This book is an introduction to integrability and conformal field theory in two dimensions using quantum groups. The book begins with a brief introduction to S-matrices, spin chains and vertex models as a prelude to the study of Yang–Baxter algebras and the Bethe ansatz. The basic ideas of integrable systems are then introduced, with particular emphasis on vertex and face models. Special attention is given to explaining the underlying mathematical tools, including braid groups, knot invariants and towers of algebras. The book then goes on to give a detailed introduction to quantum groups as a prelude to chapters on integrable models, two-dimensional conformal field theories and super-conformal field theories. The book contains many diagrams and exercises to illustrate key points in the text.

This book will be of use to graduate students and researchers in theoretical physics and applied mathematics interested in integrable systems, string theory and conformal field theory.

CAMBRIDGE MONOGRAPHS ON MATHEMATICAL PHYSICS

QUANTUM GROUPS IN TWO-DIMENSIONAL PHYSICS

CAMBRIDGE MONOGRAPHS ON MATHEMATICAL PHYSICS

A. M. Anile *Relativistic Fluids and Magneto-Fluids*
J. A. de Azcárraga and J. M. Izquierdo *Lie Groups, Lie Algebras, Cohomology and Some Applications in Physics*
J. Bernstein *Kinetic Theory in the Early Universe*
G. F. Bertsch and R. A. Broglia *Oscillations in Finite Quantum Systems*
N. D. Birrell and P. C. W. Davies *Quantum Fields in Curved Space*†
D. M. Brink *Semiclassical Methods in Nucleus–Nucleus Scattering*
J. C. Collins *Renormalization*†
P. D. B. Collins *An Introduction to Regge Theory and High Energy Physics*
M. Creutz *Quarks, Gluons and Lattices*†
F. de Felice and C. J. S. Clarke *Relativity on Curved Manifolds*†
B. DeWitt *Supermanifolds, 2nd edition*†
P. G. O. Freund *Introduction to Supersymmetry*†
F. G. Friedlander *The Wave Equation on a Curved Space-Time*
J. Fuchs *Affine Lie Algebras and Quantum Groups*†
J. A. H. Futterman, F. A. Handler and R. A. Matzner *Scattering from Black Holes*
M. Göckeler and T. Schücker *Differential Geometry, Gauge Theories and Gravity*†
C. Gómez, M. Ruiz Altaba and G. Sierra *Quantum Groups in Two-dimensional Physics*
M. B. Green, J. H. Schwarz and E. Witten *Superstring Theory, volume 1: Introduction*†
M. B. Green, J. H. Schwarz and E. Witten *Superstring Theory, volume 2: Loop Amplitudes, Anomalies and Phenomenology*†
S. W. Hawking and G. F. R. Ellis *The Large-Scale Structure of Space-Time*†
F. Iachello and A. Arima *The Interacting Boson Model*
F. Iachello and P. van Isacker *The Interacting Boson–Fermion Model*
C. Itzkyson and J.-M. Drouffe *Statistical Field Theory, volume 1: From Brownian Motion to Renormalization and Lattice Gauge Theory*†
C. Itzkyson and J.-M. Drouffe *Statistical Field Theory, volume 2: Strong Coupling, Monte Carlo Methods, Conformal Field Theory, and Random Systems*†
J. I. Kapusta *Finite-Temperature Field Theory*†
V. E. Korepin, A. G. Izergin and N. M. Boguliubov *The Quantum Inverse Scattering Method and Correlation Functions*
D. Kramer, H. Stephani, M. A. H. MacCallum and E. Herlt *Exact Solutions of Einstein's Field Equations*
N. H. March *Liquid Metals: Concepts and Theory*
I. M. Montvay and G. Münster *Quantum Fields on a Lattice*
L. O'Raifeartaigh *Group Structure of Gauge Theories*†
A. Ozorio de Almeida *Hamiltonian Systems: Chaos and Quantization*†
R. Penrose and W. Rindler *Spinors and Space-time, volume 1: Two-Spinor Calculus and Relativistic Fields*†
R. Penrose and W. Rindler *Spinors and Space-time, volume 2 : Spinor and Twistor Methods in Space-Time Geometry*†
S. Pokorski *Gauge Field Theories*†
V. N. Popov *Functional Integrals and Collective Excitations*†
R. Rivers *Path Integral Methods in Quantum Field Theory*†
R. G. Roberts *The Structure of the Proton*†
J. M. Stewart *Advanced General Relativity*
A. Vilenkin and E. P. S. Shellard *Cosmic Strings and other Topological Defects*
R. S. Ward and R. O. Wells Jr *Twistor Geometry and Field Theories*†

† Issued as a paperback

QUANTUM GROUPS IN TWO-DIMENSIONAL PHYSICS

CÉSAR GÓMEZ
Consejo Superior de Investigaciones Científicas, Madrid

MARTÍ RUIZ-ALTABA
Université de Genève

GERMÁN SIERRA
Consejo Superior de Investigaciones Científicas, Madrid

CAMBRIDGE
UNIVERSITY PRESS

Published by the Press Syndicate of the University of Cambridge
The Pitt Building, Trumpington Street, Cambridge CB2 1RP
40 West 20th Street, New York, NY 10011-4211, USA
10 Stamford Road, Oakleigh, Melbourne 3166, Australia

First published 1996

Printed in Great Britain at the University Press, Cambridge

A catalogue record for this book is available from the British Library

Library of Congress cataloguing in publication data

Gómez, César.
Quantum groups in two-dimensional physics / César Gómez,
Martí Ruiz-Altaba, Germán Sierra.
p. cm. – (Cambridge monographs on mathematical physics)
Includes bibliographical references and index.
ISBN 0 521 46065 4 (hc)
1. Quantum groups. 2. Yang–Baxter equation. 3. Conformal invariants.
4. Mathematical physics. I. Ruiz-Altaba, Martí. II. Sierra, Germán.
III. Title. IV. Series.
QC20.7.G76G66 1996
530.1′43′0151255 – dc20 95–32598 CIP

ISBN 0 521 46065 4 hardback

A nuestras niñas

Ana, Camila, Laura, Marina, Martina y Pepa

Contents

Preface xv

1 S-matrices, spin chains and vertex models 1
1.1 Factorized S-matrix models 1
1.1.1 Zamolodchikov algebra 5
1.1.2 Example 7
1.2 Bethe's diagonalization of spin chain hamiltonians 10
1.3 Integrable vertex models: the six-vertex model 14
Exercises 25
Appendix A Form factors 28
A1.1 Introduction to Smirnov's program 28
A1.2 Form factors at work: the Ising model 31
Exercise 33

2 The Yang–Baxter equation: a first look 34
2.1 The Yang–Baxter algebra 34
2.1.1 The $\mathscr{R}$-matrix and the Yang–Baxter equation 34
2.1.2 The monodromy matrix 38
2.1.3 Co-product and the Yang–Baxter algebra 39
2.1.4 Algebraic Bethe ansatz 41
2.2 Yang–Baxter algebras and braid groups 47
2.3 Yang–Baxter algebras and quantum groups 51
2.3.1 The $\mathscr{R}$-matrix as an intertwiner 53
2.3.2 A first contact with affine Hopf algebras 55
2.4 Descendants of the six-vertex model 58
2.4.1 Descent procedure 58
2.4.2 Bethe ansatz for descendant models 61
2.5 Comments 66
2.5.1 Explanation of our conventions 66
2.5.2 On parametrizations of the six-vertex weights 67
Exercises 67

3 Bethe ansatz: some examples 71
3.1 Introduction and summary 71
3.2 The phase structure of the six-vertex model 73

3.3 Low lying excitations 78
3.3.1 Strings and holes 78
3.3.2 Dispersion relations 83
3.4 Integrable higher-spin chains 83
3.4.1 Hidden symmetry 85
3.5 Integrability in a box: open boundary conditions 88
3.5.1 Vertex models: the Sklyanin equation 88
3.5.2 Spin chains: the Bethe ansatz 92
3.6 Hamiltonians with quantum group invariance 96
3.7 Spin-1 chains 98
Exercises 102

4 The eight-vertex model 108
4.1 Definitions and Yang–Baxter relations 108
4.2 Bethe ansatz solution 111
4.2.1 A smart change of basis 112
4.2.2 Eight-vertex Yang–Baxter algebra 116
4.3 Reference state and θ parameter 118
4.3.1 Further comments on the θ parameter 120
Exercises 121
Appendix B Elliptic functions 125
Exercises 128
Appendix C Sklyanin algebra 130
Exercises 132

5 Face models 134
5.1 Weights and graphs: the definitions 134
5.2 Trigonometric solutions 137
5.2.1 Bratelli diagrams 137
5.2.2 Yang–Baxter operators 138
5.2.3 The Temperley–Lieb–Jones algebra 139
5.2.4 Towers of algebras associated with a graph 140
5.2.5 The algebra of face observables 144
5.2.6 The trigonometric solution for Coxeter models 146
5.3 Elliptic solutions 148
5.3.1 An example: the Ising model 148
5.3.2 Restricted and unrestricted models 154
5.4 Fusion for face models 155
5.5 The corner transfer matrix 158
Exercises 160
Appendix D Knots and integrable models 165
D5.1 Introduction: the Jones polynomial 165
D5.2 Markov moves 168
D5.3 Markov traces for the Hecke algebra 170

D5.4 The Burau representation 172
D5.5 Extended Yang–Baxter systems 173
Exercises 176
Appendix E Spin models 177
E5.1 Factors and subfactors 177
E5.2 Spin models 180
Exercises 183

6 Quantum groups: mathematical review 184
6.1 Hopf algebras 184
6.2 Quasi-triangular Hopf algebras 186
6.3 Drinfeld's quantum double 187
6.4 The quantum group $U_q(s\ell(2))$ 189
  6.4.1 Quantum double construction 189
  6.4.2 Irreducible representations 194
6.5 Centralizer and Hecke algebra 201
  6.5.1 Representations of $H_n(q)$ 203
6.6 Link invariants from quantum groups 205
6.7 The quantum group $U_q(\mathscr{G})$ 207
6.8 R-matrices: an incomplete catalog 208
6.9 Classical Yang–Baxter equation 210
6.10 Affine quantum groups 211
6.11 Quasi-Hopf algebras 218
Exercises 219

7 Integrable models at roots of unity 228
7.1 Mathematical preliminaries 228
  7.1.1 The center of $U_q(s\ell(2))$ 228
  7.1.2 Finite-dimensional irreps 229
  7.1.3 The co-adjoint action 231
  7.1.4 Intertwiners 234
7.2 A family of R-matrices 235
  7.2.1 Highest weight intertwiner 235
  7.2.2 The nilpotent R-matrix 238
7.3 Nilpotent hamiltonians 240
7.4 Bethe ansatz 244
7.5 The limit $\ell \to \infty$ 250
  7.5.1 Quantum harmonic oscillators 251
  7.5.2 Link invariants 252
7.6 The chiral Potts model 252
  7.6.1 Star-triangle relations 255
  7.6.2 The associated spin chain hamiltonian 258
  7.6.3 Self-dual chiral Potts models 260
  7.6.4 Super-integrable chiral Potts models 262

7.6.5 The quantum symmetry 263
7.7 Solving the Yang–Baxter equation 268
Exercises 269

8 Two-dimensional conformal field theories 272
8.1 Introduction: critical phenomena 272
8.2 Renormalization group 272
8.3 Examples 275
8.3.1 The one-dimensional Ising model 275
8.3.2 The gaussian model 278
8.4 Operator algebra of a universality class 280
8.5 Conformal invariance and statistical mechanics 281
8.6 The two-dimensional conformal group 282
8.7 Representations of the Virasoro algebra 286
8.8 Decoupling of null vectors 291
8.8.1 The Kac formula 292
8.8.2 Conformal Ward identities 294
8.8.3 Minimal models 295
8.9 Fusion algebra 299
8.10 Finite-size effects 300
Exercises 304

9 Duality in conformal field theories 308
9.1 Monodromy invariance 309
9.2 Conformal blocks and chiral vertex operators 311
9.3 Sewing 314
9.4 Braiding and fusion 319
9.5 Conformal field theories and towers of algebras 323
9.6 Genus one polynomial equations 327
Exercises 336

10 Coulomb gas representation 340
10.1 Free and Feigin–Fuks scalar fields 340
10.2 Screening charges in correlation functions 344
10.2.1 Braiding matrices: an explicit example 348
10.2.2 Contour techniques 350
10.3 Lagrangian approach 353
10.4 Wess–Zumino models 355
10.4.1 The Knizhnik–Zamolodchikov equation 355
10.4.2 Free field representation of Wess–Zumino models 359
10.4.3 The Goddard–Kent–Olive construction 363
10.5 Magic corner transfer matrix 365
Exercises 366
Appendix F Vertex operators 373

11 Quantum groups in conformal field theory 376
11.1 The hidden quantum symmetry 376
11.2 Braiding matrices and quantum $6j$ symbols 381
11.3 Ribbon Hopf algebras 384
11.4 The contour representation of quantum groups 386
11.4.1 Screened vertex operators 386
11.4.2 Examples 390
11.4.3 The quantum qroup 392
11.4.4 The $\mathscr{R}$-matrix 396
11.4.5 Chiral vertex operators 400
11.5 The quantum group of $\widehat{SU(2)}_k$ 401
11.5.1 The $\mathscr{R}$-matrix 407
11.5.2 Fusion rules and chiral vertex operators 409
11.5.3 On intertwiners: a clarification 413
11.6 The quantum group of minimal models 413
Exercises 415
Appendix G Super-conformal field theories 422
G11.1 Super-conformal transformations 422
G11.2 Representations 424
G11.3 $N = 2$ super-conformal algebras 425
G11.4 $N = 2$ irreps and the chiral ring 426
G11.5 $N = 2$ topological theories 429
G11.6 Perturbed chiral ring 430
G11.7 Landau–Ginsburg description 432
G11.8 Quantum groups and solitons 434
Exercise 437

References 439

Index 452

Preface

Satius est supervacua discere quam nihil
Seneca

This book addresses the need among theoretical physicists and mathematicians for a modern, intuitive and moderately comprehensive introduction to the subject of integrable systems in two dimensions, one plus one or two plus zero. The requisite background for reading this book profitably amounts to elementary quantum field theory and statistical mechanics, in addition to basic group theory. We have tried to present all the material, both standard and new, in modern language and consistent notation.

It is perhaps still premature to evaluate the real physical impact of string theory, but it is certainly true that the current renaissance of two-dimensional physics owes much to the string wave. Traditionally, physics in two dimensions was considered a theoretical laboratory, the realm of toy models. Only after the recent work on string theory did two-dimensional quantum field theories graduate from pedagogical simplifications to serious candidates for the understanding of nature: physics in the purest aristotelian sense.

Independently of how much truth lies within string theory or elsewhere, a beautiful feature of physics in two dimensions is of course its mathematical richness. Astonishingly, almost any branch of mathematics becomes relevant in the study of two-dimensional field theories. The main physical reason for such mathematical inflation is the existence of non-trivial completely integrable two-dimensional field theories. More technically, the wonders of two dimensions have their origin in the powerful artillery of complex analysis. It is remarkable that so much of what we now understand in great generality was already contained in Onsager's solution (1944) of the two-dimensional Ising model. In particular, he discovered the star-triangle relation, now called the Yang–Baxter equation.

At the root of integrability we find a kind of trivial dynamics, described by factorizable S-matrices. This dynamics, implying infinite-dimensional symmetries, takes its most concrete expression in the wonderful Yang–Baxter equation, linking solvable two-dimensional models in statistical mechanics and quantum field theory to knot invariants and quantum groups.

The Yang–Baxter equation was originally formulated as a condition that the basic quantities of the model (be it Boltzmann weights, scattering matrices, or braiding matrices) should satisfy in order for the theory to be solvable. Later on, it was realized that there exists a hidden symmetry underlying the trigonometric and rational solutions to the Yang–Baxter equation. This hidden symmetry is captured by a new concept in mathematics, the quantum group, which unifies the framework of two-dimensional exact models. History teaches us that whenever a new kind of symmetry is discovered, a revolution is knocking at the door of knowledge. In these revolutionary times, new ideas abound and the dust has not yet begun to settle in order to distinguish more clearly what is truly revolutionary from what is fashionable opportunism; we feel nevertheless that rephrasing the Yang–Baxter equation in terms of a symmetry is of enormous epistemological relevance.

Although quantum groups were born from integrability, providing us with an algebraic explanation for the Yang–Baxter equation in terms of symmetry, they constitute such an interesting conceptual breakthrough that the whole subject of integrable models deserves a re-examination in their light. The tender age of quantum groups (about ten years old at the time of this writing) hides somewhat the fact that, albeit in disguise, they had already surfaced earlier in physics and mathematics, for example in the discrete calculus associated with q-numbers. Much work remains to be done in the development of quantum groups, now a coherent foundation upon which fancier towers may be built. A major challenge, for instance, is to understand the elliptic solutions to the Yang–Baxter equation, notably that to the eight-vertex model, in the quantum group language.

Already, the deeper understanding of integrability afforded by quantum groups has allowed the construction of new integrable models. The old and historic models, such as the Heisenberg, the sine–Gordon, or the six-vertex models, have thereby multiplied into quantum group descendants, and the growing family is still not complete. Let us stress that, for the time being, quantum groups have remained confined to two-dimensional physics: either two non-relativistic spatial dimensions (statistical mechanics) or one time and one space (quantum field theory). Higher dimensional applications of quantum groups are perhaps possible though at any event very rare, due perhaps to the difficulty of finding integrable models in dimensions higher than two. Most likely, the quantum group symmetry is intimately tied with two dimensions, and any extension to other dimensions of the quantum group technology will call for a different algebraic structure. However desirable *a priori* these extensions might appear, the perfect uniqueness of strings takes away much of the motivation for looking anywhere else than two dimensions for fundamental structures, and thus the research program around quantum groups acquires even more urgency and appeal.

The above considerations explain the ideology behind this book, which attempts to distill the structure of quantum groups from two-dimensional physics and, conversely, to frame physical questions in a formalism such that quantum groups may provide us with the answers. Were the various topics in this book not so closely linked by quantum groups, one would dare call our work an interdisciplinary effort between physics and mathematics, but given the environment of prime interest to us it is perhaps best to speak of a physics book with mathematical applications. In the spirit of the Vienna circle, mathematical physics should be considered as synonymous with formalization. Nevertheless, in our days mathematical physics is approaching criticism and shying away from the old-fashioned aim of axiomatization. For a critic, the material consists typically of works of art, and the goal is to find the clues to provide new and unifying points of view. In this sense, this book is closer in spirit to modern criticism than to canonical formalization. Our *corpus* is made out of two-dimensional theoretical physics. The chosen outlook hinges on the symmetry clue: we rely on quantum groups to extend our understanding of symmetries in physics. Recalling the French Anatolian saying that a good critic is somebody who describes their adventures among masterpieces, we can only hope that this book provides the reader with a pleasant tour.

We start with an introduction to integrable vertex models: in chapters 1 and 2 we introduce factorized S-matrices, the Bethe ansatz and the Yang–Baxter equation, along with the basic concepts about quantum groups. Chapter 3 reviews the Bethe ansatz solution to some simple spin chain hamiltonians. It also includes a few words on more general spin chains, and its last sections present boundary effects and Sklyanin's algebra. Chapters 4 and 5 serve to introduce the reader to some mathematical tools not generally known to practicing physicists, while discussing the algebraic Bethe ansatz solution to the eight-vertex model and, thus motivated, the face models. The trigonometric solution to the latter is presented in the general framework of the Temperley–Lieb–Jones algebra, complemented by an appendix on knot theory and another one on finite-dimensional subfactors. Chapter 6, the mid-point of the book, contains most of the necessary mathematical information about quantum groups, both finite and affine. In chapter 7 we turn to the representation theory of quantum groups at roots of unity and present several physical applications thereof, with an emphasis on explicit calculations. Chapter 8 introduces the reader to the universal behavior of second order phase transitions, conformal field theory, and the decoupling of null vectors. Chapter 9 exploits the concepts introduced along the book to study the duality structure of rational conformal field theories. Chapter 10 proceeds to the free field representation of these theories and introduces also the simplest Wess–Zumino models. Finally, in chapter 11, we use the free field realization of rational conformal field theories to discuss their quantum symmetries.

To keep the book to a reasonable size, we have been forced to make some painful choices. Quite a few relevant topics have been discussed only very briefly – we hope that the appendices and the exercises will fill these gaps to some extent.

Each of the chapters ends with a very brief bibliographical overview for further reading. We have not made any attempt at comprehensiveness, and we cite only essential major works that we have actually used. These references should serve the reader as an introduction to the vast literature available.

In the last few years, each of us has given various series of lectures on quantum groups, conformal field theory and integrable models in different places. This volume is an outgrowth of the course which one of us (C.G.) taught at the Troisième Cycle de Physique de la Suisse Romande during the spring of 1990, and it owes much to the feedback provided by a number of eager and inquisitive audiences.

This book was invented, as a non-existing object, by Henri Ruegg. Only thanks to his insistent prodding does it now become part of a reality, as realities occur in a Borges universe.

We have learned this physics from many colleagues. In particular, we wish to mention Luis Alvarez-Gaumé with whom we started the study of conformal field theories and quantum groups several years ago, Alexander Berkovich, who shared with us his knowledge of the Bethe ansatz, and Cupatitzio Ramírez, with whom we developed the contour picture of quantum groups. We would also like to acknowledge interesting scientific discussions with Rodolfo Cuerno, Tetsuji Miwa, Carmen Núñez and Philippe Zaugg.

Finally, we wish to acknowledge the support of the whole Theoretical Physics Department at the University of Geneva, and of the Fonds National Suisse pour la Recherche Scientifique.

Genève

1

S-matrices, spin chains and vertex models

In classical mechanics, two functions over phase space are said to be in involution if their Poisson bracket vanishes. Since Liouville's time, a dynamical system whose phase space is $2N$ dimensional is called completely integrable if there are N functions, or "hamiltonians", or "charges" in involution. A field theoretic system is called integrable if it possesses an infinity of mutually commuting conserved observables. All these mutually commuting conserved charges or hamiltonians allow us to solve the system exactly, without resorting to approximation schemes. Integrability is an unusual wealth of symmetry we might not think of requiring on realistic physical models. Rather, we should expect the complexity of nature not to be exactly solvable. However, integrability is of epistemological importance: exact solutions allow more perfect understanding. Toy models, of which physicists have been very fond since antiquity, often exist just so that exact and complete solutions can be found, in order to grasp the nature of the phenomenon being modeled. Furthermore, and quite surprisingly, physical systems with an infinite symmetry do exist: any non-linear system with soliton solutions is integrable. We shall be interested in discovering under which circumstances certain kinds of physical systems admit complete integrability, what types of systems these are, and in pointing out the physical roots of such a wonderful property. In the process, we shall have the occasion to use some of the most powerful tools elaborated by workers in mathematics.

Given our present understanding of two-dimensional models, integrability appears as a consequence of very simple dynamics, characterized by factorized scattering matrices. In this first chapter, we shall become acquainted with some of the most striking properties of factorized S-matrices, following Zamolodchikov; we shall then introduce Bethe's classical and beautiful work on the one-dimensional Heisenberg ferromagnet.

1.1 Factorized S-matrix models

Consider the scattering of relativistic massive particles in a $(1+1)$-dimensional spacetime. There is only one spatial dimension (the real line, say), and therefore the ordering of the particles is a well-defined, frame-independent concept. Equivalently, the distinction between left and right is unambiguous; in more spatial dimensions this is never so, and thus we should not expect that the interesting features depending strongly on the ordering of the particles extend to theories in higher dimensions.

It is convenient to use as kinematical variable for each particle a rapidity θ, in terms of which the momentum p^1 and energy p^0 read as follows:

$$p^0 = m\cosh\theta\ , \qquad p^1 = m\sinh\theta \tag{1.1}$$

This parametrization ensures the on shell condition $\vec{p}^{\,2} = \left(p^0\right)^2 - \left(p^1\right)^2 = m^2$.

Alternatively, we could use the lightcone momenta p and $\overline{p}$,

$$p = p^0 + p^1 = m\,\mathrm{e}^{\theta}\ , \qquad \overline{p} = p^0 - p^1 = m\,\mathrm{e}^{-\theta} \tag{1.2}$$

which transform under a Lorentz boost, $L_\alpha : \theta \to \theta + \alpha$, as

$$p \to p\,\mathrm{e}^{\alpha}\ , \qquad \overline{p} \to \overline{p}\,\mathrm{e}^{-\alpha} \tag{1.3}$$

Quite generally, an irreducible tensor Q_s of the Lorentz group in $1+1$ dimensions is labeled by its spin s according to the rule

$$L_\alpha : Q_s \to \mathrm{e}^{s\alpha} Q_s \tag{1.4}$$

so that p is of spin 1 and its parity conjugate $\overline{p}$ is of spin -1.

Suppose that Q_s is a local conserved quantity of spin $s > 0$ (the case $s < 0$ is obtained by a parity transformation) in a scattering process involving n particles A_i $(i = 1, \ldots, n)$ with masses m_i. On a one-particle state $|A_i(\theta)\rangle$ of rapidity θ, the operator Q_s acts as

$$Q_s\,|A_i(\theta)\rangle \sim p^s\,|A_i(\theta)\rangle \tag{1.5}$$

Since Q_s is local and conserved by assumption, the scattering process must satisfy

$$\sum_{i\in\{\mathrm{in}\}} p_i^s = \sum_{f\in\{\mathrm{out}\}} p_f^s \tag{1.6}$$

If the theory happens to be parity invariant, then we also have

$$\sum_{i\in\{\mathrm{in}\}} \overline{p}_i^s = \sum_{f\in\{\mathrm{out}\}} \overline{p}_f^s \tag{1.7}$$

Setting $s = 1$ in (1.6) and (1.7), we recover the usual energy and momentum conservation laws of a relativistic theory.

We are interested in theories with conserved quantities of higher spin (i.e. $|s| > 1$), leading to the conservation laws (1.6) and (1.7). In fact, integrability is synonymous with an infinity of such conserved higher spin quantities.

The physical behavior of integrable systems is quite remarkable. For instance, if (1.6) and (1.7) hold for an infinity of different spins s, it follows immediately that the incoming and outgoing momenta must be the same:

$$\left\{p_i^\mu\ ; i \in \mathrm{in}\right\} = \left\{p_f^\mu\ ; f \in \mathrm{out}\right\} \tag{1.8}$$

This means that no particle production or annihilation may ever occur.

Also, particles with equal mass may reshuffle their momenta among themselves in the scattering, but particles with different masses may not. Equivalently, we may say that the momenta are conserved individually and that particles of equal mass may interchange additional internal quantum numbers. If all the incoming particles have different masses, then the only effect of the scattering is a time delay (a phase shift) in the outgoing state with respect to the incoming one.

The infinite symmetry of the physical systems we are describing restricts tremendously the allowed processes, much to our intellectual advantage. As we shall see now, all scattering processes can be understood and pictured as a sequence of two-particle scatterings. This property is called factorizability.

By relativistic invariance, the scattering amplitude between two particles A_i and A_j may only depend on the scalar

$$p_i^\mu p_j^\nu \eta_{\mu\nu} = m_i m_j \cosh\left(\theta_i - \theta_j\right) \tag{1.9}$$

so that, in fact, it may depend only on the rapidity difference $\theta_{ij} = \theta_i - \theta_j$. Using (1.8), the most general form of the basic two-particle S-matrix in terms of which all other S-matrices will be written is thus

$$S\ |A_i(\theta_1), A_j(\theta_2)\rangle_{\text{in}} = \sum_{k,\ell} S_{ij}^{k\ell}(\theta_{12})\,|A_k(\theta_2), A_\ell(\theta_1)\rangle_{\text{out}} \tag{1.10}$$

In this notation, $|A_i(\theta_1), A_j(\theta_2)\rangle_{\text{in(out)}}$ stands for the initial (respectively, final) state of two incoming (respectively, outgoing) particles of kinds A_i and A_j and rapidities θ_1 and θ_2. This elementary process is shown diagrammatically in figure 1.1.

In addition to (1.8), the second crucial feature of a factorizable S-matrix theory, from which such models get their name, is the property of factorizability: the N-particle S-matrix can always be written as the product of $\binom{N}{2}$ two-particle S-matrices.

As in equation (1.10), we choose an initial state of N particles with rapidities $\theta_1 > \theta_2 > \cdots > \theta_N$ arranged in the infinite past in the opposite order, i.e. $x_1 < x_2 < \cdots < x_N$. This presumes simply that no scatterings have occurred before we begin studying the process, i.e. that we have been observing the system long before any particles meet. After the $N(N-1)/2$ pair collisions, the particles

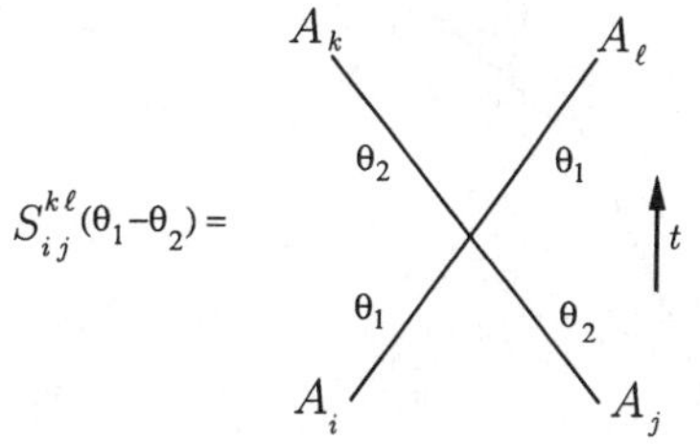

Fig. 1.1. Collision of two particles A_i and A_j with rapidities θ_1 and θ_2 ($\theta_1 > \theta_2$) into particles A_k and A_ℓ.

reach the infinite future ordered along the spatial direction in increasing rapidity. Thus we write

$$S\ |A_{i_1}(\theta_1),\ldots,A_{i_N}(\theta_N)\rangle_{\text{in}} = \sum_{j_1,\ldots,j_N} S_{i_1\cdots i_N}^{j_1\cdots j_N}\ |A_{j_1}(\theta_N),\ldots,A_{j_N}(\theta_1)\rangle_{\text{out}} \tag{1.11}$$

Factorization means that this process can be interpreted as a set of independent and consecutive two-particle scattering processes.

The spacetime picture of this multi-particle factorized scattering is obtained by associating with each particle a line whose slope is the particle's rapidity. The scattering process is thus represented by a planar diagram with N straight world-lines, such that no three ever coincide at the same point. Any world-line will therefore intersect, in general, all the others. The complete scattering amplitude associated with any such diagram is given by the (matrix) product of two-particle S-matrices. For instance, for the four-particle scattering shown in figure 1.2, we obtain

$$S_{i_1i_2i_3i_4}^{j_1j_2j_3j_4}(\theta_1,\theta_2,\theta_3,\theta_4) = \sum_{\substack{k,\ell,m,n,\\p,q,r,u}} S_{i_1i_2}^{k\ell}(\theta_{12})S_{\ell i_3}^{mn}(\theta_{13}) \times S_{km}^{pq}(\theta_{23})S_{ni_4}^{rj_4}(\theta_{14})S_{qr}^{uj_3}(\theta_{24})S_{pu}^{j_1j_2}(\theta_{34}) \tag{1.12}$$

The kinematical data (the rapidities of all the particles) do not fix a diagram uniquely. In fact, for the same rapidites, we have a whole family of diagrams, differing from each other by the parallel shift of some of the straight world-lines (figure 1.3). The parallel shift of any one line can (and should) be interpreted as a symmetry transformation. It corresponds to the translation of the (asymptotic in- and out-) x co-ordinates of the particle associated to the line. This is indeed

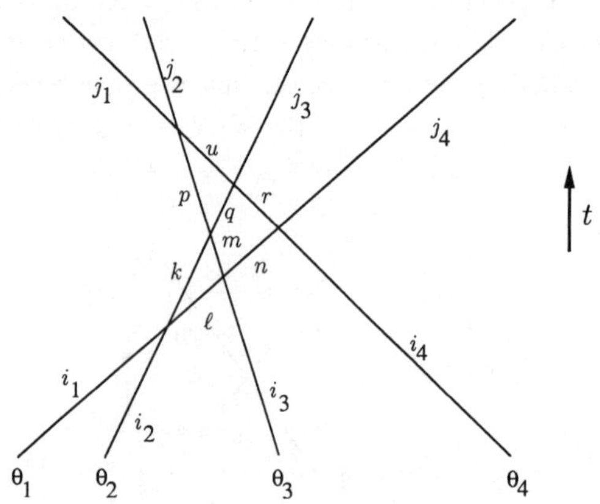

Fig. 1.2. Spacetime diagram of the scattering of four particles. Each line is the world-line of a particle.

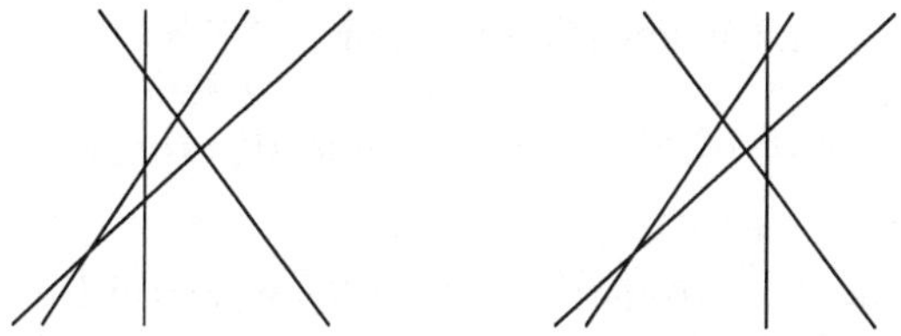

Fig. 1.3 Diagrams differing by the parallel shift of a line.

the symmetry underlying the conservation laws (1.8), which Baxter has called the **Z**-symmetry. Requiring the factorizability condition is equivalent to imposing that the scattering amplitudes of diagrams differing by such parallel shifts should be the same.

For the simple case of three particles, the condition that the factorization be independent of parallel shifts of the world-lines amounts to the following noteworthy factorization equation, which is the necessary and sufficient condition for any two diagrams differing by parallel shifts to have equal associated amplitudes (see figure 1.4):

$$\sum_{p_1,p_2,p_3} S_{i_1i_2}^{p_1p_2}(\theta_{12})S_{p_2i_3}^{p_3j_3}(\theta_{13})S_{p_1p_3}^{j_1j_2}(\theta_{23})$$
$$= \sum_{p_1,p_2,p_3} S_{i_2i_3}^{p_2p_3}(\theta_{23})S_{i_1p_2}^{j_1p_1}(\theta_{13})S_{p_1p_3}^{j_2j_3}(\theta_{12}) \tag{1.13}$$

Let us stress that the factorization equations (1.13) are a direct consequence of the postulated factorization condition and conservation laws (1.8). Equation (1.13) is the famous Yang–Baxter equation, a matrix equation.

1.1.1 Zamolodchikov algebra

Having sketched the physical meaning of the factorization or Yang–Baxter equation (1.13), we may now turn to a more mathematical interpretation of the factorization conditions from which it derives. To this end, we shall consider a set of operators $\{A_i(\theta)\}$ $(i = 1, \ldots, n)$ associated with each particle i with rapidity

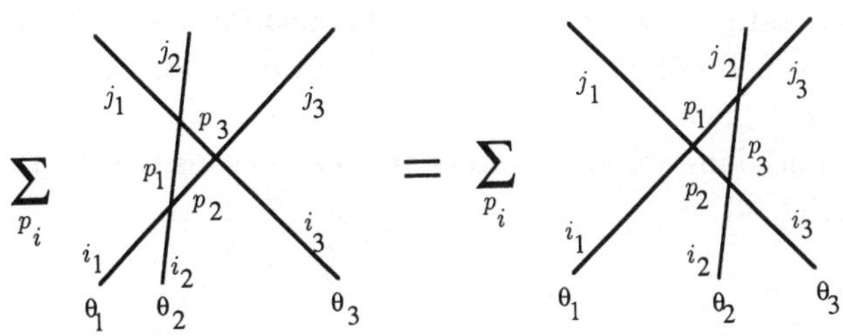

Fig. 1.4 The factorization equations (1.13).

θ, obeying the following commutation relations:

$$A_i(\theta_1)A_j(\theta_2) = \sum_{k,\ell} S_{ij}^{k\ell}(\theta_{12})A_k(\theta_2)A_\ell(\theta_1) \tag{1.14}$$

This equation encodes the two-particle scattering process (1.10), where "collision" has been replaced by "commutation". Furthermore, the relationship between (1.10) and (1.14) becomes evident if we interpret $A_i(\theta)$ as an operator (Zamolodchikov operator) which creates the particle $|A_i(\theta)\rangle$ when it acts on the Hilbert space vacuum $|0\rangle$:

$$A_i(\theta)\,|0\rangle = |A_i(\theta)\rangle \tag{1.15}$$

The factorization equation (1.13) emerges in this context as a "generalized Jacobi identity" of the algebra (1.14), assumed associative. Indeed, let us consider the product of three operators, $A_{i_1}(\theta_1)A_{i_2}(\theta_2)A_{i_3}(\theta_3)$, and let us try to reverse the order of the rapidities. This can be done in two different ways: either

$$\begin{aligned} A_{i_1}(\theta_1)A_{i_2}(\theta_2)A_{i_3}(\theta_3) &\underset{\theta_1\leftrightarrow\theta_2}{=} \sum_{p_1,p_2} S_{i_1 i_2}^{p_1 p_2}(\theta_{12})A_{p_1}(\theta_2)A_{p_2}(\theta_1)A_{i_3}(\theta_3) \\ &\underset{\theta_1\leftrightarrow\theta_3}{=} \sum_{p_1,p_2,p_3,j_3} S_{i_1 i_2}^{p_1 p_2}(\theta_{12})S_{p_2 i_3}^{p_3 j_3}(\theta_{13})A_{p_1}(\theta_2)A_{p_3}(\theta_3)A_{j_3}(\theta_1) \\ &\underset{\theta_2\leftrightarrow\theta_3}{=} \sum_{\substack{p_1,p_2,p_3\\ j_1,j_2,j_3}} S_{i_1 i_2}^{p_1 p_2}(\theta_{12})S_{p_2 i_3}^{p_3 j_3}(\theta_{13})S_{p_1 p_3}^{j_1 j_2}(\theta_{23})A_{j_1}(\theta_3)A_{j_2}(\theta_2)A_{j_3}(\theta_1) \end{aligned} \tag{1.16}$$

or else

$$\begin{aligned} A_{i_1}(\theta_1)A_{i_2}(\theta_2)A_{i_3}(\theta_3) &\underset{\theta_2\leftrightarrow\theta_3}{=} \sum_{p_2,p_3} S_{i_2 i_3}^{p_2 p_3}(\theta_{23})A_{i_1}(\theta_1)A_{p_2}(\theta_3)A_{p_3}(\theta_2) \\ &\underset{\theta_1\leftrightarrow\theta_3}{=} \sum_{p_1,p_2,p_3,j_1} S_{i_2 i_3}^{p_2 p_3}(\theta_{23})S_{i_1 p_2}^{j_1 p_1}(\theta_{13})A_{j_1}(\theta_3)A_{p_1}(\theta_1)A_{p_3}(\theta_2) \\ &\underset{\theta_1\leftrightarrow\theta_2}{=} \sum_{\substack{p_1,p_2,p_3\\ j_1,j_2,j_3}} S_{i_2 i_3}^{p_2 p_3}(\theta_{23})S_{i_1 p_2}^{j_1 p_1}(\theta_{13})S_{p_1 p_3}^{j_2 j_3}(\theta_{12})A_{j_1}(\theta_3)A_{j_2}(\theta_2)A_{j_3}(\theta_1) \end{aligned} \tag{1.17}$$

The equality between these two results is simply the factorization equations (1.13). In practice, it is easier to write down the cubic relations (1.13) with the help of a labeled diagram such as figure 1.4, rather than through the explicit computations (1.16) and (1.17).

Two more conditions are needed to guarantee the consistency of the Zamolodchikov algebra (1.14):

(i) Normalization:

$$\lim_{\theta\to 0} S_{ij}^{k\ell}(\theta) = \delta_i^k \delta_j^\ell \qquad \Longleftrightarrow \qquad \lim_{\theta\to 0} S(\theta) = \mathbf{1} \tag{1.18}$$

This condition is obtained by setting $\theta_1 = \theta_2$ in (1.14). In physical terms, it means that no scattering takes place if the relative velocity of the two particles vanishes, i.e. if the two world-lines are parallel.

(ii) Unitarity:

$$\sum_{j_1,j_2} S_{j_1 j_2}^{i_1 i_2}(\theta) S_{k_1 k_2}^{j_1 j_2}(-\theta) = \delta_{k_1}^{i_1} \delta_{k_2}^{i_2} \qquad \Longleftrightarrow \qquad S(\theta)S(-\theta) = \mathbf{1} \tag{1.19}$$

This follows from applying (1.14) twice.

The two conditions above are supplemented, on physical grounds, by two more, namely

(iii) Real analyticity:

$$S^{\dagger}(\theta) = S(-\theta^*) \tag{1.20}$$

which, together with (1.19), implies the physical unitarity condition $S^{\dagger}S = \mathbf{1}$;

(iv) Crossing symmetry:

$$S_{ij}^{k\ell}(\theta) = S_{j\bar{\ell}}^{\bar{\imath}k}(i\pi - \theta) \tag{1.21}$$

where $\bar{\imath}$ and $\bar{\ell}$ denote the antiparticles of i and ℓ, respectively.

1.1.2 Example

Let us consider a theory with only one kind of particle A and its antiparticle $\overline{A}$. According to (1.8), all the possible two-particle scattering processes are shown in figure 1.5.

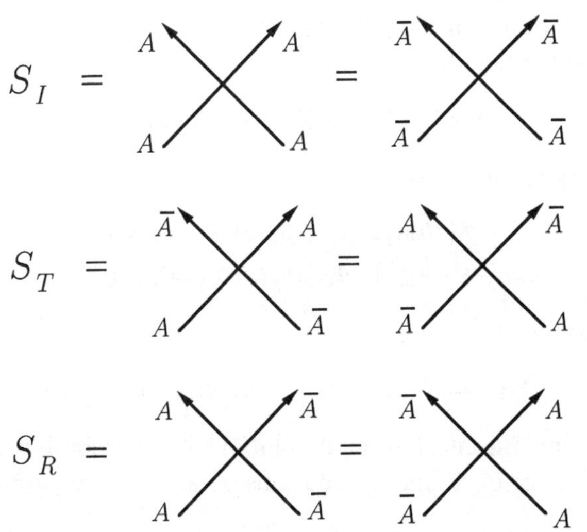

Fig. 1.5 Pair collisions among particles A and antiparticles $\overline{A}$.

Due to CPT invariance, there exist only three different amplitudes (we also assume conservation of particle number, i.e. $\mathbb{Z}_2$ invariance). The scattering amplitude between identical particles (or antiparticles) is denoted S_I, whereas S_T and S_R denote the transmission and reflection amplitudes, respectively. Notice that S_R need not vanish because the masses of a particle and its antiparticle are equal. This illustrates an important consequence of condition (1.8), namely that the redistribution of quantum numbers may take place only among particles with the same mass. In the following, we shall always consider the scattering of n different types of particles (in this example, $n = 2$) with the same mass, whose internal quantum numbers are denoted by the generic label i ($i = 1, \ldots, n$).

The scattering processes of figure 1.5 can be summarized in the following Zamolodchikov algebra:

$$\begin{aligned}
A(\theta_1)A(\theta_2) &= S_I(\theta_{12})A(\theta_2)A(\theta_1) \\
A(\theta_1)\overline{A}(\theta_2) &= S_T(\theta_{12})\overline{A}(\theta_2)A(\theta_1) + S_R(\theta_{12})A(\theta_2)\overline{A}(\theta_1) \\
\overline{A}(\theta_1)A(\theta_2) &= S_T(\theta_{12})A(\theta_2)\overline{A}(\theta_1) + S_R(\theta_{12})\overline{A}(\theta_2)A(\theta_1) \\
\overline{A}(\theta_1)\overline{A}(\theta_2) &= S_I(\theta_{12})\overline{A}(\theta_2)\overline{A}(\theta_1)
\end{aligned} \tag{1.22}$$

It is not hard to check that the factorization equations for this algebra read as

$$\begin{aligned}
S_I S'_R S''_I &= S_T S'_R S''_T + S_R S'_I S''_R \\
S_I S'_T S''_R &= S_T S'_I S''_R + S_R S'_R S''_T \\
S_R S'_T S''_I &= S_R S'_I S''_T + S_T S'_R S''_R
\end{aligned} \tag{1.23}$$

where we have set

$$S_a = S_a(\theta_{12}), \qquad S'_a = S_a(\theta_{13}), \qquad S''_a = S_a(\theta_{23}) \tag{1.24}$$

for $a \in \{I, T, R\}$ to lighten the notation.

The normalization conditions read

$$S_I(0) = 1, \quad S_T(0) = 0, \quad S_R(0) = 1 \tag{1.25}$$

whereas unitarity requires

$$\begin{aligned}
S_T(\theta)S_T(-\theta) + S_R(\theta)S_R(-\theta) &= 1 \\
S_T(\theta)S_R(-\theta) + S_R(\theta)S_T(-\theta) &= 0
\end{aligned} \tag{1.26}$$

and the crossing symmetry implies

$$S_I(\theta) = S_T(i\pi - \theta), \qquad S_R(\theta) = S_R(i\pi - \theta) \tag{1.27}$$

In order to find the most general solution to the equations (1.23), we first eliminate S''_a in favor of S_a and S'_a, and thus obtain the compatibility conditions

$$\frac{S_I^2 + S_T^2 - S_R^2}{S_I S_T} = \frac{S_I'^2 + S_T'^2 - S_R'^2}{S'_I S'_T} \tag{1.28}$$

Eliminating S'_a in favor of S_a and S''_a, or S_a in favor of S'_a and S''_a, we find similar equations which imply that the quantity

$$\Delta = \frac{S_I(\theta)^2 + S_T(\theta)^2 - S_R(\theta)^2}{2S_I(\theta)S_T(\theta)} \tag{1.29}$$

is independent of the rapidity θ, and must therefore be related to the coupling constants of the theory.

Clearly, if $\{S_a, S'_a, S''_a\}$ satisfy the Yang–Baxter equation (1.23), then so do $\{\lambda S_a, \lambda S'_a, \lambda S''_a\}$. The overall scale ambiguity can be removed by working with the ratios

$$x(\theta) = \frac{S_I(\theta)}{S_R(\theta)}, \qquad y(\theta) = \frac{S_T(\theta)}{S_R(\theta)} \tag{1.30}$$

whereby equation (1.29) becomes the quadric

$$x^2 + y^2 - 2\Delta xy = 1 \tag{1.31}$$

All three x, y and Δ may be complex. If $|\Delta| \neq 1$, we may parametrize the quadric (1.31) in terms of trigonometric functions of the rapidity θ: the factorizable S-matrix then provides a trigonometric solution to the Yang–Baxter equation. On the other hand, if $|\Delta| = 1$, the S-matrix elements are parametrized by rational functions of θ. The rapidity θ plays the role of the uniformizing parameter for the curve above.

An interesting factorized S-matrix is provided by the sine-Gordon theory, whose lagrangian density is

$$\mathscr{L} = \frac{1}{2}\left(\partial_\mu \phi\right)^2 + \frac{m_\circ^2}{\beta^2}\cos\left(\beta\phi\right) \tag{1.32}$$

If we identify the states A and $\overline{A}$ of (1.22) with the soliton and antisoliton, then the three independent amplitudes are given by

$$\begin{aligned} S_I(\theta) &= \sinh\left[\frac{8\pi}{\eta}(i\pi - \theta)\right] U(\theta) \\ S_T(\theta) &= \sinh\left(\frac{8\pi}{\eta}\theta\right) U(\theta) \\ S_R(\theta) &= i\sin\left(\frac{8\pi^2}{\eta}\right) U(\theta) \end{aligned} \tag{1.33}$$

where $U(\theta)$ is a complicated combination of Γ functions fixed by the normalization and unitarity conditions but not by the Yang–Baxter equation, and η is a renormalization of the coupling constant β of the theory:

$$\eta = \frac{\beta^2}{1 - \frac{\beta^2}{8\pi}} \tag{1.34}$$

For this model, the value of Δ in (1.29) is

$$\Delta = -\cos\left(\frac{8\pi^2}{\eta}\right) \tag{1.35}$$

Recall Coleman's result that for $\beta^2 < 8\pi$ the sine-Gordon hamiltonian is bounded below. The point $\beta^2 = 8\pi$ corresponds to a free fermion model, for which $S_T(\theta) = S_I(\theta) = 1$ and $S_R(\theta) = 0$.

1.2 Bethe's diagonalization of spin chain hamiltonians

Let us start afresh with a different kind of physical system, namely a one-dimensional spin chain. For simplicity, we restrict ourselves for the time being to a periodic one-dimensional regular lattice (a periodic chain) with L sites. At each site, the spin variable may be either up or down, so that the Hilbert space of the spin chain is simply

$$\mathscr{H}^{(L)} = \bigotimes^{L} V^{\frac{1}{2}} \tag{1.36}$$

where $V^{\frac{1}{2}}$ is the spin-$\frac{1}{2}$ irreducible representation of $SU(2)$ with basis $\{|\Uparrow\rangle, |\Downarrow\rangle\}$. By simple combinatorics, the dimension of the Hilbert space is $\dim \mathscr{H}^{(L)} = 2^L$. On $\mathscr{H}^{(L)}$, we consider a very general hamiltonian H, subject to three constraints.

First, we assume that the interaction is of short range, for example only among nearest neighbors, or next to nearest neighbors.

Next, we impose that the hamiltonian H be translationally invariant. Letting e^{iP} denote the operator which shifts the states of the chain by one lattice unit to the right, then this requirement reads as

$$\left[\mathrm{e}^{iP}, H\right] = 0 \tag{1.37}$$

It is more convenient to work with this shift operator directly than with P itself, which is just the lattice version of the momentum operator. From the periodicity of the closed chain, we must have

$$\mathrm{e}^{iPL} = \mathbf{1} \tag{1.38}$$

Finally, we demand that the hamiltonian preserve the third component of the spin:

$$\left[H, S^z_{\text{total}}\right] = \left[H, \sum_{i=1}^{L} S^z_i\right] = 0 \tag{1.39}$$

This requirement allows us to divide the Hilbert space of states into different sectors, each labeled by the third component of the spin or, equivalently, by the total number of spins down. We shall denote by $\mathscr{H}^{(L)}_M$ the subspace of $\mathscr{H}^{(L)}$ with M spins down. Obviously, $\dim \mathscr{H}^{(L)}_M = \binom{L}{M}$, so that $\dim \mathscr{H}^{(L)} = \sum_{M=0}^{L} \dim \mathscr{H}^{(L)}_M$.

We wish to study the eigenstates and spectrum of H. The zeroth sector $\mathscr{H}^{(L)}_0$ contains only one state, the "Bethe reference state" with all spins up. The most natural ansatz for the eigenvectors of H in the other sectors is some superposition of "spin waves" with different velocities. For the first sector, i.e. the subspace of states with all spins up except one down, the ansatz for the eigenvector is of the form

$$|\Psi_1\rangle = \sum_{x=1}^{L} f(x)\,|x\rangle \tag{1.40}$$

where $|x\rangle$ represents the state with all spins up but for the one at lattice site x $(1 \le x \le L)$. The unknown wave-function $f(x)$ determines the probability that the single spin down is precisely at site x.

In this simplest case, owing to the complete translational invariance due to periodic boundary conditions, it is reasonable to assume that $f(x)$ is just the wave-function for a plane wave:

$$f(x) = e^{ikx} \tag{1.41}$$

with some particular momentum k to be fixed by the boundary condition

$$f(x+L) = f(x) \tag{1.42}$$

Thus, $k = 2\pi I/L$, with $I = 0, 1, \ldots, L-1$. Hence, the eigenvectors of H with one spin down span a basis of the Hilbert space $\mathscr{H}_1^{(L)}$, by dimensionality counting.

Let us generalize the above straightforward construction to the sectors with more than one spin wave, $M > 1$.

Look first for the wave-function which solves the eigenvalue problem for the sector with two spins down,

$$H\,|\Psi_2\rangle = E_2\,|\Psi_2\rangle \tag{1.43}$$

It will be of the form

$$|\Psi_2\rangle = \sum_{x_1,x_2} f(x_1,x_2)\,|x_1,x_2\rangle \tag{1.44}$$

where now $|x_1,x_2\rangle$ stands for the state with all spins up except two spins down at positions x_1 and x_2. Since a copy of the same two-dimensional representation sits at every site, the Pauli exclusion principle implies the constraint $1 \le x_1 < x_2 \le L$, in agreement with the fact that $\dim \mathscr{H}_2^{(L)} = \binom{L}{2}$.

The analog of the periodicity condition (1.42) now reads

$$f(x_1,x_2) = f(x_2, x_1+L) \tag{1.45}$$

The order of the arguments on the right-hand side has been changed to preserve the condition that $f(x_1,x_2)$ is defined for $x_1 < x_2$.

The most naive ansatz for $f(x_1,x_2)$, which generalizes (1.41), is

$$f(x_1,x_2) = A_{12}\; e^{i(k_1x_1+k_2x_2)} \tag{1.46}$$

This ansatz is inappropriate, however, because it violates the periodicity condition (1.45). Physically, we have forgotten to include the scattering of the two "spin waves" with "quasi-momenta" k_1 and k_2. The solution to this problem was found by Bethe, who wrote the useful ansatz

$$f(x_1, x_2) = A_{12}\, e^{i(k_1 x_1 + k_2 x_2)} + A_{21}\, e^{i(k_1 x_2 + k_2 x_1)} \tag{1.47}$$

which satisfies the periodicity condition (1.45) provided the following equations hold:

$$A_{12} = A_{21}\, e^{ik_1 L}, \qquad A_{21} = A_{12}\, e^{ik_2 L} \tag{1.48}$$

Note that these two conditions imply, in particular, that $e^{i(k_1+k_2)L} = 1$, which reflects the invariance of the wave-function under a full turn around the chain, i.e. under the shift (1.38) of L units of lattice space:

$$f(x_1 + L, x_2 + L) = f(x_1, x_2) \iff e^{i(k_1+k_2)L} = 1 \tag{1.49}$$

This equation must hold if the wave-function is to be single-valued.

The ansatz (1.47) already assumes that the S-matrix for two spin waves is purely elastic. In fact, the only dynamics allowed is the permutation of the quasi-momenta. This is the first sign of an infinity of conserved charges, as we know from the discussion in the previous section.

To understand the physical meaning of equations (1.48), let us introduce the "scattering amplitudes for spin waves"

$$\hat{S}_{12} = \frac{A_{21}}{A_{12}}, \qquad \hat{S}_{21} = \frac{A_{12}}{A_{21}} \tag{1.50}$$

in terms of which equations (1.48) read as

$$e^{ik_1 L}\hat{S}_{12}(k_1, k_2) = 1, \qquad e^{ik_2 L}\hat{S}_{21}(k_2, k_1) = 1 \tag{1.51}$$

These equations tell us that the total phase shift undergone by a spin wave after traveling all the way around the closed chain is 1. This phase shift receives two contributions: one is purely kinematic ($e^{ik_1 L}$ or $e^{ik_2 L}$), and depends only on the quasi-momenta of the spin waves, while the other reflects the phase shift produced by the interchange of the two spin waves. An example of $\hat{S}_{12}$ appears in exercise 1.1 below.

Summarizing the previous discussion, we have found that the Bethe ansatz for the eigenvector of the hamiltonian H in the sector $M = 2$ is

$$f(x_1, x_2) = A_{12}\left(e^{i(k_1 x_1 + k_2 x_2)} + \hat{S}_{12}(k_1, k_2)\, e^{i(k_1 x_2 + k_2 x_1)}\right) \tag{1.52}$$

A *caveat* is in order. For a hamiltonian describing a ferromagnetic regime, the Bethe reference state is the ground state of the theory. Only in this case are the spin waves under consideration real particle-like excitations. In an antiferromagnetic regime, however, the ground state has a much more complicated structure, and thus the spin waves are not real excitations above the ground state. In fact, the study of these excitations is rather involved, and we postpone it until chapter 3. To prevent confusion between spin waves and real excitations, we refer to $\hat{S}_{12}$ as the spin wave S-matrix.

Let us continue, now considering the sector with $M > 2$ spins down. The generic form of a state $|\Psi_M\rangle \in \mathscr{H}_M^{(L)}$ is

$$|\Psi_M\rangle = \sum_{1\leq x_1<x_2<\cdots<x_M\leq L} f(x_1,\ldots,x_M)\,|x_1,\ldots,x_M\rangle \tag{1.53}$$

The Bethe ansatz is now

$$f(x_1,\ldots,x_M) = \sum_{p\in\mathscr{S}_M} A_p\, \mathrm{e}^{i(k_{p(1)}x_1+\cdots+k_{p(M)}x_M)} \tag{1.54}$$

where the sum runs over the $M!$ permutations p of the labels of the quasi-momenta k_i. The periodicity condition is now

$$f(x_1,x_2,\ldots,x_M) = f(x_2,\ldots,x_M,x_1+L) \tag{1.55}$$

Let us pause to consider explicitly the case $M = 3$, which exhibits the full structure while the expressions remain of manageable size. From (1.55) we obtain the following six equations:

$$\begin{aligned} \mathrm{e}^{ik_1L} &= \frac{A_{123}}{A_{231}} = \frac{A_{132}}{A_{321}} \\ \mathrm{e}^{ik_2L} &= \frac{A_{231}}{A_{312}} = \frac{A_{213}}{A_{132}} \\ \mathrm{e}^{ik_3L} &= \frac{A_{312}}{A_{123}} = \frac{A_{321}}{A_{213}} \end{aligned} \tag{1.56}$$

Thus, in addition to the relationships between the quasi-momenta k_i and the amplitudes A_p, there exist additional constraints between the amplitudes of three quasi-particles, which were absent in the simpler case of $M = 2$. These relations contain a Yang–Baxter relation in disguise. Indeed, from equations (1.56) it follows that

$$\frac{A_{213}}{A_{123}} = \frac{A_{321}}{A_{312}}, \quad \frac{A_{231}}{A_{213}} = \frac{A_{312}}{A_{132}}, \quad \frac{A_{321}}{A_{231}} = \frac{A_{132}}{A_{123}} \tag{1.57}$$

These equations tell us that the interchange of two particles is independent of the position of the third particle. Locality of the interactions is thus equivalent to the factorization property of the S-matrix, according to which the scattering amplitude of M quasi-particles factorizes into a product of $\binom{M}{2}$ two-point S-matrices.

The Yang–Baxter content of equations (1.57) is illustrated by the following equalities:

$$A_{321} = \begin{cases} \hat{S}_{12}\, A_{312} = \hat{S}_{12}\hat{S}_{13}\, A_{132} = \hat{S}_{12}\hat{S}_{13}\hat{S}_{23}\, A_{123} \\ \hat{S}_{23}\, A_{231} = \hat{S}_{23}\hat{S}_{13}\, A_{213} = \hat{S}_{23}\hat{S}_{13}\hat{S}_{12}\, A_{123} \end{cases} \tag{1.58}$$

With the help of (1.56) and (1.58) we arrive at the "Bethe ansatz equations"

$$\mathrm{e}^{ik_iL} = \prod_{\substack{j=1\\ j\neq i}}^{M} \hat{S}_{ji}(k_j,k_i) \qquad \text{for } i = 1,\ldots,M \tag{1.59}$$

written, in general, for a sector with arbitrary M. These equations will play a central role in future discussions.

The spin wave scattering amplitude $\hat{S}_{12}$ depends, of course, on the detailed form of the hamiltonian, and it can be computed by solving the eigenvalue equation (1.43), which reads more explicitly as

$$E_2\, f(x_1, x_2) = \sum_{1 \le y_1 < y_2 \le L} \langle x_1, x_2 | \, H \, | y_1, y_2 \rangle \, f(y_1, y_2) \tag{1.60}$$

Using (1.52) in (1.60), we would find $\hat{S}_{12}$ as a function of k_1, k_2 and the matrix elements of H. We shall give an explicit example below.

Unfortunately, there does not exist a simple criterion that would allow us to decide when a spin chain hamiltonian is integrable, i.e. when it allows the Bethe construction. As we have shown, however, the Bethe ansatz will work whenever the spin wave S-matrix satisfies the integrability condition (1.8) and factorization. We shall return to this important point later on. Let us stress that the diagonalization of a hamiltonian with the help of the Bethe ansatz does not even work for any translationally invariant and short range hamiltonian preserving the total spin. Only a very special class of such hamiltonians can be diagonalized via the Bethe procedure, namely those which describe integrable models. The first spin chain model shown to be integrable was the Heisenberg hamiltonian (also known as the XXX model):

$$H_{XXX} = J \sum_{i=1}^{L} \left(\sigma_i^x \sigma_{i+1}^x + \sigma_i^y \sigma_{i+1}^y + \sigma_i^z \sigma_{i+1}^z \right) \tag{1.61}$$

A generalization of this model, to which the Bethe ansatz technique is still applicable, is the XXZ model:

$$H_{XXZ} = J \sum_{i=1}^{L} \left(\sigma_i^x \sigma_{i+1}^x + \sigma_i^y \sigma_{i+1}^y + \Delta \sigma_i^z \sigma_{i+1}^z \right) \tag{1.62}$$

which differs from the previous one by the anisotropy parameter Δ. In fact, the even more general XYZ model,

$$H_{XYZ} = J \sum_{i=1}^{L} \left(\sigma_i^x \sigma_{i+1}^x + \Gamma \sigma_i^y \sigma_{i+1}^y + \Delta \sigma_i^z \sigma_{i+1}^z \right) \tag{1.63}$$

is also amenable to a complete solution. We will discuss the modified Bethe ansatz required by this last case in chapter 4.

1.3 Integrable vertex models: the six-vertex model

Having displayed the main feature of integrable one-dimensional quantum spin chain hamiltonians, namely their solvability via the Bethe ansatz, or, equivalently, the factorization of the spin wave amplitudes, we proceed now to explore integrability in lattice two-dimensional field theories. More specifically, we shall concentrate on classical statistical systems in two spatial dimensions (in equilibrium, so no time dimension) on a lattice. Whereas in the previous section the main problem consisted of diagonalizing a one-dimensional hamiltonian, in this section we address the computation of the partition function of the lattice system. Following Baxter, we shall introduce the concept of transfer matrices, which establishes an illuminating and fecund bridge between one-dimensional quantum mechanics and two-dimensional statistical physics. Among the plethora of integrable lattice models known today, we shall choose the very simple six-vertex model as its main representative. This model is a beautiful cornerstone because it is easy to formulate and extremely rich. In many ways, most examples in this book may be viewed as variations of the six-vertex model. Let us start, however, with some generalities.

A vertex model is a statistical model defined on a lattice $\mathscr{L}$ whose geometry is given by a (possibly infinite) set of straight lines on the plane. The intersections of these lines are called the vertices or sites, and we require that not more than two lines meet at every vertex. A generic lattice of this kind is called a Baxter lattice (see figure 1.6).

Obviously, a regular rectangular lattice is of the Baxter sort, and we shall generally restrict ourselves to it, for simplicity. We shall thus consider an $L \times L'$ lattice with L vertical lines (columns) and L' horizontal lines (rows). This fixes the geometry of the model.

A physical state on this lattice is defined by the assignment to each lattice edge of a state variable, characterized by some labels. The simplest possibility is that the labels belong to a discrete and finite set; allowing for two possible states on each link suffices for our purposes. These two possibilities may be interpreted physically in various ways. Thinking of spins up or down is most appropriate in a magnetic context. Alternatively, we may regard each edge of the lattice as representing some finite region of space, which may be either empty or occupied. Consistency of the particle picture requires that the particles which may occupy the links be fermions (exercise 1.2). Finally, if we imagine the lattice links as

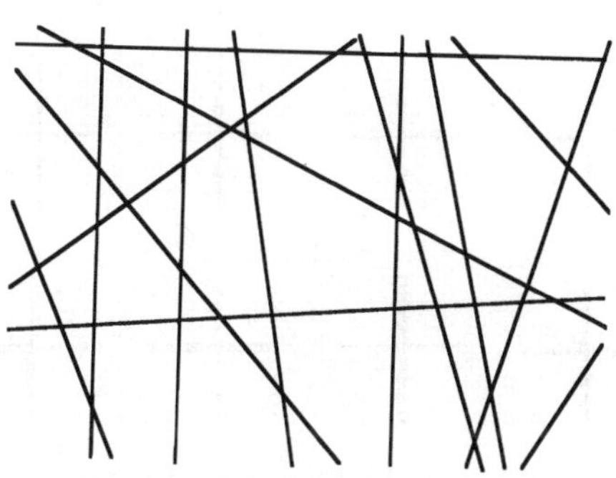

Fig. 1.6 Example of a Baxter lattice.

Table 1.1 Three interpretations of the lattice variables of a two-state model.

Spin	Occupancy	Current
$\lvert\Uparrow\rangle$	0	$\rightarrow$ or $\uparrow$
$\lvert\Downarrow\rangle$	1	$\leftarrow$ or $\downarrow$

lattice links as electric wires with a current of constant intensity running through them, then the two states are associated with the direction of the current. These alternative interpretations are summarized in table 1.1. A state of the whole system, or configuration, is the list of what variables are assigned to each and every link.

The dynamics of the model is characterized by the interactions among the lattice variables, which take place at the vertices, whence the name vertex model. The energy ε_V associated with a vertex V depends only on the four states on the edges meeting at that vertex (locality). This is also true for the $2^4 = 16$ Boltzmann weights $W_V = \exp\left(-\varepsilon_V/k_B T\right)$, which measure the probability of each local configuration. It is clear from the various interpretations of the state variables in table 1.1 that we might expect physical constraints on the values of these Boltzmann weights or, equivalently, of the vertex energies. If we impose that the interaction conserves the total spin, the fermion number, or the local current, then all but six Boltzmann weights must vanish. The six allowed vertex configurations, in the particle and current pictures, are shown in figure 1.7.

In addition to the spin, particle number or current conservation, we may also impose the $\mathbb{Z}_2$ reversal symmetry under $\lvert\Uparrow\rangle \leftrightarrow \lvert\Downarrow\rangle$, particle $\leftrightarrow$ hole or $(\rightarrow,\uparrow) \leftrightarrow (\leftarrow,\downarrow)$. Under this condition, the number of independent Boltzmann

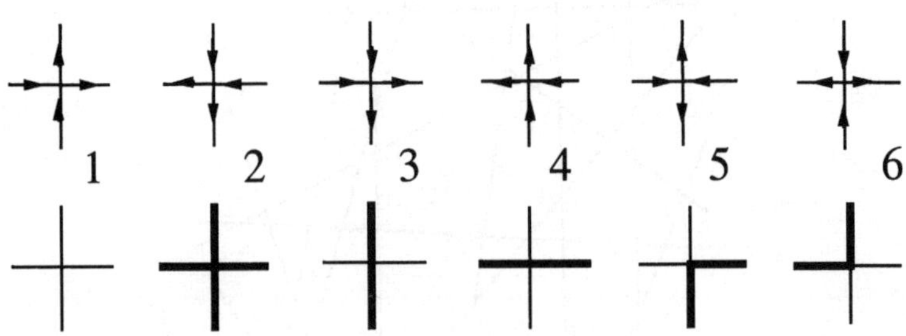

Fig. 1.7. Allowed configurations in the six-vertex model. The arrows indicate the direction of current flow; thick lines mean that the edge is occupied.

weights is reduced to three, which we shall call just a, b and c:

$$\begin{aligned} a &= \exp\frac{-\varepsilon_1}{k_B T} = \exp\frac{-\varepsilon_2}{k_B T} \\ b &= \exp\frac{-\varepsilon_3}{k_B T} = \exp\frac{-\varepsilon_4}{k_B T} \\ c &= \exp\frac{-\varepsilon_5}{k_B T} = \exp\frac{-\varepsilon_6}{k_B T} \end{aligned} \tag{1.64}$$

These weights define the symmetric or zero-field six-vertex model, which we shall call the six-vertex model for short.

Quite generally, it is convenient to represent Boltzmann weights as $W\begin{pmatrix} \beta & \nu \\ \mu & \alpha \end{pmatrix}$, where μ and ν are the horizontal edge state labels and α and β are the vertical ones:

$$W\begin{pmatrix} \beta & \nu \\ \mu & \alpha \end{pmatrix} \quad = \tag{1.65}$$

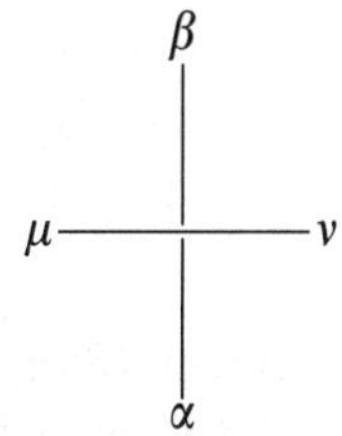

With the help of this general notation, the six-vertex model is characterized by link variables $\in \mathbb{Z}_2 = \{0, 1\}$, with Boltzmann weights subject to current conservation

$$W\begin{pmatrix} \beta & \nu \\ \mu & \alpha \end{pmatrix} = 0 \qquad \text{unless} \qquad \mu + \alpha = \nu + \beta \tag{1.66}$$

and reflection symmetry ($\overline{x} = 1 - x$)

$$W\begin{pmatrix} \beta & \nu \\ \mu & \alpha \end{pmatrix} = W\begin{pmatrix} \overline{\beta} & \overline{\nu} \\ \overline{\mu} & \overline{\alpha} \end{pmatrix} \tag{1.67}$$

A more compact way to write the weights (1.64) is readily available; for instance,

$$W\begin{pmatrix} \beta & \nu \\ \mu & \alpha \end{pmatrix} = b\,\delta_{\mu\nu}\,\delta_{\alpha\beta} + c\,\delta_{\mu\beta}\,\delta_{\nu\alpha} + (a - b - c)\,\delta_{\mu\alpha}\,\delta_{\nu\beta} \tag{1.68}$$

The statistical properties of the six-vertex model are fully characterized by the partition function

$$Z_{L\times L'}(a, b, c) = \sum_{\mathscr{C}} \exp\frac{-E(\mathscr{C})}{k_B T} = \sum_{\mathscr{C}} \prod_{V} W_V \tag{1.69}$$

where the sum runs over all possible configurations $\mathscr{C}$, of which there are $2^{LL'}$ for an $L \times L'$ periodic lattice. In the thermodynamic limit, when L and L' tend to infinity, the computation of the sum (1.69) becomes a rather formidable and apparently insurmountable problem, quite unsuited for brute force methods such as the use of electronic machines. Lieb's breakthrough in computing the partition function (1.69) of the six-vertex model relies basically on rephrasing the problem as the diagonalization of the anisotropic spin-$\frac{1}{2}$ chain, which had been solved already by the Bethe ansatz. Let us perform the sum over the horizontal variables, which involves only the Boltzmann weights on the same row of the lattice, and then carry out the sum over the vertical variables. The double sum (1.69) can thus be rearranged as follows:

$$Z_{L\times L'}(a,b,c) = \sum_{\substack{\text{vertical}\\ \text{states}}} \prod_{\text{rows}} \left(\sum_{\substack{\text{horizontal}\\ \text{states}}} \prod_{V\in\text{row}} W_V \right) \tag{1.70}$$

The quantity in parentheses depends on the two sets of vertical states above and below the row of horizontal variables. It will play a central role in future discussions: it is the (row to row) transfer matrix of the model. For conceptual clarity, it is convenient to introduce the "fixed time states" as the set of vertical link variables on the same row:

$$|\alpha\rangle =$$

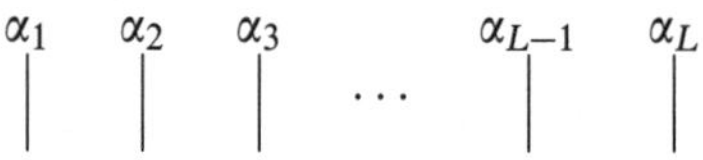

(1.71)

The transfer matrix element $\langle\beta|t|\alpha\rangle$ can then be understood as the transition probability for the state $|\alpha\rangle$ to project on the state $|\beta\rangle$ after a unit of time. We are thinking now of the horizontal direction as space and the vertical one as time:

$$\begin{aligned}\langle\beta|\, t(a,b,c)\,|\alpha\rangle = \sum_{\mu_i} W\begin{pmatrix}\beta_1 & \mu_2\\ \mu_1 & \alpha_1\end{pmatrix} W\begin{pmatrix}\beta_2 & \mu_3\\ \mu_2 & \alpha_2\end{pmatrix}\cdots\\ \cdots W\begin{pmatrix}\beta_{L-1} & \mu_L\\ \mu_{L-1} & \alpha_{L-1}\end{pmatrix} W\begin{pmatrix}\beta_L & \mu_1\\ \mu_L & \alpha_L\end{pmatrix}\end{aligned} \tag{1.72}$$

We agree with the Chinese, who think that a picture is better than a formula:

$$\langle\beta|t|\alpha\rangle = \sum_{\mu_i}$$

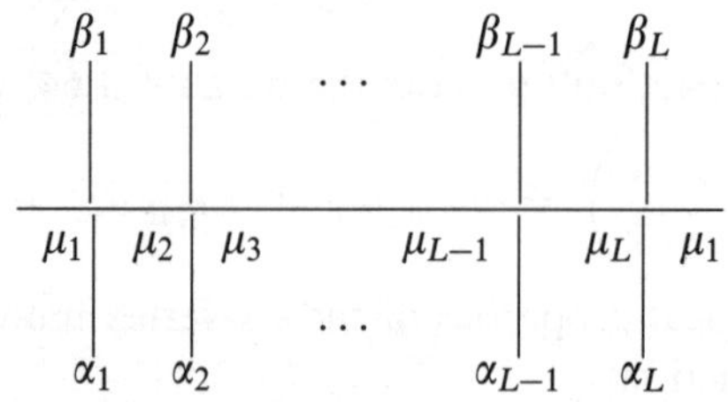

(1.73)

The transfer matrix $t(a,b,c)$ plays the role of a discrete evolution operator acting on the Hilbert space $\mathscr{H}^{(L)}$ spanned by the row states $|\alpha\rangle$ (dim $\mathscr{H}^{(L)} = 2^L$), isomorphic to the one considered in section 1.3 when diagonalizing the spin-$\frac{1}{2}$ hamiltonian. The full partition function reads thus:

$$Z_{L\times L'}(a,b,c) = \mathrm{tr}_{\mathscr{H}^{(L)}}\left(t(a,b,c)\right)^{L'} \tag{1.74}$$

The trace on $\mathscr{H}^{(L)}$ implements periodic boundary conditions in the "time" direction. This expression is just the hamiltonian formulation of the partition function (1.69). Thus, evaluating the partition function is equivalent to finding the eigenvalues of the transfer matrix. We are led, therefore, to essentially the same problem considered in the previous section, namely the diagonalization of an operator on $\mathscr{H}^{(L)}$. Let us follow the same strategy as above.

First of all, the local conservation law (1.66) translates into the conservation of particle number by the transfer matrix:

$$\langle \beta | \, t \, | \alpha \rangle = 0 \qquad \text{unless} \qquad \sum_{i=1}^{L} \alpha_i = \sum_{i=1}^{L} \beta_i \tag{1.75}$$

More technically, the number operator

$$M = \sum_{i=1}^{L} \alpha_i \tag{1.76}$$

commutes with the transfer matrix:

$$[t(a,b,c), M] = 0 \tag{1.77}$$

This is the analog of equation (1.39), and the relationship between the total spin S^z and M is simply (recall table 1.1)

$$S^z = \frac{L}{2} - M \tag{1.78}$$

Once again, the Hilbert space $\mathscr{H}^{(L)}$ can be broken down into sectors $\mathscr{H}^{(L)}_M$ labeled by $M \in \{0, 1, \ldots, L\}$. In each of these sectors, the transfer matrix can be diagonalized independently:

$$t(a,b,c) \, |\Psi_M\rangle = \Lambda_M(a,b,c) \, |\Psi_M\rangle \tag{1.79}$$

The states $|x_1, \ldots, x_M\rangle$ with 1's at the positions $x_1, \ldots, x_M$ and 0's elsewhere form a basis of $\mathscr{H}^{(L)}_M$. Expanding $|\Psi_M\rangle$ in this basis,

$$|\Psi_M\rangle = \sum_{1 \le x_1 < x_2 < \cdots < x_M \le L} f(x_1, \ldots, x_M) \, |x_1, \ldots, x_M\rangle \tag{1.80}$$

we find the equation for the eigenfunctions $f(x_1, \ldots, x_M)$:

$$\begin{aligned} \sum_{1 \le y_1 < y_2 < \cdots < y_M \le L} & \langle x_1, \ldots, x_M | \, t(a,b,c) \, | y_1, \ldots, y_M \rangle \, f(y_1, \ldots, y_M) \\ & = \Lambda_M(a,b,c) f(x_1, \ldots, x_M) \end{aligned} \tag{1.81}$$

The matrix elements $\langle x_1, \ldots, x_M | \, t(a,b,c) \, | y_1, \ldots, y_M \rangle$ of the transfer matrix can be represented graphically, as in figure 1.8. The transfer matrix connects states with the same number M of lines, whose locations may change.

The eigenvalue problem (1.81) can be solved with the help of the Bethe ansatz technique. Before attempting the general solution, it is most instructive to consider the simplest cases with $M = 0$, 1 and 2.

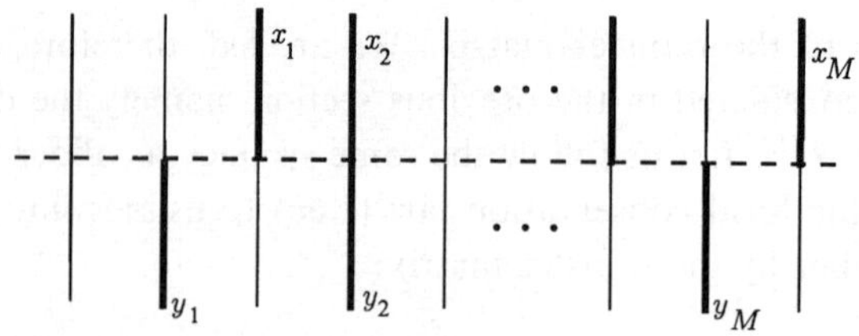

Fig. 1.8. The probability for the state $|\{y_i\}\rangle$ to evolve into $|\{x_i\}\rangle$ after one unit of discrete time is measured by the transfer matrix. Thick vertical lines represent the state 1, at the positions x_i or y_i.

The sector $M = 0$ contains only one state $|\Omega\rangle = |00\ldots0\rangle$, which is the Bethe reference state. This state plays the role of a vacuum in the construction of the other states, but it is important to realize that it need not coincide with the ground state of the model: the physical vacuum minimizes the free energy and may have nothing to do with $|\Omega\rangle$ (a similar situation arose for the ground state of the hamiltonians of section 1.2). From (1.72) and (1.81), we obtain

$$\Lambda_0 = \langle\Omega|\, t\, |\Omega\rangle = \sum_{\mu=0,1} \left[W \begin{pmatrix} 0 & \mu \\ \mu & 0 \end{pmatrix} \right]^L = a^L + b^L \tag{1.82}$$

In the sector $M = 1$, we choose $f(x) = \mathrm{e}^{ikx}$, and some elementary algebra yields (with the assumption of periodic boundary conditions)

$$\Lambda_1(k) = a^L P(k) + b^L Q(k) \tag{1.83}$$

with

$$\begin{aligned} P(k) &= \frac{ab + (c^2 - b^2)\,\mathrm{e}^{-ik}}{a^2 - ab\,\mathrm{e}^{-ik}} \\ Q(k) &= \frac{a^2 - c^2 - ab\,\mathrm{e}^{-ik}}{ab - b^2\,\mathrm{e}^{-ik}} \end{aligned} \tag{1.84}$$

The sector with $M = 2$ excitations contains more structure. The Bethe ansatz reads in this case

$$f(x_1, x_2) = A_{12}\,\mathrm{e}^{i(k_1x_1+k_2x_2)} + A_{21}\,\mathrm{e}^{i(k_2x_1+k_1x_2)} \tag{1.85}$$

subject to the periodic boundary conditions (1.45), which yield equations (1.50) and (1.51). The eigenvalue of (1.85) is given by

$$\Lambda_2 = a^L P_1 P_2 + b^L Q_1 Q_2 \tag{1.86}$$

where $P_i = P(k_i)$ and $Q_i = Q(k_i)$. The ratio of the amplitudes A_{12} and A_{21}, which again may be interpreted as a spin wave scattering matrix, is thus

$$\hat{S}_{12} = \frac{A_{21}}{A_{12}} = -\frac{1 - 2\frac{a^2+b^2-c^2}{2ab}\,\mathrm{e}^{ik_2} + \mathrm{e}^{i(k_1+k_2)}}{1 - 2\frac{a^2+b^2-c^2}{2ab}\,\mathrm{e}^{ik_1} + \mathrm{e}^{i(k_1+k_2)}} \tag{1.87}$$

For later convenience, we define the anisotropy parameter Δ as

$$\Delta = \frac{a^2 + b^2 - c^2}{2ab} \tag{1.88}$$

Notice the strong similarity between (1.88) and (1.29).

The generalization of the above results to sectors with more than two excitations proceeds through the factorization properties of the higher order Bethe amplitudes $A_{1\cdots M}$; see equations (1.58). The general formula for the eigenvalue Λ_M of a vector of the form (1.54) is

$$\Lambda_M = a^L \prod_{i=1}^{M} P(k_i) + b^L \prod_{i=1}^{M} Q(k_i) \tag{1.89}$$

and the quasi-momenta k_i ($i = 1, \ldots, M$) must satisfy the Bethe equations (1.59), which follow from the periodicity (1.55) of the wave-functions and the factorization properties of the Bethe amplitudes. In this case, they read explicitly as

$$e^{ik_i L} = (-1)^{M-1} \prod_{\substack{j=1 \\ j \neq i}}^{M} \frac{1 - 2\Delta\, e^{ik_i} + e^{i(k_i+k_j)}}{1 - 2\Delta\, e^{ik_j} + e^{i(k_i+k_j)}} \tag{1.90}$$

The final step in the computation of the eigenvalues of the transfer matrix and, ultimately, of the partition function hinges upon the solution of the Bethe equations (1.90). Their analytic solution for finite values of L and M is not known, although a fair amount of numerical progress can be made under certain assumptions. At any rate, for our purposes it suffices to consider (1.90) in the thermodynamic limit $L \to \infty$, where, surprisingly enough, they can be solved exactly. We shall consider this issue later in the book. For the time being, let us focus on the Bethe equations (1.90) themselves.

It is very important that the Bethe equations associated with the six-vertex model depend on the Boltzmann weights a, b and c only through the combination yielding the anisotropy Δ in (1.88). This is the key to understanding the integrability of the six-vertex model. The first immediate consequence from this observation is, indeed, that two different transfer matrices $t(a, b, c)$ and $t(a', b', c')$, sharing the same value for Δ, have the same eigenvectors, and thus they commute:

$$\left[\, t(a, b, c)\, ,\ t(a', b', c')\, \right] = 0 \quad \Longleftrightarrow \quad \Delta(a, b, c) = \Delta(a', b', c') \tag{1.91}$$

Therefore, given a value of Δ, through the Bethe procedure we diagonalize not just a transfer matrix but a whole continuous family of mutually commuting transfer matrices. Since each transfer matrix defines a different time evolution, with each transfer matrix with the same parameter Δ is associated a conserved quantity. So, the six-vertex model has a large number of conserved quantities, in fact an infinity of them in the thermodynamic limit.

Let us sketch now the relation between the one-dimensional hamiltonian $H_{XXZ}(\Delta) = H_\Delta$ of the anisotropic Heisenberg chain (1.62) and the two-dimensional six-vertex model. From the identification of the Bethe ansatz eigenvectors under (1.88), we see that

$$\left[\, H_{\Delta(a,b,c)}\,,\; t(a,b,c) \,\right] \;=\; 0 \tag{1.92}$$

Comparison of this expression with (1.91) leads us to suspect that the hamiltonian H_Δ must already be contained somehow in the transfer matrix $t(a,b,c)$, i.e. it should be one of the conserved quantities in the system. The same argument applies also to the translation operator e^{iP}, which commutes both with the hamiltonian and with the transfer matrix.

To make these suggestive connections explicit, let us start with the momentum operator e^{-iP}. Suppose we make the following choice of Boltzmann weights:

$$a = c = c_0, \qquad b = 0 \tag{1.93}$$

which is consistent with any value of Δ. Then, from (1.68) we obtain

$$W\begin{pmatrix} \beta & \nu \\ \mu & \alpha \end{pmatrix}\Bigg|_{\substack{a=c=c_0\\ b=0}} = c_0\; \delta_{\mu\beta}\; \delta_{\nu\alpha} \tag{1.94}$$

which can be imagined as an operator which multiplies by c_0 the incoming state $\{\mu,\alpha\}$ but otherwise leaves it untouched: the horizontal state on the left becomes the vertical state on top, and the vertical state below becomes the horizontal state to the right. Thus, the transfer matrix from these weights behaves as the shift operator e^{-iP}:

$$\begin{aligned} t_0\,|\alpha\rangle &= t(c_0,0,c_0)\,|\alpha_1,\alpha_2,\ldots,\alpha_L\rangle \\ &= c_0^L \quad |\alpha_L,\alpha_1,\ldots,\alpha_{L-1}\rangle \end{aligned} \tag{1.95}$$

and the momentum operator P is identified with

$$P = i\log\left(\frac{t_0}{c_0^L}\right) \tag{1.96}$$

This identification is easily checked on one-particle states. From (1.95), we see that $t_0\,|x\rangle = |x+1\rangle$ $(c_0 = 1)$, and thus in the Fourier transformed states $|k\rangle$ we find

$$\begin{aligned} t_0\,|k\rangle &= t_0 \sum_{x=1}^{L} \mathrm{e}^{ikx}\,|x\rangle = \sum_{x=1}^{L} \mathrm{e}^{ikx}\,|x+1\rangle \\ &= \sum_{x=1}^{L} \mathrm{e}^{ik(x+1)}\,\mathrm{e}^{-ikx}\,|x+1\rangle = \mathrm{e}^{-ik}\,|k\rangle \end{aligned} \tag{1.97}$$

Similarly, the hamiltonian H_Δ can be obtained by expanding the transfer matrix in the vicinity of the parameter point (1.93), keeping the value of Δ constant. Note that this amounts to expanding the transfer matrix about t_0, i.e. about a matrix proportional to the shift operator. So, we fix

$$\Delta = \frac{\delta a - \delta c}{\delta b} \tag{1.98}$$

and obtain

$$t_0^{-1}\,\delta t = \frac{\delta b}{2c_0}\sum_{i=1}^{L}\left\{\frac{\delta a+\delta c}{\delta b}\,\mathbf{1}+\sigma_i^x\sigma_{i+1}^x+\sigma_i^y\sigma_{i+1}^y+\Delta\sigma_i^z\sigma_{i+1}^z\right\} \tag{1.99}$$

Thus, the hamiltonian H_Δ (see (1.62)) appears in the expansion of the logarithm of the transfer matrix about the shift operator. We urge the reader to carry out the above simple and most instructive calculation explicitly.

The above derivation can be understood in more mathematical terms as follows. The parameter space of the six-vertex model consists of all triplets of Boltzmann weights (a, b, c) modulo a common scale factor, which would multiply the partition function by an overall factor:

$$Z_{L\times L'}(\lambda a, \lambda b, \lambda c) = \lambda^{LL'}\, Z_{L\times L'}(a, b, c) \tag{1.100}$$

This means that the model is defined effectively on the complex projective space $\mathbb{CP}^2$ (we allow in general for complex Boltzmann weights, although for physically meaningful statistical mechanical models they should be real and positive). This two-dimensional parameter space can be further split into integrability submanifolds, the regions of constant Δ. The transfer matrices of any two points on the same integrability region commute. Each of these regions is a complex one-dimensional submanifold, or, more precisely, an algebraic curve of genus zero. We may therefore introduce a uniformization parameter u along the curve such that (1.88) holds true for $a(u)$, $b(u)$ and $c(u)$ for any value of u.

We postpone the explanation of the parametrization that we shall use, which is

$$\begin{aligned} a(u) &= \sinh(u+i\gamma) \\ b(u) &= \sinh u \\ c(u) &= i\sin\gamma = \sinh i\gamma \end{aligned} \tag{1.101}$$

whereby Δ is related to γ by

$$\Delta = \cos\gamma \tag{1.102}$$

In order to have real Boltzmann weights for real γ, we should include an overall factor of i and restrict u to be purely imaginary.

Choosing this parametrization, we have

$$a(u=0) = c(u=0) = c_0 = i\sin\gamma, \qquad b(u=0) = 0 \tag{1.103}$$

and we can rewrite equations (1.96) and (1.99) as

$$P = i\log\frac{t(u)}{a(u)^L}\bigg|_{u=0} \tag{1.104a}$$

$$H_\Delta = i\frac{\partial}{\partial u}\log\frac{t(u)}{a(u)^L}\bigg|_{u=0} \tag{1.104b}$$

Expanding $\log t(u)$ in powers of u, we obtain a whole set of local conserved quantities involving, in general, interactions over a finite range, not just among nearest neighbors. In this sense, the transfer matrix is the generating functional for a large class of commuting conserved quantities: this follows from the integrability equation (1.91). Recall that we arrived at that equation after a long analysis involving the Bethe ansatz. Instead, we could have taken equation (1.91) as the starting point to define a vertex model, and then could have asked ourselves under what conditions the Boltzmann weights of such vertex model lead to integrability. The answer to this question is that, in order that the vertex model be integrable, the Boltzmann weights must satisfy the justly celebrated Yang–Baxter equation (see figure 1.9):

$$\sum_{\mu',\nu',\gamma} W\begin{pmatrix} \gamma & \mu' \\ \mu & \alpha \end{pmatrix} W'\begin{pmatrix} \beta & \nu' \\ \nu & \gamma \end{pmatrix} W''\begin{pmatrix} \nu'' & \mu'' \\ \nu' & \mu' \end{pmatrix}$$
$$= \sum_{\mu',\nu',\gamma} W''\begin{pmatrix} \nu' & \mu' \\ \nu & \mu \end{pmatrix} W'\begin{pmatrix} \gamma & \mu'' \\ \mu' & \alpha \end{pmatrix} W\begin{pmatrix} \beta & \nu'' \\ \nu' & \gamma \end{pmatrix} \tag{1.105}$$

where W, W' and W'' are three different sets of Boltzmann weights. We shall derive this equation in the next chapter.

Equation (1.105) is the Yang–Baxter equation for vertex models. Note that in the two problems worked out in this chapter – the diagonalization of spin chain hamiltonians and the diagonalization of the transfer matrix for vertex models – integrability is encoded in the same mathematical structure, namely the factorization of the spin wave S-matrix (1.58) and the vertex Yang–Baxter equation (1.105). We shall find several other formulations of the Yang–Baxter equation, which is always a cubic equality; it first appeared under the name of the star-triangle relation in Onsager's solution to the two-dimensional Ising model. For the time being, let us stress that the Yang–Baxter equation (1.105) summarizes the integrability properties of both the two-dimensional vertex models and of their associated one-dimensional hamiltonians. A sketchy flowchart of the basic concepts introduced in this chapter is shown in table 1.2.

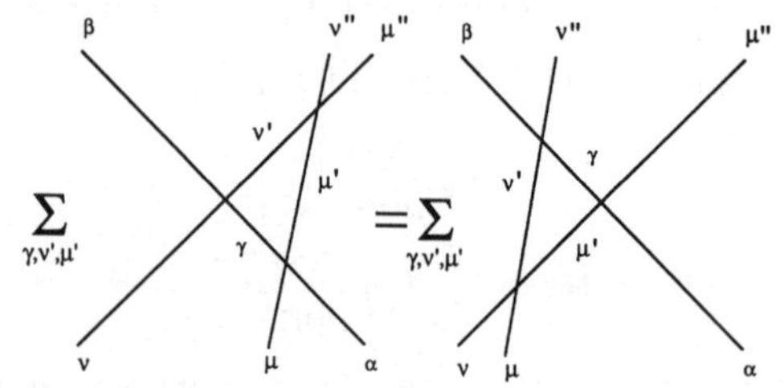

Fig. 1.9 The Yang–Baxter equation (1.105) for vertex models.

Table 1.2 Schematic summary of chapter 1.

		2-D vertex models		
		↓		
1-D hamiltonians	⟶	Bethe ansatz		
		↓	↘	factorized
		integrability		S-matrix
		↓	↙	theories
		Yang–Baxter		

Exercises

Ex. 1.1 As an example of the analysis in section 1.2, consider the anisotropic Heisenberg magnet, with hamiltonian

$$H_\Delta = H(J,\Delta) = J \sum_{i=1}^{L} \left(\sigma_i^x \sigma_{i+1}^x + \sigma_i^y \sigma_{i+1}^y + \Delta \sigma_i^z \sigma_{i+1}^z \right) \tag{1.106}$$

where $\vec{\sigma}_i$ denote the Pauli matrices acting at the ith site ($i = 1, \ldots, L$) of the lattice,

$$\sigma^x = \begin{pmatrix} 0 & 1 \\ 1 & 0 \end{pmatrix}, \qquad \sigma^y = \begin{pmatrix} 0 & -i \\ i & 0 \end{pmatrix}, \qquad \sigma^z = \begin{pmatrix} 1 & 0 \\ 0 & -1 \end{pmatrix} \tag{1.107}$$

We set $\vec{\sigma}_{L+1} = \vec{\sigma}_1$ to implement periodic boundary conditions. The parameter Δ in the XXZ hamiltonian (1.106) is called the anisotropy; when $\Delta = 1$ we recover the isotropic Heisenberg spin chain (the XXX model) with manifest $SU(2)$ symmetry. Solve equation (1.60) to obtain the following two-point spin wave S-matrix:

$$\hat{S}_{12}(k_1, k_2) = -\, \frac{1 - 2\Delta\, e^{ik_2} + e^{i(k_1+k_2)}}{1 - 2\Delta\, e^{ik_1} + e^{i(k_1+k_2)}} \tag{1.108}$$

Show then that the factorization conditions (1.57) for three spin waves do hold. Note that (1.108) coincides with (1.87) with the identification (1.88).

Ex. 1.2 Consider the XXZ hamiltonian

$$H_\Delta(J) = H(J,\gamma) = J \sum_{i=1}^{L} \left(\sigma_i^x \sigma_{i+1}^x + \sigma_i^y \sigma_{i+1}^y + \Delta \sigma_i^z \sigma_{i+1}^z \right) \tag{1.109}$$

with $\Delta = (q + q^{-1})/2$ and $q = e^{i\gamma}$. Show that, for L even, $H(J, \gamma)$ is equivalent to $-H(J, \pi - \gamma) = -H_{-\Delta}(J)$. [Hint: use the unitary transformation $U = \prod_{j \text{ odd}} \sigma_j^z$.] Therefore, introducing J in H is redundant, provided we consider both positive and negative values of Δ. Alternatively, we could restrict γ to lie in the real interval $[0, \pi/2]$ and consider both signs of J.

Ex. 1.3 Setting $\Delta = 0$ in (1.108) yields a negative sign as the result of interchanging two quasi-particles, regardless of their quasi-momenta. This immediately brings to mind fermion statistics. Therefore, we should compare the XXZ hamiltonian (1.106) with $\Delta = 0$ (known as the XX model) with the one-dimensional free fermion hamiltonian given by

$$H_F = -t \sum_{i=1}^{L} \left(c_i^\dagger c_{i+1} + c_{i+1}^\dagger c_i \right) \tag{1.110}$$

The canonical fermionic operators c_i and $c_i^\dagger$ satisfy

$$\{c_i, c_j\} = \{c_i^\dagger, c_j^\dagger\} = 0, \qquad \{c_i, c_j^\dagger\} = \delta_{i,j} \tag{1.111}$$

and t is known as the hopping parameter.

Identifying the spin down state with the presence of an electron at the site, and spin up with its absence (i.e. with a "hole"), show that the local one-dimensional operators can be represented as

$$c_i = \sigma_i^+ \prod_{j=1}^{i-1} \sigma_j^z , \qquad c_i^\dagger = \sigma_i^- \prod_{j=1}^{i-1} \sigma_j^z \tag{1.112}$$

where

$$\sigma_j^\pm = \frac{1}{2} \left(\sigma_j^x \pm i\sigma_j^y \right) \tag{1.113}$$

The non-local operator

$$K_n = \prod_{j=1}^{n-1} \sigma_j^z \tag{1.114}$$

is a kink operator.

Using the Jordan–Wigner transformation (1.112), show that H_{XX} and H_F are equivalent when the number of sites L is even, and find their mismatch when L is odd.

Ex. 1.4 Consider the XXZ hamiltonian (1.106) with $J = -1$, and take the limit $\Delta \to \pm\infty$:

$$H_\Delta \underset{|\Delta| \to \infty}{\longrightarrow} -\Delta \sum_{i=1}^{L} \sigma_i^z \sigma_{i+1}^z \tag{1.115}$$

Show that, when $\Delta \to +\infty$, the ground state is the ferromagnetic configuration $|\Uparrow\Uparrow\Uparrow\Uparrow\cdots\rangle$ or $|\Downarrow\Downarrow\Downarrow\Downarrow\cdots\rangle$. Alternatively, show that when $\Delta \to -\infty$ the ground state is the Néel state $|\Uparrow\Downarrow\Uparrow\Downarrow\cdots\rangle$ or $|\Downarrow\Uparrow\Downarrow\Uparrow\cdots\rangle$. From the previous exercise, we know that when $\Delta = 0$ the resulting XX model is equivalent to a free fermion model, whose ground state should thus be a Fermi sea at half filling. Try to figure out from these three extreme cases the phase structure of the XXZ model over the whole range of Δ. We postpone until chapter 3 a detailed analysis of this issue.

Ex. 1.5 [D.V. Chudnovsky and G.V. Chudnovsky, 'Characterization of completely X-symmetric factorized S-matrices for a special type of interaction', *Phys. Lett.* **79A** (1980) 36.]

Imagine a (1+1)-dimensional massive particle with N possible flavors, supposed to be cyclically symmetric, so that

$$S_{i,j}^{k,\ell}(\theta) = S_{i+m,j+m}^{k+m,\ell+m}(\theta)\,, \qquad \forall i,j,k,\ell,m \in \mathbb{Z}_N \tag{1.116}$$

Write down the Yang–Baxter equation and the unitarity condition, and try to solve them. (You might want to warm up with $N = 2$ and $N = 3$ first.) If this is too hard, assume furthermore that only particles with the same flavor interact, i.e. that $S_{ij}^{k\ell} = 0$ if $i \neq j$.

Ex. 1.6 Observe the close relationship between the S-matrix (1.33) of the sine-Gordon model and the Boltzmann weights (1.101) of the six-vertex model. The rapidity, θ, of the solitons and antisolitons corresponds to the uniformization parameter u. Find the relationship between the sine-Gordon coupling constant β (or η) and the anisotropy parameter γ. Study the relationship between the spectrum of the sine-Gordon model as a function of β and the phases of the six-vertex model in terms of Δ. Next, compare the factorization equation (1.13) and the Yang–Baxter equation (1.105) for the Boltzmann weights. Can one say, in general, that to any factorized S-matrix model there belongs an associated integrable vertex model?

Appendix A
Form factors

A1.1 Introduction to Smirnov's program

This appendix is rather technical, and the reader unfamiliar with S-matrix technology might want to postpone its reading until after digesting chapters 5 and 8. We start from the Zamolodchikov algebra (1.14), with the subscripts a_i labeling the internal quantum numbers of the particle with rapidity θ_i:

$$A_{a_1}(\theta_1)A_{a_2}(\theta_2) = \sum S_{a_1 a_2}^{a_2' a_1'}(\theta_{12})\, A_{a_2'}(\theta_2)A_{a_1'}(\theta_1) \tag{A1.1}$$

With the help of the vacuum (1.15), we can define a representation of the algebra (A1.1) in the Fock space $\mathscr{F}$. Consider now local operators $\mathscr{O}(\theta)$ well defined on $\mathscr{F}$. These operators are completely characterized by their form factors

$$f_{a_1 \cdots a_n}^{\mathscr{O}}(\theta; \theta_1, \theta_2, \ldots, \theta_n) = \langle 0|\, \mathscr{O}(\theta) A_{a_1}(\theta_1) \cdots A_{a_n}(\theta_n)\, |0\rangle \tag{A1.2}$$

In this section, we shall use Zamolodchikov's algebra (A1.1) to derive equations satisfied by the form factors (A1.2).

The data we shall use are the following:

(i) the spin s_{a_i} and statistics $(-1)^{\varepsilon_{a_i}}$ (with $\varepsilon_{a_i} \in \{0, 1\}$) of the particles A_{a_i};
(ii) the analytic structure of the scattering matrix;
(iii) the mutual locality index between the observable $\mathscr{O}(\theta)$ and the particle operators $A_{a_i}(\theta_i)$.

In general, the mutual locality index ω_{PQ} for two local operators $P(z)$ and $Q(z)$ is defined by

$$P(0)Q(z\, \mathrm{e}^{2\pi i}) = \mathrm{e}^{2\pi i \omega_{PQ}}\, P(0)Q(z) \tag{A1.3}$$

In this book, we shall come across many physical examples to substantiate the concept of the mutual locality index. For the time being, the reader may consider $\omega_{\mathscr{O}A_{a_i}}$ as providing extra information about the local operator algebra. The reader may think of the Bohm–Aharonov effect and imagine that P carries an electric charge e and Q a magnetic charge g. Then the mutual locality index is simply $\omega_{PQ} = eg$. From now on, we write $f_{a_1 \ldots a_n}^{\mathscr{O}}(\theta_1, \ldots, \theta_n) = f_{a_1 \ldots a_n}^{\mathscr{O}}(0; \theta_1, \ldots, \theta_n)$.

The first condition on the form factors (A1.2) imposed by the algebra (A1.1) relates the form factors before and after a two-particle scattering:

$$\begin{aligned} f^{\mathcal{O}}_{a_1\cdots a_n}(\theta_1,\ldots,\theta_n) &= (-1)^{\varepsilon_{a_i}\varepsilon_{a_{i+1}}} \sum_{b_i,b_{i+1}} S^{b_{i+1}b_i}_{a_i a_{i+1}}(\theta_{i,i+1}) \\ &\times f^{\mathcal{O}}_{a_1\cdots b_{i+1}b_i\cdots a_n}(\theta_1,\ldots\theta_{i+1},\theta_i,\ldots,\theta_n) \end{aligned} \tag{A1.4}$$

In graphical notation, this equation is shown in figure A1.1.

The spin factor $(-1)^{\varepsilon_{a_i}\varepsilon_{a_{i+1}}}$ in (A1.4) takes into account the statistics of the particles. Observe that equation (A1.4) is independent of the detailed structure of the operator $\mathcal{O}$; it depends only on its mutual locality index with the particles.

A second equation for the form factors follows on changing θ_n to $\theta_n + 2\pi i$, whereby the particle a_n turns once counter-clockwise around $\mathcal{O}$, picking up the phase appropriate to its mutual locality index with the operator $\mathcal{O}$. In addition to this obvious phase, the particle a_n is also braided through all the other particles; the result of each such braiding follows from the Zamolodchikov algebra (A1.1). The net result is thus

$$\begin{aligned} f^{\mathcal{O}}_{a_1\cdots a_n}(\theta_1,\ldots,\theta_n+2\pi i) &= \mathrm{e}^{2\pi i\left(s_{a_n}+\omega_{a_n,\mathcal{O}}\right)} S^{b_1c_1}_{a_na_1}(\theta_{n,1}) \\ &\times S^{b_2c_2}_{c_1a_2}(\theta_{n,2})\cdots S^{b_{n-1}b_n}_{c_{n-2}a_{n-1}}(\theta_{n,n-1})\, f^{\mathcal{O}}_{b_1\cdots b_n}(\theta_1,\ldots,\theta_n) \\ &= \mathrm{e}^{2\pi i\left(s_{a_n}+\omega_{a_n,\mathcal{O}}\right)}(-1)^{\varepsilon_{a_n}\sum_{j=1}^{n-1}\varepsilon_{a_j}} f^{\mathcal{O}}_{a_n,a_1,\ldots,a_{n-1}}(\theta_n,\theta_1,\ldots,\theta_{n-1}) \end{aligned} \tag{A1.5}$$

The pictorial mnemonics for a full turn of particle a_n around the operator $\mathcal{O}(0)$ are shown in figure A1.2.

Next, let us use the analytic structure of the scattering matrix appearing in (A1.1). Before applying it to the form factors, consider the pure S-matrix for $2 \to 2$ scattering. As a function of the rapidity difference θ, $S(\theta)$ is a meromorphic function on the strip $0 < \mathrm{Im}(\theta) < \pi$. The simple poles are located on the imaginary axis, $\mathrm{Re}(\theta) = 0$, and signal the existence of bound states as virtual intermediate states $a + b \to c$. The values of the rapidity where these bound state poles exist will be called $i\theta^c_{ab}$. By analyticity of the S-matrix, the residue at these poles ought

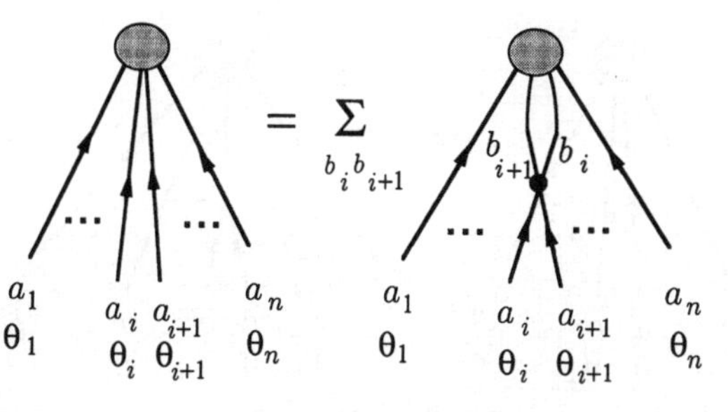

Fig. A1.1. Equation (A1.4) in graphical form. The blob stands for the operator $\mathcal{O}(0)$, and the dot for the S-matrix; the spin factor is not depicted graphically.

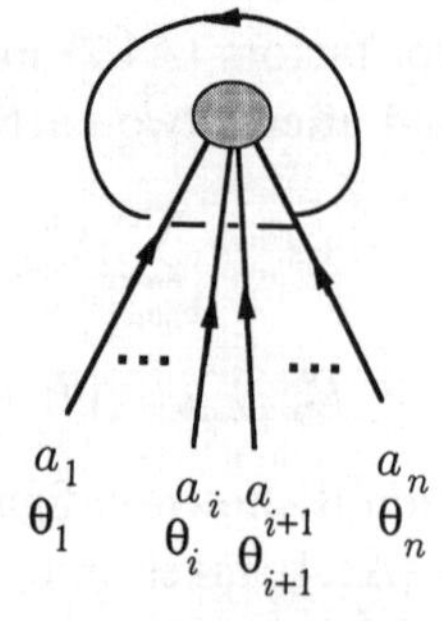

Fig. A1.2

to be related to the three-point couplings Γ^c_{ab} of the external particles to the intermediate bound states c:

$$\underset{\theta=i\theta^c_{ab}}{\mathrm{res}} \; S^{a'b'}_{ab}(\theta) = \Gamma^c_{ab}\,\Gamma^{a'b'}_c \tag{A1.6}$$

In a diagram, this is represented as

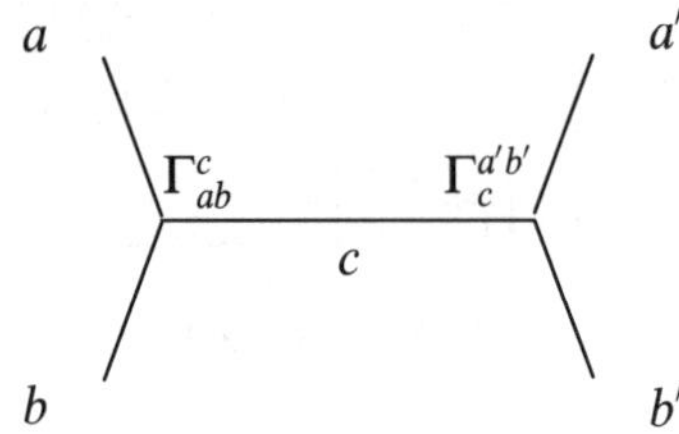

(A1.7)

From the factorization (A1.6) of the four-point S-matrix into three-point processes at its poles, we obtain a similar equation for the form factors (see exercise A1.1 and figure A1.3):

$$\begin{aligned} &\underset{\theta=i\theta^c_{i,i+1}}{\mathrm{res}} \; f^{\mathcal{O}}_{a_1\cdots a_n}(\theta_1,\ldots,\theta_{i-1},\theta_i,\theta_{i+1},\theta_{i+2},\ldots,\theta_n) \\ &= \Gamma^{a_c}_{a_i a_{i+1}}\, f^{\mathcal{O}}_{a_1\cdots a_{i-1}a_c a_{i+2}\cdots a_n}(\theta_1,\ldots,\theta_{i-1},\theta_c,\theta_{i+2},\ldots,\theta_n) \end{aligned} \tag{A1.8}$$

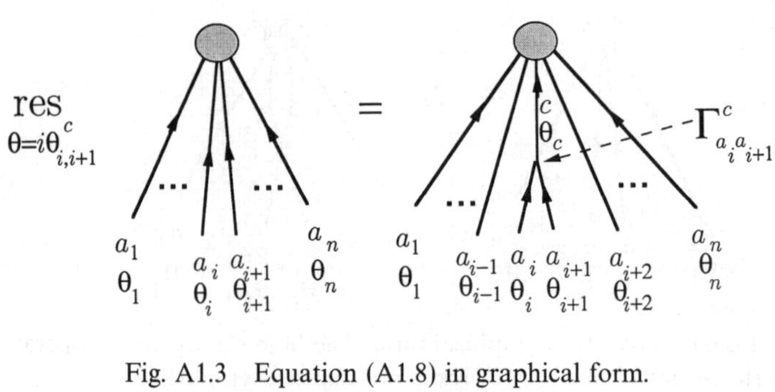

Fig. A1.3 Equation (A1.8) in graphical form.

To complete the conditions we may impose on the form factors, let us assume we have both particles and antiparticles in the theory; i.e. let us suppose the discrete time reversal and charge conjugation symmetries are defined. To distinguish between particles and antiparticles, we draw incoming or outgoing arrows in the graphical representation of the form factor. Applying now the same reasoning which led to equation (A1.8), we find

$$\underset{\overline{\theta}'=i\pi+\theta}{\text{res}} \; f^{\mathcal{O}}_{a_1\cdots a_n a\bar{a}}\left(\theta_1,\ldots,\theta_n,\theta,\overline{\theta}'\right) \tag{A1.9}$$
$$= \left[\delta^{a'_1}_{a_1}\cdots\delta^{a'_n}_{a_n} - \; e^{2\pi i\omega_{a,\mathcal{O}}}\; S^{a'_1\cdots a'_n}_{a_1\cdots a_n}\left(\theta_1,\ldots,\theta_n;\theta\right)\right] f^{\mathcal{O}}_{a'_1\cdots a'_n}\left(\theta_1,\ldots,\theta_n\right)$$

This equation reflects the zero angle scattering of a particle a with the other particles $a_1,\ldots,a_n$, as shown in figure A1.4.

Equations (A1.4), (A1.5), (A1.8) and (A1.9) are the constraints on the form factors derived from the Zamolodchikov algebra (1.14), provided the local operator $\mathcal{O}$ maps elements in the Fock space representation of this algebra into elements of the same Fock space.

A1.2 Form factors at work: the Ising model

At this early stage of the book, the construction above might puzzle the reader, or perhaps appear to be too abstract. Yet the power of this formalism comes out in splendor if we apply it to the computation of the one- and two-point correlation functions for the simple Ising model. The reader is invited to carry out this computation following the guidelines below.

The two-dimensional Ising model near the critical point $T \approx T_c$ is described by a free massive Majorana fermion ψ, whose mass is proportional to $(T - T_c)$. When $T < T_c$ (respectively, $T > T_c$), the system is in an ordered (respectively, disordered) phase. The Kramers–Wannier duality, to be discussed in chapter 8, is thus equivalent to the change $m \leftrightarrow -m$.

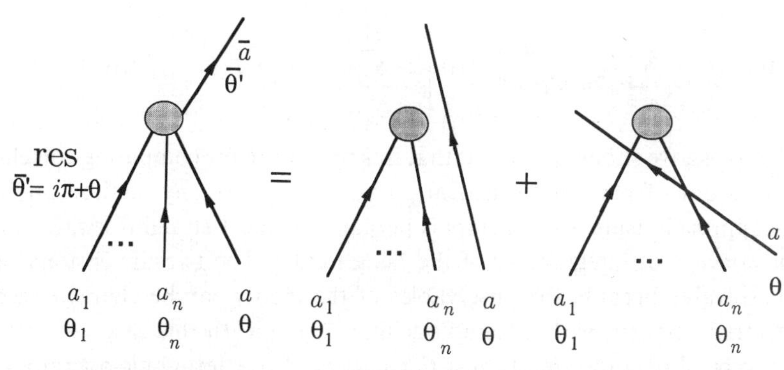

Fig. A1.4 Equation (A1.9) in graphical form.

The S-matrix for the scattering of Majorana fermions is just 1: the familiar (-1) for the interchange of two fermions is already included in the factor $(-1)^{\varepsilon_{a_i}\varepsilon_{a_{i+1}}}$ in equation (A1.4), since $\varepsilon_\psi = 1$.

Besides the local operators ψ, there exist order and disorder operators, σ and μ, whose mutual local index with the real fermions is one-half, $\omega_{\mu\psi} = \omega_{\mu\sigma} = \frac{1}{2}$.

In the high temperature or disordered phase, only the following one-point insertions of σ or μ are non-vanishing:

$$\begin{aligned} f^\sigma(\theta_1, \ldots, \theta_{2n+1}) &= \langle \sigma(0)\psi(\theta_1)\cdots\psi(\theta_{2n+1})\rangle \\ f^\mu(\theta_1, \ldots, \theta_{2n}) &= \langle \mu(0)\psi(\theta_1)\cdots\psi(\theta_{2n})\rangle \end{aligned} \tag{A1.10}$$

Thus, for instance, $\langle\sigma\rangle = 0$ and $\langle\mu\rangle \neq 0$ in the disordered phase. We shall unify the two types of correlators (A1.10) into $f(\theta_1, \ldots, \theta_n)$, with the understanding that we deal with one or the other depending on whether n is even or odd.

The bootstrap equations (A1.4), (A1.5) and (A1.9) now read as follows:

$$\begin{aligned} f(\theta_1, \ldots, \theta_i, \theta_{i+1}, \ldots, \theta_n) &= -f(\theta_1, \ldots, \theta_{i+1}, \theta_i, \ldots, \theta_n) \\ f(\theta_1, \ldots, \theta_n + 2\pi i) &= f(\theta_1, \ldots, \theta_n) \\ \operatorname*{res}_{\overline{\theta'} = i\pi + \theta} f(\theta_1, \ldots, \theta_n, \theta, \overline{\theta'}) &= (1 + S) f(\theta_1, \ldots, \theta_n) \end{aligned} \tag{A1.11}$$

The solution to these equations is simply

$$f(\theta_1, \ldots, \theta_n) = f_\pm \prod_{i>j}^{n} \tanh\left(\frac{\theta_i - \theta_j}{2}\right) \tag{A1.12}$$

where f_+ (respectively, f_-) is a constant depending for n odd (respectively, even) on the normalization of σ (respectively, μ); for instance, $\langle\mu(x)\rangle = f_-$.

To obtain the actual correlators, we use the following factorization equation (and others like it), which amounts to the usual quantum mechanical expansion over a complete set of states:

$$\langle\mu(0)\mu(x)\rangle = \sum_{n\geq 0} \frac{d\theta_1 \cdots d\theta_{2n}}{(2n)!(2\pi)^{2n}} \exp\left[x^2 \sum_{i=1}^{2n} \cosh\theta_i\right] (f^\mu(\theta_1, \ldots, \theta_n))^2 \tag{A1.13}$$

In all fairness, we should point out that this procedure for computing correlators quickly gets out of hand for practical purposes (as does any method anyway). Still, the approach using form factors is perhaps the most straightforward way to seize the concept of integrability of the Ising model. The two-dimensional Ising model is solvable because the observables of the theory can be characterized by form factors consistent with a Zamolodchikov algebra. In this case, this algebra happens to be almost too simple, just the algebra of free fermionic operators, but the general framework is well illustrated.

Exercise

Ex. A1.1 In order to work out in more detail the factorization property of the form factors, determine θ_{ab}^{c} as a function of the masses m_a, m_b and m_c subject to $m_c \leq m_a + m_b$. Find the value of the rapidity θ_c of the particle a_c in (A1.8) as a function of θ_i, θ_{i+1} and θ_c.

2

The Yang–Baxter equation: a first look

In this chapter, we shall analyze in more detail the structure of the Yang–Baxter equation. Our goal is to capture, in a general algebraic framework, the integrability properties of the models studied in chapter 1. This will improve our understanding of these basic models and open the door to powerful generalizations. The first objects we shall encounter are the $\mathcal{R}$-matrix and the monodromy matrix, which will lead us to the notion of a Yang–Baxter algebra. The elements of the Yang–Baxter algebra are the entries of the monodromy matrix of a statistical vertex model. This algebra is at the core of the quantum inverse scattering method, and it will allow us to understand the Bethe ansatz with purely algebraic techniques. Exploring the properties of Yang–Baxter algebras, we shall encounter the closely related objects known as braid groups, quantum groups and affine Hopf algebras. By the end of this chapter, the reader should be convinced of the richness of the Yang–Baxter equation.

2.1 The Yang–Baxter algebra

To establish these ideas, let us continue with the six-vertex model introduced in the previous chapter. Many of the results are easily generalizable to any integrable vertex model.

2.1.1 The $\mathcal{R}$-matrix and the Yang–Baxter equation

The main concept developed in section 1.3 was the transfer matrix, which is an endomorphism

$$t(a,b,c) : V_1 \otimes \cdots \otimes V_L \to V_1 \otimes \cdots \otimes V_L \tag{2.1}$$

where V_i stands for the spin-$\frac{1}{2}$ representation space at the ith position of the lattice. This operator is built up by multiplying Boltzmann weights on the same row and summing over the horizontal states connecting them, while keeping the vertical states above and below them fixed (see equation (1.72)). To make this distinction even clearer, we refer to the space of horizontal states as auxiliary, and denote it by V_a. The space of vertical states on which the transfer matrix acts we shall call quantum, and denote it as $\mathcal{H}^{(L)} = V_1 \otimes \cdots \otimes V_L$, as in (2.1). For the six-vertex model, V_a is also a spin-$\frac{1}{2}$ representation space, like V_i $(i = 1, \ldots, L)$.

According to these definitions, it is a natural progression to interpret the Boltzmann weights associated to the ith vertex as an operator $\mathcal{R}_{ai}$:

$$\mathcal{R}_{ai} : V_a \otimes V_i \to V_a \otimes V_i \tag{2.2}$$

where the subscripts in $\mathcal{R}$ label the vector spaces it acts upon. The operator $\mathcal{R}_{ai}$ is defined by its matrix elements:

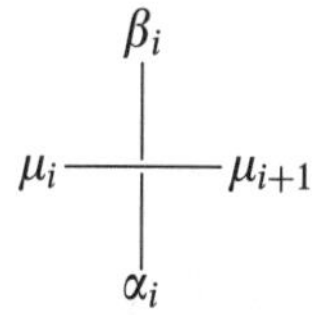

$$= W\begin{pmatrix} \beta_i & \mu_{i+1} \\ \mu_i & \alpha_i \end{pmatrix} \equiv \mathcal{R}^{\mu_{i+1}\beta_i}_{\mu_i\alpha_i} \tag{2.3}$$

$$= {}_a\langle\mu_{i+1}| \otimes {}_i\langle\beta_i|\ \mathcal{R}_{ai}\ |\mu_i\rangle_a \otimes |\alpha_i\rangle_i$$

Note that if $\mathcal{R}$ appears with two subscripts, they label the spaces $\mathcal{R}$ acts upon, whereas if $\mathcal{R}$ appears with two subscripts and two superscripts, they label the basis vectors of the spaces $\mathcal{R}$ is acting between. Using this notation, the transfer matrix (1.72) can be written as

$$\langle\beta|\, t\, |\alpha\rangle = \sum_{\mu' s} \mathcal{R}^{\mu_1\beta_L}_{\mu_L\alpha_L}\mathcal{R}^{\mu_L\beta_{L-1}}_{\mu_{L-1}\alpha_{L-1}} \cdots \mathcal{R}^{\mu_3\beta_2}_{\mu_2\alpha_2}\mathcal{R}^{\mu_2\beta_1}_{\mu_1\alpha_1} \tag{2.4}$$

We have reversed the order of multiplication of the Boltzmann weights to agree with the convention we follow (check out the conventions at the end of this chapter, just before the exercises) for multiplying matrices in the auxiliary space, namely

$$(XY)^{\mu}_{\nu} = X^{\mu}_{\lambda} Y^{\lambda}_{\nu} \tag{2.5}$$

We thus arrive finally at a label-independent expression for the transfer matrix:

$$t = \mathrm{tr}_a\left(\mathcal{R}_{aL}\mathcal{R}_{aL-1}\cdots\mathcal{R}_{a2}\mathcal{R}_{a1}\right) \tag{2.6}$$

Here, tr_a denotes the trace over the auxiliary space V_a.

After these preliminaries, we may wonder whether two transfer matrices t and t', derived from two sets of Boltzmann weights $\mathcal{R}$ and $\mathcal{R}'$, do commute. Of course, t and t' must act on the same quantum space $V_1 \otimes \cdots \otimes V_L$, but the auxiliary spaces for each of them may be different. We multiply t and t', and for clarity we label their respective auxiliary spaces as V_a and V_b, even in the case when these spaces are isomorphic:

$$t\, t' = \mathrm{tr}_{a\times b}\left(\mathcal{R}_{aL}\mathcal{R}'_{bL}\cdots\mathcal{R}_{a1}\mathcal{R}'_{b1}\right) \tag{2.7}$$

where $\mathrm{tr}_{a\times b}$ denotes the trace on $V_a \otimes V_b$. Similarly, multiplying t' and t, we obtain

$$t'\, t = \mathrm{tr}_{a\times b}\left(\mathcal{R}'_{bL}\mathcal{R}_{aL}\cdots\mathcal{R}'_{b1}\mathcal{R}_{a1}\right) \tag{2.8}$$

Hence, t commutes with t' if and only if there exists an invertible matrix X_{ab} such that

$$\mathcal{R}'_{bi}\mathcal{R}_{ai} = X_{ab}\mathcal{R}_{ai}\mathcal{R}'_{bi}X^{-1}_{ab} \qquad \forall i = 1, \ldots, L \tag{2.9}$$

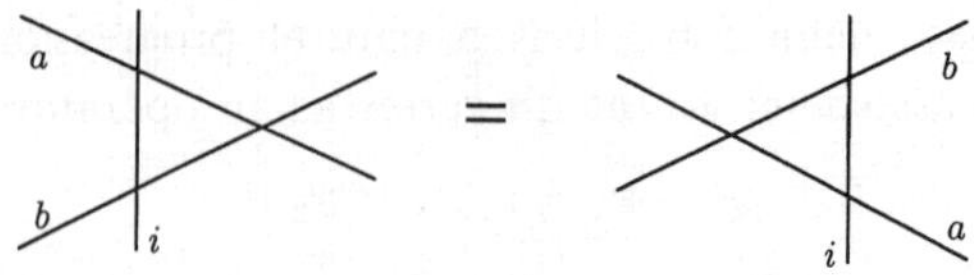

Fig. 2.1. The Yang–Baxter equation (2.11), which is an operator equality on $V_a \otimes V_b \otimes V_i$. Each of these vector spaces is represented by a straight line.

Indeed, using the cyclicity of the trace we find

$$t'\,t = \mathrm{tr}_{a\times b}\Big(X_{ab}\mathscr{R}_{aL}\mathscr{R}'_{bL}X_{ab}^{-1}X_{ab}\mathscr{R}_{aL-1}\mathscr{R}'_{bL-1}X_{ab}^{-1}\cdots$$
$$\cdots X_{ab}\mathscr{R}_{a1}\mathscr{R}'_{b1}X_{ab}^{-1}\Big) = t\,t' \tag{2.10}$$

Moreover, the matrix X_{ab} may be interpreted as arising from Boltzmann weights on the space $V_a \otimes V_b$: we shall call them $\mathscr{R}''_{ab}$. The integrability condition (2.9) is the Yang–Baxter equation in operator formalism:

$$\mathscr{R}''_{ab}\mathscr{R}_{ai}\mathscr{R}'_{bi} = \mathscr{R}'_{bi}\mathscr{R}_{ai}\mathscr{R}''_{ab} \tag{2.11}$$

which can be represented graphically as in figure 2.1. The graphical representation of the relation (2.10) is just an L-fold iteration of figure 2.1, shown in figure 2.2. The reader might want to distinguish $\mathscr{R}$ from $\mathscr{R}^{-1}$ by thinking of strings (cords, ropes, ribbons) on a table instead of lines on the plane, and thus differentiating overcrossings from undercrossings.

With some minor changes in notation, equation (2.11) can be written as

$$\mathscr{R}_{12}\mathscr{R}'_{13}\mathscr{R}''_{23} = \mathscr{R}''_{23}\mathscr{R}'_{13}\mathscr{R}_{12} \tag{2.12}$$

where $\mathscr{R}_{12}$, $\mathscr{R}'_{13}$ and $\mathscr{R}''_{23}$ are Yang–Baxter matrices acting on the spaces $V_1 \otimes V_2$, $V_1 \otimes V_3$ and $V_2 \otimes V_3$, respectively. In components, the operator Yang–Baxter

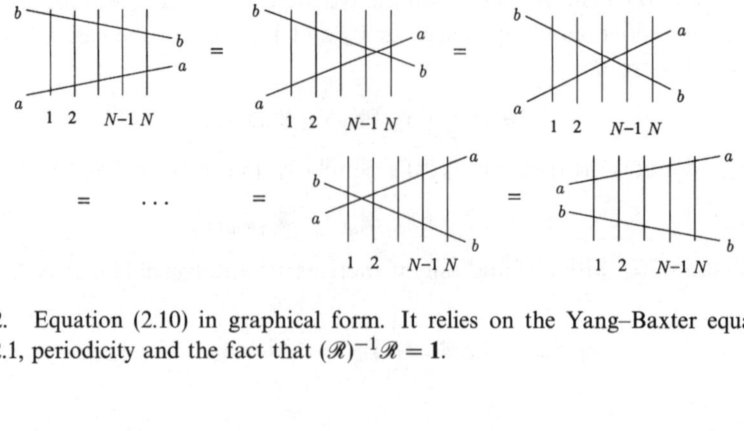

Fig. 2.2. Equation (2.10) in graphical form. It relies on the Yang–Baxter equation of figure 2.1, periodicity and the fact that $(\mathscr{R})^{-1}\mathscr{R} = \mathbf{1}$.

equation (2.12) reads as follows:

$$\sum_{j_1,j_2,j_3} \mathscr{R}^{k_1k_2}_{j_1j_2} \mathscr{R}'^{j_1k_3}_{i_1j_3} \mathscr{R}''^{j_2j_3}_{i_2i_3} = \sum_{j_1,j_2,j_3} \mathscr{R}''^{k_2k_3}_{j_2j_3} \mathscr{R}'^{k_1j_3}_{j_1i_3} \mathscr{R}^{j_1j_2}_{i_1i_2} \tag{2.13}$$

and its diagrammatic representation is shown in figure 2.3.

We need to add a note about this diagram. We will always read operator equations from right to left, thinking of the order in which these operators act as the direction of time. We choose the convention that, in our figures, time will always flow upwards.

Equation (2.13) is the most general form of the Yang–Baxter equation for vertex models, in the sense that the spaces V_1, V_2 and V_3 need not be isomorphic. We shall consider this possibility later on, but, for the time being, all these vector spaces are two-dimensional, so that the $\mathscr{R}$ operator is a 4×4 matrix which, in the case of the six-vertex model, is

$$\mathscr{R}^{(6v)}(a,b,c) = \begin{pmatrix} a & & & \\ & b & c & \\ & c & b & \\ & & & a \end{pmatrix} \tag{2.14}$$

If $\mathscr{R}^{(6v)}$ is to be invertible, then we must require $a \neq 0$ and $b \neq \pm c$.

On taking three six-vertex $\mathscr{R}$-matrices

$$\mathscr{R} = \mathscr{R}^{(6v)}(a,b,c), \quad \mathscr{R}' = \mathscr{R}^{(6v)}(a',b',c'), \quad \mathscr{R}'' = \mathscr{R}^{(6v)}(a'',b'',c'') \tag{2.15}$$

the Yang–Baxter equation holds provided

$$\Delta(a,b,c) = \Delta(a',b',c') = \Delta(a'',b'',c'') \tag{2.16}$$

in full agreement with equation (1.91). The Yang–Baxter equation captures completely the integrability of the six-vertex model, encoded in (2.16).

We discussed in section 1.3 how the equation $\Delta = (a^2 + b^2 - c^2)/(2ab)$ can be uniformized for constant Δ in terms of a variable u living on the sphere. Expressing the weights a, b and c in terms of u, we find that the Yang–Baxter matrix $\mathscr{R}(u) = \mathscr{R}^{(6v)}(a(u), b(u), c(u))$ satisfies the Yang–Baxter equation (2.12) in the form

$$\mathscr{R}_{12}(u)\mathscr{R}_{13}(u')\mathscr{R}_{23}(u'') = \mathscr{R}_{23}(u'')\mathscr{R}_{13}(u')\mathscr{R}_{12}(u) \tag{2.17}$$

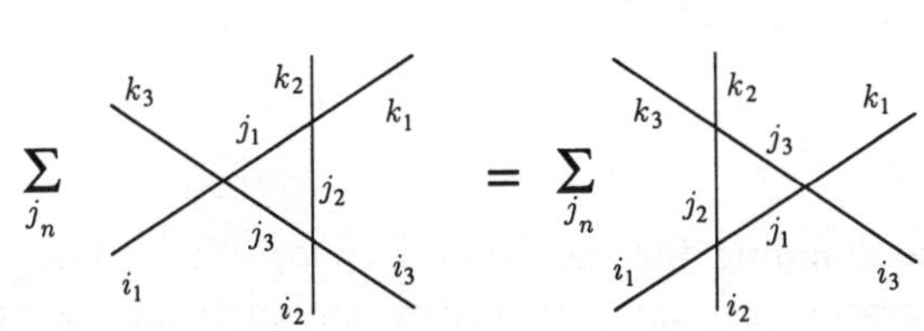

Fig. 2.3. Operator Yang–Baxter equation (2.13) in components. Each straight line stands for a vector space, and the indices label the vector components.

with u'' fixed in terms of u and u'. Now, on a sphere, all points are equivalent, in the sense that any point can be mapped to any other one by a conformal transformation. We may therefore choose the functions $a(u)$, $b(u)$ and $c(u)$ in such a way that u'' is just $u' - u$. Then (2.17) adopts the usual additive form

$$\mathscr{R}_{12}(u)\mathscr{R}_{13}(u+v)\mathscr{R}_{23}(v) = \mathscr{R}_{23}(v)\mathscr{R}_{13}(u+v)\mathscr{R}_{12}(u) \tag{2.18}$$

valid for any complex u and v. The reader is invited to check that the parametrization (1.101) does indeed satisfy this form of the Yang–Baxter equation, with the difference property.

In more general vertex models, the Boltzmann weights could be parametrized in a more complicated way. Since a genus one curve enjoys the existence of a conformal Killing vector, we could use a uniformization parameter u with the difference property ($u'' = u' - u$) in this case also. However, for a generic curve of higher genus, the absence of conformal Killing vectors destroys the difference property. This does not prevent the existence of solutions to the Yang–Baxter relation (2.12) involving Boltzmann weights sited on curves of higher genus (chapter 7). The moral of this story is that the Yang–Baxter equation is very general, far transcending the six-vertex model and its close relatives.

2.1.2 The monodromy matrix

Let us now introduce another important concept in the context of integrable models, that of the monodromy matrix $T(u)$. We define $T(u)$ in the same manner as the transfer matrix, except that we do not trace over the first (or last, due to periodic boundary conditions) horizontal states in (2.4), that is to say

$$T(u) = \mathscr{R}_{aL}\mathscr{R}_{aL-1}\cdots\mathscr{R}_{a2}\mathscr{R}_{a1} \tag{2.19}$$

The trace of the monodromy matrix on the auxiliary space is just the transfer matrix

$$t(u) = \mathrm{tr}_a T(u) \tag{2.20}$$

Using $i, j, \ldots$ as labels in the auxiliary space V_a, we see that $T(u)$ is, in fact, a matrix $T_i^j(u)$ of operator valued functions which act, in this case, on the Hilbert space $\mathscr{H}^{(L)} = V_1 \otimes \cdots \otimes V_L$. These operators will be represented graphically as

$T_i^j(u) =$ 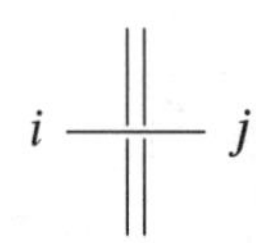

(2.21)

with the double line standing for the Hilbert space $\mathscr{H}^{(L)}$. The characteristic feature of these operators is that they satisfy an important set of quadratic relations reflecting their behavior under monodromy:

$$\mathscr{R}_{ab}(u-v)\,(T_a(u) \otimes T_b(v)) = (T_b(v) \otimes T_a(u))\,\mathscr{R}_{ab}(u-v) \tag{2.22}$$

This equation constitutes the cornerstone of the quantum inverse scattering method; it is also at the origin of the quantum group. The subscripts a and b are short-hand for the two auxiliary spaces V_a and V_b on which the operators T and $\mathcal{R}$ act. For extra clarity, we have indulged in a notational redundance when indicating the tensor product in (2.22), which is taken over these auxiliary spaces, while the quantum indices (not shown) are multiplied as ordinary matrix indices. The proof of (2.22) uses the Yang–Baxter equation (2.18) repeatedly, and elucidates the index interplay:

$$\begin{aligned}
&\mathcal{R}_{ab}(u-v)\,(T_a(u)\otimes T_b(v)) \\
&\qquad = \mathcal{R}_{ab}(u-v)\mathcal{R}_{aL}(u)\mathcal{R}_{bL}(v)\cdots\mathcal{R}_{a1}(u)\mathcal{R}_{b1}(v) \\
&\qquad = \mathcal{R}_{bL}(v)\mathcal{R}_{aL}(u)\cdots\mathcal{R}_{b1}(v)\mathcal{R}_{a1}(u)\mathcal{R}_{ab}(u-v) \\
&\qquad = (T_b(v)\otimes T_a(u))\,\mathcal{R}_{ab}(u-v)
\end{aligned} \tag{2.23}$$

For practical purposes, it is often convenient to write equation (2.22) in components:

$$\sum_{j_1,j_2} \mathcal{R}^{k_1k_2}_{j_1j_2}(u-v)T(u)^{j_1}_{i_1}\;T(v)^{j_2}_{i_2} = \sum_{j_1,j_2} T(v)^{k_2}_{j_2}\;T(u)^{k_1}_{j_1}\mathcal{R}^{j_1j_2}_{i_1i_2}(u-v) \tag{2.24}$$

Given (2.22), it is an easy task to prove the commutativity of the transfer matrices, i.e. $[\mathrm{tr}\,T_a(u), \mathrm{tr}\,T_b(u)] = 0$. Using the diagrammatic representation (2.21), this expression can be depicted as in figure 2.4. Note the similarity between equations (2.13) and (2.24), and between their corresponding pictures (figures 2.3 and 2.4).

Equation (2.24) has been derived for the six-vertex model, but it can be taken as the starting point for the construction of integrable vertex models, at least for those with the difference property. To this end, we shall introduce the formal notion of a Yang–Baxter algebra.

2.1.3 Co-product and the Yang–Baxter algebra

A Yang–Baxter algebra $\mathcal{A}$ consists of a couple $(\mathcal{R}, T)$, where $\mathcal{R}$ is an $n^2 \times n^2$ invertible matrix and $T_i^j(u)$ $(i, j \in \{1, \ldots, n\}; u \in \mathbb{C})$ are the generators of $\mathcal{A}$. They

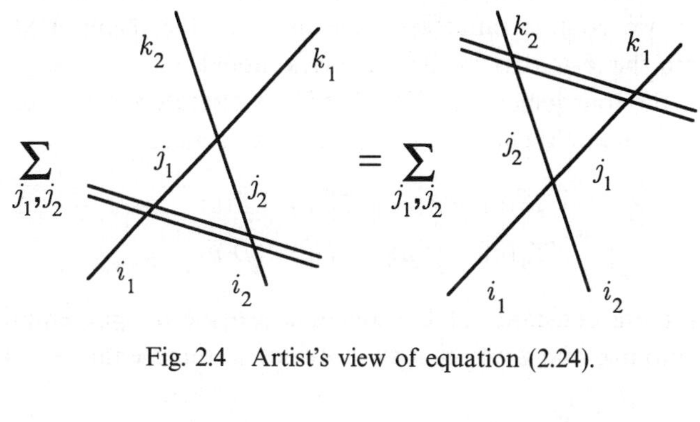

Fig. 2.4 Artist's view of equation (2.24).

must satisfy the quadratic relations (2.24), the consistency of which implies the Yang–Baxter equation (2.18) for $\mathscr{R}(u)$. The entries of the matrix $\mathscr{R}(u)$ play the role of structure constants (or rather, "structure functions") of the algebra $\mathscr{A}$. This is quite analogous to a Lie algebra, or better yet to its universal enveloping algebra, which is also defined in terms of a set of generators and structure constants. Following this analogy, the Yang–Baxter relation plays the role of the Jacobi identity: they both reflect the associativity of the corresponding algebras. We shall see later on, when we discuss quantum groups, that this similarity is quite strong.

By now, the distinction between the abstract algebra $\mathscr{A}$ itself and its representations should be clear enough. The elements of the algebra are represented as operators acting on some Hilbert space, two-dimensional in the case of the six-vertex model. The construction of representations of the Yang–Baxter algebra is equivalent to the construction of integrable vertex models. We shall return to this point in future chapters.

An important property of Yang–Baxter algebras is their "addition law", called co-multiplication or co-product Δ, which maps the algebra $\mathscr{A}$ into the tensor product $\mathscr{A} \otimes \mathscr{A}$ whilst preserving the algebraic relations of $\mathscr{A}$:

$$\begin{aligned} \Delta : \mathscr{A} &\to \mathscr{A} \otimes \mathscr{A} \\ T_i^j(u) &\mapsto \sum_k T_i^k(u) \otimes T_k^j(u) \end{aligned} \qquad (2.25)$$

The diagrammatic representation of the co-product follows from (2.21):

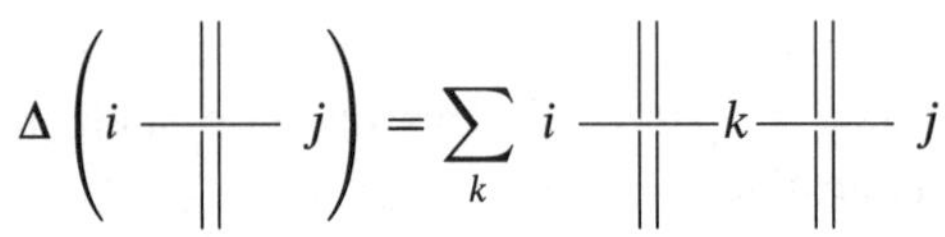

(2.26)

It is left as an exercise for the reader to check that ΔT_i^j satisfy the same relations as T_i^j in (2.24). The algebra $\mathscr{A}$ has thus both a multiplication and a co-multiplication; $\mathscr{A}$ is called a bi-algebra.

The definition of a Yang–Baxter algebra just provided is general, and it can be applied to any $\mathscr{R}$-matrix satisfying the Yang–Baxter relation. In the remainder of this chapter, we confine ourselves once more to the Yang–Baxter algebra constructed from the $\mathscr{R}$-matrix of the six-vertex model.

We represent the four generators T_i^j of $\mathscr{A}^{(6v)}$ as operators acting on a Hilbert space $\mathscr{H}$. We shall name them as follows, for convenience:

$$\begin{aligned} T_0^0(u) = A(u), \quad T_1^0(u) = B(u) \\ T_0^1(u) = C(u), \quad T_1^1(u) = D(u) \end{aligned} \qquad (2.27)$$

Just as the structure constants of Lie algebras provide a representation of the algebra (the adjoint), the $\mathscr{R}^{(6v)}$ matrix provides a representation of $\mathscr{A}^{(6v)}$ of dimension two under the identification

$$\left(T_i^j(u)\right)^k_\ell = \mathscr{R}^{jk}_{i\ell}(u) \tag{2.28}$$

or explicitly

$$\begin{aligned}
A(u) &= \begin{pmatrix} a(u) & 0 \\ 0 & b(u) \end{pmatrix} = \frac{a+b}{2}\,\mathbf{1} + \frac{a-b}{2}\,\sigma_3 \\
B(u) &= \begin{pmatrix} 0 & 0 \\ c(u) & 0 \end{pmatrix} = c\,\sigma^- \\
C(u) &= \begin{pmatrix} 0 & c(u) \\ 0 & 0 \end{pmatrix} = c\,\sigma^+ \\
D(u) &= \begin{pmatrix} b(u) & 0 \\ 0 & a(u) \end{pmatrix} = \frac{a+b}{2}\,\mathbf{1} - \frac{a-b}{2}\,\sigma_3
\end{aligned} \tag{2.29}$$

In diagrams, this can be represented as

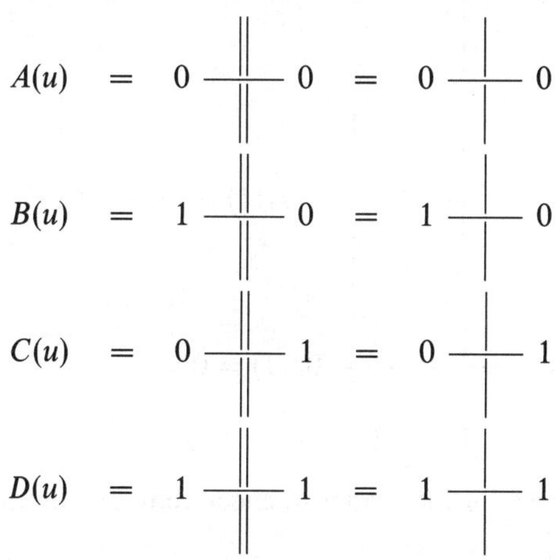

(2.30)

The single vertical line in (2.30) denotes the two-dimensional spin-$\frac{1}{2}$ irrep (recall equation (2.21)).

2.1.4 Algebraic Bethe ansatz

Equations (2.30) yield what we might call the spin-$\frac{1}{2}$ representation of the algebra (2.24). This nomenclature is appropriate since $C(u)$ and $B(u)$ act as raising and lowering operators, respectively, whereas $A(u)$ and $D(u)$ span the Cartan subalgebra of $SU(2)$. Using now the bi-algebra structure of $\mathscr{A}^{(6v)}$ defined by the co-multiplication (2.25), we may obtain a representation of $\mathscr{A}^{(6v)}$ on the space

$\mathscr{H}^{(L)} = \bigotimes^L V_{\frac{1}{2}}$. In particular, for $L = 2$ we find

$$\begin{aligned}\Delta(A(u)) &= A(u) \otimes A(u) + C(u) \otimes B(u)\\ \Delta(B(u)) &= B(u) \otimes A(u) + D(u) \otimes B(u)\\ \Delta(C(u)) &= C(u) \otimes D(u) + A(u) \otimes C(u)\\ \Delta(D(u)) &= D(u) \otimes D(u) + B(u) \otimes C(u)\end{aligned} \tag{2.31}$$

To see how this works in detail, let us present one simple diagrammatic computation, the action of ΔB on $\mathscr{H}^{(2)}$:

$\Delta B(u)\,|00\rangle = 1$ 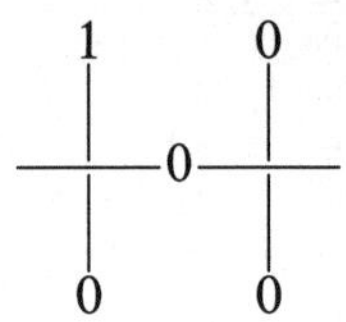

$0\,|10\rangle + 1$ 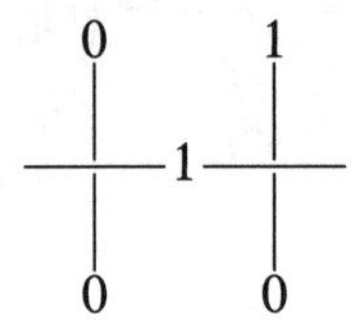

$0\,|01\rangle$

$= c(u)a(u)\,|10\rangle + b(u)c(u)\,|01\rangle$

$\Delta B(u)\,|10\rangle = 1$ 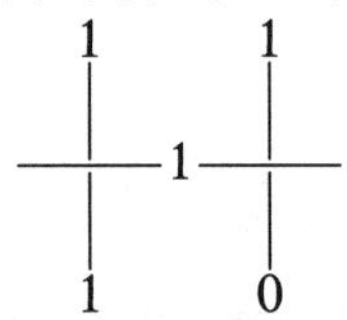

$0\,|11\rangle = a(u)c(u)\,|11\rangle$

$\Delta B(u)\,|01\rangle = 1$ 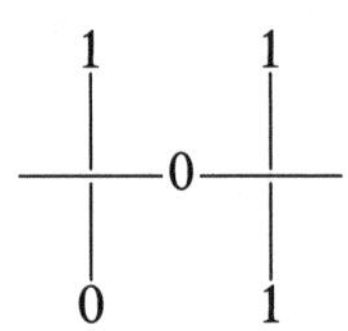

$0\,|11\rangle = b(u)c(u)\,|11\rangle$ (2.32)

$\Delta B(u)\,|11\rangle = 1$ 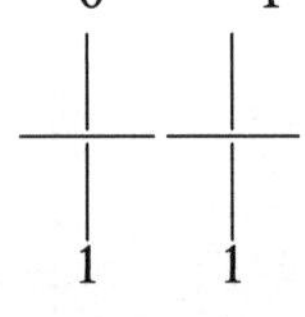

$0\,|11\rangle = 0$

We leave it as an exercise for the reader to check that ΔC annihilates the reference state $|\Omega\rangle \equiv |00\rangle = |\Uparrow\Uparrow\rangle$:

$$\Delta C(u)\,|00\rangle = 0 \tag{2.33}$$

Using this interpretation of $B(u)$ and $C(u)$ as creation and annihilation operators, it follows from (2.32) that $\Delta B(u)$ acting on the reference state $|\Omega\rangle$ yields a state in the sector with the number of spins down equal to one. Using the notation of section 1.3, we can rewrite this state as

$$\Delta B(u)\,|\Omega\rangle = |\Psi_1\rangle = \sum_x f(x)\,|x\rangle = f(1)\,|10\rangle + f(2)\,|01\rangle \tag{2.34}$$

with

$$f(1) = c(u)a(u)\,, \qquad f(2) = b(u)c(u) \tag{2.35}$$

Comparing (2.34) and (2.35) with (1.40) and (1.41), we deduce the relationship between Boltzmann weights and quasi-momenta:

$$\frac{b(u)}{a(u)} = \mathrm{e}^{ik} \tag{2.36}$$

This method for lowering spins (i.e. creating 1's) from a reference state by means of the B operators can be extended to a lattice with $L > 2$ sites. To do so, we recall the definition (2.27) of the operator B as the entry T_1^0 of the monodromy matrix; thanks to the co-product (2.25), it can be made to act on the space $\mathscr{H} = \otimes^L V_{\frac{1}{2}}$:

$$B(u) = \Delta^{L-1}\left(T_1^0(u)\right) \tag{2.37}$$

where

$$\Delta^{L-1} : \mathscr{A} \to \overbrace{\mathscr{A} \otimes \cdots \otimes \mathscr{A}}^{L \text{ times}} \tag{2.38}$$

is the associative generalization of (2.25), $\Delta^{L-1} = (\mathbf{1} \otimes \Delta)\Delta^{L-2}$ with $L \geq 2$. Hence, a state with M spins down can be constructed as follows:

$$|\Psi_M\rangle = \prod_{i=1}^{M} B(u_i)\,|00\cdots 0\rangle = \prod_{i=1}^{M} B(u_i)\,|\Omega\rangle \tag{2.39}$$

The states (2.39) are called algebraic Bethe ansatz states ("algebraic" in contrast with the "co-ordinate" description (1.53)), and constitute a very good starting point for solving the eigenvalue problem of the transfer matrix. In order to show this, let us work out more explicitly the relations satisfied by the generators of the six-vertex Yang–Baxter algebra.

From (2.14) and (2.24) we obtain, for arbitrary u and v,

$$B(u)B(v) = B(v)B(u) \tag{2.40a}$$

$$A(u)B(v) = \frac{a(v-u)}{b(v-u)}B(v)A(u) - \frac{c(v-u)}{b(v-u)}B(u)A(v) \tag{2.40b}$$

$$D(u)B(v) = \frac{a(u-v)}{b(u-v)}B(v)D(u) - \frac{c(u-v)}{b(u-v)}B(u)D(v) \tag{2.40c}$$

$$C(u)B(v) - B(v)C(u) = \frac{c(u-v)}{b(u-v)}\left(A(v)D(u) - A(u)D(v)\right) \tag{2.40d}$$

Equation (2.40a) implies that the algebraic Bethe ansatz state (2.39) is independent of the ordering in which the B operators are multiplied.

The transfer matrix of the six-vertex model can be written from equations (2.20) and (2.27) as

$$t^{(6v)}(u) = \mathrm{tr}_a T^{(6v)}(u) = A(u) + D(u) \tag{2.41}$$

Therefore, the problem of diagonalizing the transfer matrix (2.41) in the algebraic Bethe ansatz basis (2.39) amounts to finding parameters $\{u_i, i = 1, \ldots, M\}$ such that

$$\begin{aligned} t^{(6v)}(u)\,|\Psi_M\rangle &= [A(u)+B(u)]\prod_{i=1}^{M} B(u_i)\,|\Omega\rangle \\ &= \Lambda_M(u;\{u_i\})\prod_{i=1}^{M} B(u_i)\,|\Omega\rangle \end{aligned} \tag{2.42}$$

The advantage of using the algebraic Bethe ansatz states is that the whole computation involved in (2.42) reduces to a systematic use of the commutation relations (2.40), in addition to the obvious relations

$$A(u)\,|\Omega\rangle = a(u)^L\,|\Omega\rangle\,, \qquad D(u)\,|\Omega\rangle = b(u)^L\,|\Omega\rangle \tag{2.43}$$

Indeed, using (2.40b), (2.40c) and (2.43), we find

$$\begin{aligned} (A(u)+D(u))\prod_{i=1}^{M} B(u_i)\,|\Omega\rangle &= \text{unwanted terms} \\ + \left[a^L(u)\prod_{i=1}^{M}\frac{a(u_i-u)}{b(u_i-u)} + b^L(u)\prod_{i=1}^{M}\frac{a(u-u_i)}{b(u-u_i)}\right] &\prod_{i=1}^{M} B(u_i)\,|\Omega\rangle \end{aligned} \tag{2.44}$$

The first term on the right-hand side of this equation gives us the eigenvalue of the transfer matrix:

$$\Lambda_M(u;\{u_i\}) = a^L(u)\prod_{i=1}^{M}\frac{a(u_i-u)}{b(u_i-u)} + b^L(u)\prod_{i=1}^{M}\frac{a(u-u_i)}{b(u-u_i)} \tag{2.45}$$

From the second summands in (2.40b) and (2.40c), however, we also obtain terms which are not of the desirable form $\prod_{i=1}^{M} B(u_i)\,|\Omega\rangle$. If these terms are present, the algebraic Bethe ansatz does not work. The unwanted terms actually cancel, for the six-vertex model, under a judicious choice of the u_i parameters. The condition that the parameters u_i should satisfy to guarantee the cancellation of the unwanted terms is precisely the Bethe equations, written in the form

$$\left(\frac{a(u_i)}{b(u_i)}\right)^L = \prod_{\substack{j=1\\ j\neq i}}^{M}\frac{a(u_i-u_j)b(u_j-u_i)}{a(u_j-u_i)b(u_i-u_j)} \tag{2.46}$$

Equations (2.45) and (2.46) are the final outcome of the diagonalization of the transfer matrix through the algebraic Bethe ansatz, which we can now compare with equations (1.89) and (1.90) from chapter 1. Matching the eigenvalues (1.89) and (2.45) yields

$$\frac{a(u_i-u)}{b(u_i-u)} = P(k_i) = \frac{a(u)b(u) + \left(c^2(u)-b^2(u)\right)\,\mathrm{e}^{-ik_i(u_i)}}{a^2(u)-a(u)b(u)\,\mathrm{e}^{-ik_i(u_i)}} \tag{2.47a}$$

$$\frac{a(u-u_i)}{b(u-u_i)} = Q(k_i) = \frac{a^2(u)-c^2(u)-a(u)b(u)\,\mathrm{e}^{-ik_i(u_i)}}{a(u)b(u)-b^2(u)\,\mathrm{e}^{-ik_i(u_i)}} \tag{2.47b}$$

Choosing $u = 0$ in (2.47a) and recalling from (1.103) that $a(0) = c(0) \neq 0$, $b(0) = 0$, we obtain

$$\frac{b(u_i)}{a(u_i)} = \mathrm{e}^{ik_i(u_i)} \tag{2.48}$$

Hence, the comparison between (2.46) and (1.90) finally produces

$$\hat{S}_{ji} = \frac{a(u_j - u_i)b(u_i - u_j)}{a(u_i - u_j)b(u_j - u_i)} = -\frac{1 - 2\Delta\,\mathrm{e}^{ik_i} + \mathrm{e}^{i(k_i+k_j)}}{1 - 2\Delta\,\mathrm{e}^{ik_j} + \mathrm{e}^{i(k_i+k_j)}} \tag{2.49}$$

which confirms the result (2.36). Equations (2.47), (2.48) and (2.49) provide the map between the quasi-momenta k_i and the uniformization variables u_i used in the algebraic Bethe ansatz construction.

Let us introduce the following explicit uniformization of the Boltzmann weights of the six-vertex model (see (1.101) and the comments in section 2.5.2):

$$\begin{aligned} a(u) &= \mathscr{R}^{00}_{00}(u) = \mathscr{R}^{11}_{11}(u) = \sinh(u + i\gamma) \\ b(u) &= \mathscr{R}^{10}_{10}(u) = \mathscr{R}^{01}_{01}(u) = \sinh u \\ c(u) &= \mathscr{R}^{10}_{01}(u) = \mathscr{R}^{01}_{10}(u) = i \sin\gamma \end{aligned} \tag{2.50}$$

where the parameter γ is related to the anisotropy Δ by the relation

$$\Delta = \cos\gamma \tag{2.51}$$

Note that $u = 0$ does indeed satisfy the conditions (1.103), which we used to define the hamiltonian. Moreover, the uniformization (2.50) satisfies the Yang–Baxter equations (2.18) identically. The solution (2.50) to the Yang–Baxter equation is called trigonometric. For the eight-vertex model, as we shall see in chapter 4, the spectral variety of Boltzmann weights is a curve of genus one, and thus the uniformization calls for elliptic functions. Accordingly, the eight-vertex solution to Yang–Baxter is called elliptic.

Using the map

$$\mathrm{e}^{ik_j} = \frac{\sinh u_j}{\sinh(u_j + i\gamma)} \tag{2.52}$$

it is easy to check that equations (2.47) and (2.49) are satisfied with the six-vertex $\mathscr{R}$-matrix (2.50), and that the Bethe equation can be written as

$$\left(\frac{\sinh(u_j + i\gamma)}{\sinh u_j}\right)^L = \prod_{\substack{k=1\\k\neq j}}^{M} \frac{\sinh(u_j - u_k + i\gamma)}{\sinh(u_j - u_k - i\gamma)} \tag{2.53}$$

There is yet another parametrization of k_i or u_i in terms of the rapidity λ_i, given by

$$u_j = \frac{\gamma}{2}(\lambda_j - i) \quad \Rightarrow \quad \mathrm{e}^{ik_j} = \frac{\mathrm{e}^{\gamma\lambda_j} - \mathrm{e}^{i\gamma}}{\mathrm{e}^{\gamma\lambda_j + i\gamma} - 1} \tag{2.54}$$

whereby the Bethe ansatz equation (2.53) takes the form

$$\left(\frac{\sinh\frac{\gamma}{2}(\lambda_j+i)}{\sinh\frac{\gamma}{2}(\lambda_j-i)}\right)^L = \prod_{\substack{k=1\\k\neq j}}^{M} \frac{\sinh\frac{\gamma}{2}(\lambda_j-\lambda_k+2i)}{\sinh\frac{\gamma}{2}(\lambda_j-\lambda_k-2i)} \tag{2.55}$$

Sometimes, u_i is also called rapidity; the $\mathcal{R}$-matrix depends only on the differences $u_i - u_j$ or $\lambda_i - \lambda_j$.

We may calculate the energy of the Bethe ansatz state (2.39) from the eigenvalue of the transfer matrix (2.45) if we recall that the hamiltonian is defined as

$$\begin{aligned} H &= i\frac{\partial}{\partial u}\log\left(\frac{t(u)}{a(u)^L}\right)\bigg|_{u=0} \\ &= \frac{1}{2\sin\gamma}\sum_{j=1}^{L}\left[\sigma_j^x\sigma_{j+1}^x + \sigma_j^y\sigma_{j+1}^y + \cos\gamma\left(\sigma_j^z\sigma_{j+1}^z - 1\right)\right] \end{aligned} \tag{2.56}$$

Indeed, carrying out the computation explicitly, we find

$$\begin{aligned} E_M\left(\{u_j\}\right) &= i\frac{\partial}{\partial u}\log\left(\frac{\Lambda_M\left(u,\{u_j\}\right)}{a(u)^L}\right)\bigg|_{u=0} \\ &= -\sum_{j=1}^{M}\frac{\sin\gamma}{\sinh u_j\,\sinh(u_j+i\gamma)} \\ &= -\sum_{j=1}^{M}\frac{\sin\gamma}{\sinh\frac{\gamma}{2}\left(\lambda_j+i\right)\,\sinh\frac{\gamma}{2}\left(\lambda_j-i\right)} \end{aligned} \tag{2.57}$$

Similarly, the total momentum of the same Bethe state is

$$\begin{aligned} P_M\left(\{u_j\}\right) &= i\log\left(\frac{t(u)}{a(u)^L}\right)\bigg|_{u=0} = i\sum_{j=1}^{M}\log\frac{\sinh\left(u_j+i\gamma\right)}{\sinh u_j} \\ &= i\sum_{j=1}^{M}\log\left(\frac{\sinh\frac{\gamma}{2}\left(\lambda_j+i\right)}{\sinh\frac{\gamma}{2}\left(\lambda_j-i\right)}\right) \\ &= \sum_{j=1}^{M}\left(2\arctan\left[\operatorname{cotan}\frac{\gamma}{2}\,\tanh\frac{\gamma\lambda_j}{2}\right]+\pi\right)\bmod(2\pi) \end{aligned} \tag{2.58}$$

The advantage of the parametrization (2.54) is that the isotropic limit $\gamma \to 0$ (i.e. $\Delta \to 1$) of equation (2.55) reproduces the historic original Bethe equation for the isotropic Heisenberg model:

$$\left(\frac{\lambda_j+i}{\lambda_j-i}\right)^L = \prod_{\substack{k=1\\k\neq j}}^{M}\frac{\lambda_j-\lambda_k+2i}{\lambda_j-\lambda_k-2i} \tag{2.59}$$

To calculate the Heisenberg hamiltonian from (2.56), we multiply (2.56) by $(\sin\gamma)/2$ and take the limit $\gamma \to 0$:

$$H_{XXX} = \frac{1}{4}\sum_{j=1}^{L}\left(\vec{\sigma}_j \cdot \vec{\sigma}_{j+1} - 1\right) \tag{2.60}$$

where $\vec{\sigma}$ are the usual Pauli matrices. In this isotropic limit, the energy and momentum of a Bethe state are given by

$$\begin{aligned} E_M &= \sum_{j=1}^{M} \frac{-2}{1+\lambda_j^2} \\ P_M &= \sum_{j=1}^{M} \left[2 \arctan\left(\lambda_j\right) + \pi\right] \quad \mathrm{mod}\,(2\pi) \end{aligned} \tag{2.61}$$

Equations (2.47)–(2.49) allow us to link up with the description we have presented in chapter 1 and to explain the integrability content of the Bethe equations of the six-vertex model, which is codified in the Yang–Baxter equation. The factorization properties discussed in chapter 1 simply reflect the consistency of the Yang–Baxter algebra. We have thus reinterpreted the factorization properties as integrability of the six-vertex model.

2.2 Yang–Baxter algebras and braid groups

The basic relation studied in the previous section is the Yang–Baxter equation (2.18). In components, it reads as

$$\begin{aligned} &\sum_{j_1,j_2,j_3} \mathcal{R}_{j_1 j_2}^{k_1 k_2}(u)\mathcal{R}_{i_1 j_3}^{j_1 k_3}(u+v)\mathcal{R}_{i_2 i_3}^{j_2 j_3}(v) \\ &\qquad = \sum_{j_1,j_2,j_3} \mathcal{R}_{j_2 j_3}^{k_2 k_3}(v)\mathcal{R}_{j_1 i_3}^{k_1 j_3}(u+v)\mathcal{R}_{i_1 i_2}^{j_1 j_2}(u) \end{aligned} \tag{2.62}$$

where all the indices run from 1 to $n = \dim V$ ($n = 2$ for the six-vertex model). An interesting way to write this equation calls for the permuted R-matrix,

$$R = P\mathcal{R} : V_1 \otimes V_2 \to V_2 \otimes V_1 \tag{2.63}$$

where P is the permutation map

$$\begin{aligned} P : V_1 \otimes V_2 &\to V_2 \otimes V_1 \\ e_i^{(1)} \otimes e_j^{(2)} &\mapsto e_j^{(2)} \otimes e_i^{(1)} \end{aligned} \tag{2.64}$$

with $\{e_r^{(i)}, r = 1, \ldots, n\}$ a basis of V_i. The relationship between the entries of R and $\mathcal{R}$ is straightforward:

$$R = P\mathcal{R} \quad \Longleftrightarrow \quad R_{ij}^{k\ell} = \mathcal{R}_{ij}^{\ell k} \tag{2.65}$$

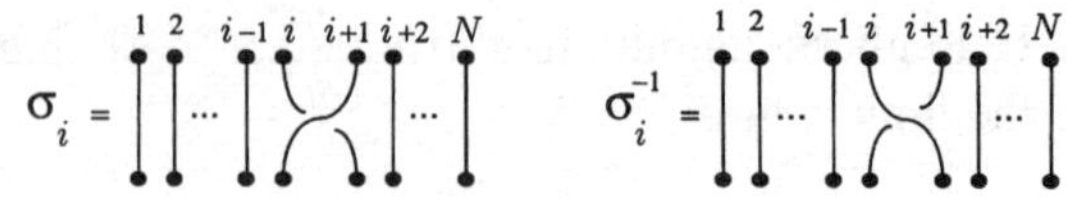

Fig. 2.5 Action of the braid group generator σ_i on L strands.

With the help of the permuted R-matrix, the Yang–Baxter equation (2.62) can be written as

$$(\mathbf{1}\otimes R(u))\,(R(u+v)\otimes\mathbf{1})\,(\mathbf{1}\otimes R(v))$$
$$= (R(v)\otimes\mathbf{1})\,(\mathbf{1}\otimes R(u+v))\,(R(u)\otimes\mathbf{1}) \tag{2.66}$$

Every operator in parentheses acts on the space $V\otimes V\otimes V$:

$$(R(u)\otimes\mathbf{1})\,e_{i_1}\otimes e_{i_2}\otimes e_{i_3} = R^{j_2j_1}_{i_1i_2}(u)\,e_{j_2}\otimes e_{j_1}\otimes e_{i_3}$$
$$(\mathbf{1}\otimes R(u))\,e_{i_1}\otimes e_{i_2}\otimes e_{i_3} = R^{j_3j_2}_{i_2i_3}(u)\,e_{i_1}\otimes e_{j_3}\otimes e_{j_2} \tag{2.67}$$

The alert reader will scrutinize the difference between equations (2.66) and (2.18), noting that R is very close in form to the factorizable S-matrices of chapter 1.

The reason for writing the Yang–Baxter equation in the form (2.66) comes from its relationship with the braid group. The braid group B_L on L strands is generated by $L-1$ elements σ_i ($i=1,\ldots,L-1$) subject to the relations

$$\sigma_i\sigma_{i+1}\sigma_i = \sigma_{i+1}\sigma_i\sigma_{i+1} \tag{2.68a}$$
$$\sigma_i\sigma_j = \sigma_j\sigma_i \qquad |i-j|\geq 2 \tag{2.68b}$$
$$\sigma_i\sigma_i^{-1} = \sigma_i^{-1}\sigma_i = \mathbf{1} \tag{2.68c}$$

The generator σ_i braids the ith strand under the $(i+1)$th strand, as shown in figure 2.5, whereas σ_i^{-1} effects the inverse braiding, i.e. it takes the ith strand over the $(i+1)$th strand. Relation (2.68a) is shown in figure 2.6.

Trying to make (2.66) look more like (2.68), we define the operators $R_i(u)$ (for $i=1,\ldots,L-1$) on $\bigotimes_{i=1}^{L} V_i$, which act on the spaces $V_i\otimes V_{i+1}$ as $R(u)$, and as the identity elsewhere:

$$R_i(u) = \mathbf{1}\otimes\cdots\otimes\mathbf{1}\otimes \overset{(i,i+1)}{R(u)} \otimes\mathbf{1}\otimes\cdots\otimes\mathbf{1} \tag{2.69}$$

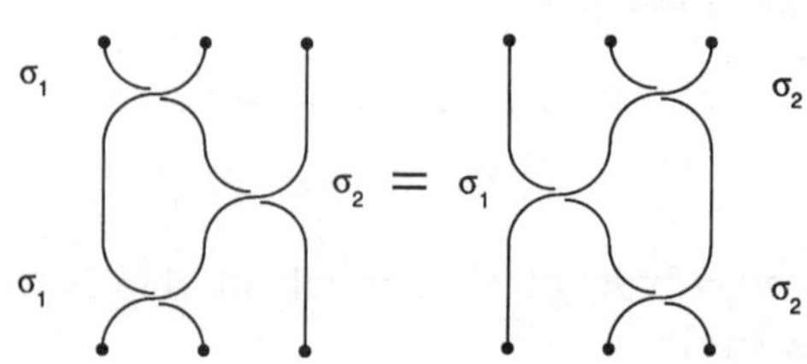

Fig. 2.6 The braid group relation (2.68a) in graphical form.

Then equation (2.66) becomes

$$R_{i+1}(u)R_i(u+v)R_{i+1}(v) = R_i(v)R_{i+1}(u+v)R_i(u) \tag{2.70}$$

and obviously

$$R_i(u)R_j(v) = R_j(v)R_i(u) \qquad |i-j|>1 \tag{2.71}$$

The identification of the Yang–Baxter equation in the form (2.70) with the braid group relation (2.68a) cannot be realized yet due to the presence of the rapidity variable u. This should be no problem, however, in a situation where $u = v = u+v$, which has two solutions:

(i) $u = v = 0$,
(ii) $u = v$, $|u| = \infty$.

The first solution is trivial, since from (2.50) and (2.65) we get merely

$$R(u=0) = i\sin\gamma\ \mathbf{1} \tag{2.72}$$

and thus $\mathscr{R}$ is just proportional to a permutation P. Solution (*ii*) is known as the braid limit. Up to constant factors,

$$\lim_{u\to\pm\infty} \mathrm{e}^{-|u|}R(u) \sim P\exp\left[\pm\frac{i\gamma}{2}\sigma^z\otimes\sigma^z\right] \tag{2.73}$$

where we have assumed u to be real. This limit provides us with a representation of the braid group in terms of, essentially, permutations. It is well known that the permutation group of L elements, $\mathscr{S}_L$, satisfies the same defining relations as the braid group B_L, except for the crucial difference that the square of a transposition is the identity, and therefore one cannot distinguish overcrossings from undercrossings. It goes without saying that a good representation of the braid group should be able to distinguish between σ and σ^{-1}. To find it, let us first observe the following. In the limit $u\to+\infty$, the Boltzmann weights behave as

$$a(u)\to\frac{1}{2}\mathrm{e}^u\,\mathrm{e}^{i\gamma}\,, \qquad b(u)\to\frac{1}{2}\mathrm{e}^u\,, \qquad c(u) = i\sin\gamma \tag{2.74}$$

Hence, the information contained in the weight $c(u)$ is washed out in the limit, which accounts for the "triviality" of the result (2.73). In order not to lose information in the limits $u\to\pm\infty$, we follow Jimbo and perform a u-dependent rescaling of the basis elements, i.e. a u-dependent "diagonal" change of basis:

$$\tilde{e}_r(u) = f_r(u)e_r(u) \qquad r = 1,\ldots,n \tag{2.75}$$

Recalling that the definition of R is

$$R\left(e_{r_1}(u_1)\otimes e_{r_2}(u_2)\right) = R^{r'_2r'_1}_{r_1r_2}(u_1-u_2)e_{r'_2}(u_2)\otimes e_{r'_1}(u_1) \tag{2.76}$$

we deduce that the R-matrix in the new basis $\tilde{e}_r$ is given by

$$\tilde{R}^{r'_2r'_1}_{r_1r_2}(u_1,u_2) = \frac{f_{r_1}(u_1)f_{r_2}(u_2)}{f_{r'_1}(u_1)f_{r'_2}(u_2)}R^{r'_2r'_1}_{r_1r_2}(u_1-u_2) \tag{2.77}$$

The trick is to preserve the difference property of the R-matrix under this change of basis, and thereby fix the scaling functions $f_r(u)$. Indeed, if $\tilde{R}(u_1, u_2)$ is to still depend only on the difference $u_1 - u_2$, the functions in the change of basis must be

$$f_r(u) = \mathrm{e}^{\alpha u r} \tag{2.78}$$

Since $\tilde{R}$ conserves the total quantum number, that is

$$\tilde{R}_{r_1 r_2}^{r_2' r_1'}(u_1, u_2) = 0 \qquad \text{unless} \qquad r_1 + r_2 = r_1' + r_2' \tag{2.79}$$

we may write the rescaled $\tilde{R}$-matrix explicitly as

$$\tilde{R}_{r_1 r_2}^{r_2' r_1'}(u_1 - u_2, \alpha) = \mathrm{e}^{\alpha(u_1 - u_2)(r_1 - r_1')} R_{r_1 r_2}^{r_2' r_1'}(u_1 - u_2) \tag{2.80}$$

The most important property of the R-matrix, namely that it satisfies Yang–Baxter (2.66), is preserved for arbitrary α.

We have thus shown that R-matrices satisfying the conservation law (2.79) possess a “scaling” property with the rapidity as the scaling parameter. This is most suitable for our investigation of the six-vertex R-matrix. We take the braid limit of (2.80) at a special value of α:

$$R \equiv 2\,\mathrm{e}^{-i\gamma/2} \lim_{u \to +\infty} \mathrm{e}^{-u} \tilde{R}(u, \alpha = 1) \tag{2.81}$$

obtaining

$$\begin{aligned} R_{00}^{00} &= R_{11}^{11} = \mathrm{e}^{i\gamma/2} \\ R_{10}^{01} &= R_{01}^{10} = \mathrm{e}^{-i\gamma/2} \\ R_{10}^{10} &= \mathrm{e}^{-i\gamma/2}\left(\mathrm{e}^{i\gamma} - \mathrm{e}^{-i\gamma}\right) \\ R_{01}^{01} &= 0 \end{aligned} \tag{2.82}$$

or, in matrix form,

$$R = \begin{array}{c} \\ 00 \\ 01 \\ 10 \\ 11 \end{array} \begin{array}{c} \begin{array}{cccc} 00 & 01 & 10 & 11 \end{array} \\ \begin{pmatrix} q^{\frac{1}{2}} & 0 & 0 & 0 \\ 0 & 0 & q^{-\frac{1}{2}} & 0 \\ 0 & q^{-\frac{1}{2}} & q^{-\frac{1}{2}}\left(q - q^{-1}\right) & 0 \\ 0 & 0 & 0 & q^{\frac{1}{2}} \end{pmatrix} \end{array} \tag{2.83}$$

with

$$q = \mathrm{e}^{i\gamma} \tag{2.84}$$

These R-matrices, first derived by Jimbo, satisfy the Yang–Baxter relation (2.66) without spectral parameter, and will appear often in this book. Note that, in deriving these scaling limits, the choice $\alpha = 1$ is crucial. It can be checked explicitly that $(R)^2 \neq \mathbf{1}$ for $\gamma \neq 0$, so that we have indeed obtained a genuine representation of the braid group. In the isotropic case ($\gamma = 0$), we obtain the previous result (2.73). The inverse matrix R^{-1} can be obtained as the other real infinite limit of the same rescaled R:

$$R^{-1} = -2\,e^{i\gamma/2} \lim_{u\to-\infty} e^{u} R(u, \alpha = 1) \tag{2.85}$$

Summarizing, we see that the braid limit $u \to \pm\infty$ of the appropriately scaled Boltzmann weights of the six-vertex model yields a non-trivial representation of the braid group:

$$\begin{aligned} \pi : B_L &\to \mathrm{End}\left(\bigotimes^{L} V_{\frac{1}{2}}\right) \\ \sigma_i^{\pm 1} &\mapsto \mathbf{1}\otimes\cdots\otimes\mathbf{1}\otimes \underset{(i,i+1)}{R^{\pm 1}} \otimes\mathbf{1}\otimes\cdots\otimes\mathbf{1} \end{aligned} \tag{2.86}$$

The properties of this representation will be studied in greater detail in chapter 5. For the time being, we return to the Yang–Baxter algebras in order to explore some other territory.

2.3 Yang–Baxter algebras and quantum groups

Any Yang–Baxter algebra $\mathscr{A}$ is a quantum group, in the sense that quantum groups are generally defined as some kind of Hopf algebras, and the latter are bi-algebras satisfying some axioms that are verified in the case of $\mathscr{A}$. Instead of giving the formal mathematical definition of quantum groups, let us go on with physical constructions for the time being.

In the previous section, we derived an interesting braid group representation in the limit $u \to \pm\infty$ of the Boltzmann weights of the six-vertex model, written in the rescaled basis $\tilde{e}_r$ of equation (2.75) with $f_r(u) = e^{ur}$. We shall use this basis again to disentangle new structures hidden in the basic equation (2.24). In the new basis $\{\tilde{e}_r\}$, the $\mathscr{R}$-matrix (without the permutation in R) and the monodromy matrix T are related to those in the basis $\{e_r\}$ by the equations

$$\tilde{\mathscr{R}}_{i_1 i_2}^{j_1 j_2}(u) = e^{u(i_1 - j_1)} \mathscr{R}_{i_1 i_2}^{j_1 j_2}(u) \tag{2.87a}$$

$$\tilde{T}(u)_i^j = e^{u(i-j)} T(u)_i^j \tag{2.87b}$$

The reader may check that $\tilde{\mathscr{R}}(u)$ and $\tilde{T}(u)$ do indeed satisfy equation (2.24).

Now, just as we took the limit $u \to \pm\infty$ of the matrix $\tilde{\mathscr{R}}(u)$, we may take the limit of the monodromy matrix $\tilde{T}(u)$. To get a feeling for what this limit may yield, let us consider the spin-$\frac{1}{2}$ representation given by equations (2.29):

$$\begin{aligned} \lim_{u\to+\infty} \tilde{A}(u) &= \lim_{u\to+\infty} \frac{1}{2} e^{u} e^{i\gamma/2} \begin{pmatrix} e^{i\gamma/2} & \\ & e^{-i\gamma/2} \end{pmatrix} = \frac{1}{2} e^{u} q^{\frac{1}{2}} q^{S^z} \\ \lim_{u\to-\infty} \tilde{A}(u) &= \lim_{u\to-\infty} -\frac{1}{2} e^{-u} e^{-i\gamma/2} \begin{pmatrix} e^{-i\gamma/2} & \\ & e^{i\gamma/2} \end{pmatrix} \\ &= -\frac{1}{2} e^{-u} q^{-\frac{1}{2}} q^{-S^z} \end{aligned} \tag{2.88}$$

where $S^z = \frac{1}{2}\sigma^z$ is the Cartan generator in the spin-$\frac{1}{2}$ representation of $SU(2)$, and we recall that $q = e^{i\gamma}$. The limits $u \to \pm\infty$ of $\tilde{B}(u)$, $\tilde{C}(u)$ and $\tilde{D}(u)$ can be evaluated similarly, whereby the braid limits of the monodromy matrix $\tilde{T}(u)$ are

$$\begin{aligned} T_+ &\equiv 2q^{-\frac{1}{2}} \lim_{u\to+\infty} e^{-u} \begin{pmatrix} \tilde{T}_0^0 & \tilde{T}_0^1 \\ \tilde{T}_1^0 & \tilde{T}_1^1 \end{pmatrix} \\ &= \begin{pmatrix} q^{S^z} & 0 \\ q^{-\frac{1}{2}}(q-q^{-1})S^- & q^{-S^z} \end{pmatrix} \end{aligned} \tag{2.89a}$$

$$\begin{aligned} T_- &\equiv -2q^{\frac{1}{2}} \lim_{u\to-\infty} e^{u} \begin{pmatrix} \tilde{T}_0^0 & \tilde{T}_0^1 \\ \tilde{T}_1^0 & \tilde{T}_1^1 \end{pmatrix} \\ &= \begin{pmatrix} q^{-S^z} & -q^{\frac{1}{2}}(q-q^{-1})S^+ \\ 0 & q^{S^z} \end{pmatrix} \end{aligned} \tag{2.89b}$$

where $S^\pm = (\sigma^x \pm i\sigma^y)/2$ are the off-diagonal generators of $SU(2)$ in the spin-$\frac{1}{2}$ irrep.

The fun starts when we take the various limits $u \to \pm\infty$, $v \to \pm\infty$ in the $\mathscr{R}TT = TT\mathscr{R}$ equation (2.22). All the extra factors work out nicely so that the result is

$$\begin{aligned} \mathscr{R}_{12}\,(T_+)_1\,(T_+)_2 &= (T_+)_2\,(T_+)_1\,\mathscr{R}_{12} \\ \mathscr{R}_{12}\,(T_+)_1\,(T_-)_2 &= (T_-)_2\,(T_+)_1\,\mathscr{R}_{12} \\ \mathscr{R}_{12}^{-1}\,(T_-)_1\,(T_-)_2 &= (T_-)_2\,(T_-)_1\,\mathscr{R}_{12}^{-1} \end{aligned} \tag{2.90}$$

with $\mathscr{R} = PR$ (R given by (2.82)) and $T_\pm$ as in (2.89).

This system of equations appears rather complicated at first sight. The equations are, in fact, equivalent to the following algebraic relationships between S^z, S^+ and S^-:

$$\left[S^z, S^\pm\right] = \pm S^\pm \tag{2.91a}$$

$$\left[S^+, S^-\right] = \frac{q^{2S^z} - q^{-2S^z}}{q - q^{-1}} \tag{2.91b}$$

These are the defining relations for the quantum group $U_q(s\ell(2))$, which is some kind of deformation of the Lie algebra $s\ell(2)$, with $q = e^{i\gamma}$ acting as deformation parameter.

The notation $U_q(s\ell(2))$ clarifies the fact that the quantum group consists of all the formal powers and linear combinations of S^+, S^- and S^z, subject to the relations (2.91). Traditionally, $U(s\ell(2))$ denotes the universal enveloping algebra of $s\ell(2)$, i.e. all the formal powers and linear combinations of $S^\pm$ and S^z modulo the standard Lie algebra relations. In the isotropic limit $\gamma \to 0$, i.e. $q \to 1$, we recover from (2.91) the usual $s\ell(2)$ algebra. The limit $\gamma \to 0$ is called classical, in the sense that the "quantum" group $U_q(s\ell(2))$ becomes the "classical" universal enveloping algebra $U(s\ell(2))$. From the viewpoint of rigid nomenclature, it is perhaps unfortunate that the classical limit of a quantum group is (the universal enveloping algebra of) a Lie algebra; beware of erroneous distinctions between quantum groups and quantum algebras!

Finally, we may take the braid limits $u \to \pm\infty$ in the co-multiplication rule (2.25) to find the co-multiplication for the generators of the quantum group $U_q(s\ell(2))$:

$$\Delta(q^{S^z}) = q^{S^z} \otimes q^{S^z} \tag{2.92a}$$
$$\Delta(S^{\pm}) = S^{\pm} \otimes q^{S^z} + q^{-S^z} \otimes S^{\pm} \tag{2.92b}$$

The co-multiplication preserves the algebraic relations (2.91), as can be checked by using the fact that Δ is a homomorphism, that is to say $\Delta(ab) = \Delta(a)\Delta(b)$. Note that the non-trivial addition rule (2.92b) is consistent with the non-trivial commutator (2.91b), and vice versa. Compare also with $\Delta B(u)$ in equation (2.31).

2.3.1 The $\mathcal{R}$-matrix as an intertwiner

In the preceding section we derived an interesting algebraic structure, the quantum group $U_q(s\ell(2))$, by letting the rapidities become infinite in the Yang–Baxter elements $T_i^j(u)$. More precisely, the quadratic relations between the monodromy matrices ($\mathcal{R}TT$ equations of the Yang–Baxter algebra) give us the defining relations of $U_q(s\ell(2))$, while the co-multiplication of $\mathcal{A}^{(6v)}$ implies that of $U_q(s\ell(2))$. Let us now study some other important uses of the $\mathcal{R}$-matrix of the Yang–Baxter algebra $\mathcal{A}^{(6v)}$.

Let us return once again to the Yang–Baxter equation satisfied by the $\mathcal{R}$-matrix:

$$\begin{aligned}\sum_{j_1,j_2,j_3} &\mathcal{R}^{k_1k_2}_{j_1j_2}(u-v)\mathcal{R}^{j_1k_3}_{i_1j_3}(u)\mathcal{R}^{j_2j_3}_{i_2i_3}(v) \\ &= \sum_{j_1,j_2,j_3} \mathcal{R}^{k_2k_3}_{j_2j_3}(v)\mathcal{R}^{k_1j_3}_{j_1i_3}(u)\mathcal{R}^{j_1j_2}_{i_1i_2}(u-v)\end{aligned} \tag{2.93}$$

We may rephrase section 2.1 to say that the $\mathcal{R}TT$ equation is based on the identification of T_i^j in the representation of dimension two with the $\mathcal{R}$-matrix itself:

$$\left(T_i^j(u)\right)_\alpha^\beta = \mathcal{R}^{j\beta}_{i\alpha} = \tag{2.94}$$

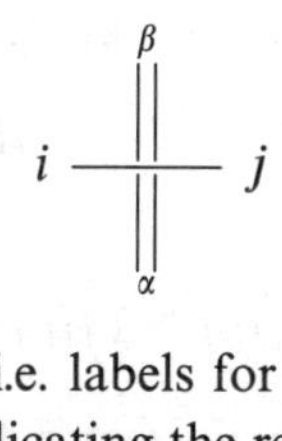

where i, j are indices of the auxiliary space (i.e. labels for the elements of $\mathcal{A}^{(6v)}$) and α, β are indices of the quantum space (indicating the representation of $\mathcal{A}^{(6v)}$). We wish to emphasize that, in all this construction, the $\mathcal{R}$-matrix has played a rather auxiliary role, as indeed its indices in $\mathcal{R}TT = TT\mathcal{R}$ are auxiliary: we would like to see the $\mathcal{R}$-matrix playing a role in quantum space as well.

Let us take advantage of an interesting property satisfied by the $\mathcal{R}$-matrix of the six-vertex model, the parity symmetry:

$$\mathcal{R}^{j_1j_2}_{i_1i_2}(u) = \mathcal{R}^{j_2j_1}_{i_2i_1}(u) \tag{2.95}$$

More intrinsically, equation (2.95) can be written as

$$P\mathcal{R}(u)P = \mathcal{R}(u) \tag{2.96}$$

with P the permutation operator (2.64). The amazing consequence of equation (2.95) is that we may actually interchange the roles of the auxiliary and quantum spaces (provided they are of the same dimensionality). Physically, this corresponds to the arbitrariness in choosing the vertical direction of the euclidean plane as time. With the help of equation (2.95), we may rewrite (2.93) as

$$\begin{aligned}\sum_{j_1,j_2,j_3} &\mathcal{R}^{k_1k_2}_{j_1j_2}(u-v)\mathcal{R}^{k_3j_1}_{j_3i_1}(u)\mathcal{R}^{j_3j_2}_{i_3i_2}(v)\\ &= \sum_{j_1,j_2,j_3} \mathcal{R}^{j_3k_1}_{i_3j_1}(u)\mathcal{R}^{k_3k_2}_{j_3j_2}(v)\mathcal{R}^{j_1j_2}_{i_1i_2}(u-v)\end{aligned} \tag{2.97}$$

Note that in (2.93) the space $V_1 \otimes V_2$ is auxiliary and V_3 is quantum, whereas now V_3 is auxiliary but both V_1 and V_2 are quantum. We have thus transferred $\mathcal{R}$ into quantum space. Using (2.94), we rewrite (2.97) as

$$\mathcal{R}(u-v)\left(T^k_j(u)\otimes T^j_i(v)\right) = \left(T^j_i(u)\otimes T^k_j(v)\right)\mathcal{R}(u-v) \tag{2.98}$$

where the tensor product takes place in $V_1 \otimes V_2$, and $\mathcal{R}(u-v) \in \mathrm{End}(V_1 \otimes V_2)$.

Let us now explore the consequences of the $\mathcal{R}TT$ equation (2.98) with $\mathcal{R}$ in quantum space. First of all, for it to be consistent with the parity symmetry (2.96), there must exist some function $\rho(u)$ such that

$$\mathcal{R}(u)\mathcal{R}(-u) = \rho(u)\rho(-u)\mathbf{1} \tag{2.99}$$

This can be derived by acting on both sides of (2.98) with the permutation operator P and then using (2.96). Equation (2.99) can also be derived from the unitarity condition (1.21), namely $\mathcal{R}(u)P\mathcal{R}(u)P \sim \mathbf{1}$ for parity invariant $\mathcal{R}$-matrices, i.e. satisfying (2.98). For the $\mathcal{R}$-matrix of the six-vertex model, $\rho(u) = a(u)$ in (2.99).

Letting $u = v$ in (2.98), and knowing that

$$\mathcal{R}(0) = \rho(0)P \tag{2.100}$$

we obtain

$$T^j_i(u)\otimes T^k_j(u) = P\left(T^k_j(u)\otimes T^j_i(u)\right)P \tag{2.101}$$

which establishes the equivalence between the co-multiplication (2.25) and its transpose. More generally, when $u \neq v$, equation (2.98) establishes an equivalence between the two ways of co-multiplying arbitrary elements of $\mathcal{A}$. Note that u and v are labels of the representation spaces V_1 and V_2, respectively. If, recalling (2.25), we define

$$\Delta_{u,v}\left(T^k_i\right) = T^j_i(u)\otimes T^k_j(v) \tag{2.102a}$$

$$\Delta'_{u,v}\left(T^k_i\right) = T^k_j(u)\otimes T^j_i(v) \tag{2.102b}$$

we can write (2.98) as

$$\Delta_{u,v}\left(T_i^k\right) = \mathcal{R}(u-v)\Delta'_{u,v}\left(T_i^k\right)\mathcal{R}^{-1}(u-v) \tag{2.103}$$

This equation means that $\mathcal{R}(u-v)$ intertwines the two possible co-multiplications $\Delta_{u,v}$ and $\Delta'_{u,v}$. In Drinfeld's definition of quantum groups as quasi-triangular Hopf algebras, equation (2.103) is one of the basic postulates.

To understand how the $\mathcal{R}$-matrix intertwines between a co-product and its transpose, let us take a closer look at the braid limit $u \to \infty$ of equation (2.98), or rather of its analog for the rescaled $\tilde{\mathcal{R}}(u)$ and $\tilde{T}(u)$ introduced in (2.87), namely

$$\tilde{\mathcal{R}}(v-u)\left(\tilde{T}_i^j(u)\otimes\tilde{T}_j^k(v)\right) = \left(\tilde{T}_j^k(u)\otimes\tilde{T}_i^j(v)\right)\tilde{\mathcal{R}}(v-u) \tag{2.104}$$

Using (2.81) and (2.89), we find the braid limit of (2.104), which is the braid limit of (2.103):

$$\Delta'(g) = \mathcal{R}\Delta(g)\mathcal{R}^{-1} \qquad g \in \left\{q^{\pm S^z}, S^\pm\right\} \tag{2.105}$$

Here, the $\mathcal{R}$-matrix is $\mathcal{R} = PR$, with R the Jimbo matrix (2.82), the co-product $\Delta(g)$ is given by (2.92) and the transposed co-product Δ' is, explicitly,

$$\begin{aligned}\Delta'(q^{S^z}) &= q^{S^z}\otimes q^{S^z}\\ \Delta'(S^\pm) &= S^\pm\otimes q^{-S^z} + q^{S^z}\otimes S^\pm\end{aligned} \tag{2.106}$$

To avoid confusion, equation (2.105) should really be written as

$$\Delta'_{\frac{1}{2}\frac{1}{2}}(g) = \mathcal{R}^{\frac{1}{2}\frac{1}{2}}\Delta_{\frac{1}{2}\frac{1}{2}}(g)\left(\mathcal{R}^{\frac{1}{2}\frac{1}{2}}\right)^{-1} \tag{2.107}$$

where $\Delta_{\frac{1}{2}\frac{1}{2}}$ denotes the restriction of $\Delta(g)$ to the irrep $\frac{1}{2}\otimes\frac{1}{2}$ of $U_q(s\ell(2))$, and the indices on $\mathcal{R}$ remind us that we are using the representation of $\mathcal{R}$ on the vector space $\frac{1}{2}\otimes\frac{1}{2}$. It is nevertheless worth stressing that (2.105) makes sense even at the level of the quantum group $U_q(s\ell(2))$, prior to the construction of its representations. By this we mean that $\mathcal{R}$ in (2.105) can be viewed as an element of $U_q(s\ell(2))\otimes U_q(s\ell(2))$, rather than as a numerical matrix as in (2.107). For this reason, the matrix $\mathcal{R}$ is called the "universal R-matrix", and its existence guarantees that the co-multiplications Δ and Δ' of $U_q(s\ell(2))$ are equivalent at the purely algebraic level.

2.3.2 *A first contact with affine Hopf algebras*

We have found that the integrability of the six-vertex model in the braid limit (when the dependence of the Boltzmann weights on the rapidity drops out) is encoded in the quantum group $U_q(s\ell(2))$. Motivated by these results, we may ask ourselves whether a mathematical structure similar to the quantum group could be associated with the rapidity-dependent R-matrix solutions to the Yang–Baxter equation. This will lead us to affine quantum groups.

For convenience, we shall always use the Chevalley basis. The $s\ell(2)$ Lie algebra (A_1 in Cartan's classification) has three Chevalley generators E, F and H with the following non-vanishing commutators:

$$\begin{aligned} [E,F] &= H \\ [H,E] &= 2E \\ [H,F] &= -2F \end{aligned} \tag{2.108}$$

The usual spin generators are related to these by

$$E = S^+, \quad F = S^-, \quad H = 2S^z \tag{2.109}$$

The affine extension of A_1, called $A_1^{(1)}$ in Kac's classification, has six Chevalley generators E_i, F_i and H_i ($i = 0, 1$). Suppose we have an irreducible representation of A_1. It can be affinized, i.e. promoted to an irreducible representation of $A_1^{(1)}$, through the identifications

$$\begin{aligned} E_0 &= \mathrm{e}^u F & \qquad E_1 &= \mathrm{e}^u E \\ F_0 &= \mathrm{e}^{-u} E & \qquad F_1 &= \mathrm{e}^{-u} F \\ H_0 &= -H & \qquad H_1 &= H \end{aligned} \tag{2.110}$$

where $x = \mathrm{e}^u$ is a complex affinization parameter.

Different irreps of $A_1^{(1)}$ (of zero central extension) may be labeled by the affine parameter e^u and the Casimir of the corresponding representation of A_1. For example, the irreducible $(\mathrm{e}^u, \frac{1}{2})$ representation of $A_1^{(1)}$ derives from the usual spin-$\frac{1}{2}$ irrep of A_1:

$$\begin{aligned} E_0 &= \begin{pmatrix} 0 & 0 \\ \mathrm{e}^u & 0 \end{pmatrix} & \qquad E_1 &= \begin{pmatrix} 0 & \mathrm{e}^u \\ 0 & 0 \end{pmatrix} \\ F_0 &= \begin{pmatrix} 0 & \mathrm{e}^{-u} \\ 0 & 0 \end{pmatrix} & \qquad F_1 &= \begin{pmatrix} 0 & 0 \\ \mathrm{e}^{-u} & 0 \end{pmatrix} \\ H_0 &= \begin{pmatrix} -1 & 0 \\ 0 & 1 \end{pmatrix} & \qquad H_1 &= \begin{pmatrix} 1 & 0 \\ 0 & -1 \end{pmatrix} \end{aligned} \tag{2.111}$$

Let us turn to the quantum deformations of A_1 and $A_1^{(1)}$, which we shall denote by $U_q(A_1)$ and $U_q(A_1^{(1)})$, respectively. If we define the operator K as

$$K = q^H \tag{2.112}$$

we may rewrite equations (2.91) in terms of E, F and K as

$$\begin{aligned} KE &= q^2 EK \\ KF &= q^{-2} FK \\ [E,F] &= \frac{K - K^{-1}}{q - q^{-1}} \end{aligned} \tag{2.113}$$

We shall denote by $U_q(A_1) = U_q(s\ell(2))$ the algebra generated by E, F and H subject to (2.91), and by $U'_q(A_1) = U'_q(s\ell(2))$ that generated by E, F and K subject to (2.113). The subtle difference between these two algebras will be discussed in chapter 6.

Affinization of the spin-$\frac{1}{2}$ irrep of $U_q(A_1)$ or $U'_q(A_1)$ yields an irrep $(\mathrm{e}^u, \frac{1}{2})$ of $U_q(A_1^{(1)})$:

$$E_0 = \begin{pmatrix} 0 & 0 \\ \mathrm{e}^u & 0 \end{pmatrix} \qquad E_1 = \begin{pmatrix} 0 & \mathrm{e}^u \\ 0 & 0 \end{pmatrix}$$
$$F_0 = \begin{pmatrix} 0 & \mathrm{e}^{-u} \\ 0 & 0 \end{pmatrix} \qquad F_1 = \begin{pmatrix} 0 & 0 \\ \mathrm{e}^{-u} & 0 \end{pmatrix} \tag{2.114}$$
$$K_0 = \begin{pmatrix} q^{-1} & 0 \\ 0 & q \end{pmatrix} \qquad K_1 = \begin{pmatrix} q & 0 \\ 0 & q^{-1} \end{pmatrix}$$

which is the same as representation (2.111) of the classical group $A_1^{(1)}$, provided we take into account relation (2.112). For the fundamental irrep, as the doublet of $s\ell(2)$, it is always true that the classical and quantum representations coincide.

We expect that the irrep $(\mathrm{e}^u, \frac{1}{2})$ of $U_q(A_1^{(1)})$ should be intimately related to the spin-$\frac{1}{2}$ representation (2.29) of the generators $A(u)$, $B(u)$, $C(u)$ and $D(u)$ of $\mathscr{A}^{(6v)}$. This is so; indeed

$$A(u) = \frac{1}{2}\left(\mathrm{e}^u \sqrt{q} K^{\frac{1}{2}} - \frac{\mathrm{e}^{-u}}{\sqrt{q}} K^{-\frac{1}{2}}\right)$$
$$B(u) = \frac{q - q^{-1}}{2\sqrt{q}}\ F\ K^{\frac{1}{2}}$$
$$C(u) = \frac{q - q^{-1}}{2\sqrt{q}}\ E\ K^{\frac{1}{2}} \tag{2.115}$$
$$D(u) = \frac{1}{2}\left(\mathrm{e}^u \sqrt{q} K^{-\frac{1}{2}} - \frac{\mathrm{e}^{-u}}{\sqrt{q}} K^{\frac{1}{2}}\right)$$

Please check that the algebraic relations (2.40) follow from those of the quantum group $U_q(A_1)$, equations (2.113).

The quantum affine algebra $U_q(A_1^{(1)})$ enjoys also a bi-algebra structure, determined by the co-product

$$\Delta(E_i) = E_i \otimes K_i + \mathbf{1} \otimes E_i$$
$$\Delta(F_i) = F_i \otimes \mathbf{1} + K_i^{-1} \otimes F_i \tag{2.116}$$
$$\Delta(K_i) = K_i \otimes K_i$$

It is thus natural to look for an intertwiner $\mathscr{R}$-matrix for the tensor product of two spin-$\frac{1}{2}$ irreps $(\mathrm{e}^{u_1}, \frac{1}{2}) \otimes (\mathrm{e}^{u_2}, \frac{1}{2})$ of $U_q(A_1^{(1)})$. It should satisfy the condition

$$\mathscr{R}(\mathrm{e}^{u_1}, \mathrm{e}^{u_2})\Delta_{\mathrm{e}^{u_1},\mathrm{e}^{u_2}}(g) = \Delta'_{\mathrm{e}^{u_1},\mathrm{e}^{u_2}}(g)\mathscr{R}(\mathrm{e}^{u_1}, \mathrm{e}^{u_2}) \tag{2.117}$$

for any $g \in U_q(A_1^{(1)})$. This is nothing but the affinized version of the intertwiner condition (2.107) for $U_q(A_1)$. At the risk of offending the reader, we show the transposed co-multiplication Δ' of the generators of $U_q(A_1^{(1)})$:

$$\begin{aligned}\Delta'(E_i) &= E_i \otimes \mathbf{1} + K_i \otimes E_i \\ \Delta'(F_i) &= F_i \otimes K_i^{-1} + \mathbf{1} \otimes F_i \\ \Delta'(K_i) &= K_i \otimes K_i\end{aligned} \tag{2.118}$$

Comparing (2.116) with (2.92), we see that the relationship between E and F and S^+ and S^- is

$$E = K^{\frac{1}{2}} S^+ , \qquad F = K^{-\frac{1}{2}} S^- \tag{2.119}$$

After some straightforward manipulations, from (2.117) we obtain the affine $\mathscr{R}^{\frac{1}{2}\frac{1}{2}}(\mathrm{e}^{u_1}, \mathrm{e}^{u_2})$ matrix:

$$\begin{aligned}\frac{\mathscr{R}_{01}^{01}(\mathrm{e}^{u_1}, \mathrm{e}^{u_2})}{\mathscr{R}_{00}^{00}(\mathrm{e}^{u_1}, \mathrm{e}^{u_2})} &= \frac{\mathrm{e}^{u_1-u_2} - \mathrm{e}^{u_2-u_1}}{q\,\mathrm{e}^{u_1-u_2} - q^{-1}\,\mathrm{e}^{u_2-u_1}} \\ \frac{\mathscr{R}_{01}^{10}(\mathrm{e}^{u_1}, \mathrm{e}^{u_2})}{\mathscr{R}_{00}^{00}(\mathrm{e}^{u_1}, \mathrm{e}^{u_2})} &= \frac{q - q^{-1}}{q\,\mathrm{e}^{u_1-u_2} - q^{-1}\,\mathrm{e}^{u_2-u_1}}\end{aligned} \tag{2.120}$$

with $\mathscr{R}_{00}^{00} = \mathscr{R}_{11}^{11}$, $\mathscr{R}_{01}^{01} = \mathscr{R}_{10}^{10}$, $\mathscr{R}_{01}^{10} = \mathscr{R}_{10}^{01}$, and all other matrix elements of $\mathscr{R}$ equal to zero.

Identifying now $q = \mathrm{e}^{i\gamma}$ and $\mathrm{e}^{u} = \mathrm{e}^{u_1-u_2}$, we see that $\mathscr{R}(\mathrm{e}^{u_1}, \mathrm{e}^{u_2})$ is just the six-vertex $\mathscr{R}$-matrix (2.50). With this happy result, we conclude our first taste of the relationship between the six-vertex model and the affine quantum group $U_q(A_1^{(1)})$.

2.4 Descendants of the six-vertex model

In standard group theory, we are used to finding all irreducible representations in the decomposition of the tensor product of the fundamental representation. Since the six-vertex model is just the vertex model associated with the fundamental irreducible representation of $U_q(A_1^{(1)})$, we ought to be able to derive from it other vertex models, associated with higher-spin irreps. In the discussion below, we shall try to stress the physical underpinnings of the construction, shying away from purely quantum group theoretic arguments.

2.4.1 Descent procedure

Starting with the six-vertex $\mathscr{R}$-matrix $\mathscr{R}^{(6v)}(u) = \mathscr{R}^{(\frac{1}{2}\frac{1}{2})}(u)$, let us define representations of the associated Yang–Baxter algebra by considering various explicit solutions to (2.24). We will denote these solutions by $L^{\frac{1}{2},\rho}$, letting the label ρ stand for an irrep. We thus require

$$\begin{aligned}&\sum_{j_1,j_2,\beta} \left(\mathscr{R}^{(\frac{1}{2},\frac{1}{2})}(u-v)\right)_{j_1 j_2}^{k_1 k_2} \left(L^{(\frac{1}{2},\rho)}(u)\right)_{i_1\beta}^{j_1\gamma} \left(L^{(\frac{1}{2},\rho)}(v)\right)_{i_2\alpha}^{j_2\beta} \\ &\quad = \sum_{j_1,j_2,\beta} \left(L^{(\frac{1}{2},\rho)}(v)\right)_{j_2\beta}^{k_2\gamma} \left(L^{(\frac{1}{2},\rho)}(u)\right)_{j_1\alpha}^{k_1\beta} \left(\mathscr{R}^{(\frac{1}{2},\frac{1}{2})}(u-v)\right)_{i_1 i_2}^{j_1 j_2}\end{aligned} \tag{2.121}$$

Clearly, $L^{(\frac{1}{2},\frac{1}{2})} = T$ is a solution, but we want more. We may use the notation $L_1 = L \otimes \mathbf{1}$ and $L_2 = \mathbf{1} \otimes L$ to write this equation more simply as follows:

$$\begin{aligned} &\mathscr{R}^{(\frac{1}{2},\frac{1}{2})}(u-v)\, L_1^{(\frac{1}{2},\rho)}(u)\, L_2^{(\frac{1}{2},\rho)}(v) \\ &\qquad = L_2^{(\frac{1}{2},\rho)}(v)\, L_1^{(\frac{1}{2},\rho)}(u)\, \mathscr{R}^{(\frac{1}{2},\frac{1}{2})}(u-v) \end{aligned} \tag{2.122}$$

In graphical notation, letting a single line represent the spin-$\frac{1}{2}$ representation and a double one the ρ representation, we have

$$\left(L^{(\frac{1}{2},\rho)}(u)\right)_{i\alpha}^{j\beta} \quad = \tag{2.123}$$

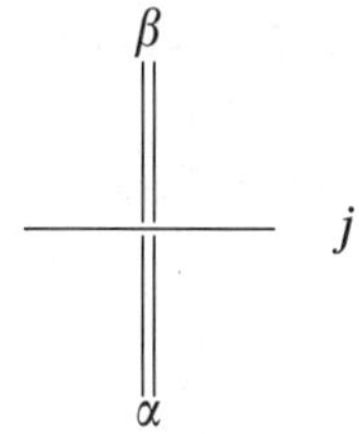

Accordingly, the defining relation (2.121) becomes figure 2.7.

Every solution to (2.121) can be interpreted formally as yielding a representation ρ of the Yang–Baxter algebra, i.e.

$$L^{(\frac{1}{2},\rho)}(u)_i^j = \rho\left(T_i^j(u)\right) \tag{2.124}$$

We will be interested mostly in matrix solutions to (2.121) where $L^{(\frac{1}{2},\rho)}{}_i^j$ is a finite $m \times m$ square matrix, i.e. a situation where the Yang–Baxter algebra is represented as endomorphisms in $\mathbb{C}^m$. The double line in (2.123) thus stands for the vector space $\mathbb{C}^m$.

Assuming we had found two solutions $L^{(\frac{1}{2},\rho_1)}(u)$ and $L^{(\frac{1}{2},\rho_2)}(u)$ of (2.121), we may now look for a new $\mathscr{R}$-matrix $\mathscr{R}^{(\rho_1,\rho_2)}$ subject to

$$\begin{aligned} &\sum_{j,\beta_1,\beta_2} \left(L^{(\frac{1}{2},\rho_1)}(v)\right)_{j\beta_1}^{k\gamma_1} \left(L^{(\frac{1}{2},\rho_2)}(u)\right)_{i\beta_2}^{j\gamma_2} \left(\mathscr{R}^{(\rho_1,\rho_2)}(u-v)\right)_{\alpha_1\alpha_2}^{\beta_1\beta_2} \\ &\quad = \sum_{j,\beta_1,\beta_2} \left(\mathscr{R}^{(\rho_1,\rho_2)}(u-v)\right)_{\beta_1\beta_2}^{\gamma_1\gamma_2} \left(L^{(\frac{1}{2},\rho_2)}(u)\right)_{j\alpha_2}^{k\beta_2} \left(L^{(\frac{1}{2},\rho_1)}(v)\right)_{i\alpha_1}^{j\beta_1} \end{aligned} \tag{2.125}$$

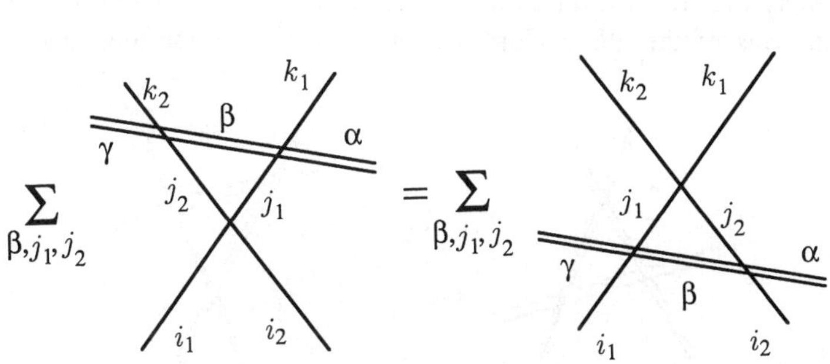

Fig. 2.7 Graphical representation of (2.121).

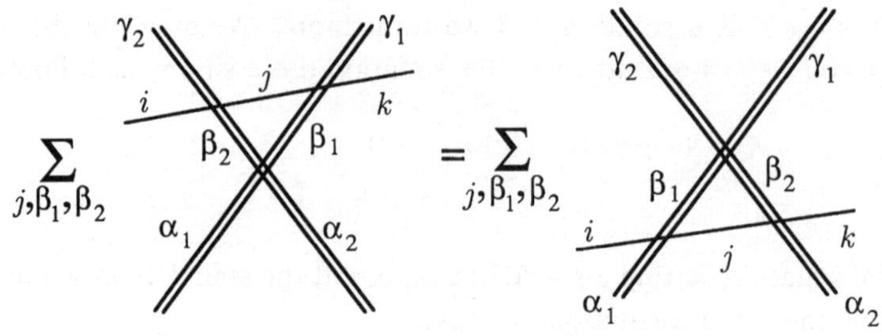

Fig. 2.8 Graphical representation of (2.125).

which in graphical form is shown in figure 2.8, with

$$\left(\mathscr{R}^{(\rho_1,\rho_2)}\right)_{\alpha_1\alpha_2}^{\beta_1\beta_2} \quad = \quad \alpha_1 \qquad\qquad \beta_1 \tag{2.126}$$

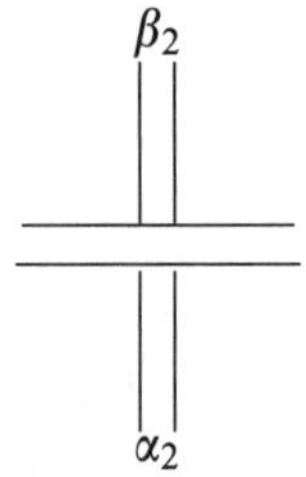

Solving (2.125), for every pair of solutions to (2.121) we find a collection of $\mathscr{R}$-matrices satisfying the Yang–Baxter equation

$$\begin{aligned}\mathscr{R}_{12}^{(\rho_1,\rho_2)}(u)\mathscr{R}_{13}^{(\rho_1,\rho_3)}(u+v)\mathscr{R}_{23}^{(\rho_2,\rho_3)}(v)\\ = \mathscr{R}_{23}^{(\rho_2,\rho_3)}(v)\mathscr{R}_{13}^{(\rho_1,\rho_3)}(u+v)\mathscr{R}_{12}^{(\rho_1,\rho_2)}(u)\end{aligned} \tag{2.127}$$

shown schematically in figure 2.9.

It is not obvious that for any two solutions $L^{(\frac{1}{2},\rho_1)}$ and $L^{(\frac{1}{2},\rho_2)}$ to (2.121) there exists an intertwiner $\mathscr{R}^{(\rho_1,\rho_2)}$ as a solution to (2.125). In chapter 7, we will encounter such a problem, and surmount it by relaxing the difference property we assumed for the $\mathscr{R}$-matrix in (2.125): the more complicated intertwiners $\mathscr{R}^{(\rho_1,\rho_2)}(u_1,u_2)$ will satisfy the corresponding generalization of (2.127).

In order to achieve a clearer mathematical understanding of the descent equations (2.121) and (2.125), let us return to the result presented in the previous section, namely that the six-vertex $\mathscr{R}$-matrix is the intertwiner between two spin-$\frac{1}{2}$ representations of the affine Hopf algebra $U_q(A_1^{(1)})$. Extending this result, we

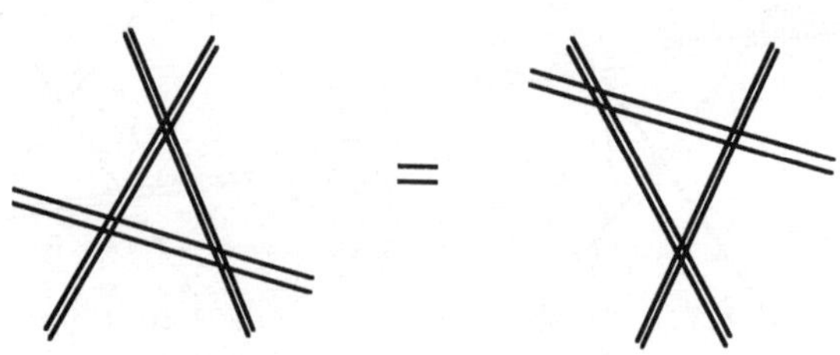

Fig. 2.9 Graphical representation of the Yang–Baxter equation (2.127).

should expect to interpret the various solutions to (2.121) as intertwiners between the spin-$\frac{1}{2}$ representation and a new representation ρ of the affine Hopf algebra $U_q(A_1^{(1)})$. Moreover, the solutions to (2.125) should be interpreted accordingly as intertwiners between two irreps ρ_1 and ρ_2 of $U_q(A_1^{(1)})$. From this point of view, equations (2.121), (2.125) and (2.127) are all the same basic equation reflecting the quasi-triangularity of $U_q(A_1^{(1)})$, projected or particularized to different representations. We shall pursue this mathematical interpretation in chapter 6, but, for the time being, we may view equations (2.121) and (2.125) simply as a descent procedure for defining new representations of the six-vertex Yang–Baxter algebra, and thus finding candidate solutions to (2.127) starting from the six-vertex $\mathcal{R}$-matrix. We leave it to the reader to wonder whether infinite-dimensional solutions $L^{(\frac{1}{2},\rho)}$ could be envisaged in this framework, and limit ourselves in what follows to finite-dimensional irreps.

2.4.2 *Bethe ansatz for descendant models*

Let $V^{(\rho)}$ denote the vector space associated with a representation ρ (i.e. with a solution $L^{(\frac{1}{2},\rho)}$), assumed finite-dimensional ($\mathbb{C}^m$). Using the operators (2.123) we may define a monodromy operator

$T(u)^{(\rho)} \quad = \quad \frac{1}{2}$

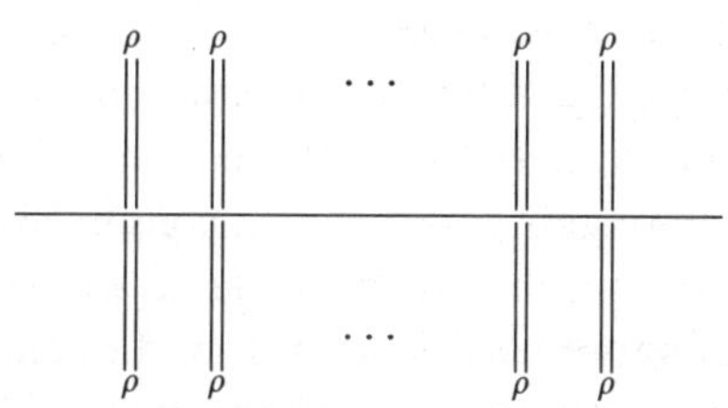

$\frac{1}{2}$ (2.128)

As a matrix over the initial and final spin-$\frac{1}{2}$ representation space, $T(u)^{(\rho)}$ is a 2×2 matrix,

$\left(T(u)^{(\rho)}\right)_i^j \quad = \quad i$

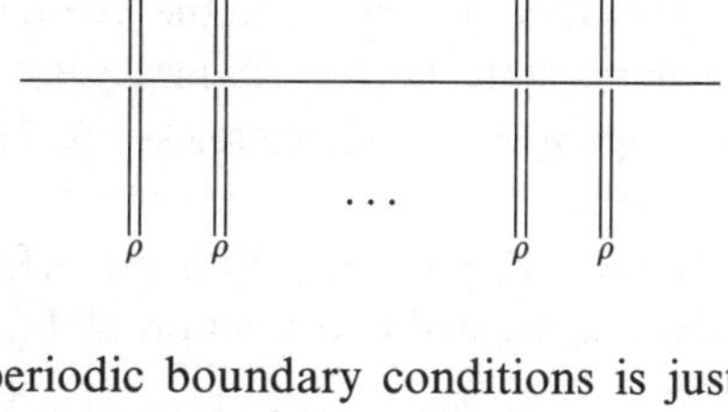

j (2.129)

The transfer matrix with periodic boundary conditions is just the trace of the monodromy matrix:

$$t(u)^{(\rho)} = \text{tr}\left(T(u)^{(\rho)}\right) = \left(T(u)^{(\rho)}\right)_0^0 + \left(T(u)^{(\rho)}\right)_1^1 \qquad (2.130)$$

where we have followed the notation introduced earlier, whereby the spin-$\frac{1}{2}$ irrep has states $\left|\frac{1}{2},0\right\rangle$ and $\left|\frac{1}{2},1\right\rangle$. Explicitly,

$$\langle\beta_1,\ldots,\beta_L|\, t(u)^{(\rho)}\, |\alpha_1,\ldots,\alpha_L\rangle = \sum_{i_1,\ldots,i_L\in\{0,1\}} \left(L^{(\frac{1}{2},\rho)}(u)\right)_{i_1\alpha_1}^{i_2\beta_1} \times \left(L^{(\frac{1}{2},\rho)}(u)\right)_{i_2\alpha_2}^{i_3\beta_2} \cdots \left(L^{(\frac{1}{2},\rho)}(u)\right)_{i_L\alpha_L}^{i_1\beta_L} \tag{2.131}$$

and this transfer matrix describes a vertex model with two possible lattice variables on the horizontal lines and $m = \dim V^{(\rho)}$ lattice variables on the vertical lines. The Boltzmann weights of the model are given by (2.123).

We may now discuss the operator $B(u)$ of the six-vertex algebra in the representation ρ:

$B(u)^{(\rho)} \quad =$ 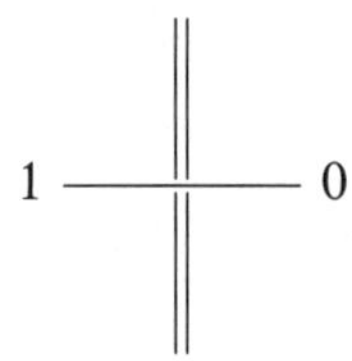

(2.132)

Recall that this operator was crucial for the algebraic Bethe ansatz solution to the six-vertex model. There, both the vertical double line (quantum space) and the single horizontal one (auxiliary space) represented a two-dimensional Hilbert space, and the B operator increased by one unit the total number of spins down. Recall also that, using the arrow representation for the variables on the links rather than the notation $a \in \{0,1\}$, the conservation of arrow number in the six-vertex Boltzmann weights gave rise to a conserved quantity, the total spin of a state. In order to carry out the algebraic Bethe ansatz for the new model defined by $L^{(\frac{1}{2},\rho)}$, the first thing we need is a conserved quantum number allowing us to split the Hilbert space into orthogonal subspaces labeled by the eigenvalue of that conserved quantity. The generalized arrow conservation law should read as

$$\left(L^{(\frac{1}{2},\rho)}(u)\right)_{i\alpha}^{j\beta} = \delta_{i+\alpha}^{j+\beta} \left(L^{(\frac{1}{2},\rho)}(u)\right)_{i\alpha}^{j\beta} \tag{2.133}$$

Representations ρ satisfying (2.133) will be called "regular representations": they are both highest weight and lowest weight representations. For these, we may define, as in the six-vertex model, a Bethe reference state $\Omega_\circ = |0,0,\ldots,0\rangle$. Moreover, for a generic state $|\alpha_1,\alpha_2,\ldots\rangle$ in $\otimes^L V^{(\rho)}$ the number $M = \sum \alpha_i$ is conserved and the B operator (2.132) increases the value of M by one unit.

We may therefore proceed with the model $L^{(\frac{1}{2},\rho)}$, subject to (2.133), along the same lines as for the six-vertex model. With the help of the co-multiplication of the Yang–Baxter algebra, we define the action of $B(u)^{(\rho)}$ on $\mathscr{H} = \otimes^L V^{(\rho)}$ as

$$B(u)^{(\rho)} = \Delta^{L-1}\left[\left(T(u)^{(\rho)}\right)_1^0\right] \tag{2.134}$$

with

$$\Delta^2 = (\Delta\otimes\mathbf{1})\,\Delta,\ldots,\Delta^n = (\Delta\otimes\mathbf{1})\,\Delta^{n-1} \tag{2.135}$$

Therefore, the Bethe states are of the form

$$\prod_{i=1}^{M} B(u_i)^{(\rho)} |\Omega\rangle \tag{2.136}$$

To diagonalize the transfer matrix (2.130), we use the Bethe states (2.136) and the algebraic relations (2.40) in the appropriate representation ρ. Explicitly,

$$\left[A(u)^{(\rho)}\right]_{\alpha}^{\beta} \quad = $$

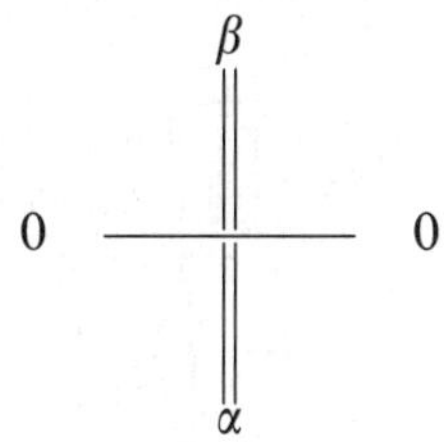

$$= \begin{pmatrix} \left[A(u)^{(\rho)}\right]_{0}^{0} & & \\ & \ddots & \\ & & \left[A(u)^{(\rho)}\right]_{m-1}^{m-1} \end{pmatrix} \tag{2.137a}$$

$$\left[D(u)^{(\rho)}\right]_{\alpha}^{\beta} \quad = $$

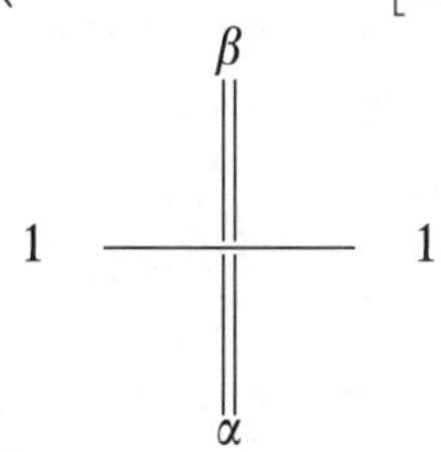

$$= \begin{pmatrix} \left[D(u)^{(\rho)}\right]_{0}^{0} & & \\ & \ddots & \\ & & \left[D(u)^{(\rho)}\right]_{m-1}^{m-1} \end{pmatrix} \tag{2.137b}$$

with

$$\left[A(u)^{(\rho)}\right]_{\alpha}^{\alpha} = \left[L^{(\frac{1}{2},\rho)}(u)\right]_{0\alpha}^{0\alpha}, \quad \left[D(u)^{(\rho)}\right]_{\alpha}^{\alpha} = \left[L^{(\frac{1}{2},\rho)}(u)\right]_{1\alpha}^{1\alpha} \tag{2.138}$$

In these expressions, we assume explicitly that ρ is an m-dimensional representation satisfying the conservation law (2.133). The Bethe reference state is thus

$$|\Omega\rangle = \bigotimes^{L} |e_0\rangle \qquad \text{with} \qquad |e_0\rangle = \begin{pmatrix} 1 \\ 0 \\ \vdots \\ 0 \end{pmatrix} \in \mathbb{C}^m \tag{2.139}$$

Putting it all together, we find

$$\begin{aligned} A(u)^{(\rho)} |\Omega\rangle &= \Delta^L \left(A(u)^{(\rho)}\right) |\Omega\rangle = \left[\left(L^{(\frac{1}{2},\rho)}(u)\right)_{00}^{00}\right]^{L} |\Omega\rangle \\ &\equiv \left[a^{(\rho)}(u)\right]^{L} |\Omega\rangle \end{aligned} \tag{2.140}$$

$$D(u)^{(\rho)}\,|\Omega\rangle = \Delta^L\left(D(u)^{(\rho)}\right)|\Omega\rangle = \left[\left(L^{(\frac{1}{2},\rho)}(u)\right)_{10}^{10}\right]^L |\Omega\rangle$$
$$\equiv \left[b^{(\rho)}(u)\right]^L |\Omega\rangle$$

From the algebraic relations (2.40) and (2.140), we obtain

$$\begin{aligned}\left[A(u)^{(\rho)} + D(u)^{(\rho)}\right]\prod_{i=1}^{M} B(u_i)^{(\rho)}\,|\Omega\rangle \;&=\; \text{unwanted terms}\\ &+\left(\left[a^{(\rho)}(u)\right]^L\prod_{j=1}^{M}\frac{a(u_j-u)}{b(u_j-u)}\right.\\ &\left.+\left[b^{(\rho)}(u)\right]^L\prod_{j=1}^{M}\frac{a(u-u_j)}{b(u-u_j)}\right)\prod_{i=1}^{M}B(u_i)^{(\rho)}\,|\Omega\rangle\end{aligned} \tag{2.141}$$

The structure of the unwanted terms is exactly as for the six-vertex case, and therefore the Bethe equations for descendants (2.46) are valid, provided we replace on its left-hand side $a(u)$ and $b(u)$ by $a^{(\rho)}(u)$ and $b^{(\rho)}(u)$. The Bethe ansatz equations fix the rapidities u_i of the spin waves, which superimpose to yield the physical states; here, they read as

$$\left[\frac{a^{(\rho)}(u_i)}{b^{(\rho)}(u_i)}\right]^L = \prod_{\substack{j=1\\ j\neq i}}^{M}\frac{a(u_i-u_j)b(u_j-u_i)}{a(u_j-u_i)b(u_i-u_j)} \tag{2.142}$$

Note that the functions $a(u)$ and $b(u)$ on the right-hand side remain the same as for the six-vertex model. In conclusion, if the representation ρ enjoys the generalized arrow number conservation (2.133), the only difference between the Bethe equations for the six-vertex model and for its descendants is a minor change on the left-hand side. The change is very important, however, because it modifies the map (2.48) between rapidities and momenta. The eigenvalues of the transfer matrix are

$$\Lambda^{(\rho)}(\{u_i\}|u) = \left[a^{(\rho)}(u)\right]^L\prod_{i=1}^{M}\frac{a(u_i-u)}{b(u_i-u)} + \left[b^{(\rho)}(u)\right]^L\prod_{i=1}^{M}\frac{a(u-u_i)}{b(u-u_i)} \tag{2.143}$$

We may now proceed to use solutions of (2.125) to generalize the vertex model with spin-$\frac{1}{2}$ irreps on the horizontal lines, ρ representation on the vertical ones, and weights (2.123), to another one with ρ representations on both vertical and horizontal lines. The Boltzmann weights of this descendant model will be

$$\left(\mathcal{R}(u)^{(\rho,\rho)}\right)_{\alpha\beta}^{\gamma\delta} \;=\; \alpha \quad \tag{2.144}$$

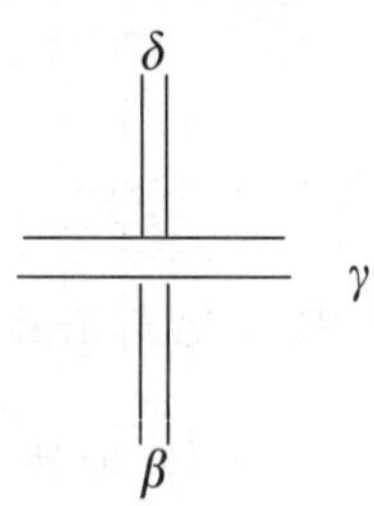

With the help of these weights, we may define a new monodromy matrix

$$T(u)^{(\rho,\rho)} : \bigotimes^{L} V^{(\rho)} \longrightarrow \bigotimes^{L} V^{(\rho)} \tag{2.145}$$

graphically represented by

$$\left(T(u)^{(\rho,\rho)}\right)_{\alpha}^{\beta} \quad = \quad \alpha$$

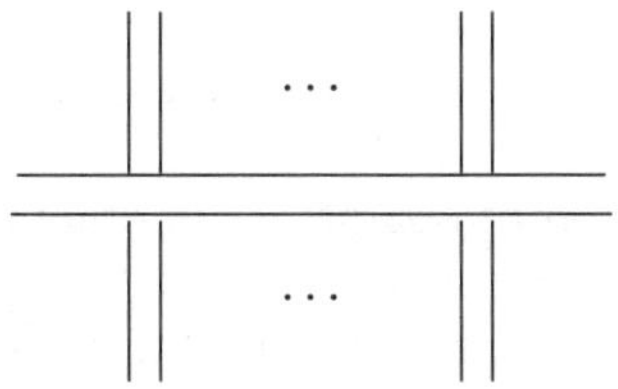

β (2.146)

This is now an $m \times m$ matrix, and its trace is the transfer matrix for periodic boundary conditions:

$$t(u)^{(\rho,\rho)} = \mathrm{tr}\left(T(u)^{(\rho,\rho)}\right) \tag{2.147}$$

We wish to diagonalize this new transfer matrix, which is an easy task because the $\mathscr{R}$-matrix $\mathscr{R}(u)^{(\rho,\rho)}$ used to define this transfer matrix is a solution to (2.125). From that descent equation it follows immediately that

$$\left[t(u)^{(\rho,\rho)} \, , \, t(v)^{(\rho)}\right] = 0 \tag{2.148}$$

whereby the transfer matrices $t(u)^{(\rho,\rho)}$ and $t(v)^{(\rho)}$ of two different statistical mechanical models share the same eigenvectors. Therefore, the Bethe states (2.136) diagonalizing $t(u)^{(\rho)}$ diagonalize $t(u)^{(\rho,\rho)}$ as well, and thus the Bethe equations (2.142) just described remain valid for the more general model with weights (2.144). Both these models are called descendants of the six-vertex model, via regular representations of the six-vertex Yang–Baxter algebra.

In the next chapter, we will consider some explicit examples of regular representations and work out the corresponding Bethe equations. Two comments are in order before that, though. First, the one-dimensional hamiltonians associated with the descendant models are given by the customary formula

$$H^{(\rho)} \; = \; i \, \frac{d}{du} \log \, t(u)^{(\rho,\rho)}\Big|_{u=0} \tag{2.149}$$

and describe spin chains whose spin variables take values in the representation $V^{(\rho)}$. If ρ is a standard representation, then $H^{(\rho)}$ preserves $M = \sum \alpha_i$. Secondly, we may consider a slightly more general model with irrep ρ_1 on the horizontal lines and irrep ρ_2 on the vertical ones. The Boltzmann weights would be given by

$$W\left(\begin{matrix} j & \beta \\ \alpha & i \end{matrix}\middle| u\right) \; = \; \left(\mathscr{R}(u)^{(\rho_1,\rho_2)}\right)_{\alpha i}^{\beta j} \tag{2.150}$$

and again equation (2.125) would imply that $t(u)^{(\rho_1,\rho_2)}$ and $t(v)^{(\rho_2)}$ commute. Only very minor changes are needed in the Bethe program to find the physical states of these models.

2.5 Comments

2.5.1 Explanation of our conventions

Let A be a linear operator acting on an n-dimensional vector space V, and let $\{e_i\}$ $(i = 1, \dots, n)$ be a basis of V. We write the action of A on this basis as

$$A\ e_i = A_i^j\ e_j \tag{2.151}$$

Upon multiplication of operators, we obtain

$$(A\ B)_i^j = A_k^j\ B_i^k \tag{2.152}$$

Hence, if e_i are assembled into a column, the upper index j of a matrix A_i^j denotes the row and the lower index i denotes the column. If the space V has the metric $\eta_{ij} = \langle e_i | e_j \rangle$, we may use it to lower indices and obtain

$$A_{ij} = \langle e_i | \, A \, | e_j \rangle = \eta_{ik} A_j^k \tag{2.153}$$

The metric used in section 2.1 is just $\eta_{ij} = \delta_{ij}$.

Throughout this book, we use the notation $\mathcal{R}$ for an $\mathcal{R}$-matrix which does not interchange the spaces it acts upon. The universal $\mathcal{R}$-matrix, belonging to $\mathcal{A} \otimes \mathcal{A}$ with $\mathcal{A}$ a finite or affine quantum group, is of this sort. The representation of the universal $\mathcal{R}$-matrix as an element of End $V^{(\rho_1)} \otimes$End $V^{(\rho_2)}$ is thus denoted $\mathcal{R}^{(\rho_1,\rho_2)}$, although sometimes we suppress or simplify the superscripts if no confusion can arise. On the other hand, the notation R always implies a permutation of the spaces, so, for instance, the matrix $R^{(\rho_1,\rho_2)}$ acts between $V^{(\rho_1)} \otimes V^{(\rho_2)}$ and $V^{(\rho_2)} \otimes V^{(\rho_1)}$. Whereas $\mathcal{R}$ is most convenient in formal algebraic discussions, R has a direct braid group interpretation. The factorizable S-matrices of chapter 1 effect a braiding, so they are essentially R-matrices. Always, then, $R = P\mathcal{R}$. In the literature, $\mathcal{R}$ and R are often denoted R and $\check{R}$, respectively.

Table 2.1 Parametrization of the six-vertex weights in the three different regimes. The domain of the rapidity θ is the real axis.

	Antiferromagnetic	Critical	Ferromagnetic
	$\Delta < -1$	$-1 < \Delta < 1$	$1 < \Delta$
a	$\sinh(\eta - \theta)$	$\sin(\eta - \theta)$	$\sinh(\theta + \eta)$
b	$\sinh \theta$	$\sin \theta$	$\sinh \theta$
c	$\sinh \eta$	$\sin \eta$	$\sinh \eta$
Δ	$-\cosh \eta$	$-\cos \eta$	$\cosh \eta$
γ	$\pi + i\eta$	$\pi - \eta$	$-i\eta$

2.5.2 On parametrizations of the six-vertex weights

In statistical mechanics, the local energies must be real, and thus the Boltzmann weights must be real and positive. It is nevertheless useful to allow the Boltzmann weights to be complex in general. This freedom is useful from the technical viewpoint, but it is also physically meaningful. In particular, the region of parameter space in which the one-dimensional spin chain hamiltonian is hermitian need not coincide with that in which the two-dimensional Boltzmann weights are real and positive. Thus, the physical spin chain hamiltonian is indeed an analytic extension of the hamiltonian derived from realistic Boltzmann weights.

If we focus on the paradigmatic two-dimensional classical statistical mechanical six-vertex model, the parametrization (2.50) is not manifestly real and positive. To exhibit this property, we need to use a different parametrization in each of the three phases determined by the anisotropy Δ (see section 3.2). These parametrizations are shown in table 2.1. The last row in the table allows us to translate from the parameter η in the table to the parameter γ in (2.50).

Exercises

Ex. 2.1 Using the discussion up to equation (2.39), construct $\Delta B(u)$ acting on the space $\bigotimes^3 V_{\frac{1}{2}}$, and expand the state $\Delta B(u)\Omega$ in the basis $|x\rangle$ as in equation (2.34). What is the general pattern of the wave-function $f(x)$ for the state $|\Psi_M\rangle$ in (2.39)? What is the relationship between the quasi-momenta k of the Bethe ansatz construction in chapter 1 and the Boltzmann weights a, b and c?

Ex. 2.2 Show that the cancellation of unwanted terms in (2.44) yields the Bethe ansatz equation (2.46).

Ex. 2.3 Plug the Boltzmann factors a, b and c given by the uniformization (2.50) into the expression (2.45) for the eigenvalue Λ_M of the transfer matrix. Show that $\Lambda_M(u;\{u_i\})$ is analytic in u if and only if the Bethe ansatz equation (2.46) is satisfied. This is to be expected since, by construction, the transfer matrix $t(u)$ is itself analytic in u. This alternative derivation of the Bethe ansatz equation is equivalent to the cancellation of unwanted terms, although, in practice, it is safer to check both the analyticity and the cancellation.

Ex. 2.4 A 2×2 matrix has two invariants: its trace and its determinant. The trace of the monodromy matrix $T(u)$ yields the transfer matrix $t(u)$. Find the analog of the determinant, called the q-determinant $\det_q$, of the monodromy matrix, and elucidate its meaning.

Ex. 2.5 The relationship between quantum spin chains in one dimension and statistical models in two dimensions becomes more transparent with the help of

Zamolodchikov operators $Z(u)$, which may act repeatedly on the reference state $|\Omega\rangle$ to create Bethe states. We write, for instance,

$$\begin{aligned} Z(u)\,|\Omega\rangle &= |\Psi_1;k(u)\rangle \\ Z(u_1)Z(u_2)\,|\Omega\rangle &= |\Psi_2;k_1(u_1),k_2(u_2)\rangle \end{aligned} \tag{2.154}$$

The physical requirement on the Zamolodchikov operators is that their commutator reproduce the phase factor in the interchange between two pseudo-particles (2.49):

$$Z(u_1)Z(u_2) = \hat{S}_{12}Z(u_2)Z(u_1) \tag{2.155}$$

Find the explicit expression for the Zamolodchikov operators $Z(u)$ in equations (2.154) and (2.155) for the case of the six-vertex model.

Ex. 2.6 The spectral manifold Γ_Δ of the six-vertex model is a complex algebraic curve of genus zero, i.e. it is the Riemann sphere $\mathbb{C} \cup \{\infty\} = S^2$. Show that the Boltzmann weights of the six-vertex model are three holomorphic vector fields on S^2, which can be chosen as

$$a(z) = qz^2 - q^{-1}, \quad b(z) = z^2 - 1, \quad c(z) = \left(q - q^{-1}\right) z \tag{2.156}$$

where $q = e^{i\gamma}$ and $z \in S^2$. Find the relationship between z and u. This shows that a, b and c are globally well defined on the Riemann sphere, indeed that they are precisely the three conformal Killing vectors of S^2 (provided $q \neq \pm 1$). With the help of this geometrical interpretation of the weights a, b and c, work through the following:

(i) The replacement $q \to q^{-1}$ is equivalent to the conformal transformation $z \to 1/z$.
(ii) The hamiltonian (2.153) is defined at the zeroes of a and b. What about the zeroes of c?
(iii) Find a relationship between the anisotropy parameter $\Delta = \left(q + q^{-1}\right)/2$ and the harmonic ratio between the zeroes of a and b.
(iv) Along these lines, what is the geometric interpretation of the higher-spin R-matrices obtained through the descent procedure?

Ex. 2.7 On the vector space $W = V_a \otimes V_1 \otimes \cdots \otimes V_L$ of chapter 1 (with V_a and all the V_i isomorphic), define the permutation operators P_{ai} as

$$\begin{aligned} P_{ai} : V_a \otimes V_i &\to V_a \otimes V_i \\ v \otimes w &\mapsto w \otimes v \end{aligned} \tag{2.157}$$

and similarly for P_{ij} $(i \neq j)$. Show that they satisfy

$$P_{ai}P_{ij} = P_{aj}P_{ij} = P_{ij}P_{ai}\,, \qquad P_{ki}P_{kj} = P_{kj}P_{ij} = P_{ij}P_{ki} \tag{2.158}$$

Use these relations to show that P_{ij} satisfy the Yang–Baxter relation (2.66), namely

$$P_{i,i+1}P_{i+1,i+2}P_{i,i+1} = P_{i+1,i+2}P_{i,i+1}P_{i+1,i+2} \tag{2.159}$$

From the definition of the transfer matrix (2.6), and taking into account that

$$\mathscr{R}_{ai}(u=0) = c_0 P_{ai} \tag{2.160}$$

for some constant c_0, show that

$$t(0) = c_0^L \operatorname{tr}_a (P_{aL} \cdots P_{a2}P_{a1}) \tag{2.161}$$

With the help of equations (2.158) and the normalization $\operatorname{tr}_a(P_{aj}) = \mathbf{1}_j$, deduce that

$$t(0) = c_0^L P_{1L} \cdots P_{13}P_{12} = c_0^L P_{12}P_{23} \cdots P_{L-1,L} \tag{2.162}$$

Show that $t(0)/c_0^L$ coincides with the shift operator e^{-iP} (see equations (1.96)).

Using (2.158) together with

$$\operatorname{tr}_a(P_{ai}M_{aj}) = M_{ij} \qquad i \neq j \tag{2.163}$$

derive the hamiltonian

$$H = i\frac{\partial}{\partial u}\log\left(\frac{t(u)}{\left(\mathscr{R}(u)_{00}^{00}\right)^L}\right)\Bigg|_{u=0} = i\sum_{j=1}^{L} P_{j,j+1}\frac{\partial}{\partial u}\left(\frac{\mathscr{R}_{j,j+1}(u)}{a(u)}\right)\Bigg|_{u=0} \tag{2.164}$$

Scaling $\mathscr{R}(u) \to \mathscr{R}(u)/a(u)$ so that $\mathscr{R}_{00}^{00} = 1$, and using the braid-like permuted R-matrix (2.65), show that

$$H = i\sum_{j=1}^{L}\frac{d}{du}R_{j,j+1}(u)\Bigg|_{u=0} = \sum_{j=1}^{L} h_{j,j+1} \tag{2.165}$$

Apply this formula to the R-matrix of the six-vertex model, and thus find the XXZ hamiltonian in the form (2.56).

Since $t(u)$ and $t(v)$ commute for all u and v, construct the following conserved current:

$$\begin{aligned} H_2 &= i\frac{\partial^2}{\partial u^2}\log t(u)\Bigg|_{u=0} \qquad (2.166)\\ &= i\sum_{j=1}^{L}\left\{R''_{j,j+1}(0) - \left(R'_{j,j+1}(0)\right)^2 + h_{j,j+1}h_{j+1,j+2} - h_{j+1,j+2}h_{j,j+1}\right\} \end{aligned}$$

You will need the identity $\operatorname{tr}_a\left(P_{ai}M_{aj}N_{ak}\right) = M_{ij}N_{ik}$, which is a generalization of (2.163). Actually, the first two terms on the right-hand side of (2.166) cancel due to the unitarity condition $R(u)R(-u) = \mathbf{1}\otimes\mathbf{1}$, where we have implicitly adopted the normalization $R_{00}^{00}(u) = \mathbf{1}$. Hence, in general,

$$H_2 = i\sum_{j=1}^{L}\left[h_{j,j+1}\ ,\ h_{j+1,j+2}\right] \tag{2.167}$$

Observe that this conserved quantity involves interactions between nearest and also next to nearest neighbors. Higher derivatives of the transfer matrix are local conserved quantities involving interactions among spins even further away.

Show that, for the six-vertex model,

$$H_2 = \frac{1}{2(\sin\gamma)^2}\sum_{j=1}^{L}\left\{\sigma_j^y\sigma_{j+1}^z\sigma_{j+2}^x - \sigma_j^x\sigma_{j+1}^z\sigma_{j+2}^y + \right. \tag{2.168}$$

$$\left. + \Delta\left(\sigma_j^x\sigma_{j+1}^y\sigma_{j+2}^z - \sigma_j^y\sigma_{j+1}^x\sigma_{j+2}^z + \sigma_j^z\sigma_{j+1}^x\sigma_{j+2}^y - \sigma_j^z\sigma_{j+1}^y\sigma_{j+2}^x\right)\right\}$$

which for the isotropic Heisenberg model becomes simply

$$H_2 \sim \sum_{j=1}^{L}\left(\vec{\sigma}_j \times \vec{\sigma}_{j+1}\right)\cdot\vec{\sigma}_{j+2} \tag{2.169}$$

Ex. 2.8 [P.P. Kulish and E.K. Sklyanin, *Lecture Notes in Physics* **151** (1981) 61.]

It is interesting to note that H_2 in the previous exercise is constructed using only the nearest neighbor hamiltonian $h_{j,j+1}$. Since H and H_2 commute, we find the Reshetikhin criterion for integrability of a local hamiltonian, namely

$$\left[h_{j,j+1} + h_{j+1,j+2}\,,\,\left[h_{j,j+1}, h_{j+1,j+2}\right]\right] = \mathcal{O}_{j,j+1} + \mathcal{O}_{j+1,j+2} \tag{2.170}$$

with $\mathcal{O}_{j,j+1}$ some nearest neighbor operator. Show that this criterion is a sufficient but not necessary condition for integrability. Study whether (2.170) holds for the spin-1 Heisenberg chain $H = \sum \vec{S}_j \cdot \vec{S}_{j+1}$, where $\vec{S}$ are the spin-1 matrices of $SU(2)$.

Ex. 2.9 Take the logarithm of (2.54) in order to analyze the conformal mapping $\lambda \to k(\lambda)$. Consider separately γ as real and purely imaginary. Study the limit $\gamma \to 0$.

Ex. 2.10 Show that R in (2.82) has only two different eigenvalues. From this fact, fix the coefficients of the characteristic equation of R

$$R^2 = \alpha R + \beta \mathbf{1} \tag{2.171}$$

in terms of q. This relation is analogous to that satisfied by the elements of the Hecke algebra $H_L(q)$, to be discussed in chapter 5.

3

Bethe ansatz: some examples

3.1 Introduction and summary

In this chapter, we present a more detailed study of some simple one-dimensional models. The objective is to show how the Bethe ansatz technology works and to develop further some insight into its physical meaning. This introductory section serves the dual purpose of providing an overview of the relevant concepts and of summarizing the main results.

It is most surprising that it took 50 years to understand such a simple system as the one-dimensional antiferromagnetic Heisenberg model (the XXX model)

$$H = J \sum_{i=1}^{L} (\vec{\sigma}_i \cdot \vec{\sigma}_{i+1} - 1) \qquad J > 0 \tag{3.1}$$

Only in 1981 was the spectrum of the low lying excitations found, by the Leningrad school.

The first issue we should address concerning a one-dimensional spin system such as (3.1) is whether a particle interpretation exists. By this, we mean whether it is possible to define a Fock representation of the Hilbert space of the model such that the vacuum $|0\rangle$ of the Fock space $\mathscr{F}$ corresponds to the ground state and the many-particle states correspond to the low lying excitations. A particle interpretation is readily available if we establish the correspondence

$$\mathscr{H}_{\infty}^{\ell.\ell.} \to |0\rangle \oplus \mathscr{H}_1 \oplus \mathscr{H}_2 \oplus \cdots = \mathscr{F} \tag{3.2}$$

where $\mathscr{H}_{\infty}^{\ell.\ell.}$ represents the Hilbert space of the low lying excitations of the hamiltonian (3.1), in the limit $L \to \infty$, and $\mathscr{H}_n$ is the Hilbert space of n elementary excitations. The Fock vacuum $|0\rangle$ represents the antiferromagnetic ground state. Notice that the Hilbert space $\mathscr{H}_L$ on which (3.1) acts is isomorphic to $\overset{L}{\otimes}\, \mathbb{C}^2$, which, in the limit $L \to \infty$, is not separable. The non-triviality of (3.2) is precisely to extract a separable Hilbert space $\mathscr{H}_{\infty}^{\ell.\ell.}$ from the non-separable $\mathscr{H}_{\infty}$.

The two crucial facts to elucidate concerning the elementary excitations are

(i) the dispersion relation $\epsilon(k)$, from which we obtain information about the existence of a mass gap: if the mass gap is zero, then the theory may correspond in the continuum limit to a massless or critical field theory, i.e. to a conformal field theory;

(ii) the internal quantum numbers (spin) of the elementary excitations.

To begin to comprehend why these two facts are so difficult to compute – indeed, it took the collective efforts of many workers over half a century – it is necessary to reflect for a moment on the richness of the antiferromagnetic vacuum. (Here and below, the word "ferromagnetic" is used to characterize the positive values of J in (3.1), and not a possible antiferromagnetic order of the vacuum. In fact, the ground state of (3.1) is disordered.) Recall from previous chapters that to solve the model (3.1) we must diagonalize the hamiltonian in the basis of spin waves. A state with M spin waves is achieved by flipping M spins down from the reference state with all spins up. For the trivial case of only one spin wave in a periodic chain, the dispersion relation for the spin wave of the XXX model is given by

$$E(k) = 4J(\cos k - 1) \tag{3.3}$$

and thus the difference between the physical behavior of the ferromagnetic phase and that of the antiferromagnetic one can be easily distinguished.

In the ferromagnetic regime, the coupling constant J is negative and the energy of the spin wave is positive, so the Bethe reference state coincides with the ground state of minimal energy; this is the ordered phase, and spins tend to align. The solution to the physical problem is relatively straightforward.

The antiferromagnetic regime, with a positive coupling $J > 0$, is trickier. The energy of the spin wave is negative, and flipping one spin down is energetically favored over keeping all spins up. The Bethe reference state has nothing to do with the ground state, which we expect to be a singlet of the global $SU(2)$ symmetry, in fact a state with $S^z_{\text{total}} = 0$. To obtain such a state in our picture, we need a "condensate" of spin waves. This physical intuition can be combined with the fact that the energy of the spin wave for $J > 0$ is negative thanks to one of the most fruitful ideas in theoretical physics, the concept of a Dirac sea. Identifying the vacuum as the Dirac sea filled up to the Fermi surface, the elementary excitations can be thought of as holes in the Dirac sea. The integrability of the model, which amounts to the factorizability of the scattering matrix for spin waves, allows us to construct the sea starting from the Bethe equations. We shall describe this construction in this chapter.

For the isotropic Heisenberg model (3.1) in the antiferromagnetic phase $J > 0$, points (*i*) and (*ii*) above were finally clarified by Faddeev and Takhtajan in the early 1980s. The dispersion relation for the low lying excitations turns out to be of the form

$$\epsilon(k) = 2\pi J \sin k \qquad 0 \leq k \leq \pi \tag{3.4}$$

which means that the system has no mass gap. This is consistent with a continuum limit described by a free massless scalar field. The big surprise has to do with the spin of the low lying excitations. Since an excitation corresponds to flipping one local spin up into a spin down, with a net change of one unit of angular momentum, you might have guessed from the one-particle hamiltonian that the elementary excitations would have spin 1. Instead, it turns out that the particle-like excitations over the antiferromagnetic Dirac sea have spin $\frac{1}{2}$. Thus, the Fock space of the model is, for a chain with an even number of sites,

$$\mathscr{F} = \bigoplus_{n=0}^{\infty} \int_0^{\pi} \cdots \int_0^{\pi} dk_1 \cdots dk_{2n} \otimes^{2n} \mathbb{C}^2 \tag{3.5}$$

where the integrations run over the possible values of the momenta, and $\mathbb{C}^2$ represents the internal spin-$\frac{1}{2}$ space of dimension two. Proper symmetrization of the states in (3.5) must be taken into account as well. The excitations come in pairs (whence the $2n$ in (3.5)), for otherwise the total spin of a chain with an even number of sites would not be an integer.

To end this introduction, let us stress the physical relevance of the study of low lying excitations in one-dimensional magnets. The construction sketched out above provides us with a new way of thinking about the difficult issue of whether a quantum system admits a particle interpretation. Also, and more pragmatically, the internal quantum numbers of the elementary excitations are a completely unexpected collective result, impossible to predict *a priori*. This is the motivation for invoking particles with strange statistics (anyons) in attempts to understand high temperature super-conductors, the copper oxide planes of which can be modeled by two-dimensional antiferromagnets.

This chapter is devoted to an explanation of the above introduction. To highlight various important features, we will study the simplest representative systems at hand. In section 3.2, we will consider in detail the antiferromagnetic ground state of the anisotropic Heisenberg spin chain. In section 3.3, we will concentrate on the antiferromagnetic isotropic spin chain, and find the spin and dispersion relations of the low lying excitations. Then, in section 3.4, we will generalize these results to other integrable spin chains. Sections 3.5 and 3.6 are devoted to open spin chains. Finally, section 3.7 provides the only analysis in this book of a general two-dimensional system, not necessarily integrable.

3.2 The phase structure of the six-vertex model

Following Baxter, we will first establish a qualitative picture of the phase structure of the six-vertex model, using heuristic arguments. We restrict ourselves to the case of a real anisotropy parameter Δ, for otherwise the spin chain hamiltonian is not hermitian.

In the thermodynamic limit $L \to \infty$ (L the number of sites in the chain), the ground state is completely determined by the maximum eigenvalue of the transfer matrix. Some information about the maximum eigenvalue can be obtained directly from the Perron–Frobenius theorem, which states that, given a finite matrix whose elements are all positive or zero, the components of its eigenvector of highest eigenvalue are all strictly positive.

Let us consider the sector with two spins down. The Bethe equations (1.51) for $M = 2$ reduce to

$$e^{ik_1 L} e^{ik_2 L} = 1 \tag{3.6}$$

Setting $z_i = e^{ik_i}$, this equation implies that $z_1 z_2$ is an Lth root of unity ε. The components of the eigenvector are given by the wave-function

$$f(x_1, x_2) = A_{12} \left(z_1^{x_1} z_2^{x_2} + \hat{S} z_1^{x_2} z_2^{x_1} \right) \tag{3.7}$$

Using (3.6), we obtain

$$f(x_1 + 1, x_2 + 1) = \varepsilon f(x_1, x_2) \tag{3.8}$$

All the components of the eigenvector of maximal eigenvalue must be strictly positive, which already implies $\varepsilon = 1$. Thus $z_2 = z_1^{-1}$, and therefore $f(x_1, x_2)$ is proportional to $\cos k(x_1 - x_2 + L/2)$, i.e. $k = k_1 = -k_2$. Since $0 < x_1 < x_2 \leq L$, it follows that $x_1 - x_2 + L/2$ ranges over the interval $[-L/2+1, L/2-1] = [-r, r]$. From the form of $f(x_1, x_2)$, it follows that the value of k defining the ground state of the $M = 2$ sector must satisfy $0 \leq |k|r \leq \pi/2$, i.e. $|k| \in [0, \pi/(2r)]$, or else it is a purely imaginary number ($\cos k(x_1 - x_2 + L/2) = \cosh |k|(x_1 - x_2 + L/2) > 0$). Using the Bethe equations (1.90), the explicit expression for $\hat{S}$ (1.108) and the condition $z_1 z_2 = 1$, we obtain

$$\Delta \left(z_1^{L-1} + z_1 \right) = z_1^L + 1 \tag{3.9}$$

which implies for the anisotropy parameter Δ the relation

$$\Delta = \frac{\cos k(r+1)}{\cos kr} \tag{3.10}$$

For $\Delta < 1$, there is one solution with real k in the allowed interval $[0, \pi/2r]$ and none for pure imaginary k. On the other hand, for $\Delta > 1$ the only solution to (3.10) calls for k to be pure imaginary and positive. From (1.84) and (1.86), it is clear that if k is pure imaginary, the eigenvalue in the sector M is always smaller than the eigenvalue in the sector $M - 1$, and therefore the maximal eigenvalue corresponds to the state in the 0 sector, which means ferromagnetic ordering. This very heuristic argument, based on the educated extrapolation to arbitrary M of the results for the $M = 2$ sector, leads us to the conclusion that hamiltonians with $\Delta > 1$ correspond to systems in a ferromagnetic phase, whereas those with $\Delta < 1$ describe some kind of antiferromagnetic situation.

Let us now consider with some care the case $\Delta < 1$, where two different regions must be distinguished, with $\Delta = -1$ at their boundary.

In the region $-1 < \Delta < 1$, we parametrize

$$\Delta = \frac{q + q^{-1}}{2} \tag{3.11}$$

with

$$q = e^{i\gamma} \tag{3.12}$$

and $0 < \gamma < \pi$. In terms of the rapidity λ defined by (2.54), we obtain

$$e^{ik} = \frac{e^{\gamma\lambda} - e^{i\gamma}}{e^{\gamma(\lambda+i)} - 1} = \frac{\sinh\frac{\gamma}{2}(\lambda - i)}{\sinh\frac{\gamma}{2}(\lambda + i)} \tag{3.13}$$

When the momentum k is real, the rapidity λ is also real. When $\lambda \to \pm\infty$, we see that $e^{ik} \to e^{\mp i\gamma}$, which agrees with the fact that the Bethe ansatz creation operator $\Delta(B)$ on a two-particle state becomes, in this limit, $\Delta(S^-)$ (see equation (2.92)). Taking the logarithm of (3.13),

$$k = 2\arctan\left[\left(\text{cotan}\frac{\gamma}{2}\right)\tanh\frac{\gamma\lambda}{2}\right] + \pi \qquad \text{mod}(2\pi) \tag{3.14}$$

we see that $k(\lambda)$ is a monotonically increasing function over the whole real interval $\lambda \in [-\infty, +\infty]$. Actually, it will prove convenient to shift the momentum from k to $k + \pi$ in all these expressions, so that the range of shifted quasi-momenta is $[\gamma - \pi, \pi - \gamma]$. The Bethe equation for the new momentum corresponds to a parametrization with $\Delta = -\cos\gamma$.

For $\Delta < -1$, we first replace γ by $i\gamma$ in (3.13), which becomes

$$e^{ik} = \frac{e^{i\gamma\lambda} - e^{-\gamma}}{e^{\gamma(i\lambda-1)} - 1} = \frac{\sin\frac{\gamma}{2}(\lambda - i)}{\sin\frac{\gamma}{2}(\lambda + i)} \tag{3.15}$$

Now, for real k the rapidity λ is real. The logarithm of (3.15) yields

$$k = 2\arctan\left[\left(\text{cotanh}\frac{\gamma}{2}\right)\tan\frac{\gamma\lambda}{2}\right] + \pi \tag{3.16}$$

Shifting the quasi-momentum by π, we see that the range for the rapidity is $\lambda \in [-\pi/\gamma, \pi/\gamma]$, and therefore the anisotropy Δ in the Bethe equations is effectively $\Delta = -\cosh\gamma$.

In summary, by extrapolating to arbitrary M the results about the ground state of the $M = 2$ sector, we have found the rapidity range contributing to the antiferromagnetic ground state of the XXZ model ($\Delta < 1$):

$$\begin{array}{lll} -1 < \Delta < 1 & \lambda \in [-\infty, \infty] & k \in [\gamma - \pi, \pi - \gamma] \\ \Delta < -1 & \lambda \in [-\pi/\gamma, \pi/\gamma] & k \in [-\pi, +\pi] \end{array} \tag{3.17}$$

We may now turn to the thermodynamic limit. The limit is defined by letting both L and M tend to infinity while keeping the ratio M/L constant.

Let us study the case $-1 < \Delta < 1$ first. Taking the logarithm of the Bethe equations (1.90), we find

$$Lk_i = 2\pi I_i + \sum_{j=1}^{M} \Theta(k_i, k_j) \tag{3.18}$$

where $\Theta(k_i, k_j) = i\log\left(-\hat{S}(k_i, k_j)\right)$, and I_i are integers when M is odd, or half-odd when M is even. When $L, M \to \infty$, we assume that the solutions to (3.18) are distributed on the real axis in such a way that they can be described by a density function $\rho(k)$:

$$\int_{-Q}^{+Q} \rho(k)dk = \frac{M}{L} \tag{3.19}$$

with the interval $[-Q, +Q] \subset [\gamma - \pi, \pi - \gamma]$; see (3.17). We assume also that the rapidities in the ground state are packed as closely as possible. This is how we fill the "Dirac sea". Thus, we take I_i as

$$I_i = i - \frac{1}{2}(M+1) \qquad i = 1, \ldots, M \tag{3.20}$$

Using the density function $\rho(k)$, we obtain from (3.18)

$$Lk = -\pi(M+1) + 2\pi L \int_{-Q}^{k} \rho(k')dk' + L \int_{-Q}^{+Q} \Theta(k,k')\rho(k')dk' \tag{3.21}$$

After differentiating with respect to k, this yields

$$2\pi\rho(k) = 1 - \int_{-Q}^{+Q} \frac{\partial\Theta(k,k')}{\partial k}\rho(k')dk' \tag{3.22}$$

Solutions to (3.22) produce the ground state distribution of roots for the various sectors, which are now characterized by the value of M/L. Note that the Fermi interval $[-Q, +Q]$ is determined in terms of this value through equation (3.19).

Before solving (3.22), it is convenient to trade the momenta k for the rapidity variables λ, because in these variables the kernel of the integral equation (3.22) depends only on the difference $(\lambda - \lambda')$. The functions $k(\lambda_i)$ and $\Theta(k_i, k_j)$ can be expressed in terms of the functions

$$\begin{aligned} f(x,n) &= i \log\left[-\frac{\sinh\frac{\gamma}{2}(x+in)}{\sinh\frac{\gamma}{2}(x-in)}\right] \\ &= 2\arctan\left[\left(\operatorname{cotan}\frac{n\gamma}{2}\right)\tanh\frac{\gamma x}{2}\right] \end{aligned} \tag{3.23}$$

as follows:

$$\begin{aligned} k(\lambda) &= f(\lambda, 1) \\ \Theta(k(\lambda), k(\lambda')) &= f(\lambda - \lambda', 2) \end{aligned} \tag{3.24}$$

Introducing now the density $\rho(\lambda)$ such that

$$\int_{-\widehat{Q}}^{+\widehat{Q}} \rho(\lambda)d\lambda = \int_{-Q}^{+Q} \rho(k)dk = \frac{M}{L} \tag{3.25}$$

and following the same steps as above, we obtain

$$a(\lambda) \equiv \frac{1}{2\pi}\frac{d}{d\lambda}f(\lambda,1) = \rho(\lambda) + \int_{-\widehat{Q}}^{+\widehat{Q}} \rho(\lambda')T(\lambda-\lambda')d\lambda' \tag{3.26}$$

where

$$T(\lambda) \equiv \frac{1}{2\pi}\frac{d}{d\lambda}f(\lambda,2) \tag{3.27}$$

Instead of fixing the value of M/L, let us choose the simplest $\widehat{Q} = \infty$, and then find out what value of M/L it corresponds to (this is possible for $-1 < \Delta < 1$, because $\lambda \in [-\infty, +\infty]$). For $\widehat{Q} = \infty$, we introduce the Fourier transform

$$\hat{F}(x) = \int_{-\infty}^{+\infty} F(\lambda)\, \mathrm{e}^{-i\lambda x} d\lambda \tag{3.28}$$

whereby (3.26) becomes

$$\hat{a}(x) = \left(1 + \hat{T}(x)\right) \hat{\rho}(x) \tag{3.29}$$

Furthermore, using the fact that

$$\frac{1}{2\pi}\hat{f}'(x, n) = \frac{\sinh x\left(\frac{\pi}{\gamma} - n\right)}{\sinh x\frac{\pi}{\gamma}} \tag{3.30}$$

we find the solution

$$\hat{\rho}(x) = \frac{1}{2\,\cosh\frac{\pi x}{\gamma}} \tag{3.31}$$

or equivalently

$$\rho(\lambda) = \frac{1}{2\pi}\int_{-\infty}^{+\infty} dx\ \mathrm{e}^{i\lambda x}\hat{\rho}(x) = \frac{\gamma}{4\pi}\frac{1}{\cosh\frac{\gamma\lambda}{2}} \tag{3.32}$$

Integrating,

$$\int_{-\infty}^{+\infty} \rho(\lambda) d\lambda = \hat{\rho}(0) = \frac{1}{2} \tag{3.33}$$

which means that $M/L = 1/2$. The total angular momentum S^z, in the hamiltonian picture of this state, is $S^z_{\mathrm{tot}} = 0$, which is consistent with our intuition about the antiferromagnetic phase.

The analysis for $\Delta < -1$ differs from the one above in the integration limits. In this regime, equation (3.26) still holds provided we effect the change $\gamma \to i\gamma$, so we use

$$f(x, n)\Big|_{\Delta<-1} = 2\arctan\left[\left(\mathrm{cotanh}\frac{\gamma n}{2}\right)\tan\frac{\gamma x}{2}\right] \tag{3.34}$$

Note that now the functions $a(\lambda)$, $\rho(\lambda)$ and $T(\lambda)$ are periodic in λ with period $2\pi/\gamma$. We choose $\hat{Q} = \pi/\gamma$, and define the Fourier series in the finite interval $[-\pi/\gamma, +\pi/\gamma]$:

$$\begin{aligned} \hat{F}_m &= \int_{-\pi/\gamma}^{\pi/\gamma} d\lambda\, \mathrm{e}^{-i\gamma\lambda m} F(\lambda) \\ F(\lambda) &= \frac{\gamma}{2\pi}\sum_{m\in\mathbb{Z}} \mathrm{e}^{i\gamma\lambda m}\hat{F}_m \end{aligned} \tag{3.35}$$

The Fourier transform of (3.26) is now

$$\hat{a}_m = \left(1 + \hat{T}_m\right) \hat{\rho}_m \tag{3.36}$$

Since the mth Fourier component of

$$\frac{1}{2\pi}\frac{d}{d\lambda}f(\lambda,n)\Big|_{\Delta<-1}$$

is $e^{-|m|\gamma n}$, we obtain

$$\hat{\rho}_m = \frac{1}{2\cosh m\gamma} \tag{3.37}$$

which is a discretized version of the distribution (3.31) for $-1 < \Delta < 1$. The density $\rho(\lambda)$ is then given from (3.35) and (3.37) by

$$\begin{aligned} \rho(\lambda) &= \frac{\gamma}{4\pi}\sum_{m\in\mathbb{Z}}\frac{e^{i\gamma\lambda m}}{\cosh(m\gamma)} \\ &= \frac{\gamma K(k)}{2\pi^2}\,\mathrm{dn}\left(\frac{K(k)\gamma\lambda}{\pi},k\right) \end{aligned} \tag{3.38}$$

where dn is a Jacobi elliptic function (see appendix B), whose modulus k is fixed by the condition

$$\frac{K'(k)}{K(k)} = \frac{\gamma}{\pi} \tag{3.39}$$

3.3 Low lying excitations

In the previous section, we learned that the ground state in the regimes $-1 < \Delta < 1$ and $\Delta < -1$ of the anisotropic Heisenberg chain is a condensate of spin waves characterized by vanishing total angular momentum, ($S^z_{\rm tot} = 0$) and the distribution of real roots (3.31) or (3.38). The real roots in the ground state are packed as tightly as possible. It is traditional to call this state the Dirac sea of real roots.

In this section, we investigate fluctuations above the ground state. In particular, we wish to find out the mass and spin of the low lying excitations.

3.3.1 Strings and holes

For a finite chain of length L, any physical state of spin $L/2-M$ is characterized by a set of M roots $\{\lambda_i\}$ satisfying the Bethe equations (2.55). These roots may be complex. Recall that the condition that the roots should be real was obtained (in the $M = 2$ sector) with the help of the Perron–Frobenius theorem, which applies only to the eigenvector of maximal eigenvalue, i.e. to the ground state. It is not very hard to prove that, in the thermodynamic limit, non-real roots come in conjugate pairs, i.e. that if z is a Bethe root then so is $\bar{z}$. It turns out, moreover, that sometimes more than one or two roots have the same real part. If p roots share the same real component, we say that there is a string of length p at that point on the real "root axis". Physically, the roots assembling into a string in the thermodynamic limit are associated with a bound state of spin waves. An arbitrary Bethe state can be characterized by the location of its roots;

in the thermodynamic limit it suffices to know the location of the strings (real numbers) and their length (integer numbers). The string hypothesis states that the roots in a string are all equally spaced in the imaginary direction. There exist deviations from the string hypothesis, but they will not be of any relevance for our discussion: we shall always assume the string hypothesis to be true. If a state in the sector M contains V_n strings of length n, then

$$\sum_{n\geq 0} nV_n = M \tag{3.40}$$

The ground state discussed in the previous section is a Dirac sea of Bethe strings of length one, $V_n = M\delta_{n,1}$.

Instead of considering complex roots, we can imagine a simpler perturbation of the ground state as just a modification of the distribution of (real) roots. In the "Dirac sea" picture, we may allow for some holes (i.e. particles, i.e. strings of zero length). A hole in the Dirac sea appears whenever the labels I_i and I_{i+1} in (3.20) associated with two consecutive roots are not consecutive themselves:

$$I_{i+1} - I_i = 1 + (\text{number of holes between the roots } i \text{ and } i+1) \tag{3.41}$$

A generic low lying Bethe state can thus carry holes and strings, as shown graphically below:

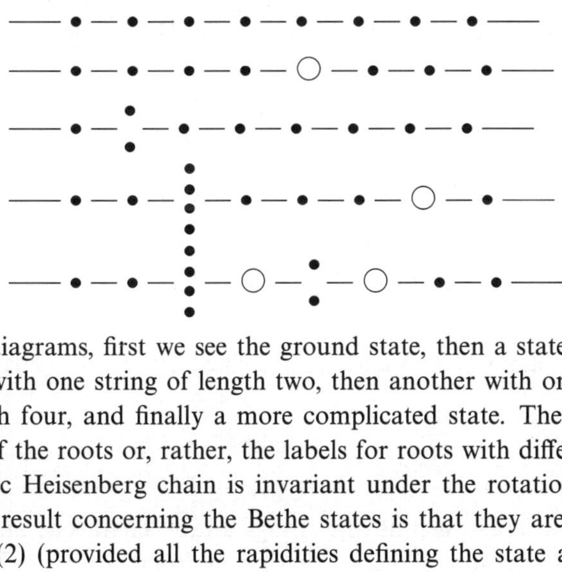

(3.42)

In the above diagrams, first we see the ground state, then a state with one hole, then another with one string of length two, then another with one hole and one string of length four, and finally a more complicated state. The axis represents the real part of the roots or, rather, the labels for roots with different real parts.

The isotropic Heisenberg chain is invariant under the rotation group $SU(2)$. An important result concerning the Bethe states is that they are highest weight vectors of $SU(2)$ (provided all the rapidities defining the state are finite). This implies, using the $SU(2)$ symmetry, that the total number of Bethe states with M spin waves on a chain of length L is given by

$$Z(L, M) = \binom{L}{M} - \binom{L}{M-1} \tag{3.43}$$

Taking into account that each Bethe vector of spin $S^z = L/2 - M$ generates an $SU(2)$ multiplet, we deduce that the Bethe basis is complete and spans the whole

Hilbert space $\mathscr{H}_L$ of the spin chain:

$$\dim \mathscr{H}_L = \sum_{M=0}^{L/2} (L - 2M + 1) Z(L, M) = 2^L \tag{3.44}$$

We have taken L to be even, a technical particularization which we keep from now on.

To lighten the presentation, we shall continue considering the simple isotropic chain with $\Delta = 1$ (and $J > 0$, so we are in the antiferromagnetic regime), although the results below are rather general. The Bethe equations for the real roots are given by (2.59), namely:

$$\left(\frac{\lambda_j + i}{\lambda_j - i} \right)^L = - \prod_{k=1}^{M} \frac{\lambda_j - \lambda_k + 2i}{\lambda_j - \lambda_k - 2i} \tag{3.45}$$

A Bethe string of length n consists of a set of n Bethe roots $\lambda_{j,\alpha}^{(n)}$ sharing the same real part $\lambda_j^{(n)} \in \mathbb{R}$ and equally spaced along the imaginary axis:

$$\lambda_{j,\alpha}^{(n)} = \lambda_j^{(n)} + i(n + 1 - 2\alpha) \qquad \alpha = 1, \ldots, n \tag{3.46}$$

Introducing (3.46) into (3.45) and multiplying the Bethe equations for all the roots of the same string, we obtain the following Bethe equation for the real part of the roots:

$$L f_n \left(\lambda_j^{(n)} \right) = 2\pi I_j^{(n)} + \sum_{m \geq 1} \sum_{k=1}^{V_m} f_{n,m} \left(\lambda_j^{(n)} - \lambda_k^{(m)} \right) \tag{3.47}$$

where

$$f_{n,m}(x) = \sum_{\ell = |n-m|/2}^{(n+m-2)/2} \left(f_{2\ell}(x) + f_{2\ell+2}(x) \right) \tag{3.48}$$

with $-\pi/2 \leq \arctan x \leq \pi/2$ and $f_n(x)$ defined by (3.23) in the isotropic limit $\gamma \to 0$:

$$f_n(x) = \begin{cases} 2 \arctan \frac{x}{n} & n \neq 0 \\ 0 & n = 0 \end{cases} \tag{3.49}$$

Note, as a check, that when all the strings are of length one and no holes are present, i.e. for the ground state, we recover from (3.47) the limit $\gamma \to 0$ of equation (3.18).

A reasonable theorem states that any array of labels $\{I_j^{(n)}\}_{j=1,\ldots,V_n}$ such that

$$I_k^{(n)} \neq I_j^{(n)} \quad ; \qquad k \neq j \quad (\forall n) \tag{3.50}$$

defines a unique Bethe vector. This condition can be interpreted as the Pauli exclusion principle for Bethe roots.

The numbers $I_j^{(n)}$ in (3.47) vary in the interval $|I_j^{(n)}| \leq I_{\max}^{(n)}$, and they are integer or half-odd depending on $I_{\max}^{(n)}$. In order to compute $I_{\max}^{(n)}$, we introduce the auxiliary functions

$$\begin{aligned} F_n(\lambda) &\equiv F_n\left(\lambda, \{\lambda_j^{(m)}\}\right) \\ &= \frac{L}{2\pi} f_n(\lambda) - \frac{1}{2\pi} \sum_{m\geq 1} \sum_{k=1}^{V_m} f_{n,m}\left(\lambda - \lambda_k^{(m)}\right) \end{aligned} \tag{3.51}$$

which depend on the whole set of roots $\{\lambda_j^{(m)}\}$. The Bethe equation (3.47) can thus be rewritten as

$$F_n\left(\lambda_j^{(n)}\right) = I_j^{(n)} \tag{3.52}$$

From (3.49), we see that $f_n(\lambda)$ is a monotonically increasing function of λ. We expect the same behavior for $F_n(\lambda)$, so we deduce that

$$-F_n(+\infty) = F_n(-\infty) \leq I_j^{(n)} \leq F_n(+\infty) \tag{3.53}$$

Finally, since the total number of allowed $I_j^{(n)}$ is $2I_{\max}^{(n)} + 1 = 2F_n(\infty)$, we obtain

$$2I_{\max}^{(n)} + 1 = 2F_n(\infty) \tag{3.54}$$

whereby

$$I_{\max}^{(n)} = \frac{L-1}{2} - \frac{1}{2\pi} \sum_{m\geq 1} V_m f_{n,m}(\infty) \tag{3.55}$$

The function $F_n(\lambda)$ serves also to illustrate the fact mentioned above (see (3.41)) that there may exist holes in the distribution of the numbers $I_j^{(n)}$. It is convenient to unify the concepts of roots and holes into the notion of vacancy. A vacancy can be either occupied (roots) or empty (hole). The number of vacancies corresponding to strings of length n is equal to $2I_{\max}^{(n)} + 1$, and they are the quantum numbers at our disposal to construct the Bethe states. Denoting by H_n the number of holes of length n, we have the relation

$$2I_{\max}^{(n)} + 1 = V_n + H_n \tag{3.56}$$

Let us study various solutions to (3.55), using directly the number, V_n, of "particles" and the number, H_n, of holes. To start with, we can derive a very interesting equation from the $n = 1$ case of (3.55). Using the result

$$\frac{1}{2\pi} f_{n,m}(\infty) = \begin{cases} n - \frac{1}{2} & \text{for } n = m \\ \min(n,m) & \text{if } n \neq m \end{cases} \tag{3.57}$$

we find

$$H_1 = L - 2\sum_{n\geq 1} V_n \tag{3.58}$$

For L even, therefore, the number of holes in the sea of one-strings is always even.

Recalling that the ground state is characterized by

$$M = \frac{L}{2} \qquad V_1 = \frac{L}{2} \qquad V_n = 0, \quad n = 2, 3, \ldots \tag{3.59}$$

it follows from (3.58) that the ground state has no holes ($H_1 = 0$). In other words, the $I_j^{(1)}$ fill up the whole allowed interval $[-I_{\max}^{(1)}, I_{\max}^{(1)}]$ in the ground state. Hence, the antiferromagnetic ground state is unique.

Let us march forward. The next simplest Bethe state with zero total angular momentum is characterized by

$$M = \frac{L}{2} \qquad V_1 = \frac{L}{2} - 2 \qquad V_2 = 1 \qquad V_n = 0\,, \; n = 3, 4, \ldots \tag{3.60}$$

Equation (3.58) implies the existence of two holes, $H_1 = 2$, and the two-string is fixed at $I_j^{(2)} = 0$. This state, a singlet ($S^z = 0$), is labeled by the rapidities θ_1 and θ_2 of the two holes, whereas the two-string has zero energy (see exercise 3.6).

Carrying on, the simplest triplet state ($S^z = 1$) would have two holes ($H_1 = 2$) and no strings of higher length:

$$M = \frac{L}{2} - 1 \qquad V_1 = \frac{L}{2} - 1 \qquad V_n = 0\,, \; n = 2, 3, 4, \ldots \tag{3.61}$$

Again the state is labeled by the rapidities of the two holes.

These simple examples suggest that the holes in fact behave like spin-$\frac{1}{2}$ particles. For instance, we have just constructed the singlet and triplet combinations of two holes. This is further supported by the following inequality for the total spin of a Bethe state:

$$S^z = \frac{L}{2} - \sum nV_n = \frac{1}{2}H_1 - \sum_{n\geq 1}(n-1)V_n \quad \Rightarrow \quad S^z \leq \frac{H_1}{2} \tag{3.62}$$

where (3.58) has been put to good use once more.

The above considerations demonstrate that the number of holes H_1 of the Dirac sea of one-strings plays a fundamental role in the analysis of the low lying excitations: let us call H_1 simply N_h. We will prove next that the dimension of the Hilbert space of solutions with N_h holes is precisely 2^{N_h}, which together with (3.62) shows the important result that holes are effectively spin-$\frac{1}{2}$ particles.

To obtain solutions to equation (3.55) with more than two holes, it is convenient to rewrite (3.55) in terms of the number of holes H_n. Using the relations

$$\begin{aligned} &2f_{1,m}(\infty) - f_{2,m}(\infty) = \pi\, \delta_{2,m} \\ &2f_{n,m}(\infty) - f_{n-1,m}(\infty) - f_{n+1,m}(\infty) = \pi \left(\delta_{n+1,m} + \delta_{n+1,m}\right) \quad (n > 1) \end{aligned} \tag{3.63}$$

which are valid for any $m \geq 1$, we derive from (3.55) the following equalities:

$$
\begin{aligned}
V_1 + H_1 &= \frac{L + H_2}{2} \\
V_2 + H_2 &= \frac{H_1 + H_3}{2} \\
&\vdots \\
V_n + H_n &= \frac{H_{n-1} + H_{n+1}}{2}
\end{aligned}
\tag{3.64}
$$

The low lying excitations over the ground state (3.59) correspond to configurations where $\frac{L}{2} - V_1$ is fixed and finite as $L \to \infty$. Thus, V_n $(n \geq 2)$ are finite in the thermodynamic limit, and the Hilbert space of the low lying excitations appearing on the left-hand side of equation (3.2) in the introduction to this chapter is meaningfully defined. With the help of (3.58) and (3.64), we may show (see exercise 3.9) that

$$
\dim \mathscr{H}^{(N_h)} = 2^{N_h} \tag{3.65}
$$

3.3.2 Dispersion relations

Let us now turn to the energy and momentum of a Bethe state characterized by the roots $\{\lambda_i\}$. We have already found the energy and momentum of Bethe states in the isotropic case, equation (2.61), which we reproduce for convenience:

$$
P = \sum_{i=1}^{M} (2 \arctan \lambda_i + \pi) \quad , \quad E = -\sum_{i=1}^{M} \frac{2}{1 + \lambda_i^2} \tag{3.66}
$$

Here, E is the eigenvalue of the Heisenberg hamiltonian and e^{-iP} is the eigenvalue of the shift operator.

The energy of the singlet (3.60) and triplet (3.61) states relative to that of the ground state allows us to obtain the energy and momentum of a hole, reinforcing the particle interpretation of holes in the Dirac sea of one-strings:

$$
\begin{aligned}
\epsilon(\theta) &= \frac{\pi}{2 \cosh \frac{\pi\theta}{2}} \\
k(\theta) &= \frac{\pi}{2} - \arctan \sinh \frac{\pi\theta}{2}
\end{aligned}
\tag{3.67}
$$

Eliminating the rapidity θ from these two expressions, we recover the advertised dispersion relation (3.4), i.e. $\epsilon(k) = (\pi/2) \sin k$.

Note the difference between the dispersion relations for spin waves (3.66) and those for the "elementary excitations" (3.67). Note also that the relations (3.67) were obtained in the thermodynamic limit $L \to \infty$, where the two-string appearing in the singlet (3.60) is of zero energy.

3.4 Integrable higher-spin chains

To model physical systems where the variables at the lattice sites may take more than two values, we could imagine spin chains with representations of $SU(2)$

of spin $> \frac{1}{2}$, so that the operators acting on them are matrices $\vec{S}$ of dimension higher than two. An immediate physical motivation for such a model comes from a metal, where the sites may either carry an electron with spin up, an electron with spin down, or no electron: three possible states which can be assembled into a vector of $SU(2)$. Integrable higher-spin chains can be easily obtained from the six-vertex solution through the descent procedure of section 2.4.1. Essentially, that procedure allows us to find the intertwiner R-matrices of $U_q(\widehat{s\ell}(2))$ for spin-S irreps, which are then interpreted as Boltzmann weights for the two-dimensional model. The explicit construction of these R-matrices is postponed until section 7.3. The integrable hamiltonians for generic spin and anisotropy derived from these R-matrices have been written down explicitly in a few cases only, for spins of $\frac{1}{2}$ and 1. A general formula is known, nevertheless, for the isotropic case. It takes the form

$$H_S(\gamma = 0) = -\sum_{j=1}^{L} Q_{2S}\left(\vec{S}_j \cdot \vec{S}_{j+1}\right) \tag{3.68}$$

where $\vec{S}$ is the spin-S matrix representation of $SU(2)$ and $Q_{2S}(x)$ is a polynomial of degree $2S$ in the $SU(2)$ invariant quantity $x_j = \vec{S}_j \cdot \vec{S}_{j+1}$. This polynomial is given by

$$Q_{2S}(x) = \sum a_k \, P_k(x) \tag{3.69}$$

with

$$\begin{aligned} a_k &= \sum_{\ell=1}^{k} \frac{1}{\ell} \\ P_k(x) &= \prod_{\substack{\ell=0 \\ \ell \neq k}}^{2S} \frac{x - x_\ell}{x_k - x_\ell} \\ x_\ell &= \frac{1}{2}\left[\ell(\ell+1) - 2S(S+1)\right] \end{aligned} \tag{3.70}$$

In fact, $P_k(\vec{S} \cdot \vec{S}')$ is just the projector onto the spin-k irrep of the tensor product $S \otimes S'$.

Here, we will concentrate on the Bethe equations for descendant models, which are very similar to the six-vertex Bethe equations. In fact, the only change in the equations is the appearance of the spin, S, on the left-hand side, in a very natural way:

$$\left(\frac{\sinh \frac{\gamma}{2}(\lambda_j + 2iS)}{\sinh \frac{\gamma}{2}(\lambda_j - 2iS)}\right)^L = \prod_{\substack{k=1 \\ k \neq j}}^{M} \frac{\sinh \frac{\gamma}{2}(\lambda_j - \lambda_k + 2i)}{\sinh \frac{\gamma}{2}(\lambda_j - \lambda_k - 2i)} \tag{3.71}$$

When $S = \frac{1}{2}$, we recover the six-vertex Bethe equations. In the isotropic limit $\gamma \to 0$, (3.71) becomes the more manageable

$$\left(\frac{\lambda_j + 2iS}{\lambda_j - 2iS}\right)^L = \prod_{\substack{k=1 \\ k \neq j}}^{M} \frac{\lambda_j - \lambda_k + 2i}{\lambda_j - \lambda_k - 2i} \tag{3.72}$$

Repeating the analysis of the previous section for (3.72), and using the same strings (3.46) of complex roots, we obtain the higher-spin generalization of (3.47), namely

$$L f_n^S\left(\lambda_j^{(n)}\right) = 2\pi I_j^{(n)} + \sum_m \sum_k f_{n,m}\left(\lambda_j^{(n)} - \lambda_k^{(m)}\right) \tag{3.73}$$

where

$$f_n^S(\lambda) = \sum_{\ell=1}^{\min(n,2S)} f_{n+1+2S-2\ell}(\lambda) \tag{3.74}$$

Note that $f_n^{\frac{1}{2}}(\lambda)$ reproduces $f_n(\lambda)$ in (3.49).

The considerations for $S = \frac{1}{2}$ concerning the allowed interval $I_{\max}^{(n)}$ can be repeated in this more general case rather effortlessly. The analog of (3.55) for any spin S is

$$I_{\max}^{(n)} = \frac{1}{2}\left(\frac{L}{\pi} f_n^S(\infty) - 1\right) - \frac{1}{2\pi} \sum_{m \geq 1} V_m f_{n,m}(\infty) \tag{3.75}$$

The ground state properties and the spin and dispersion relations for the low lying excitations are summarized in table 3.1. Let us stress a remarkable feature of the higher-spin chains: for any S, the particle-like excitations are always of spin one-half. The skeleton of the relevant derivations is addressed in exercises 3.4 and 3.5.

3.4.1 Hidden symmetry

Although the spin of the particle-like excitations is always $\frac{1}{2}$, some mysterious extra quantum numbers for the low lying excitations of the spin-S chain recall the original multiplet structure of the local lattice variables.

For any spin S, both the singlet and triplet low lying states are composed of two holes, and this is the reason why we attribute spin $\frac{1}{2}$ to the elementary excitation of a hole in the Dirac sea. For the spin-$\frac{1}{2}$ chain, this is pretty much the end of the story, because the dimension of the space $\mathscr{H}^{(N_h)}$ of Bethe states with N_h holes is precisely 2^{N_h}. By a state with N_h holes we mean a Bethe state with N_h holes in the Dirac sea of one-strings, and possibly also any number of ℓ-strings ($\ell > 1$), of zero energy in the thermodynamic limit. For instance, the space of states with two holes contains the state (3.60) with one two-string.

Table 3.1 Some results from the Bethe ansatz for the isotropic XXX Heisenberg model ($S = \frac{1}{2}$) and its higher-spin descendants, in the antiferromagnetic phase.

	$S = \frac{1}{2}$	$S > \frac{1}{2}$
n-string of roots	$\lambda_{j,\alpha}^{(n)} = \lambda_j^{(n)} + i(n + 1 - 2\alpha)$	$\alpha \in \{1, \ldots, n\}$
Energy of Bethe state $\{\lambda_i\}_{i=1}^M$	$\sum_{i=1}^{M} \frac{-2}{1 + \lambda_i^2}$	$\sum_{i=1}^{M} \frac{-4S}{4S^2 + \lambda_i^2}$
Momentum of Bethe state $\{\lambda_i\}_{i=1}^M$	$\sum_{i=1}^{M} (2 \arctan\lambda_i + \pi)$	$\sum_{i=1}^{M} \left(2 \arctan\frac{\lambda_i}{2S} + \pi\right)$
Dispersion relation	$\epsilon = \frac{\pi}{2} \sin k \, , \; 0 \leq k \leq \frac{\pi}{2}$	
Ground state	Dirac sea of one-strings	Dirac sea of $(2S)$-strings
	non-degenerate, with $S_{\text{total}}^z = 0$	
Low lying excitation	one hole in Dirac sea	
Spin of a hole	$\frac{1}{2}$	
Energy of a hole	$\epsilon(\theta) = \frac{\pi}{2} \left[\cosh \frac{\pi\theta}{2}\right]^{-1}$	
Momentum of a hole	$k(\theta) = \frac{\pi}{2} - \arctan \sinh \frac{\pi\theta}{2}$	
	$\frac{L}{2} - 2$ one-strings	$\frac{L}{2} - 2$ $(2S)$-strings
Low lying singlet	one two-string	one $(2S - 1)$-string
		one $(2S + 1)$-string
Low lying triplet	$\frac{L}{2} - 1$ one-strings	$\frac{L}{2} - 1$ $(2S)$-strings
		one $(2S - 1)$-string
Mass gap	no	no

For the spin-S chain, on the other hand, the situation changes drastically. An important step towards the formulation and understanding of this problem was taken by the Leningrad school. Using the string hypothesis (complex roots assemble at common real values), Faddeev and Reshetikhin found the following result for $S >> N_h$:

$$\lim_{S \to \infty} \dim \mathscr{H}^{(N_h)} = 2^{N_h} \left[\binom{N_h}{N_h/2} - \binom{N_h}{N_h/2 - 1} \right] \tag{3.76}$$

The correction factor in brackets points to the existence of some other, hidden, quantum numbers. Fortunately, the extra factor has a very nice combinatoric interpretation. In fact, it is the multiplicity of the identity (spin-0 irrep) in the tensor product of N_h fundamental irreps of $SU(2)$. Denoting the decomposition of this tensor product by

$$\overbrace{V^{(\frac{1}{2})} \otimes \cdots \otimes V^{(\frac{1}{2})}}^{N_h \text{ times}} = \bigoplus_j W(N_h, j) \otimes V^{(j)} \tag{3.77}$$

it is not hard to check that

$$\dim W(N_h, j=0) = \binom{N_h}{N_h/2} - \binom{N_h}{N_h/2 - 1} \tag{3.78}$$

A clear diagrammatic representation of the states with N_h holes is the following:

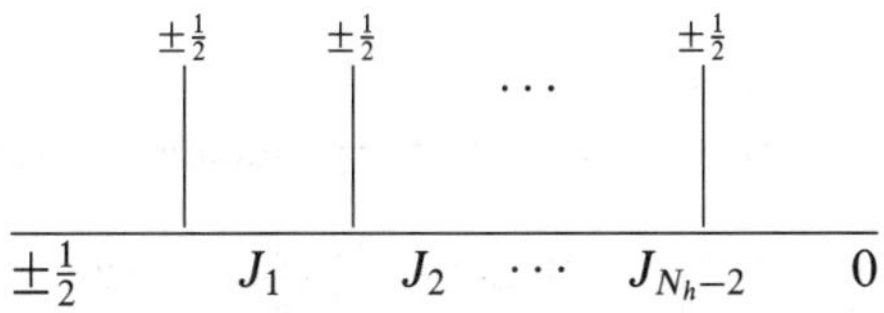

(3.79)

To read this diagram, start on the left and multiply two spin-$\frac{1}{2}$ irreps, yielding the irrep J_1, then $\frac{1}{2} \times J_1 = J_2$, and so on until $\frac{1}{2} \times J_{N_h-2} = 0$. The external spin-$\frac{1}{2}$ legs are labeled $\pm\frac{1}{2}$ because they have two "polarization states"; this yields the factor 2^{N_h} in (3.76). The various possible sequences of intermediate channels $(J_1, J_2, \ldots, J_{N_h-2})$ label the basis vectors of $W(N_h, 0)$.

We have interpreted the extra multiplicity factor in (3.76) to be the requirement that physical states with N_h holes form, altogether, a singlet. But this nice interpretation holds only, as the Leningrad formula does, for very large S. By direct computation, it is possible to evaluate dim $\mathscr{H}^{(N_h)}$ for small N_h and S. It turns out that the dimensionality of the space of Bethe states with N_h holes is still given by the multiplicity of the identity in the product of N_h spin-$\frac{1}{2}$ irreps, provided we modify the $SU(2)$ decomposition rules appropriately. The suitable modification of the decomposition rules calls for postulating the existence of a maximal spin. Normally, if we multiply $\frac{1}{2} \times \frac{1}{2} \times \cdots$ many times, we may reach irreps of arbitrarily high spin. By requiring that any spin J_i in (3.79) be at most S, we implement a truncation in the decomposition rule for the tensor product of two irreps of $SU(2)$:

$$J_1 \otimes J_2 = \sum_{J=|J_1-J_2|}^{\min(J_1+J_2, 2S-J_1-J_2)} J \tag{3.80}$$

This truncated tensor product reduces the multiplicity of the identity in $\otimes^{N_h} V^{(\frac{1}{2})}$ such that the wonderful formula

$$\dim \mathscr{H}^{(N_h)} = 2^{N_h} \times \dim W(N_h, 0) \tag{3.81}$$

remains valid. Note that the truncation disappears in the limit $S \to \infty$, as indeed it should.

We shall encounter such truncated decomposition rules several times in the chapters ahead. Mathematically, they have to do with the representation theory of quantum groups when q is a root of unity (chapters 6 and 7). Let us mention in passing two instances where we shall encounter such a truncation phenomenon again.

(i) In chapter 5, we will study generalizations of the two-dimensional Ising model called face models on Coxeter graphs. The decomposition rule (3.80) corresponds to a type A graph with Coxeter number $2S + 2$.

(ii) In chapter 10, we will study $SU(2)$ Wess–Zumino models of level k. The decomposition rule (3.80) is precisely the fusion algebra of the Wess–Zumino model with level $k = 2S$.

3.5 Integrability in a box: open boundary conditions

Until now, in our study of both integrable vertex models and one-dimensional spin chains, we have always dealt with periodic boundary conditions. Physically, the condition of integrability turned out to be equivalent to the factorizability of the spin wave S-matrix. To find the integrability conditions in the more general case of open boundary conditions, we should consider the interplay between the scattering S-matrix for the interchange of two spin waves, and the reflection coefficients characterizing the scattering of a spin wave off the boundary or wall.

3.5.1 Vertex models: the Sklyanin equation

For vertex models, the existence of boundary conditions amounts to the introduction of two new sets of Boltzmann weights describing the boundary of the lattice:

$K^-(u)_i^j =$ 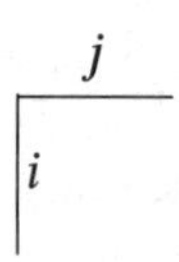

, $K^+(u)_i^j =$ 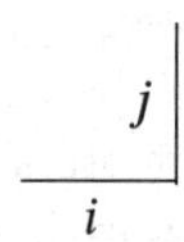

(3.82)

These new Boltzmann weights for the boundary sites, in addition to the usual

$\mathscr{R}_{ij}^{k\ell}(u) = W\begin{pmatrix} \ell & k \\ i & j \end{pmatrix}\!\!\bigg|\, u \bigg) =$ 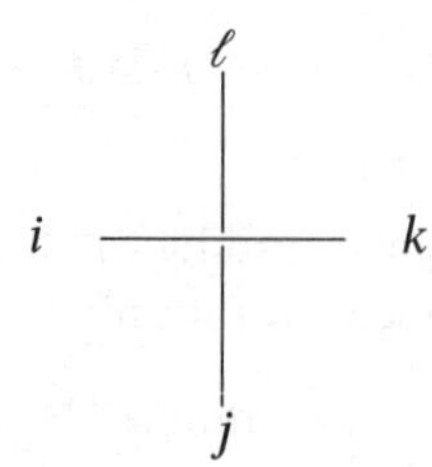

(3.83)

for the internal sites, define a vertex model with open boundary conditions. Each set of boundary conditions is associated with a choice of boundary operators $K^{\pm}(u)$.

What conditions must the Boltzmann weights W and $K^{\pm}$ satisfy so that the vertex model is integrable? This question was answered in full generality by Sklyanin. In the following, we present both his result and some simple examples.

For simplicity, let us start with an integrable vertex model whose internal Boltzmann weights W satisfy the Yang–Baxter equation. Following the notation of previous chapters, we may assemble these weights into an R-matrix satisfying

$$\mathscr{R}_{12}(u)\mathscr{R}_{13}(u+v)\mathscr{R}_{23}(v) = \mathscr{R}_{23}(v)\mathscr{R}_{13}(u+v)\mathscr{R}_{12}(u) \tag{3.84}$$

To establish these ideas, let us restrict ourselves to the six-vertex R-matrix, $R(u) : V^{\frac{1}{2}} \otimes V^{\frac{1}{2}} \rightarrow V^{\frac{1}{2}} \otimes V^{\frac{1}{2}}$, so the boundary operators act as $K^{\pm}(u) : V^{\frac{1}{2}} \rightarrow V^{\frac{1}{2}}$. We know that $R(u)$ can be interpreted as the intertwiner between the representations $(\frac{1}{2}, u_1)$ and $(\frac{1}{2}, u_2)$ of the affine quantum group $U_q(A_1^{(1)})$, with $u = u_1 - u_2$. In a more mathematically precise way, then, the operators $K^{\pm}(u)$ can be viewed as maps between level zero representation spaces of the affine quantum group $U_q(A_1^{(1)})$:

$$K^{\pm}(u) : V^{(\frac{1}{2}, u_1)} \longrightarrow V^{(\frac{1}{2}, u_2)} \; ; \qquad u = u_1 - u_2 \tag{3.85}$$

Given $R(u)$ and $K^{\pm}(u)$, let us construct the transfer matrix of the vertex model with open boundary conditions. Using the Boltzmann weights (3.82) and (3.83), we construct a lattice model with a boundary. We define first an auxiliary operator $\mathscr{U}_{ij}(u)$ in terms of the monodromy matrices $T_i^j(u)$, shown pictorially in figure 3.1:

$$\mathscr{U}_i^j(u) = \sum_{\ell, m} T(u)_m^j K^-(u)_\ell^m \left(T^{-1}(-u)\right)_i^\ell \tag{3.86}$$

The operator $\mathscr{U}_i^j$ includes a monodromy matrix towards the left, a reflection from the left wall, and a monodromy matrix towards the right (we have used the difference property of the six-vertex R-matrix). The monodromy matrix $T(u)$ and its inverse are defined by equation (2.19), which can be rewritten as

$$\begin{aligned} T(u) &= L_L(u) \cdots L_2(u) L_1(u) \\ T^{-1}(-u) &= L_1^{-1}(-u) \cdots L_L^{-1}(-u) \end{aligned} \tag{3.87}$$

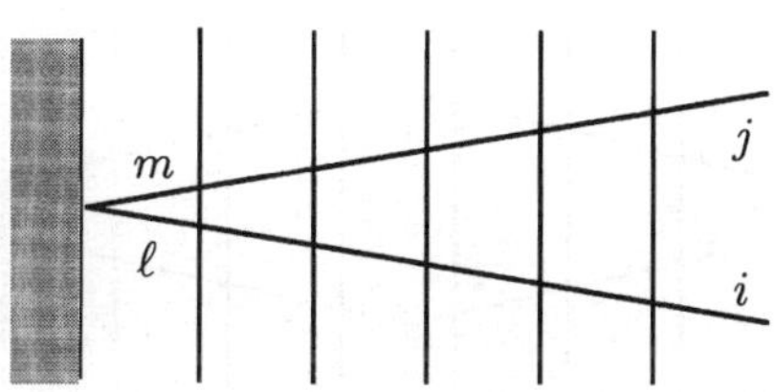

Fig. 3.1 Graphical representation of the operator $\mathscr{U}_{ij}(u)$ in (3.86).

where the L_i operators are simply $\mathcal{R}_{ai}$. The transfer matrix for the open chain is obtained by closing the operator $\mathcal{U}_i^j(u)$ with a reflection on the right wall (figure 3.2):

$$\begin{aligned}\hat{t}(u) &= \sum_{i,j} K^+(u)^i_j \mathcal{U}^j_i(u) \\ &= \sum_{i,j,\ell,m} K^+(u)^i_j T(u)^j_m K^-(u)^m_\ell T^{-1}(-u)^\ell_i \\ &= \mathrm{tr}_{V^{\frac{1}{2}}} \left(K^+(u) T(u) K^-(u) T^{-1}(-u)\right)\end{aligned} \quad (3.88)$$

All the products and traces in (3.88) are understood in the auxiliary space, using the terminology of section 2.1, whereas the labels $1, \ldots, L$ denote the quantum space.

Integrability of the vertex model with open boundary conditions means that the transfer matrices commute for any rapidity:

$$\left[\, \hat{t}(u)\, ,\, \hat{t}(v)\, \right] \quad = \quad 0 \quad (3.89)$$

Just as we found in chapter 2 the Yang–Baxter equation for $\mathcal{R}(u)$ as the necessary and sufficient condition for $[t(u), t(v)] = 0$, we will now derive the compatibility conditions among $\mathcal{R}(u)$ and $K^{\pm}(u)$ for (3.89) to hold. To simplify the discussion, we shall assume that the $\mathcal{R}$-matrix satisfies the following four properties:

(i) P invariance

$$P\mathcal{R}_{12}(u)P = \mathcal{R}_{12}(u) \quad (3.90)$$

(ii) T invariance

$$\mathcal{R}_{12}(u)^{t_1 t_2} = \mathcal{R}_{12}(u) \quad (3.91)$$

(iii) unitarity

$$\mathcal{R}_{12}(u)\mathcal{R}_{12}(-u) = \rho(u)\mathbf{1} \quad (3.92)$$

(iv) crossing unitarity

$$\mathcal{R}_{12}(u)^{t_1}\mathcal{R}_{12}(-u-2i\gamma)^{t_1} = \tilde{\rho}(u)\mathbf{1} \quad (3.93)$$

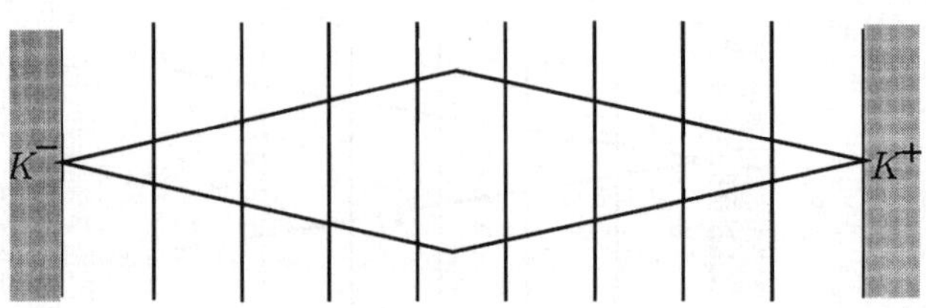

Fig. 3.2 Graphical representation of the transfer matrix $\hat{t}(u)$ defined in (3.88).

where t_1 and t_2 denote the transposition on the first or second spaces, respectively. The six-vertex R-matrix satisfies all the requirements above.

Under these assumptions, the integrability conditions for (3.89) to hold, known as the Sklyanin equations, are the following:

$$\begin{aligned}&\mathscr{R}_{12}(u_1-u_2)K_1^-(u_1)\mathscr{R}_{12}(u_1+u_2)K_2^-(u_2)\\&\quad=K_2^-(u_2)\mathscr{R}_{12}(u_1+u_2)K_1^-(u_1)\mathscr{R}_{12}(u_1-u_2)\\&\mathscr{R}_{12}(-u_1+u_2)K_1^{+t_1}(u_1)\mathscr{R}_{12}(-u_1-u_2-2i\gamma)K_2^{+t_2}(u_2)\\&\quad=K_2^{+t_2}(u_2)\mathscr{R}_{12}(-u_1-u_2-2i\gamma)K_1^{+t_1}(u_1)\mathscr{R}_{12}(-u_1+u_2)\end{aligned} \tag{3.94}$$

We view the Yang–Baxter equation as reflecting Baxter's **Z** invariance of the three-point amplitude, i.e. its invariance under the parallel shift of one line across the intersection of the other two. Similarly, the Sklyanin equations express the invariance of the scattering amplitude of two particles with different rapidites off a boundary, with respect to the parallel shift of their two world-lines. This is shown in figure 3.3.

Let us give the example of the six-vertex model. Its R-matrix (2.50) can be nicely written in terms of Pauli matrices and the identity $\mathbf{1}=\sigma^0$ as

$$R^{(6v)}=\sum_{a=0}^{3}\alpha_a\,\sigma^a\otimes\sigma^a \tag{3.95}$$

where

$$\begin{aligned}\alpha_0&=\sinh\left(u+i\frac{\gamma}{2}\right)\,\cos\frac{\gamma}{2}\\\alpha_1&=\alpha_2=i\,\sin\frac{\gamma}{2}\,\cos\frac{\gamma}{2}\\\alpha_3&=i\,\sin\frac{\gamma}{2}\,\cosh\left(u+i\frac{\gamma}{2}\right)\end{aligned} \tag{3.96}$$

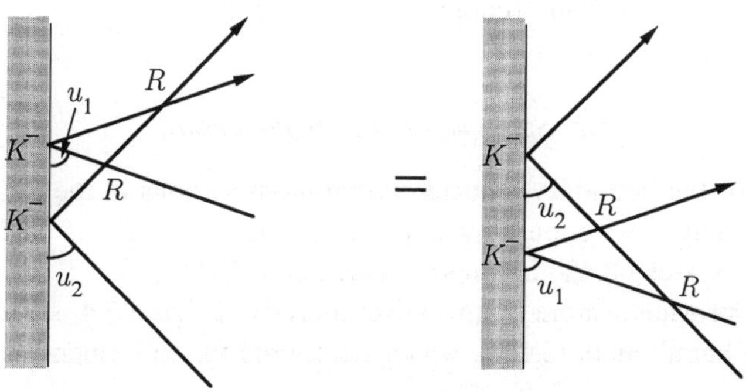

Fig. 3.3. The Sklyanin equation (3.94) reflects **Z** invariance in the scattering process of two particles off a wall.

A solution to (3.94) with R given by (3.96) was found by Cherednik; it depends on two arbitrary constants ξ_+ and ξ_-:

$$K^-(u) = \frac{1}{\sinh \xi_-} \begin{pmatrix} \sinh(u+\xi_-) & 0 \\ 0 & -\sinh(u-\xi_-) \end{pmatrix}$$
$$K^+(u) = \frac{1}{\sinh \xi_+} \begin{pmatrix} \sinh(u+i\gamma+\xi_+) & 0 \\ 0 & -\sinh(u+i\gamma-\xi_+) \end{pmatrix} \tag{3.97}$$

The one-dimensional spin chain hamiltonian H with open boundary conditions derived from the vertex model above is given by

$$H = \sum_{j=1}^{L-1} h_{j,j+1} + \frac{i}{2} K_1^-(0)' + \frac{\mathrm{tr}_0 K_0^+(0)\, h_{L0}}{\mathrm{tr} K_0^+(0)} \tag{3.98}$$

where we have assumed that $\mathscr{R}_{12}(0) = P_{12}$ and $K^-(0) = \mathbf{1}$. By $h_{j,j+1}$, we mean the two-body hamiltonian of the closed chain,

$$h_{j,j+1} = i \frac{d}{du} P \mathscr{R}_{j,j+1}(u) \Big|_{u=0} \tag{3.99}$$

Equation (3.98) can be easily verified by proving

$$i\hat{t}'(0) = 2H \,\mathrm{tr} K^+(0) \; + \; i \,\mathrm{tr} K^{+\prime}(0) \tag{3.100}$$

and recalling that the uniformization parameter u is arranged so that $\hat{t}(0) = \mathbf{1}$.

To illustrate the formula (3.98), we may apply it to the solution (3.97) for the six-vertex model, obtaining the hamiltonian

$$H(\gamma, \xi_+, \xi_-) = \frac{1}{2 \sin \gamma} \left[\sum_{j=1}^{L-1} \left(\sigma_j^x \sigma_{j+1}^x + \sigma_j^y \sigma_{j+1}^y + \cos\gamma\, \sigma_j^z \sigma_{j+1}^z \right) + i \sin\gamma \left(\mathrm{cotanh}\, \xi_- \; \sigma_1^z + \mathrm{cotanh}\, \xi_+ \; \sigma_L^z \right) \right] \tag{3.101}$$

3.5.2 Spin chains: the Bethe ansatz

The presence of the boundary induces several modifications in the Bethe ansatz. In the whole analysis we should include the reflection coefficients describing the bounce of spin waves off the boundaries (see figure 3.4).

In order to provide a more quantitative analysis of figure 3.4, let us consider the spin chain hamiltonian (3.101), which we rewrite for convenience as

$$H = \frac{1}{2} \left[\sum_{i=1}^{L-1} \left(\sigma_i^x \sigma_{i+1}^x + \sigma_i^y \sigma_{i+1}^y + \Delta \sigma_i^z \sigma_{i+1}^z \right) + K \sigma_1^z + K' \sigma_L^z \right] \tag{3.102}$$

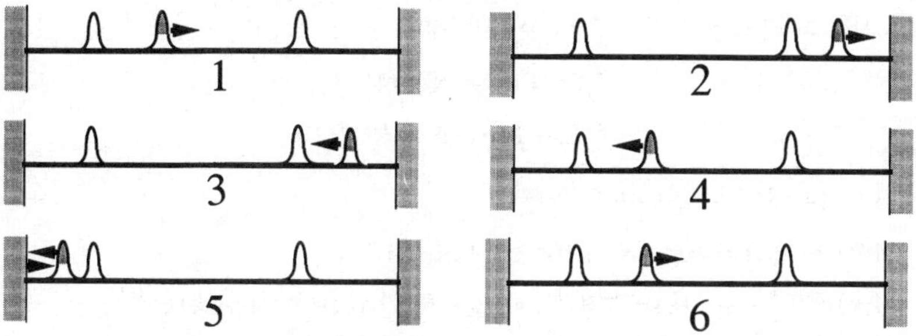

Fig. 3.4. The Bethe ansatz for a spin chain with boundary conditions amounts to requiring that the wave-function be invariant when a particle completes a whole period of motion, taking into account scatterings with other particles and reflections off the boundaries.

which is obtained from the XXZ hamiltonian by the addition of some particular boundary terms. So far, Δ, K and K' are arbitrary constants. The hamiltonian (3.102) still commutes with

$$S^z_{\rm tot} = \sum_{i=1}^{L} \sigma^z_i \tag{3.103}$$

and therefore we may still diagonalize it separately in each sector of fixed angular momentum or, equivalently, fixed number M of spins down.

The Bethe reference state is the state with all spins up. In the $M = 1$ sector, the wave-function of the spin wave $f(x)$ with eigenvector $\Psi_1 = \sum_x f(x)\,|x\rangle$ must be a solution to the equations

$$Ef(x) = f(x-1) + f(x+1) + \frac{(L-5)\Delta + K + K'}{2} f(x) \tag{3.104}$$

with $2 \le x \le L-1$, whereas the equations satisfied by $f(x)$ are slightly different if x is at the ends of the open chain, due to the boundary terms in the hamiltonian:

$$\begin{aligned} Ef(1) &= f(2) + \frac{(L-3)\Delta - K + K'}{2} f(1) \\ Ef(L) &= f(L-1) + \frac{(L-3)\Delta + K - K'}{2} f(L) \end{aligned} \tag{3.105}$$

The Bethe ansatz for the wave-function is

$$f(x) = A(k)\,\mathrm{e}^{ikx} + \tilde{A}(k)\,\mathrm{e}^{-ikx} \tag{3.106}$$

which already reflects the existence of a boundary. Substituting (3.106) into (3.104) and (3.105), we obtain the following dispersion relation for a spin wave:

$$E(k) = 2\cos k + \frac{1}{2}\left[(L-5)\Delta + K + K'\right] \tag{3.107}$$

In order to find the factor $A(k)$, it is convenient to simulate the boundary conditions at $x = 1$ and $x = L$, using two fictitious extra spins, in such a way that (3.104) holds for $1 \le x \le L$. We obtain

$$\begin{aligned} f(0) &= (\Delta - K)f(1) \\ f(L+1) &= (\Delta - K')f(L) \end{aligned} \tag{3.108}$$

The Bethe ansatz (3.106) then implies

$$\begin{aligned} &A(k) + \tilde{A}(k) = (\Delta - K)\left[A(k)\,\mathrm{e}^{ik} + \tilde{A}(k)\,\mathrm{e}^{-ik}\right] \\ &A(k)\,\mathrm{e}^{ik(L+1)} + \tilde{A}(k)\,\mathrm{e}^{-ik(L+1)} = (\Delta - K')\left[A(k)\,\mathrm{e}^{ikL} + \tilde{A}(k)\,\mathrm{e}^{-ikL}\right] \end{aligned} \tag{3.109}$$

The functions

$$\begin{aligned} \alpha(k) &= 1 + (K - \Delta)\,\mathrm{e}^{-ik} \\ \beta(k) &= \left[1 + (K' - \Delta)\,\mathrm{e}^{-ik}\right]\,\mathrm{e}^{i(L+1)k} \end{aligned} \tag{3.110}$$

allow us to rewrite (3.109) more compactly:

$$\begin{aligned} A(k)\alpha(-k) + \tilde{A}(k)\alpha(k) &= 0 \\ A(k)\beta(k) + \tilde{A}(k)\beta(-k) &= 0 \end{aligned} \tag{3.111}$$

Compatibility of these two equations requires

$$\alpha(k)\beta(k) = \alpha(-k)\beta(-k) \tag{3.112}$$

The solution to (3.111) is simply

$$A(k) = -\tilde{A}(-k) = \beta(-k) \tag{3.113}$$

Equivalently, if we undo the change of variables (3.110) we can write (3.112) in terms of the original parameters as

$$\mathrm{e}^{i2(L-1)k}\quad \frac{\left(\mathrm{e}^{ik} + K - \Delta\right)\left(\mathrm{e}^{ik} + K' - \Delta\right)}{\left(\mathrm{e}^{-ik} + K - \Delta\right)\left(\mathrm{e}^{-ik} + K' - \Delta\right)} = 1 \tag{3.114}$$

For the special values

$$\Delta = \frac{q + q^{-1}}{2} \qquad K = \frac{q - q^{-1}}{2} \qquad K' = -K = \frac{q^{-1} - q}{2} \tag{3.115}$$

the contribution of the reflection coefficients in (3.114) cancels, and equation (3.114) becomes simply

$$\mathrm{e}^{i2Lk} = 1 \tag{3.116}$$

The energy spectrum of the $M = 1$ sector is thus given by (3.107) with discrete momenta

$$k = \frac{\pi m}{L} \qquad m \in \mathbb{Z} \tag{3.117}$$

In this $M = 1$ sector, the state with minimum energy has momentum label $m = 0$, whereas the state with $m = L$ is the most energetic. Note already the difference between (3.116) and the corresponding equation for the periodic chain, namely $\mathrm{e}^{ikL} = 1$.

Let us proceed to the sector with two spins down, i.e. $M = 2$. The Bethe ansatz now reads as

$$f(x_1, x_2) = \sum \epsilon_p A(k_{p(1)}, k_{p(2)})\, \mathrm{e}^{i(k_{p(1)}x_1 + k_{p(2)}x_2)} \tag{3.118}$$

The summation runs over all permutations p of the momenta k_1 and k_2 and of their opposites $-k_1$ and $-k_2$. The signature ϵ_p is ± 1, and this changes with every transposition or negation, as in (3.113). For $M = 2$, we have two momenta, each of which may be negated (four possibilities), and there is only one permutation: eight terms in all.

Repeating exactly the same steps as for $M = 1$, we obtain the energy in terms of the momenta in the ansatz wave-function:

$$E(k_1, k_2) = 2\cos k_1 + 2\cos k_2 + \frac{1}{2}\left[(L-9)\Delta + K + K'\right] \tag{3.119}$$

The compatibility equations for $A(k_1, k_2)$ read as follows:

$$A(k_1, k_2)\beta(k_2) - A(k_1, -k_2)\beta(-k_2) = 0 \tag{3.120a}$$

$$A(k_1, k_2)\alpha(-k_1) - A(-k_1, k_2)\alpha(k_1) = 0 \tag{3.120b}$$

$$A(k_1, k_2)s(k_1, k_2) - A(k_2, k_1)s(k_2, k_1) = 0 \tag{3.120c}$$

with $\alpha(k)$ and $\beta(k)$ as in (3.110) and

$$s(k_1, k_2) = 1 - 2\Delta\, \mathrm{e}^{ik_2} + \mathrm{e}^{i(k_1+k_2)} \tag{3.121}$$

The novelty of the $M = 2$ sector is the constraint (3.120c), involving two down spins on neighboring sites. From (3.120), we obtain the analog of (3.114) for the $M = 2$ sector, i.e. the equation restricting the values of the momenta k_1 and k_2:

$$\mathrm{e}^{i2(L-1)k_1} \frac{\left(\mathrm{e}^{ik_1} + K - \Delta\right)\left(\mathrm{e}^{ik_1} + K' - \Delta\right)}{\left(\mathrm{e}^{-ik_1} + K - \Delta\right)\left(\mathrm{e}^{-ik_1} + K' - \Delta\right)} = \frac{B(-k_1, k_2)}{B(k_1, k_2)} \tag{3.122}$$

where we have set

$$B(k_1, k_2) = s(k_1, k_2)s(k_2, -k_1) \tag{3.123}$$

For the special parameters (3.115), we find

$$\mathrm{e}^{i2Lk_1} = \frac{B(-k_1, k_2)}{B(k_1, k_2)} \tag{3.124}$$

and the solution for the coefficients of the plane waves is

$$A(k_1, k_2) = \beta(-k_1)\beta(-k_2)B(-k_1, k_2)\mathrm{e}^{-ik_2} \tag{3.125}$$

Inspired by figure 3.4, equations (3.121)–(3.124) can be given a very neat physical interpretation. In fact, (3.122) contains explicitly the reflection coefficients and the contributions from the scattering of two spin waves. As expected, the scattering of spin waves is the same as for the periodic chain. For the special boundaries (3.115), the reflection contributions disappear, and the open chain behaves as if it were effectively closed and periodic.

For generic M, the Bethe ansatz is

$$f(x_1,\dots,x_M) = \sum \epsilon_p A(k_{p(1)},\dots,k_{p(n)})\, \mathrm{e}^{i(k_{p(1)}x_1\cdots+k_{p(M)}x_M)} \tag{3.126}$$

with

$$A(k_1,\dots,k_M) = \prod_{j=1}^{M} \beta(-k_j) \prod_{1\le j<\ell\le n} B(-k_j,k_\ell)\,\mathrm{e}^{-ik_\ell} \tag{3.127}$$

and the energy is given by

$$E(\{k_j\}) = \frac{1}{2}\left[(L-1)\Delta + K + K'\right] + 2\sum_{j=1}^{M}\left(\cos k_j - \Delta\right) \tag{3.128}$$

The Bethe equations for the special choice of parameters (3.115) are

$$\mathrm{e}^{2Lk_j} = \prod_{\substack{\ell=1\\ \ell\neq j}}^{M} \frac{B(-k_j,k_\ell)}{B(k_j,k_\ell)} \qquad 1\le j\le M \tag{3.129}$$

More explicitly, this equation is

$$\mathrm{e}^{2iLk_j} = \prod_{\substack{\ell=1\\ \ell\neq j}}^{M} \frac{1-2\Delta\,\mathrm{e}^{ik_j} + \mathrm{e}^{i\pi(k_j+k_\ell)}}{1-2\Delta\,\mathrm{e}^{ik_\ell} + \mathrm{e}^{i\pi(k_j+k_\ell)}} \times \frac{1-2\Delta\,\mathrm{e}^{ik_\ell} + \mathrm{e}^{i\pi(-k_j+k_\ell)}}{1-2\Delta\,\mathrm{e}^{-ik_j} + \mathrm{e}^{i\pi(-k_j+k_\ell)}} \tag{3.130}$$

Compare this expression with that for the closed spin chain, namely equation (1.90).

3.6 Hamiltonians with quantum group invariance

In the previous section, we dealt with some aspects of the Bethe ansatz for the XXZ model on an open chain. The boundary terms in the hamiltonian (3.102) were chosen to be as simple as possible, and then we saw that, for the special values (3.115) of the boundary parameters K and K' in terms of Δ, several formulae were simplified throughout. What are all these nice algebraic cancellations telling us?

Let us consider the hamiltonian

$$H = \frac{1}{2}\left[\sum_{i=1}^{L-1}\left(\sigma_i^x\sigma_{i+1}^x + \sigma_i^y\sigma_{i+1}^y + \frac{q+q^{-1}}{2}\sigma_i^z\sigma_{i+1}^z\right) + \frac{q-q^{-1}}{2}\left(\sigma_1^z - \sigma_L^z\right)\right] \tag{3.131}$$

The main peculiarity of this hamiltonian is that the contribution of the reflection coefficients of spin waves to the Bethe equations is invisible. Particularly noteworthy are some spin waves, with momentum $\tilde{k}$ such that

$$\mathrm{e}^{i\tilde{k}} = q \tag{3.132}$$

From (3.113), it follows that for such a spin wave there is no reflected wave: $A(-\tilde{k}) = 0$. Moreover, on comparing equations (3.107) and (3.119), we discover an unexpected degeneracy in energy of the low lying excitations: for any value of k, the state with $M = 1$ spin wave of momentum k has the same energy as the state with $M = 2$ spin waves of momenta k and $\tilde{k} = -i \log q$:

$$\begin{aligned} E(k) &= 2\cos k + \frac{L-5}{2} \frac{q+q^{-1}}{2} \\ &= 2\cos k + 2\cos \tilde{k} + \frac{L-9}{2} \frac{q+q^{-1}}{2} \end{aligned} \tag{3.133}$$

This degeneracy in the energy of the states $|k\rangle$ and $|k, \tilde{k}\rangle$ is a strong indication of some additional symmetry in the hamiltonian (3.131).

Note that the degeneracy occurs between states with different numbers of spins down, i.e. between states with different S^z_{total}. Whatever the length of the chain, the eigenvalue of S^z_{total} on $|k, \tilde{k}\rangle$ is one unit less than on $|k\rangle$. It would seem natural to look for some lowering operator $\Delta'(S^-)$ which would create the state $|k, -i \log q\rangle$ from $|k\rangle$. (Do not confuse the co-product operators Δ or Δ' with the anisotropy parameter Δ!) This lowering operator would change the spin of a state but, since it acts among states degenerate in energy, it would commute with the hamiltonian. Clearly, the standard co-product of the usual $SU(2)$ lowering operator

$$\Delta'^{(L-1)}\left(S^-\right) = \sum_{i=1}^{L} \mathbf{1} \otimes \cdots \otimes \mathbf{1} \otimes S_i^- \otimes \mathbf{1} \otimes \cdots \otimes \mathbf{1} \tag{3.134}$$

is not appropriate, because the state $|k, 0\rangle$ obtained by acting with it on the state $|k\rangle$ is degenerate with the latter only in the isotropic ($\Delta = 1$) case. This is reasonable, since the isotropic Heisenberg spin chain is $SU(2)$ invariant (the three directions x, y and z of spinor space are equivalent). But we are looking for a symmetry valid even for $\Delta \neq 1$. Furthermore, the operator $\Delta'^{(L-1)}(S^-)$ should create from the one spin wave state $|k\rangle$ a two spin wave state with momenta k and $\tilde{k} = q$, not $\tilde{k} = 0$. As the reader might have guessed, the symmetry is not classical $SU(2)$ but its quantum deformation. From (2.92), we read

$$\Delta'^{(L-1)}\left(S^-\right) = \sum_{i=1}^{L} q^{S^z} \otimes \cdots \otimes q^{S^z} \otimes S_i^- \otimes q^{-S^z} \otimes \cdots \otimes q^{-S^z} \tag{3.135}$$

Changing the sign of the boundary term in (3.131) would force us to use the transposed co-multiplication $\Delta = \sigma \circ \Delta'$. It is a non-trivial but easy exercise to check that the operator $\Delta'^{(L-1)}\left(S^-\right)$ commutes with the hamiltonian (3.131). In fact, the hamiltonian (3.131) commutes with the full quantum group $U_q(s\ell(2))$.

Having discovered that the XXZ hamiltonian with suitable boundary terms is invariant under the quantum group $U_q(s\ell(2))$, it is no surprise to find that the energy eigenstates organize into $U_q(s\ell(2))$ multiplets, with no energy splitting among members of the same multiplet. In the critical regime $-1 < \Delta < 1$, the ground state is a Dirac sea of one-strings, and the low lying excitations are identified with holes in this Dirac sea. Also, it is natural to expect that the Fock space representation (3.2) of the Hilbert space of this model is still valid, provided we consider the elementary excitations as "quantum" doublets, i.e. spin-$\frac{1}{2}$ irreps of $U_q(s\ell(2))$. The reader is invited to prove or disprove the following conjecture: the low lying excitations of the quantum group invariant XXZ chain constitute quantum group representations of spin $\frac{1}{2}$ in the very same way as the holes of the isotropic chain define spin-$\frac{1}{2}$ irreps of the $SU(2)$ classical symmetry.

As a final comment on the hamiltonian (3.131), note that it can be written as

$$H = -\sum_{i=1}^{L-1} \left(e_i - \frac{q+q^{-1}}{4} \right) \tag{3.136}$$

with the matrices e_i acting on the lattice variables at sites i and $i+1$ as

$$e_i = \begin{pmatrix} 0 & 0 & 0 & 0 \\ 0 & q^{-1} & -1 & 0 \\ 0 & -1 & q & 0 \\ 0 & 0 & 0 & 0 \end{pmatrix} \tag{3.137}$$

The interesting fact concerning the version (3.136) is that the e_i operators satisfy the Temperley–Lieb–Jones algebra

$$e_i^2 = e_i \quad , \quad e_i e_{i\pm1} e_i = e_i \quad , \quad e_i e_j = e_j e_i \ \ |i-j| \geq 2 \tag{3.138}$$

which we shall study in great detail in chapter 5.

3.7 Spin-1 chains

This section is devoted to some generalizations of the spin-1 Heisenberg hamiltonian

$$H = \sum_i \vec{S}_i \cdot \vec{S}_{i+1} \tag{3.139}$$

Despite its simplicity, this hamiltonian is not integrable for spin 1 or > 1, and thus the Bethe ansatz cannot be applied in the study of its properties. The main physical motivation for considering such hamiltonians comes from Haldane's conjecture, which predicts a mass gap in the spectrum of the hamiltonian (3.139) for spin 1. This conjecture, which enjoys ample theoretical and experimental support, was a big surprise since, as we have learned, the spin-$\frac{1}{2}$ Heisenberg antiferromagnet has no mass gap, i.e. there exist local excitations with arbitrarily low energy above the ground state. To make our discussion more general, we shall consider the most general isotropic spin-1 hamiltonians with only nearest neighbor interactions, namely

$$H_\beta = \sum_i \left[\vec{S}_i \cdot \vec{S}_{i+1} - \beta \left(\vec{S}_i \cdot \vec{S}_{i+1} \right)^2 \right] \tag{3.140}$$

The case $\beta = 1$ corresponds to the integrable spin-1 model studied in the previous section with $\Delta = 1$.

Particularly interesting is the value $\beta = -\frac{1}{3}$, in which case (3.140) may be written as

$$H_{-\frac{1}{3}} = \sum_i P_2(\vec{S}_i, \vec{S}_{i+1}) \tag{3.141}$$

where

$$P_2(\vec{S}_i, \vec{S}_{i+1}) = \frac{1}{2}\left[\vec{S}_i \cdot \vec{S}_{i+1} + \frac{1}{3}\left(\vec{S}_i \cdot \vec{S}_{i+1}\right)^2 + \frac{2}{3}\right] \tag{3.142}$$

is the orthogonal projector on spin 2 in the decomposition $1 \times 1 = 0 + 1 + 2$. As we will show in a moment, the ground state $|\Omega_\circ\rangle$ of this hamiltonian provides us with an example of what is known as a valence bond solid ground state. It is constructed as follows.

Introduce first the following basis for the spin variables at each site:

$$\psi_{\alpha\beta} = \frac{1}{\sqrt{2}}\left[\psi_\alpha \otimes \psi_\beta + \psi_\beta \otimes \psi_\alpha\right] \tag{3.143}$$

with ψ_α a basis of the spin-$\frac{1}{2}$ irrep. Explicitly,

$$\begin{pmatrix} \psi_{11} & \psi_{12} \\ \psi_{21} & \psi_{22} \end{pmatrix} = \begin{pmatrix} \sqrt{2}\,|1,+1\rangle & |1,0\rangle \\ |1,0\rangle & \sqrt{2}\,|1,-1\rangle \end{pmatrix} \tag{3.144}$$

Introduce next the antisymmetric tensor $\epsilon^{\alpha\beta}$, with $\epsilon^{12} = +1$, whereby the singlet is $\psi_\alpha \otimes \psi_\beta \epsilon^{\alpha\beta}$.

The valence bond solid ground state is then

$$|\Omega_\circ\rangle_{\alpha\beta} = \psi_{\alpha\beta_1} \otimes \psi_{\alpha_2\beta_2} \otimes \psi_{\alpha_3\beta_3} \otimes \cdots \otimes \psi_{\alpha_L\beta} \quad \epsilon^{\beta_1\alpha_2}\epsilon^{\beta_2\alpha_3}\cdots\epsilon^{\beta_{L-1}\alpha_L} \tag{3.145}$$

which admits the nice pictorial representation

$$\underbrace{\underset{\alpha}{\bullet}\ \ \underset{\beta_1}{\bullet}}\ \overset{\epsilon^{\beta_1\alpha_2}}{\rule{3em}{0.4pt}}\ \underbrace{\underset{\alpha_2}{\bullet}\ \ \underset{\beta_2}{\bullet}}\ \overset{\epsilon^{\beta_2\alpha_3}}{\rule{3em}{0.4pt}}\ \underbrace{\underset{\alpha_3}{\bullet}\ \ \underset{\beta_3}{\bullet}}\ \rule{2em}{0.4pt}\ \cdots \tag{3.146}$$

Note that, for the open chain, there exist four ground states

$$|\Omega_\circ\rangle_{11}\ , \quad |\Omega_\circ\rangle_{21}\ , \quad |\Omega_\circ\rangle_{12}\ , \quad |\Omega_\circ\rangle_{22} \tag{3.147}$$

with total spin (value of S^z_{total}) equal to 1, 0, 0 and -1, respectively. There exist, hence, two ground states: a singlet and a triplet.

The proof that the valence bond solid states are ground states is indeed trivial from the construction and the fact that the hamiltonian (3.141) is positive definite. In fact, the maximum spin we may achieve for two adjacent spins, having contracted two spin-$\frac{1}{2}$ pieces with the ϵ tensor, is 1. Therefore,

$$H_{-\frac{1}{3}}\,|\Omega_\circ\rangle_{\alpha,\beta} = \sum_i P_2(\vec{S}_i, \vec{S}_{i+1}) \left| \underset{\alpha_i}{\bullet}\ \ \underset{\beta_i}{\bullet}\ \overset{\epsilon}{\rule{2em}{0.4pt}}\ \underset{\alpha_{i+1}}{\bullet}\ \ \underset{\beta_{i+1}}{\bullet} \right\rangle = 0 \tag{3.148}$$

Clearly, valence bond solid states are possible only if the chain carries integer spin variables. This is the first indication of an important difference between chains with integer and half-odd spins. We will sketch now the proof that correlation functions over a valence bond solid ground state decay exponentially with distance and thus display a mass gap. In general, isotropic integer spin chains exhibit a mass gap, whereas half-odd spin chains do not. The main lines of the proof, for the spin-1 case at hand, are as follows.

To evaluate scalar products of valence bond states, we shall use a simple diagrammatic procedure. The scalar product of two states of the form (3.143), for example, is given by

$$\langle \psi_{\gamma\delta} | \psi_{\alpha\beta} \rangle = \delta_{\alpha\gamma}\, \delta_{\beta\delta} \,+\, \delta_{\alpha\delta}\, \delta_{\beta\gamma} \tag{3.149}$$

and can be represented graphically by

$\langle \gamma\delta | \alpha\beta \rangle \;=$

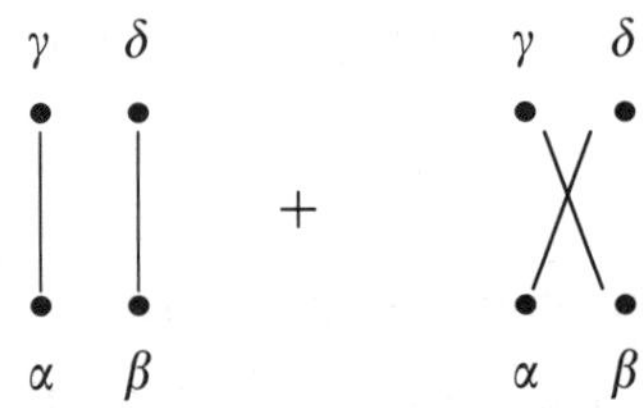

(3.150)

To compute the scalar product $\langle \Omega_{\gamma\delta} | \Omega_{\alpha\beta} \rangle$ of two ground states, we draw two figures (3.146) on top of each other, and apply the scalar product rule (3.150) to each site of the chain. This gives a total of 2^L diagrams. After disentangling the lines, the result is either $\delta_{\alpha\gamma}\delta_{\beta\delta}$ or $\delta_{\alpha\delta}\delta_{\beta\gamma}$ with some integer coefficients. In the application of the rules, one encounters loops, which represent the trace δ^{α}_{α} and contribute a factor of two.

Using the rules just indicated, we may read off from the following diagrammatic computation (with $L = 2$)

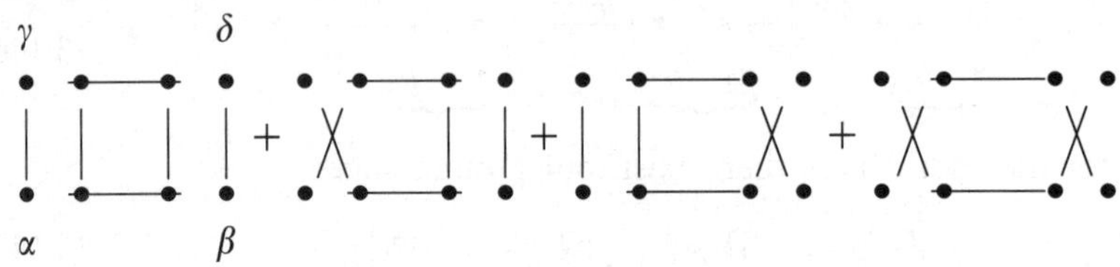

(3.151)

the scalar product between two valence bond solid ground states for a two-site chain:

$$\begin{aligned} \langle \Omega_{\delta\gamma} | \Omega_{\alpha\beta} \rangle &= 2\delta_{\alpha\gamma}\, \delta_{\beta\delta} + \delta_{\alpha\gamma}\, \delta_{\beta\delta} + \delta_{\alpha\gamma}\, \delta_{\beta\delta} + \delta_{\alpha\beta}\, \delta_{\gamma\delta} \\ &= 4\delta_{\alpha\gamma}\, \delta_{\beta\delta} + \delta_{\alpha\beta}\, \delta_{\gamma\delta} \end{aligned} \tag{3.152}$$

The diagrams for the chain with L even sites contributing to the scalar product of two valence bond solid ground states can be classified according to the number, ℓ, of closed loops they contain. The contribution from diagrams with $\ell \geq 1$ loops is $2^{\ell}\delta_{\alpha_\gamma}\delta_{\beta\delta}$, whereas that from the single diagram with no loops is $\delta_{\alpha\beta}\delta_{\gamma\delta}$. After some elementary combinatorics, we find

$$\langle \Omega_{\gamma\delta} | \Omega_{\alpha\beta} \rangle = \frac{3^L - 1}{2} \delta_{\alpha\gamma}\, \delta_{\beta\delta} + \delta_{\alpha\beta}\, \delta_{\gamma\delta} \tag{3.153}$$

(Note that, if L was even, the single diagram with no loops would contribute instead $\delta_{\alpha\delta}\delta_{\beta\gamma}$, and thus the last term in (3.153) would take this form.) Hence, the normalization of the ground state with periodic boundary conditions is

$$\langle \Omega_{\beta\beta} | \Omega_{\alpha\alpha} \rangle = 3^L + 3 \tag{3.154}$$

Next, we compute the spin-spin correlation function. For periodic boundary conditions, we are interested in the matrix element

$$F(x, y) = \langle \Omega_{\gamma\delta} |\, S_x^a S_y^b \,| \Omega_{\alpha\beta} \rangle \tag{3.155}$$

where we should expect $F(x, y)$ to depend only on $r = |x - y|$, so it is enough to compute $F(0, r)$. The effect of the spin operator $\vec{S}$ acting on the basis $\psi_{\alpha\beta}$ is

$$\vec{S}\psi_{\alpha\beta} = -\frac{1}{2}\,\vec{\sigma}_\alpha^\gamma \psi_{\gamma\beta} - \frac{1}{2}\,\vec{\sigma}_\beta^\gamma \psi_{\alpha\gamma} \tag{3.156}$$

with σ the Pauli matrices. From (3.145), we see that the action of $\vec{S}_x$ on the periodic ground state breaks the link between site x and $x - 1$ or $x + 1$, and contracts the two dangling indices with $\pm\frac{1}{2}\vec{\sigma}$ (for x even or odd). Thus, the action of $\vec{S}_0\vec{S}_x$ on Ω yields four terms, one of which is

(3.157)

To obtain the correlation function (3.155), we must compute the overlap of this state with the ground state. Using the same diagrammatic rules as above, we see that the contribution from (3.157) is

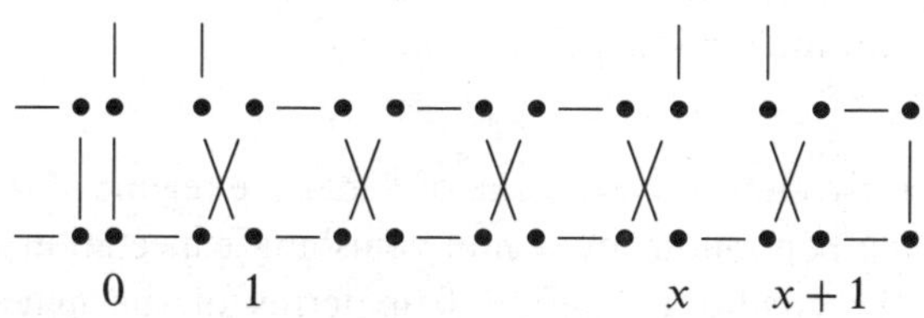

(3.158)

After summing over all such contributions to the correlation function, dividing the result by the ground state normalization (3.154) and taking the limit $L \to \infty$, we reach Haldane's predicted exponential decay:

$$\langle \Omega |\, S_0^a S_r^b \,| \Omega \rangle = \frac{4}{3}(-1)^r 3^{-r} \delta^{ab} \tag{3.159}$$

The phase structure in regard to the parameter β in (3.140) is not completely understood yet. From the study of $\beta = -\frac{1}{3}$ it is plausible to conjecture the following phases:

(i) $-1 < \beta < 1$ – massive region containing the Heisenberg antiferromagnet ($\beta = 0$) and the valence bond solid model ($\beta = -\frac{1}{3}$);

(ii) $\beta = 1$ – massless and integrable; the spectrum can be found exactly with the help of the Bethe ansatz;

(iii) $\beta = -1$ – expected to be massless; the system enjoys $SU(3)$ symmetry.

Exercises

Ex. 3.1 Using the Jordan–Wigner transformation of exercise 1.3, show that the XXZ hamiltonian can be written as follows, up to an additive constant:

$$H_{XXZ} = \sum_{i=1}^{L-1} \left(c_i^\dagger c_{i+1} + \text{h.c.} \right) - \Delta \sum_{i=1}^{L-1} \left(c_i^\dagger c_i - c_{i+1}^\dagger c_{i+1} \right)^2 = H_0 + \Delta H_1 \tag{3.160}$$

With periodic boundary conditions and the notation $\hat{c}_n = i^n c_n$, show that

$$H_0 = \sum_{n=1}^{L/2} i\hat{c}_{2n}^\dagger \left(\hat{c}_{2n+1} - \hat{c}_{2n}\right) + \hat{c}_{2n+1}^\dagger \left(\hat{c}_{2n+2} - \hat{c}_{2n}\right) \tag{3.161}$$

To take the continuum limit, introduce the two-component spinor field

$$\hat{\psi}_\alpha(n) = \begin{cases} \psi_1(n) = \hat{c}_{2n} & \text{if } n \text{ is even} \\ \psi_2(n) = \hat{c}_{2n+1} & \text{if } n \text{ is odd} \end{cases} \tag{3.162}$$

If a is the lattice spacing and $\psi(x) = \left(1/\sqrt{2a}\right) \hat{\psi}(x)$, show that

$$H_0 = \int dx\, \overline{\psi}(x) i\gamma^1 \partial_x \psi(x) \tag{3.163}$$

where $\overline{\psi} = \psi^\dagger \gamma^0$ and γ^i are Dirac matrices in 1+1 dimensions, $\{\gamma^i, \gamma^j\} = 2g^{ij}$. Using the same techniques, write the continuum limit of the interaction hamiltonian H_1 as Fermi's four-point interaction $\int dx\, [\overline{\psi}(x)\psi(x)]^2$.

Ex. 3.2 Qualitatively, show the existence of a massive regime of the XXZ model, characterized in the continuum by a non-vanishing expectation value of $\langle \overline{\psi}\psi \rangle$. Try to interpret the condition $\langle \overline{\psi}\psi \rangle \neq 0$ in terms of the magnetization, and compare it with the Néel state.

Ex. 3.3 Bosonize the continuum limit of the XXZ hamiltonian and prove its equivalence with the sine-Gordon model,

$$\mathscr{L} = (\partial\phi)^2 + \lambda : \cos\beta\phi : \tag{3.164}$$

Find the explicit relationship between β and the anisotropy Δ, and find the region of anisotropy for which the constant λ is dimensionful.

Ex. 3.4 In the thermodynamic limit $L \to \infty$, the equations (3.73) become integral equations. Let $\rho_n(\lambda)$ and $\rho_n^{(h)}(\lambda)$ denote the densities of roots and holes of length n, respectively, so that $\rho_n(\lambda)d\lambda$ is the number of n-strings in the interval $[\lambda, \lambda + d\lambda]$ and $\rho_n^{(h)}(\lambda)d\lambda$ counts the number of n-strings absent in the interval $[\lambda, \lambda + d\lambda]$. Introduce the functions

$$\begin{aligned} a_n^S(\lambda) &= \frac{1}{2\pi}\frac{d}{d\lambda}f_n^S(\lambda) \\ A_{n,m}(\lambda) &= \frac{1}{2\pi}\frac{d}{d\lambda}f_{n,m}(\lambda) + \delta_{n,m}\delta(\lambda) \end{aligned} \tag{3.165}$$

and prove that the $L \to \infty$ limit of the equations (3.73) is equivalent to

$$a_n^S(\lambda) = \rho_n^{(h)}(\lambda) + \sum_m A_{n,m} * \rho_m(\lambda) \tag{3.166}$$

where

$$(A * f)(\lambda) = \int_{-\infty}^{+\infty} A(\lambda - \mu) f(\mu) d\mu \tag{3.167}$$

Ex. 3.5 In the thermodynamic limit $L \to \infty$, a Bethe state is characterized by the densities $\rho_n(\lambda)$. Prove that the energy per unit length of a Bethe state is

$$E(\rho) = -J \int_{-\infty}^{+\infty} \sum_n a_n^S(\lambda)\rho_n(\lambda) d\lambda \tag{3.168}$$

where J is the overall coupling constant of the hamiltonian ($J > 0$ in the antiferromagnetic phase). The free energy is defined by $F = E - T\mathcal{S}$, with T the temperature and $\mathcal{S}$ the entropy

$$\begin{aligned} \mathcal{S} = \sum_n \int_{-\infty}^{+\infty} \Big[&\left(\rho_n(\lambda) + \rho_n^{(h)}(\lambda)\right) \log\left(\rho_n(\lambda) + \rho_n^{(h)}(\lambda)\right) \\ &- \rho_n(\lambda)\log\rho_n(\lambda) - \rho_n^{(h)}(\lambda)\log\rho_n^{(h)}(\lambda) \Big] \end{aligned} \tag{3.169}$$

Using this expression and equations (3.166) and (3.168), check that the minimum of F is determined by the conditions

$$\frac{J}{T}a_n^S(\lambda) + \log\left(1 + \exp\frac{\epsilon_n}{T}\right) - \sum_m A_{n,m} * \log\left(1 + \exp\frac{-\epsilon_m}{T}\right) = 0 \tag{3.170}$$

where the string energy $\epsilon_n(\lambda)$ is defined by

$$\exp -\frac{\epsilon_m}{T} = \frac{\rho_n(\lambda)}{\rho_n^{(h)}(\lambda)} \tag{3.171}$$

Solve equation (3.170) in the zero temperature limit and check that all the Bethe strings have zero energy except those of length $2S$, whose energy is negative. This result explains why the ground state is a "Dirac sea" of $2S$-strings.

Ex. 3.6 Check that the energy degeneracy between the singlet (3.60) and triplet (3.61) states in the spin-$\frac{1}{2}$ isotropic closed chain holds only in the thermodynamic limit $L \to \infty$. This means that some additional symmetry arises in this limit. To prove this result, check that the two-string of the singlet does not contribute to the energy only in the thermodynamic limit.

Ex. 3.7 Using the results of exercise 3.5, the free energy at zero temperature becomes

$$F(T=0) = \int_{-\infty}^{+\infty} a_{2S}^{S}(\lambda)\epsilon_{2S}(\lambda)d\lambda \tag{3.172}$$

with ϵ_{2S} the (negative) energy of $2S$-strings. For a gapless system with rotational invariance, the dispersion relation is

$$\epsilon \sim vk \tag{3.173}$$

where v is the speed of sound. The central extension c is defined by the leading term of the heat capacity (see section 8.10)

$$C(T) = T\frac{\partial \mathscr{S}}{\partial T} = -T\frac{\partial^2 F}{\partial T^2} \tag{3.174}$$

in the limit $T \to 0$ as follows:

$$C(T) \underset{T\to 0}{\longrightarrow} \frac{\pi c}{3v} T + o(T) \tag{3.175}$$

Compute the value of c for the isotropic spin-S chain and check that it is equal to $3S/(S+1)$.

Ex. 3.8 Solve the equations (3.64) and (3.58) for the spin-$\frac{1}{2}$ Heisenberg chain with $N_h = H_1 = 4$, and obtain $\dim \mathscr{H}^{(N_h)} = 2^4$. [Hint: you must take into account the degeneracy $2S^z + 1$ for a Bethe state of spin $S^z = \frac{L}{2} - M$.] More generally, prove that the dimension of the space of Bethe states with N_h holes is 2^{N_h}. [Hint: solution $\{V_n, H_n\}$ to the equations has another degeneracy due to the number of choices for assigning the particles V_n into $2I_{\max}^{(n)} + 1$ numbers for $n \geq 2$. It is given by the product of combinatorial numbers $\prod_{n\geq 2} \binom{V_n+H_n}{V_n}$. Thus the equality to be proven is $\sum_{\{V_n,H_n\}}(2S^z+1)\prod_{n\geq 2}\binom{V_n+H_n}{V_n} = 2^{N_h}$.]

Ex. 3.9 Allowing for N_c pairs of complex roots $(z_\ell, \bar{z}_\ell)$ and N_h holes in the Dirac sea of roots, the Bethe equations (2.53) for the XXZ model of length L in the regime $\Delta < -1$ become

$$\left(\frac{\sinh(u_j + i\gamma)}{\sinh u_j}\right)^L = \prod_{\substack{k=1\\k\neq j}}^{M} \left(\frac{\sinh(u_j - u_k + i\gamma)}{\sinh(u_j - u_k)}\right) \times \prod_{\ell=1}^{N_c} \left(\frac{\sinh(u_j - z_\ell + i\gamma)}{\sinh(u_j - z_\ell)}\right)\left(\frac{\sinh(u_j - \overline{z}_\ell + i\gamma)}{\sinh(u_j - \overline{z}_\ell)}\right) \tag{3.176a}$$

$$\left(\frac{\sinh(z_j + i\gamma)}{\sinh z_j}\right)^L = \left(\frac{\sinh(z_j - \overline{z}_j + i\gamma)}{\sinh(z_j - \overline{z}_j)}\right) \prod_{k=1}^{M} \left(\frac{\sinh(z_j - u_k + i\gamma)}{\sinh(z_j - u_k)}\right) \times \prod_{\substack{\ell=1\\ \ell\neq j}}^{N_c} \left(\frac{\sinh(z_j - z_\ell + i\gamma)}{\sinh(z_j - z_\ell)}\right)\left(\frac{\sinh(z_j - \overline{z}_\ell + i\gamma)}{\sinh(z_j - \overline{z}_\ell)}\right) \tag{3.176b}$$

Fix γ to some numerical value of your choice and use a computer to search for solutions to these equations (actually, their logarithm) with small values for L, M and N_c. Once the program is debugged, ask the computer to write out the spin and energy of the Bethe states it finds. Do you observe any degeneracy in energies between Bethe states of different spin? Are there states you expected by number counting which do not seem to show up? What happens if you increase L? Into what multiplets does the spectrum assemble?

Ex. 3.10 Try to find the statistics of the elementary excitations (holes) of the spin-S chain. In other words, try to find the symmetry of a system of holes under the interchange of two holes. If the wave-function is symmetric (respectively, antisymmetric) under the interchange of any two holes, we would say that the holes are bosonic (respectively, fermionic). As a hint, consider the diagram below, where holes a and b are interchanged. If this exercise is too hard right now, return to it after reading chapters 6, 9 and 10.

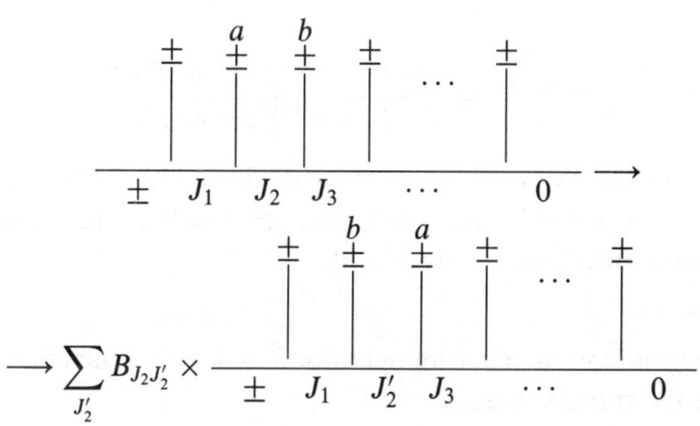

(3.177)

Ex. 3.11 Consider the Majumdar–Ghosh spin-$\frac{3}{2}$ hamiltonian

$$H = \sum_i P_{\frac{3}{2}}\left(\vec{S}_i, \vec{S}_{i+1}, \vec{S}_{i+2}\right) \tag{3.178}$$

with $\vec{S}_i$ the spin-$\frac{1}{2}$ operator at site i, and $P_{\frac{3}{2}}$ the projector of $\frac{1}{2}\times\frac{1}{2}\times\frac{1}{2}$ onto $\frac{3}{2}$. Find the ground states of the model and show that there is a mass gap.

Ex. 3.12 Using the representation theory of the quantum group $U_q(s\ell(2))$, we may decompose 1×1 into $0+1+2$, where 0, 1 and 2 are the spin labels of the irreps. In analogy with (3.178), define the "quantum" hamiltonian

$$H_{-\frac{1}{3}}(q) = \sum_i P_2^{(q)}\left(\vec{S}_i, \vec{S}_{i+1}\right) \tag{3.179}$$

where $\vec{S}$ are the quantum spin-1 matrices. What are the ground states? What is the role, if any, of the q-deformed Levi–Civita tensor $\epsilon^{12} = q^{\frac{1}{2}}$, $\epsilon^{21} = -q^{-\frac{1}{2}}$, $\epsilon^{11} = \epsilon^{22} = 0$, in the construction of the presumed valence bond solid ground state?

Ex. 3.13 Consider the XXZ spin chain with boundary terms given by (3.115) and $q = \exp\frac{\pi i}{p}$, $p \in \mathbb{N}$, $p \geq 3$. Check first that the central extension is given by the following limit (the factor $-1/12$ is conventional):

$$-\frac{c}{12} = \lim_{\substack{L,M\to\infty \\ 2L-M=0}} L\left[\frac{1}{2}(L-1)\frac{q+q^{-1}}{2} + 2\sum_{j=1}^{M}\left(\cos k_j - \frac{q+q^{-1}}{2}\right)\right] \tag{3.180}$$

where k_j is given by

$$Lk_j = \pi j + \frac{1}{2}\sum_{\substack{\ell=1 \\ \ell\neq j}}^{M}\left[\Phi(k_j, k_\ell) + \Phi(k_\ell, -k_j)\right] \tag{3.181}$$

with

$$e^{i\Phi(k_j,k_\ell)} = \frac{1-\left(q+q^{-1}\right)e^{ik_j} + e^{i(k_j+k_\ell)}}{1-\left(q+q^{-1}\right)e^{ik_\ell} + e^{i(k_j+k_\ell)}} \tag{3.182}$$

Evaluate the central extension c above with the usual ζ function trick, namely $\sum_{n\geq 1} n = \zeta(-1) = -1/12$, and check that the result is the central charge of a unitary minimal model (see section 8.10).

Ex. 3.14 Suppose that we have an operator $K \in \mathrm{End}(V)$ such that $[\mathscr{R}(u), K\otimes K] = 0$. Show that the transfer matrix

$$t_K(u) = \mathrm{tr}_V\left(K L_n(u)\cdots L_1(u)\right) \tag{3.183}$$

describes a closed integrable quantum chain with quasi-periodic or "twisted" boundary conditions determined by the matrix K. Derive the eigenvalues and the Bethe ansatz equation for an XXZ chain modified by the operator twisted boundary conditions

$$K = \begin{pmatrix} e^{i\phi} & \\ & e^{-i\phi} \end{pmatrix} \tag{3.184}$$

Ex. 3.15 Taking the limit $\xi_+ = -\xi_- \to -\infty$ in $H(\gamma, \xi_+, \xi_-)$ yields the quantum group invariant hamiltonian (3.131). In this limit, the boundary operators become

$$K^-(u) = \begin{pmatrix} e^{u} & \\ & e^{-u} \end{pmatrix}, \qquad K^+(u) = \begin{pmatrix} e^{-u-i\gamma} & \\ & e^{u+i\gamma} \end{pmatrix} \tag{3.185}$$

Show that there exists a gauge transformation of the form (2.75) such that the new $\tilde{K}$ and $\tilde{R}$ matrices become

$$\tilde{K}^-(u) = \mathbf{1}, \qquad \tilde{K}^+(u) = \begin{pmatrix} e^{-i\gamma} & \\ & e^{i\gamma} \end{pmatrix} \tag{3.186}$$

while the hamiltonian is given by $\tilde{H} = i \sum P_{j,j+1} \tilde{\mathscr{R}}'(0)_{j,j+1}$. Thus, the quantum group structure of the six-vertex model becomes more transparent in the new gauge.

Ex. 3.16 [V. Pasquier and H. Saleur, *Nucl. Phys.* **B330** (1990) 523.]

Note that the quantum group invariant hamiltonian (3.131) is not hermitian. Prove that its eigenvalues are, nevertheless, real. Diagonalize this hamiltonian for a baby chain with $L = 3$ sites, with a generic value of q. Note that for the special value $q^3 = 1$, something singular happens to the eigenvectors and eigenvalues of the hamiltonian, which cannot be fully diagonalized. Compare this with the representation theory of $U_q(s\ell(2))$ at $q^3 = 1$.

Ex. 3.17 Try to solve the integrable spin-1 hamiltonian (3.140) with $\beta = 1$, namely

$$H_1 = \sum_i \left[\vec{S}_i \cdot \vec{S}_{i+1} - \left(\vec{S}_i \cdot \vec{S}_{i+1} \right)^2 \right] \tag{3.187}$$

using the co-ordinate Bethe ansatz as we did for the spin-$\frac{1}{2}$ case in chapter 1. First of all, find the phase shift for the scattering of two spin waves, and secondly show that the scattering amplitudes for three spin waves factorizes according to the principles exposed in chapter 1. [Hint: write the states with M excitations as

$$|\Psi_M\rangle = \sum_{1 \le x_1 \le x_2 \le \cdots \le x_m \le L} f(x_1, \ldots, x_M) \, |x_1, \ldots, x_M\rangle \tag{3.188}$$

where the coincidence $x_i = x_{i+1}$ means that state "2" occupies the ith site, whereas, if $x_{i-1} < x_i < x_{i+1}$, then the ith site is occupied by state "1".]

4

The eight-vertex model

4.1 Definitions and Yang–Baxter relations

The eight-vertex model, introduced and solved by Baxter, is a simple generalization of the six-vertex model, and is obtained by relaxing the condition of current conservation on the Boltzmann weights, but still preserving the $\mathbb{Z}_2$ reflection symmetry. In addition to the six non-zero weights of the six-vertex model, the eight-vertex model has two non-zero weights associated with current sources and sinks, equal by virtue of the reflection symmetry. Thus, we start with the Boltzmann weights

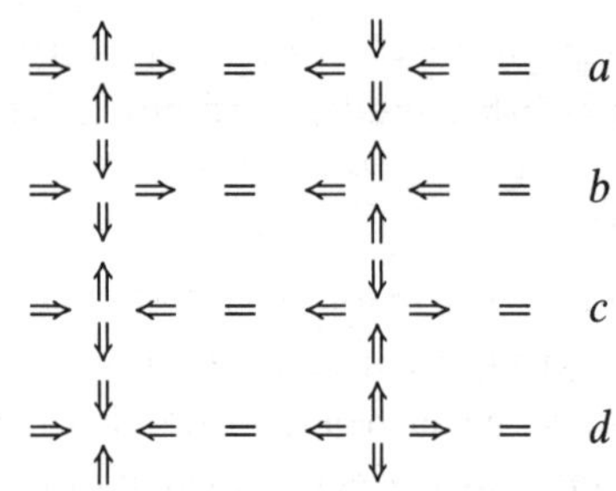

(4.1)

In terms of current conservation (we keep the convention that time flows diagonally upwards and rightwards), it is clear that the new weight d conserves the current only up to $\mathbb{Z}_2$ transformations. We shall see that this apparently minor relaxation of the symmetries of the model will lead us to new and unsuspected features.

Following Zamolodchikov, we may interpret the previous set of vertices in the language of S-matrices (see exercise 4.1). We picture again, as in chapter 1 for the six-vertex model, a theory with particles A and antiparticles $\overline{A}$ evolving on the line, with non-vanishing matrix elements for the following two-particle scattering processes:

$$\begin{aligned} AA &\to S_1\, AA + S_4\, \overline{A}\,\overline{A} \\ A\overline{A} &\to S_2\, \overline{A}A + S_3\, A\overline{A} \end{aligned} \tag{4.2}$$

along with their CPT conjugates, $A \leftrightarrow \overline{A}$, and we use the identifications

$$\begin{aligned} S_1 &= a \qquad & S_2 &= b \\ S_3 &= c \qquad & S_4 &= d \end{aligned} \tag{4.3}$$

In the same spirit as our previous study of the six-vertex model, we may collect the Boltzmann weights (4.1) in the form of a 4×4 $\mathscr{R}$-matrix:

$$\mathscr{R}^{(8v)} = \begin{pmatrix} a & 0 & 0 & d \\ 0 & b & c & 0 \\ 0 & c & b & 0 \\ d & 0 & 0 & a \end{pmatrix} \tag{4.4}$$

As usual, this $\mathscr{R}$-matrix can be interpreted as an operator $\mathscr{R}^{(8v)} : V^{\frac{1}{2}} \otimes V^{\frac{1}{2}} \to V^{\frac{1}{2}} \otimes V^{\frac{1}{2}}$ with matrix elements

$$\mathscr{R}^{(8v)} |\alpha\rangle \otimes |\beta\rangle = \sum_{\delta,\gamma} \left(\mathscr{R}^{(8v)}\right)^{\delta\gamma}_{\alpha\beta} |\delta\rangle \otimes |\gamma\rangle \qquad (\alpha, \beta, \gamma, \delta \in \{0, 1\}) \tag{4.5}$$

We have learned that integrability of a model arises if in the space of couplings a submanifold exists, such that the transfer matrices corresponding to any two points P and P' on it commute:

$$[t(P), t(P')] = 0 \tag{4.6}$$

Baxter's remarkable result for the eight-vertex model can be summarized as follows. Let $p = (a, b, c, d)$ be a point in $\mathbb{CP}^3$. Define the elliptic curve Γ_{YB} by the intersection of the two quadrics

$$\alpha_1 = \frac{cd}{ab}, \qquad \alpha_2 = \frac{a^2 + b^2 - c^2 - d^2}{2ab} \tag{4.7}$$

with $\alpha_1, \alpha_2 \in \mathbb{C}$. Then, for two arbitrary points P and P' in Γ_{YB}, the integrability condition (4.6) is satisfied.

We know, from the discussion in previous chapters, that (4.6) is equivalent to finding three sets of weights (a, b, c, d), (a', b', c', d') and (a'', b'', c'', d'') that satisfy the star-triangle or Yang–Baxter equation for vertex models, shown again in figure 4.1. We can write the eight-vertex weights more conveniently as

$$\left(\mathscr{R}^{(8v)}\right)^{\mu'\alpha'}_{\mu\alpha} = \sum_{j=1}^{4} w_j \left(\sigma^j\right)_{\mu\mu'} \left(\sigma^j\right)_{\alpha\alpha'} \tag{4.8}$$

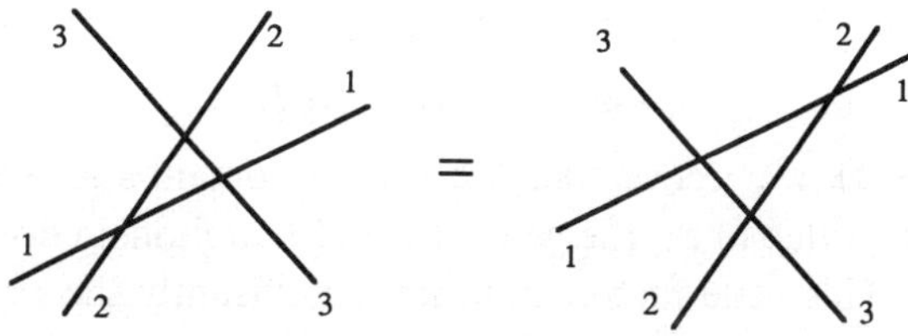

Fig. 4.1. Diagrammatic mnemotechnics for the Yang–Baxter equation, (12) ↔ (a, b, c, d), (13) ↔ (a', b', c', d') and (23) ↔ (a'', b'', c'', d'').

where σ^1, σ^2 and σ^3 are the usual Pauli matrices and σ^4 is the identity. This change of variables reads explicitly as follows:

$$\begin{aligned} w_1 &= \frac{1}{2}(c+d), & w_2 &= \frac{1}{2}(c-d) \\ w_3 &= \frac{1}{2}(a-b), & w_4 &= \frac{1}{2}(a+b) \end{aligned} \tag{4.9}$$

In the new variables, the Yang–Baxter equations read as

$$w_n w'_\ell w''_j + w_k w'_j w''_\ell = w_\ell w'_n w''_k + w_j w'_k w''_n \tag{4.10}$$

for all cyclic permutations (j,k,ℓ,n) of $(1,2,3,4)$.

The solution to (4.10) found by Baxter is

$$\begin{aligned} a(u) &= \rho\ \Theta(\gamma)\ \Theta(u)\ H(u+\gamma) \\ b(u) &= \rho\ \Theta(\gamma)\ H(u)\ \Theta(u+\gamma) \\ c(u) &= \rho\ H(\gamma)\ \Theta(u)\ \Theta(u+\gamma) \\ d(u) &= \rho\ H(\gamma)\ H(u)\ H(u+\gamma) \end{aligned} \tag{4.11}$$

where Θ and H are Jacobi elliptic functions and ρ is an overall factor. (For definitions and a brief review of elliptic functions see appendix B.) The three different sets of Boltzmann weights appearing in (4.10) are associated with three different points u, u' and u'' in Γ_{YB} such that $u'' = u' - u$. Writing the elliptic curve Γ_{YB} as $\mathbb{C}/L$, where L is the lattice generated by $4K$ and $2iK'$ with $q = \exp(-\pi K'/K)$ the elliptic nome, the rapidity conservation is relaxed to become $u'' = u' - u \bmod L$, which preserves nevertheless the difference property modulo the lattice L.

The parameter γ in (4.11) and the modulus of the elliptic functions are determined by (4.7) in terms of the two constants α_1 and α_2. The physical meaning of these constants is unveiled with the help of the associated spin chain hamiltonian

$$H = \sum_j \sigma_j^x \sigma_{j+1}^x + \frac{1-\alpha_1}{1+\alpha_1} \sigma_j^y \sigma_{j+1}^y + \frac{\alpha_2}{1+\alpha_1} \sigma_j^z \sigma_{j+1}^z \tag{4.12}$$

which is the XYZ model. The standard notation is

$$\begin{aligned} \Gamma &= \frac{1-\alpha_1}{1+\alpha_1} = \frac{ab-cd}{ab+cd} \\ \Delta &= \frac{\alpha_2}{1+\alpha_1} = \frac{a^2+b^2-c^2-d^2}{2(ab+cd)} \end{aligned} \tag{4.13}$$

When $\alpha_1 = 0$ ($\Gamma = 1$), we recover the XXZ hamiltonian, and Δ becomes $\cos\gamma$. Due to the non-zero value of α_1, the general XYZ hamiltonian does not conserve the total spin $S^z_{\rm tot}$. This little fact complicates significantly the resolution of the model via the Bethe ansatz

The symmetry pattern of the spin chains with which we are becoming acquainted is as follows. The isotropic Heisenberg chain, or XXX model, is invariant under $SU(2)$ because the three directions x, y and z are equivalent: it is associated with a sphere. The XXZ model is at first sight invariant only under the $U(1)$ rotations in the x–y plane, about the z axis: it is associated with a revolutional ellipsoid. The XYZ model, finally, has no obvious invariance, and it is associated with a general two-dimensional ellipsoid in three-space. In chapter 3, we saw that by adding a boundary term to the XXZ model, we can make it $U_q(s\ell(2))$ invariant. This apparently minor change enhances $U(1)$ to $U_q(s\ell(2))$. Moreover, the infinite XXZ chain enjoys the infinite-dimensional affine quantum symmetry $U_q(A_1^{(1)})$, and integrability is thus assured. Along these lines, the natural question to ask is what additional deformation of $U_q(s\ell(2))$ or $U_q(A_1^{(1)})$ describes the symmetry of the integrable XYZ chain? The deformation parameter q is related to the anisotropy as in chapter 2, namely $q = e^{i\gamma}$ with $\Delta = \cos\gamma$, but the extra anisotropy in the x–y plane is somehow measured by the elliptic modulus k. A synopsis of this discussion appears in table 4.1.

The last two columns in table 4.1 refer to perhaps the most intriguing aspect of the eight-vertex solution, at least from a mathematical point of view. In chapter 2, we noticed that the $\mathcal{R}$-matrix solution to the six-vertex Yang–Baxter equation admits an interpretation as an intertwiner of $U_q(\widehat{s\ell}(2))$. In chapter 7, we will show that intertwiners of finite-dimensional irreps of the quantum affine algebra $U_q(\widehat{s\ell}(2))$ provide, quite generally, trigonometric solutions to the Yang–Baxter equations. The space containing question marks in table 4.1 should thus be filled with some deformation of the quantum affine algebra, whose intertwiners yield elliptic solutions to the Yang–Baxter equation. At the the time of writing, this hypothetical mathematical structure is not yet known.

4.2 Bethe ansatz solution

Neither the algebraic nor the analytic Bethe ansatzes studied in previous chapters can be extended naturally to the eight-vertex case, because the current is not conserved. Since we may not classify the Hilbert space into subspaces of constant S^z_{tot}, we lack a good starting point. However, not everyting is lost, and the general ideas of the algebraic Bethe ansatz presented in chapters 2 and 3 can still be put to good use. To proceed, we will present Baxter's solution of the eight-vertex

Table 4.1 Symmetries underlying some integrable models and the kind of solutions to the Yang–Baxter equation that they produce.

$\Delta = 1$	$\Gamma = 1$	six-vertex	XXX chain	$SU(2)$	rational
$\Delta \neq 1$	$\Gamma = 1$	six-vertex	XXZ chain	$U_q(s\ell(2))$	trigonometric
$\Delta \neq 1$	$\Gamma \neq 1$	eight-vertex	XYZ chain	? ? ?	elliptic

model in the version due to Faddev and Takhtadjan. The strategy is to make the eight-vertex model as similar as possible to the six-vertex model.

4.2.1 A smart change of basis

In the usual basis $\{e_0, e_1\} = \{|\Uparrow\rangle, |\Downarrow\rangle\}$, the eight-vertex $\mathscr{R}$-matrix is given by (4.4) and (4.5). The first step is to find a new basis $\{X^+, X^-\}$ which can depend on the spectral parameter

$$X^{\pm}(u) = \begin{pmatrix} X_1^{\pm}(u) \\ X_2^{\pm}(u) \end{pmatrix} \tag{4.14}$$

such that the eight-vertex $\mathscr{R}$-matrix in this new basis has the texture of the six-vertex $\mathscr{R}$-matrix, i.e. of the form

$$\mathscr{R}^{(8v)} \text{ in basis } \{X^+, X^-\} = \begin{pmatrix} \times & 0 & 0 & 0 \\ 0 & \times & \times & 0 \\ 0 & \times & \times & 0 \\ 0 & 0 & 0 & \times \end{pmatrix} \tag{4.15}$$

Quite remarkably, the problem of such a change of basis admits a solution. Moreover, it actually admits an infinity of solutions $\{X_\ell^{\pm}(u)\}$ labeled by an integer index $\ell \in \mathbb{Z}$:

$$\begin{aligned} X_\ell^+(u) &= H(s + \gamma\ell - u)\,|\Uparrow\rangle + \Theta(s + \gamma\ell - u)\,|\Downarrow\rangle \\ X_\ell^-(u) &= H(t + \gamma\ell + u)\,|\Uparrow\rangle + \Theta(t + \gamma\ell + u)\,|\Downarrow\rangle \end{aligned} \tag{4.16}$$

where s and t are free parameters.

Using the addition theorems for theta functions (see appendix B), the reader can prove the following six-vertex-like relations involving the permuted matrix $R^{(8v)} = P\mathscr{R}^{(8v)}$:

$$\begin{aligned}
& R^{(8v)}(u_1 - u_2)\ X_\ell^+(u_1) \otimes X_{\ell+1}^+(u_2) \\
&\qquad = B\left(\begin{matrix} \ell - 1 & \ell \\ \ell & \ell + 1 \end{matrix}\,\middle|\, u_1 - u_2\right)\ X_\ell^+(u_2) \otimes X_{\ell+1}^+(u_1) \\
& R^{(8v)}(u_1 - u_2)\ X_\ell^-(u_1) \otimes X_{\ell-1}^-(u_2) \\
&\qquad = B\left(\begin{matrix} \ell + 1 & \ell \\ \ell & \ell - 1 \end{matrix}\,\middle|\, u_1 - u_2\right)\ X_\ell^-(u_2) \otimes X_{\ell-1}^-(u_1) \\
& R^{(8v)}(u_1 - u_2)\ X_{\ell+1}^+(u_1) \otimes X_\ell^-(u_2) \\
&\qquad = B\left(\begin{matrix} \ell & \ell + 1 \\ \ell + 1 & \ell \end{matrix}\,\middle|\, u_1 - u_2\right)\ X_{\ell+1}^+(u_2) \otimes X_\ell^-(u_1) \\
&\qquad + B\left(\begin{matrix} \ell & \ell - 1 \\ \ell + 1 & \ell \end{matrix}\,\middle|\, u_1 - u_2\right)\ X_{\ell-1}^-(u_2) \otimes X_\ell^+(u_1) \\
& R^{(8v)}(u_1 - u_2)\ X_{\ell-1}^-(u_1) \otimes X_\ell^+(u_2)
\end{aligned} \tag{4.17}$$

$$= B\begin{pmatrix} \ell & \ell-1 \\ \ell-1 & \ell \end{pmatrix}\!\!\left|\, u_1-u_2 \right) X^-_{\ell-1}(u_2)\otimes X^+_\ell(u_1)$$

$$+ B\begin{pmatrix} \ell & \ell+1 \\ \ell-1 & \ell \end{pmatrix}\!\!\left|\, u_1-u_2 \right) X^+_{\ell+1}(u_2)\otimes X^-_\ell(u_1)$$

where the B-matrices are given by

$$\begin{aligned}
B\left(\begin{matrix} \ell\pm1 & \ell \\ \ell & \ell\mp1 \end{matrix}\,\middle|\, u\right) &= h(u+\gamma) \\
B\left(\begin{matrix} \ell & \ell\pm1 \\ \ell\pm1 & \ell \end{matrix}\,\middle|\, u\right) &= \frac{h(\gamma)h(\tau+\gamma\ell\mp u)}{h(\tau+\gamma\ell)} \\
B\left(\begin{matrix} \ell & \ell\pm1 \\ \ell\mp1 & \ell \end{matrix}\,\middle|\, u\right) &= \frac{h(u)h(\tau+\gamma\ell\mp\gamma)}{h(\tau+\gamma\ell)}
\end{aligned} \tag{4.18}$$

In these expressions,

$$\tau = \frac{s+t}{2} - K \tag{4.19}$$

where K is the real half period of $\Theta(u)$ and

$$h(u) = \Theta(0)H(u)\Theta(u) \tag{4.20}$$

In deriving (4.18), we have used the theorems

$$\begin{aligned}
\Theta(u)\Theta(v)H(w)H(u+v+w) + H(u)H(v)\Theta(w)\Theta(u+v+w) \\
= \Theta(0)\Theta(u+v)H(u+w)H(v+w) \\
H(u-v)\Theta(u+v) - \Theta(u-v)H(u+v) = 2\frac{g(u-K)g(v)}{g(K)}
\end{aligned} \tag{4.21}$$

where $g(u) = \Theta(u)H(u)$. Notice that the six non-vanishing entries of the R-matrix in the basis $X^\pm_\ell(u)$ reproduce the six-vertex texture. In the new basis $X^\pm_\ell(u)$, the R-matrix conserves current exactly.

The four equations (4.17) may be written in a more compact way as:

$$\begin{aligned}
&\left(R^{(8v)}\right)(u_1-u_2)X^{(\ell_1,\ell_2)}(u_1)\otimes X^{(\ell_2,\ell_3)}(u_2) \\
&= \sum_{\ell_4} B\left(\begin{matrix} \ell_1 & \ell_4 \\ \ell_2 & \ell_3 \end{matrix}\,\middle|\, u_1-u_2\right) X^{(\ell_1,\ell_4)}(u_2)\otimes X^{(\ell_4,\ell_3)}(u_1)
\end{aligned} \tag{4.22}$$

where the only non-zero B's are those in (4.18), and we use the notation

$$\begin{aligned}
X^{(\ell-1,\ell)}(u) &= X^+_\ell(u) \\
X^{(\ell+1,\ell)}(u) &= X^-_\ell(u) \\
X^{(\ell,\ell')}(u) &= 0 \qquad \text{if} \quad |\ell-\ell'| \neq 1
\end{aligned} \tag{4.23}$$

In components, (4.22) reads as follows:

$$\sum_{\alpha,\beta} \left(R^{(8v)}\right)_{\alpha\beta}^{\gamma\delta}(u_1-u_2) X_\alpha^{(\ell_1,\ell_2)}(u_1)\, X_\beta^{(\ell_2,\ell_3)}(u_2) \tag{4.24}$$

$$= \sum_{\ell_4} B\left(\begin{matrix}\ell_1 & \ell_4\\ \ell_2 & \ell_3\end{matrix}\middle| u_1-u_2\right) X_\gamma^{(\ell_1,\ell_4)}(u_2)\, X_\delta^{(\ell_4,\ell_3)}(u_1)$$

with the Greek indices labeling the components in the basis $\{|\Uparrow\rangle, |\Downarrow\rangle\} = \{0,1\}$ of X_ℓ^+ and X_ℓ^-.

It will prove to be convenient to use some diagrams. We write

$$X_\ell^+ = \left(X^{(\ell-1,\ell)}\right)_\alpha =$$

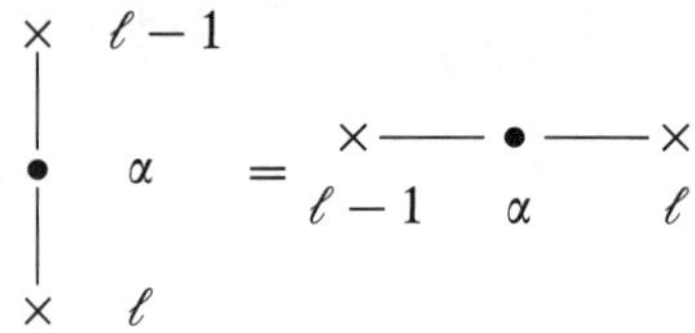

(4.25a)

$$X_\ell^- = \left(X^{(\ell+1,\ell)}\right)_\alpha =$$

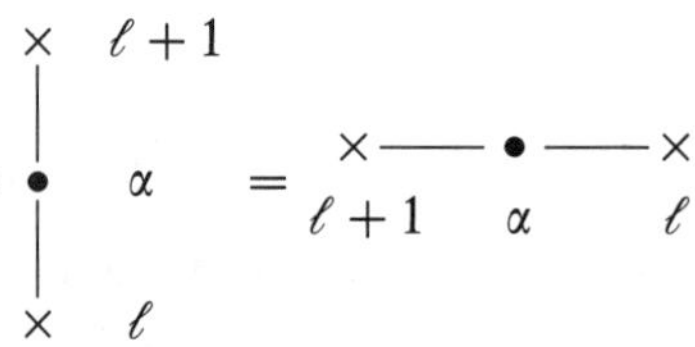

(4.25b)

We may also represent the coefficients B as

$$B\left(\begin{matrix}\ell_1 & \ell_4\\ \ell_2 & \ell_3\end{matrix}\middle| u\right) =$$

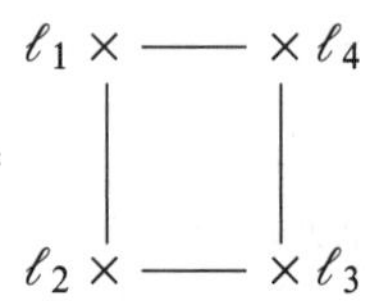

(4.26)

In diagrams, equation (4.24) becomes the very pleasing

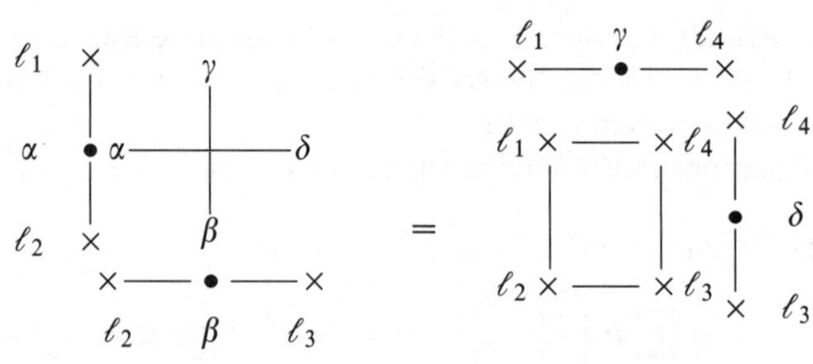

(4.27)

It is clear from (4.22) that since $R^{(8v)}$ satisfies the Yang–Baxter equation, the coefficients B satisfy another Yang–Baxter equation. To derive it, apply both sides of the Yang–Baxter equation

$$R_1^{(8v)}(u_{23})R_2^{(8v)}(u_{13})R_1^{(8v)}(u_{12}) = R_2^{(8v)}(u_{12})R_1^{(8v)}(u_{13})R_2^{(8v)}(u_{23}) \tag{4.28}$$

to the space

$$X^{(\ell_1,\ell_2)}(u_1) \otimes X^{(\ell_2,\ell_3)}(u_2) \otimes X^{(\ell_3,\ell_4)}(u_3) \tag{4.29}$$

and use the fact that the vectors $X_\ell^{\pm}$ are linearly independent. The resulting Yang–Baxter equation

$$\begin{aligned} &\sum_{\ell_0} B\left(\begin{matrix}\ell_1 & \ell_0\\ \ell_2 & \ell_3\end{matrix}\middle| u_{12}\right) B\left(\begin{matrix}\ell_0 & \ell_5\\ \ell_3 & \ell_4\end{matrix}\middle| u_{13}\right) B\left(\begin{matrix}\ell_1 & \ell_6\\ \ell_0 & \ell_5\end{matrix}\middle| u_{23}\right) \\ &\quad = \sum_{\ell_0} B\left(\begin{matrix}\ell_2 & \ell_0\\ \ell_3 & \ell_4\end{matrix}\middle| u_{23}\right) B\left(\begin{matrix}\ell_1 & \ell_6\\ \ell_2 & \ell_0\end{matrix}\middle| u_{13}\right) B\left(\begin{matrix}\ell_6 & \ell_5\\ \ell_0 & \ell_4\end{matrix}\middle| u_{12}\right) \end{aligned} \tag{4.30}$$

is known as the hexagon identity, and is shown in figure 4.2.

In chapter 5, we shall consider the Yang–Baxter equation (4.30) in more detail. There, we shall see that B are the Boltzmann weights of a "face model"; equation (4.22) and its graphical representation (4.27) is called the "vertex–face map". In figure 4.3, we have represented the vertex–face map for the case of a vertex matrix $R_{m_1m_2}^{m_4m_3}$ which preserves the total third component of isospin, $m_1 + m_2 = m_3 + m_4$. The face degrees of freedom are identified with the total spin j_a ($a = 1, 2, 3, 4$). We take $m_i \in \{+\frac{1}{2}, -\frac{1}{2}\}$, and identify

$$m_1 = j_2 - j_1,\ m_2 = j_3 - j_2,\ m_3 = j_3 - j_4,\ m_4 = j_4 - j_1 \tag{4.31}$$

whereby

$$|j_1 - j_2| = |j_2 - j_3| = |j_3 - j_4| = |j_4 - j_1| = \frac{1}{2} \tag{4.32}$$

Note the parallel with equation (4.23). To see more explicitly the relationship between the hexagon equation (4.30) and the usual "scattering" Yang–Baxter equation (1.105), it is enough to apply the vertex–face map repeatedly, as shown in figure 4.4.

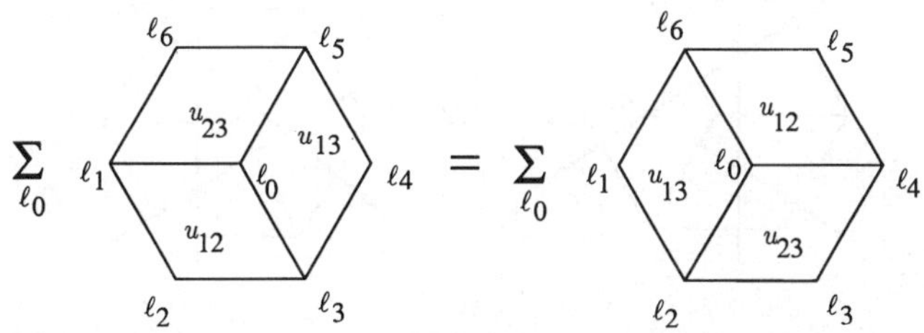

Fig. 4.2 Diagram for the the Yang–Baxter equation for the coefficients B, equation (4.30).

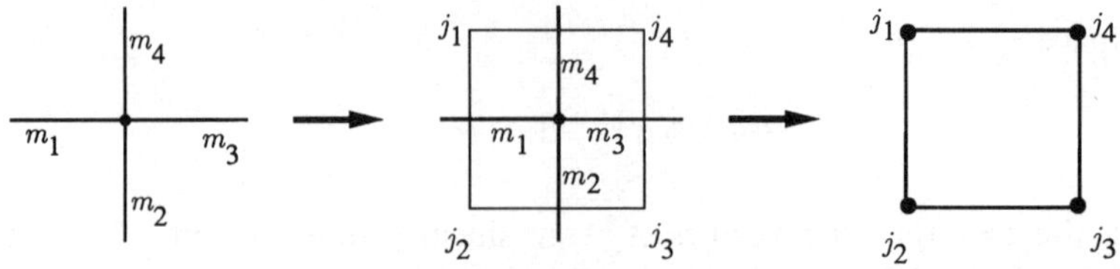

Fig. 4.3. The vertex–face map when the vertex R-matrix preserves current number. The label identifications are as in equations (4.31).

4.2.2 Eight-vertex Yang–Baxter algebra

Having found the new basis $\{X_\ell^+(u), X_\ell^-(u)\}$ in which the eight-vertex matrix of weights $R^{(8v)}$ has the six-vertex texture, we proceed to generalize the Yang–Baxter algebra. The change of basis (4.16) can be described by the 2×2 matrix

$$M_\ell(u) = \left(X_\ell^+(u), X_\ell^-(u)\right) = \begin{pmatrix} H(s+\gamma\ell-u) & H(t+\gamma\ell+u) \\ \Theta(s+\gamma\ell-u) & \Theta(t+\gamma\ell+u) \end{pmatrix} \tag{4.33}$$

and its inverse

$$M_\ell^{-1}(u) = \begin{pmatrix} \tilde{X}_\ell^-(u) \\ \tilde{X}_\ell^+(u) \end{pmatrix} \tag{4.34}$$
$$= \frac{1}{\det M_\ell(u)} \begin{pmatrix} \Theta(t+\gamma\ell+u) & -H(t+\gamma\ell+u) \\ -\Theta(s+\gamma\ell-u) & H(s+\gamma\ell-u) \end{pmatrix}$$

where $X_\ell^\pm$ are column vectors and their contragradients $\tilde{X}_\ell^\pm$ are row vectors.

We can now define a family of monodromy matrices $T_{k,\ell}^{(8v)}$ which are related to the "canonical" monodromy matrix $T^{(8v)}$, defined in equation (2.19), by the "gauge transformation"

$$T^{(8v)}(u) \quad \longrightarrow \quad T_{k,\ell}^{(8v)} = M_k^{-1}(u)\, T^{(8v)}(u)\, M_\ell(u) \tag{4.35}$$

In the new basis, the four entries $A(u)$, $B(u)$, $C(u)$ and $D(u)$ of the transfer matrix $T^{(8v)}$ (see chapter 2 for the detailed definitions) become the following entries of

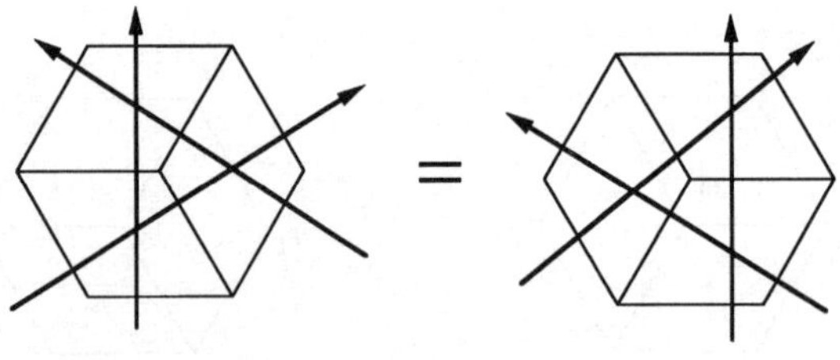

Fig. 4.4. The vertex–face map allows us to go from the vertex version of the Yang–Baxter equation to its face version.

$T^{(8v)}_{k,\ell}$:

$$\begin{aligned}
A_{k,\ell}(u) &= \tilde{X}^-_k(u)T^{(8v)}(u)X^+_\ell(u) \\
B_{k,\ell}(u) &= \tilde{X}^-_k(u)T^{(8v)}(u)X^-_\ell(u) \\
C_{k,\ell}(u) &= \tilde{X}^+_k(u)T^{(8v)}(u)X^+_\ell(u) \\
D_{k,\ell}(u) &= \tilde{X}^+_k(u)T^{(8v)}(u)X^-_\ell(u)
\end{aligned} \tag{4.36}$$

The diagrammatics (4.25), supplemented by

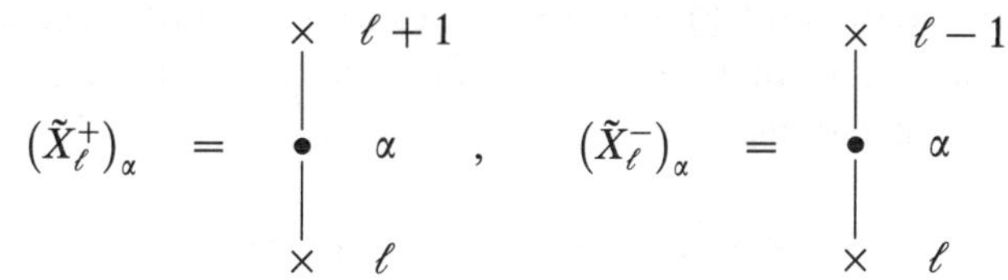

(4.37)

imply the following graphical representation of equations (4.36):

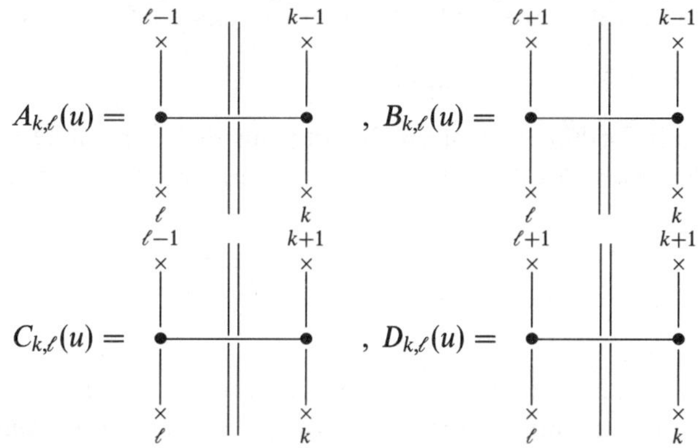

(4.38)

Note that the index ℓ (respectively, k) appears on the left (respectively, right) because the monodromy matrix $T^{(8v)} = L_L \cdots L_1$ in (2.19), with $L_i = \mathscr{R}_{ai}$, is represented as in (2.21). The four operators (4.38), which are the entries of $T^{(8v)}_{k,\ell}$, should be understood as acting on the vertical double-lined space, i.e. on the quantum space of the two-dimensional model or the Hilbert space of the one-dimensional chain.

From (4.35) and (4.36), it is clear that the transfer matrix, which is the trace of the monodromy matrix, is given by

$$t^{(8v)} = A_{\ell\ell}(u) + D_{\ell\ell}(u) \tag{4.39}$$

and it is independent of ℓ. The algebra of the operators (4.36) is given by

$$B_{k,\ell+1}(u)B_{k+1,\ell}(v) = B_{k,\ell+1}(v)B_{k+1,\ell}(u) \tag{4.40a}$$

$$\begin{aligned}
A_{k,\ell}(u)B_{k+1,\ell-1}(v) &= \alpha(u-v)B_{k,\ell-2}(v)A_{k+1,\ell-1}(u) \\
&\quad -\beta_{\ell-1}(u-v)B_{k,\ell-2}(u)A_{k+1,\ell-1}(v)
\end{aligned} \tag{4.40b}$$

$$\begin{aligned}
D_{k,\ell}(u)B_{k+1,\ell-1}(v) &= \alpha(v-u)B_{k+2,\ell}(v)D_{k+1,\ell-1}(u) \\
&\quad +\beta_{k+1}(u-v)B_{k+2,\ell}(u)D_{k+1,\ell-1}(v)
\end{aligned} \tag{4.40c}$$

with

$$
\begin{aligned}
\alpha(u) &= \frac{h(u-\gamma)}{h(u)} \\
\beta_k(u) &= \frac{h(\gamma)}{h(u)} \frac{h\left(\frac{s+t}{2} + \gamma k - K + u\right)}{h\left(\frac{s+t}{2} + \gamma k - K\right)}
\end{aligned}
\tag{4.41}
$$

and $h(u)$ as in (4.20). It is a good exercise to check that the labels k and ℓ in these equations match the graphical representation (4.38). The reader will profit from comparing equations (4.40) with the equivalent ones for the six-vertex model (2.40); note, in particular, that only the combination $(s+t)$ of the free parameters appears in these expressions.

4.3 Reference state and θ parameter

We have already collected most of the ingredients necessary to apply the algebraic Bethe ansatz. The last item we need is the reference state $|\Omega\rangle$ with no spin waves

$$C_{k,\ell}\,|\Omega\rangle = 0 \tag{4.42}$$

It is natural to take for $|\Omega\rangle$ the state with "all spins up", i.e. a state made out of $X_\ell^+(u)$'s only. The new index ℓ must be carefully arranged, however, according to the pattern

$$
\begin{array}{ccccccccc}
\ell & & \ell+1 & & \ell+2 & & \ell+3 & & \\
\times & \!\!-\!\!\bullet\!\!-\!\! & \times & \!\!-\!\!\bullet\!\!-\!\! & \times & \!\!-\!\!\bullet\!\!-\!\! & \times & & \cdots \\
 & \alpha_1 & & \alpha_2 & & \alpha_3 & & &
\end{array}
\tag{4.43}
$$

Again, the Greek indices α_i refer to the standard $\{\Uparrow, \Downarrow\}$ basis. Note that the increase of the "quantum label" ℓ is fully consistent with the graphical representation (4.25). In order to satisfy (4.42), the Bethe reference state must be of the form (4.43) with the rapidity $u = \gamma$:

$$|\Omega_\ell\rangle = X_\ell^+(\gamma) \otimes X_{\ell+1}^+(\gamma) \otimes X_{\ell+2}^+(\gamma) \otimes \cdots \tag{4.44}$$

Using this construction, we find a whole family of reference states $|\Omega_\ell\rangle$. Since we start with X_ℓ^+ and the chain has L sites, the last label is $\ell + L$. It is a good exercise to check that

$$C_{L+\ell,\ell}\,|\Omega_\ell\rangle = 0 \tag{4.45}$$

Also,

$$
\begin{aligned}
A_{L+\ell,\ell}(u)\,|\Omega_\ell\rangle &= h^L\!\left(u + \frac{\gamma}{2}\right)|\Omega_{\ell-1}\rangle \\
D_{L+\ell,\ell}(u)\,|\Omega_\ell\rangle &= h^L\!\left(u - \frac{\gamma}{2}\right)|\Omega_{\ell+1}\rangle
\end{aligned}
\tag{4.46}
$$

Next, we look for the Bethe eigenvectors which are required to satisfy

$$\left(A_{\ell\ell}(u) + D_{\ell\ell}(u)\right)|\Psi\rangle = \Lambda(u)\,|\Psi\rangle \tag{4.47}$$

with the help of equations (4.40) and (4.46).

To simplify matters as much as possible, let us start with a chain of $L = 2$ sites and consider the states

$$|\Psi_\ell(u)\rangle = B_{\ell+1,\ell-1}(u)\,|\Omega_{\ell-1}\rangle \tag{4.48}$$

Using (4.40b) and (4.40c) we obtain

$$\begin{aligned}\Big[A_{\ell\ell}(u) + D_{\ell\ell}(u)\Big]\ |\Psi_\ell(u_1)\rangle \\ &= \frac{h(u_1 - u + \gamma)}{h(u_1 - u)} h^2\left(u + \frac{\gamma}{2}\right) |\Psi_{\ell-1}(u_1)\rangle \\ &+ \frac{h(u - u_1 + \gamma)}{h(u - u_1)} h^2\left(u - \frac{\gamma}{2}\right) |\Psi_{\ell+1}(u_1)\rangle \\ &+ \text{ unwanted terms}\end{aligned} \tag{4.49}$$

To reach equation (4.47), we must now introduce some coherent sum of $\Psi_\ell(u)$ states. This will introduce some non-perturbative parameter into the game. By analogy with quantum chromodynamics, we call it angle θ:

$$\Psi_\theta(u) = \sum_{\ell\in\mathbb{Z}} e^{i\ell\theta}\Psi_\ell(u) \tag{4.50}$$

Then we obtain

$$\begin{aligned}\Big[A_{\ell\ell}(u) + D_{\ell\ell}(u)\Big]\, |\Psi_\theta(u_1)\rangle &= \text{unwanted terms} \\ &+ \left(\frac{h(u_1 - u + \gamma)}{h(u_1 - u)}\, h^2\left(u + \frac{\gamma}{2}\right)\, e^{i\theta}\right. \\ &+ \left.\frac{h(u - u_1 + \gamma)}{h(u - u_1)}\, h^2\left(u - \frac{\gamma}{2}\right)\, e^{-i\theta}\right) |\Psi_\theta(u_1)\rangle\end{aligned} \tag{4.51}$$

which is essentially the desired result. As expected, the eigenvalue is independent of ℓ. As for the six-vertex case, the Bethe ansatz equations for the rapidities follow from the requirement that the unwanted terms vanish. In the specific $L = 2$ situation under consideration, the Bethe equations are simply

$$\frac{h^2(u_1 + \frac{\gamma}{2})}{h^2(u_1 - \frac{\gamma}{2})} = e^{-2i\theta} \tag{4.52}$$

Before jumping to a chain with arbitrary L, let us point out some peculiarities of the Bethe states that are apparent at first sight. We observe from (4.40b), (4.40c) and (4.46) that the only candidates for Bethe states are the ones with $L/2$ insertions of B operators. By analogy with the six-vertex model, we say that these states have zero total Bethe spin. For the shortest chain with $L = 2$ sites, the reader will find it easy to convince herself that, in sharp contrast with the six-vertex model, there are no Bethe states with no B operator insertions.

Quite generally, if L is even, the Bethe states are obtained from

$$\Big|\Psi_\ell(u_1, \ldots, u_{L/2})\Big\rangle = B_{\ell+1,\ell-1}(u_1)\cdots B_{\ell+L/2,\ell-L/2}(u_{L/2})\,\Big|\Omega_{\ell-L/2}\Big\rangle \tag{4.53}$$

These states have zero Bethe spin. The eigenvalue in this general case is

$$\Lambda(u; u_1, \ldots, u_{L/2}) = e^{i\theta} h^L(u + \frac{\gamma}{2}) \prod_{i=1}^{L/2} \frac{h(u_i - u + \gamma)}{h(u_i - u)} + e^{-i\theta} h^L(u - \frac{\gamma}{2}) \prod_{i=1}^{L/2} \frac{h(u - u_i + \gamma)}{h(u - u_i)} \tag{4.54}$$

and the corresponding Bethe equations are

$$e^{-2i\theta} \prod_{\substack{k=1 \\ j \neq k}}^{L/2} \frac{h(u_j - u_k + \gamma)}{h(u_j - u_k - \gamma)} = \frac{h^L(u_j + \frac{\gamma}{2})}{h^L(u_j - \frac{\gamma}{2})} \tag{4.55}$$

Note that the eigenvalues (4.54) of the transfer matrix are independent of the parameters s and t, which appear only in the expressions for the eigenvectors. Also, note that the parameter $\theta \in [0, 2\pi]$ remains free in the whole analysis. We focus on θ more closely in the next section, and summarize this short but dense chapter in table 4.2.

4.3.1 Further comments on the θ parameter

In all the expressions until now, the dependence on the integers ℓ is always weighted by the anisotropy through the combination $\gamma\ell$. Moreover, the "quantum numbers" ℓ may range over all the integers, $\ell \in \mathbb{Z}$. Nevertheless, it turns out that

Table 4.2 Comparison of the main concepts underlying the algebraic Bethe ansatz for the six- and eight-vertex models.

Model	Six-vertex	Eight-vertex
Basis	$\{\lvert\Uparrow\rangle, \lvert\Downarrow\rangle\}$	$\{X_\ell^+(u), X_\ell^-(u)\}$ $(\ell \in \mathbb{Z})$
Reference state	$\lvert\Omega\rangle = \lvert\Uparrow\Uparrow\cdots\Uparrow\rangle$	$\lvert\Omega_\ell\rangle = X_\ell^+(\gamma)\cdots X_{\ell+L}^+(\gamma)$
$\mathscr{A}_{YB}$ generators	$A(u)$, $B(u)$, $C(u)$, $D(u)$	$A_{k,\ell}(u)$, $B_{k,\ell}(u)$, $C_{k,\ell}(u)$, $D_{k,\ell}(u)$ $(k, \ell \in \mathbb{Z})$
Reference state condition	$C(u)\lvert\Omega\rangle = 0$	$C_{\ell+L,\ell}(u)\lvert\Omega_\ell\rangle = 0$
Bethe states	$\prod_{i=1}^{m} B(u_i)\lvert\Omega\rangle$	$\sum_{\ell\in\mathbb{Z}} e^{i\ell\theta} \prod_{k=1}^{L/2} B_{\ell+k,\ell-k}(u_k)\lvert\Omega_{\ell-L/2}\rangle$
Bethe spin	$\frac{L}{2} - m$	0

for the special values of the parameters

$$\tau = 0\,, \qquad \gamma = \frac{2K}{r} \quad \text{with} \quad r \in \mathbb{N} - \{0,1\} \tag{4.56}$$

then the labels ℓ may be restricted to a finite interval:

$$1 \leq \ell \leq r - 1 \tag{4.57}$$

We will study such restricted models in greater detail in chapter 5; the case $r = 4$ corresponds to the Ising model. Here, we wish to emphasize the "elliptic" origin of such restricted models.

The choice (4.56) can be justified using the periodicity of the elliptic functions (see appendix B). When the anisotropy γ takes a value such that $r\gamma = 2K$, with r an integer, the Bethe eigenvectors (4.53) become periodic of period r in the subscript ℓ. This periodicity restricts the allowed values of $\theta \in [0, 2\pi]$ to the discrete choices

$$\theta = \frac{2\pi m}{r} \qquad m = 0, 1, \ldots, r - 1 \tag{4.58}$$

In this case, the Bethe spin of the eigenvectors will be zero only modulo r.

The trigonometric limit of the eight-vertex model is defined by degenerating the torus into a strip, i.e. $K' \to \infty$ and $K \to \pi/2$. In this limit, the anisotropy becomes $\gamma = \pi/r$, leading to a deformation parameter $q = e^{i\gamma}$, which is a $2r$th root of unity. This last comment inspires several interesting exercises.

Exercises

Ex. 4.1 [A.B. Zamolodchikov, 'Irreversibility of the flux of the renormalization group in a 2D field theory', *Pisma ZhETF* **43** (1986) 565.]

Some years ago, Zamolodchikov suggested the possibility of using the elliptic solution (4.11) to define a scattering matrix (4.2). The elliptic nature of the solution seems to imply an oscillatory behavior for the cross-section similar to the one expected for limiting cycles of the renormalization group equation. Using Zamolodchikov's c theorem (exercise 8.8), prove that such a behavior is impossible in $(1+1)$ quantum field theories.

Ex. 4.2 Consider the trigonometric limit $K \to \pi/2$, and choose $r\gamma = \pi$. Taking the value $m = 0$ in (4.58) and solving the Bethe ansatz equations should presumably yield the ground state of the six-vertex model. Check that this is indeed the case by computing the free energy of the ground state. Since this expression is a function of the lattice size L, the central extension of the continuum effective theory can be deduced from the coefficient of L^{-1}. Check that the central extension is the same as that for a free boson, $c = 1$.

Ex. 4.3 [M. Karowski, in *Quantum Groups, Proceedings of the 1989 Clausthal Workshop*, ed. H.D. Doebner (1990) Springer.]

Repeat exercise 4.2 but with $m = 1$, i.e. $\theta = 2\pi/r$. You should now find a ground state energy of the form

$$E_0 = -\frac{\pi c}{6L} + \text{constant} + O(L) \tag{4.59}$$

with

$$c = 1 - \frac{6}{r(r-1)} \tag{4.60}$$

This is the central extension for conformal minimal models, which we shall study in chapter 8.

Ex. 4.4 Repeat exercise 4.2 again for $m = 2, \ldots, r-1 = Q$. We shall meet the formulae for the energies thus obtained in chapter 8, in association with the primary fields $\Phi_{m,1}$ of the unitary minimal models.

Ex. 4.5 Since elliptic functions are doubly periodic, we might expect that the Bethe eigenvectors (4.53) become periodic whenever $r\gamma = 2m_1 K + i m_2 K'$, with m_1 and m_2 integers (note that (4.56) is recovered setting $m_1 = 1$ and $m_2 = 0$). Investigate whether this is indeed the case.

Ex. 4.6 Observe that the degenerate (trigonometric) limit of the eight-vertex Bethe ansatz equations is almost the same as that for the six-vertex model, with two major differences:

- the Bethe spin is 0, or 0 mod r if $\gamma = \pi/r$;
- the θ parameter need not vanish.

To introduce the θ parameter into the six-vertex model in a natural way, modify the trace in the horizontal space as follows:

$$\mathrm{tr}'\,[T(u)] = \mathrm{tr}\left[T(u)\,\mathrm{e}^{i\pi m S^z/r}\right] \tag{4.61}$$

Can the new trace tr′ be interpreted as a Markov trace?

Ex. 4.7 [C. Fan and F.Y. Wu, 'General lattice model of phase transitions', *Phys. Rev.* **B2** (1970) 723.]

Given the most general eight-vertex $\mathscr{R}$-matrix

$$\mathscr{R} = \begin{pmatrix} a & 0 & 0 & d \\ 0 & b & c & 0 \\ 0 & c' & b' & 0 \\ d' & 0 & 0 & a' \end{pmatrix} \tag{4.62}$$

the free fermion eight-vertex solution is characterized by the Fan–Wu condition

$$aa' + bb' = cc' + dd' \tag{4.63}$$

(i) Check that the free fermion condition (4.63) is satisfied, in the symmetric case studied in this chapter, with the choice $\gamma = K(k)$.

(ii) Taking the logarithmic derivative of the free fermion eight-vertex model, we obtain the hamiltonian of the XY model,

$$H_{XY} = \sum_i J_x \sigma_i^x \sigma_{i+1}^x + J_y \sigma_i^y \sigma_{i+1}^y \tag{4.64}$$

Find the values of the coupling constants J_x and J_y.

(iii) Note that the free fermion condition (4.63) can be written as

$$\det \begin{pmatrix} a & d \\ d' & a' \end{pmatrix} = \det \begin{pmatrix} c & b \\ b' & b' \end{pmatrix} \tag{4.65}$$

Give an interpretation of why this condition amounts to freedom.

(iv) To understand the name "free fermion" of the condition (4.63), apply a Jordan–Wigner transformation

$$c_i^{x,y} = \sigma_i^{x,y} \left(\prod_{\ell < i} \sigma_\ell^z \right) \tag{4.66}$$

to the hamiltonian (4.64), and express it in terms of the operators $c_i^{x,y}$.

Ex. 4.8 [E.H. Lieb, D.C. Mattis and T.D. Schultz, *Ann. Phys.* **16** (1961) 407.]
To diagonalize the XY hamiltonian (4.64), define the operators

$$B_{x,y}(\lambda) = \sum_j \mathrm{e}(i\lambda)^{j-1} c_j^{x,y} \tag{4.67}$$

where $\mathrm{e}(u)$ is the elliptic exponential defined in appendix B. Check the equation

$$\left[H_{XY}, B_\pm(\lambda)\right] = \pm \Lambda(\lambda) B_\pm(\lambda) \tag{4.68}$$

where

$$\begin{aligned} B_\pm(\lambda) &= \left[D(\lambda)B_x(\lambda) \pm C(\lambda)B_y(\lambda)\right] / \sqrt{2} \\ D(\lambda) &= k' \, \frac{ik \,\mathrm{sn}\,\lambda \,\mathrm{cn}\,\lambda + \mathrm{dn}\,\lambda}{1 - k^2 \mathrm{sn}^2 \lambda} \\ C(\lambda) &= k' \, \frac{i \,\mathrm{sn}\,\lambda \,\mathrm{dn}\,\lambda - \mathrm{cn}\,\lambda}{1 - k^2 \mathrm{sn}^2 \lambda} \end{aligned} \tag{4.69}$$

and the eigenvalue $\Lambda(\lambda)$ is given by

$$\Lambda(\lambda) = \sqrt{(1+k^2)\cos^2 P(\lambda) + (1-k^2)\sin^2 P(\lambda)} \tag{4.70}$$

where we have introduced the logarithm of the elliptic exponential with the notation $\mathrm{e}(i\lambda) = \exp iP(\lambda)$.

Ex. 4.9 [V.V. Bazhanov and Yu.G. Stroganov, 'Hidden symmetry of free fermion model', *Teor. Mat. Fiz.* **62** (1985) 377; H.B. Thacker and H. Itoyama, 'Integrability and Virasoro symmetry of the non-critical Baxter–Ising model', *Nucl. Phys.* **B320** (1989) 541.]

There exist more general free fermion solutions than the one analyzed in the preceding two exercises, namely $\gamma = K$. Allowing for asymmetric weights, a free fermion solution to the eight-vertex model is given by

$$\begin{aligned} a &= 1 - \mathrm{e}(u)\,\mathrm{e}(\psi_1)\,\mathrm{e}(\psi_2) \quad , \qquad a' = \mathrm{e}(u) - \mathrm{e}(\psi_1)\,\mathrm{e}(\psi_2) \\ b &= \mathrm{e}(\psi_1) - \mathrm{e}(u)\,\mathrm{e}(\psi_2) \quad , \qquad b' = \mathrm{e}(\psi_2) - \mathrm{e}(u)\,\mathrm{e}(\psi_1) \\ c &= c' = \sqrt{\mathrm{e}(\psi_1)\mathrm{sn}(\psi_1)\,\mathrm{e}(\psi_2)\mathrm{sn}(\psi_2)}\frac{1 - \mathrm{e}(u)}{\mathrm{sn}\frac{u}{2}} \\ d &= d' = -ik\sqrt{\mathrm{e}(\psi_1)\mathrm{sn}(\psi_1)\,\mathrm{e}(\psi_2)\mathrm{sn}(\psi_2)}\,(1 + \mathrm{e}(u))\,\mathrm{sn}\frac{u}{2} \end{aligned} \tag{4.71}$$

where the elliptic exponential is a linear combination of Jacobi elliptic functions, $\mathrm{e}(x) = \mathrm{cn}\,x + i\,\mathrm{sn}\,x$ (see appendix B), and all the functions share the same elliptic modulus k. The R-matrix defined by these weights depends on the usual spectral parameter u, living on a torus, and two "external rapidities" ψ_1 and ψ_2, associated with the external incoming (or outgoing) lines. The Yang–Baxter equation satisfied by this R-matrix is of a generalized type:

$$\begin{aligned} &\mathscr{R}_{12}(u;\psi_1,\psi_2)\mathscr{R}_{13}(u+v;\psi_1,\psi_3)\,\mathscr{R}_{23}(v;\psi_2,\psi_3) \\ &\qquad = \mathscr{R}_{23}(v;\psi_2,\psi_3)\,\mathscr{R}_{13}(u+v;\psi_1,\psi_3)\,\mathscr{R}_{12}(u;\psi_1,\psi_2) \end{aligned} \tag{4.72}$$

Show that the one-dimensional hamiltonian which this R-matrix gives rise to (with $\psi = \psi_1 = \psi_2$) is that of the XY model in the presence of an external field:

$$H = \sum_{j=1}^{L} (1+\xi)\sigma_j^x\sigma_{j+1}^x + (1-\xi)\sigma_j^y\sigma_{j+1}^y + h\left(\sigma_j^z + \sigma_{j+1}^z\right) \tag{4.73}$$

where

$$\begin{aligned} \xi &= \frac{2cd}{ab + a'b'} = k\,\mathrm{sn}\,\psi \\ h &= \frac{a^2 + b^2 - (a')^2 - (b')^2}{2\,(ab + a'b')} = \mathrm{cn}\,\psi \end{aligned} \tag{4.74}$$

Appendix B
Elliptic functions

Let τ be a complex number with positive imaginary part, $\operatorname{Im}\tau > 0$, and set $q = e^{i\pi\tau}$. (This q should not be confused with the deformation parameter of quantum groups.) We may consider the entire function of z

$$\theta_4(z) = \sum_{n\in\mathbb{Z}} (-1)^n q^{n^2} e^{2\pi i n z} \tag{B4.1}$$

which is periodic, of period 1, and quasi-periodic, of quasi-period τ:

$$\begin{aligned} \theta_4(z+1,q) &= \theta_4(z,q) \\ \theta_4(z+\tau,q) &= -q^{-1} e^{-2\pi i z} \theta_4(z,q) \end{aligned} \tag{B4.2}$$

This function can also be expressed as an infinite product:

$$\theta_4(z,q) = \prod_{n=1}^{\infty} \left(1 - 2q^{2n-1}\cos 2u + q^{4n-2}\right)\left(1 - q^{2n}\right) \tag{B4.3}$$

It is convenient to give names to θ_4 with shifted arguments:

$$\begin{aligned} \theta_1(z,q) &= -iq^{1/4} e^{i\pi z} \theta_4\left(z + \frac{\tau}{2}, q\right) \\ \theta_2(z,q) &= q^{1/4} e^{i\pi z} \theta_4\left(z + \frac{1}{2} + \frac{\tau}{2}, q\right) \\ \theta_3(z,q) &= \theta_4\left(z + \frac{1}{2}, q\right) \end{aligned} \tag{B4.4}$$

Where confusion cannot arise, we omit the second argument of all theta functions, $\theta_i(z) = \theta_i(z,q)$ $(i = 1,\ldots,4)$.

Whereas $\theta_1(z)$ is odd and has zeroes at $m + n\tau$ $(m,n \in \mathbb{Z})$, the functions $\theta_2(z)$, $\theta_3(z)$ and $\theta_4(z)$ are even.

Many surprising relationships between elliptic functions and number theory exist; they serve to illustrate the fundamental nature of these functions and their properties. In particular, the theta functions θ_1 and θ_4 satisfy the following functional relations:

$$\begin{aligned}
&\theta_1(u+x)\theta_1(u-x)\theta_1(v+y)\theta_1(v-y)\\
&\qquad -\theta_1(u+y)\theta_1(u-y)\theta_1(v+x)\theta_1(v-x)\\
&\qquad\quad =\theta_1(u+v)\theta_1(u-v)\theta_1(x+y)\theta_1(x-y)\\
&\theta_4(u+x)\theta_4(u-x)\theta_4(v+y)\theta_4(v-y)\\
&\qquad -\theta_4(u+y)\theta_4(u-y)\theta_4(v+x)\theta_4(v-x)\\
&\qquad\quad =-\theta_1(u+v)\theta_1(u-v)\theta_1(x+y)\theta_1(x-y)\\
&\theta_4(u+x)\theta_4(u-x)\theta_1(v+y)\theta_1(v-y)\\
&\qquad -\theta_4(u+y)\theta_4(u-y)\theta_1(v+x)\theta_1(v-x)\\
&\qquad\quad =\theta_4(u+v)\theta_4(u-v)\theta_1(x+y)\theta_1(x-y)
\end{aligned} \tag{B4.5}$$

The Jacobi theta functions used in the text are related to the ones above by the following:

$$H(u)=\theta_1\left(\frac{u}{2K},q\right), \qquad \Theta(u)=\theta_4\left(\frac{u}{2K},q\right) \tag{B4.6}$$

where K is the complete elliptic integral of the first kind of modulus k:

$$K=K(k)=\int_0^{\pi/2}\frac{d\phi}{\sqrt{1-k^2\sin^2\phi}}=\frac{\pi}{2}\theta_3^2(0) \tag{B4.7}$$

and

$$k=\frac{\theta_2^2(0)}{\theta_3^2(0)} \tag{B4.8}$$

It is convenient to also introduce k', the supplementary modulus of k:

$$k^2+k'^2=1 \qquad \Rightarrow \qquad k'=\frac{\theta_4^2(0)}{\theta_3^2(0)} \tag{B4.9}$$

and

$$K'=K'(k)=K(k')=\int_0^{\pi/2}\frac{d\phi}{\sqrt{1-k'^2\sin^2\phi}}=\frac{-i\pi\tau}{2}\theta_3^2(0) \tag{B4.10}$$

The complete integrals K and K' are related to q by

$$q=\exp(i\pi\tau)=\exp\left(-\pi\frac{K'}{K}\right) \tag{B4.11}$$

The Jacobi theta functions $H(u)$ and $\Theta(u)$ behave nicely under reflection and shifts in the lattice:

$$
\begin{aligned}
H(-u) &= -H(u)\,, \qquad\qquad H(u+2K) = -H(u) \\
\Theta(-u) &= \Theta(u)\,, \qquad\qquad \Theta(u+2K) = \Theta(u) \\
H(u+iK') &= iq^{-1/4}\exp\left(\frac{-i\pi u}{2K}\right)\Theta(u) \\
\Theta(u+iK') &= iq^{-1/4}\exp\left(\frac{-i\pi u}{2K}\right)H(u)
\end{aligned}
\tag{B4.12}
$$

Owing to these properties, $K(k)$ and $K'(k)$ are called the real and imaginary quarter periods of the elliptic functions of modulus k.

The addition theorems analogous to (B4.5) are

$$
\begin{aligned}
&H(u)H(v)H(w)H(u+v+w) + \Theta(u)\Theta(v)\Theta(w)\Theta(u+v+w) \\
&\qquad = \Theta(0)\Theta(u+v)\Theta(u+w)\Theta(v+w) \\
&H(u)H(v)\Theta(w)\Theta(u+v+w) + \Theta(u)\Theta(v)H(w)H(u+v+w) \\
&\qquad = \Theta(0)\Theta(u+v)H(u+w)H(v+w)
\end{aligned}
\tag{B4.13}
$$

and

$$
H(u-v)\Theta(u+v) - \Theta(u-v)H(u+v) = 2\frac{g(u-K)g(v)}{g(K)} \tag{B4.14}
$$

where

$$
g(u) = H(u)\Theta(u) \tag{B4.15}
$$

The doubly periodic Jacobi elliptic functions are also interesting. Starting from the incomplete elliptic integral of the first kind

$$
u = \int_0^{\xi} \frac{d\phi}{\sqrt{1-k^2\sin^2\phi}} \tag{B4.16}
$$

we define

$$
\begin{aligned}
\operatorname{sn} u &= \operatorname{sn}(u,k) = \sin\xi \\
\operatorname{cn} u &= \operatorname{cn}(u,k) = \cos\xi \\
\operatorname{dn} u &= \operatorname{cn}(u,k) = \sqrt{1-k^2\operatorname{sn}^2 u}
\end{aligned}
\tag{B4.17}
$$

They are related to each other

$$
\operatorname{sn}^2 u + \operatorname{cn}^2 u = 1, \qquad k^2\operatorname{sn}^2 u + \operatorname{dn}^2 u = 1 \tag{B4.18}
$$

and also to the theta functions above:

$$
\begin{aligned}
\operatorname{sn} u &= \frac{1}{\sqrt{k}}\frac{H(u)}{\Theta(u)} = \frac{1}{\sqrt{k}}\frac{\theta_1(v)}{\theta_4(v)} \\
\operatorname{cn} u &= \sqrt{\frac{k'}{k}}\frac{\theta_2(v)}{\theta_4(v)} \\
\operatorname{dn} u &= \sqrt{k'}\frac{\theta_3(v)}{\theta_4(v)}
\end{aligned}
\tag{B4.19}
$$

where $v = u/(2K)$.

The functions cn and dn are even, but sn is odd, and also

$$\begin{aligned} \operatorname{sn}(2K-u) &= \operatorname{sn} u \qquad & \operatorname{sn}(2iK'-u) &= -\operatorname{sn} u \\ \operatorname{cn}(2K-u) &= -\operatorname{cn} u \qquad & \operatorname{cn}(2iK'-u) &= -\operatorname{cn} u \\ \operatorname{dn}(2K-u) &= \operatorname{dn} u \qquad & \operatorname{dn}(2iK'-u) &= -\operatorname{dn} u \end{aligned} \tag{B4.20}$$

The elementary periods of sn are $4K$ and $2iK'$; of cn they are $4K$ and $2K+2iK'$; and of dn they are $2K$ and $4iK'$.

Their addition theorems read as follows:

$$\begin{aligned} \operatorname{sn}(u+v) &= \frac{\operatorname{sn} u \operatorname{cn} v \operatorname{dn} v + \operatorname{cn} u \operatorname{dn} u \operatorname{sn} v}{1-k^2 \operatorname{sn}^2 u \operatorname{sn}^2 v} \\ \operatorname{cn}(u+v) &= \frac{\operatorname{cn} u \operatorname{cn} v - \operatorname{sn} u \operatorname{dn} u \operatorname{sn} v \operatorname{dn} v}{1-k^2 \operatorname{sn}^2 u \operatorname{sn}^2 v} \\ \operatorname{dn}(u+v) &= \frac{\operatorname{dn} u \operatorname{dn} v - k^2 \operatorname{sn} u \operatorname{cn} u \operatorname{sn} v \operatorname{cn} v}{1-k^2 \operatorname{sn}^2 u \operatorname{sn}^2 v} \end{aligned} \tag{B4.21}$$

In the limit $k \to 0$ ($k' \to 1$), the elliptic functions become trigonometric:

$$\begin{aligned} &K \to \frac{\pi}{2}, \qquad && K' \to \infty \\ &\frac{1}{\sqrt{k}} H(u) \to \sin u, \qquad && \Theta(u) \to 1 \\ &\operatorname{sn} u \to \sin u, \qquad \operatorname{cn} u \to \cos u, \qquad && \operatorname{dn} u \to 1 \end{aligned} \tag{B4.22}$$

The elliptic exponential $\mathrm{e}(u) = \operatorname{cn}(u) + i \operatorname{sn}(u)$ is an intriguing object; its derivative is just $\operatorname{dn}(u)\,\mathrm{e}(u)$, and $\mathrm{e}(-u) = \mathrm{e}(u)^{-1}$.

Exercises

Ex. B4.1 Another parametrization of the eight-vertex Boltzmann weights is given by

$$a(u) = \operatorname{sn}(u+\gamma), \quad b(u) = \operatorname{sn} u, \quad c(u) = \operatorname{sn} \gamma, \quad d(u) = k\,abc \tag{B4.23}$$

As an exercise, the reader might want to relate this parametrization to the one in equation (4.11) and may find, in particular, that

$$J_x = 1 + k \operatorname{sn}^2\gamma, \quad J_y = 1 - k \operatorname{sn}^2\gamma, \quad J_z = \operatorname{cn}\gamma \operatorname{dn}\gamma \tag{B4.24}$$

Ex. B4.2 It is often useful to use the notation $\mathrm{pq}(u) = \mathrm{pn}(u)/\mathrm{qn}(u)$ with $\mathrm{p,q} \in \{\mathrm{n, s, c, d}\}$ and the convention nn=1. A third parametrization of the eight-vertex Boltzmann weights is given by

$$a = 1 + \alpha_3, \quad b = 1 - \alpha_3, \quad c = \alpha_1 + \alpha_2, \quad d = \alpha_1 - \alpha_2 \tag{B4.25}$$

with

$$\alpha_i = \frac{\omega_i(u+\tilde{\gamma})}{\omega_i(\tilde{\gamma})} \qquad i = 1, 2, 3 \tag{B4.26}$$

so that the *R*-matrix can be written as

$$\mathscr{R}^{8v}(u) = \mathbf{1} \otimes \mathbf{1} + \sum_{i=1}^{3} \alpha_i \, \sigma^i \otimes \sigma^i \tag{B4.27}$$

which reduces, in the trigonometric limit, to the six-vertex parametrization (3.96). Verify that the choice

$$\omega_1(u) = \mathrm{ns}\,(u,k)\,, \quad \omega_2(u) = \mathrm{ds}\,(u,k)\,, \quad \omega_3(u) = \mathrm{cs}\,(u,k) \tag{B4.28}$$

yields a solution of the Yang–Baxter equation. From the definition of the Jacobi elliptic functions (B4.16) and (B4.17), find their derivatives and thus compute the hamiltonian associated with the *R*-matrix (B4.28). Check that $\Gamma = \mathrm{dn}\,\tilde{\gamma}$, $\Delta = \mathrm{cn}\,\tilde{\gamma}$, and find the change of basis from this parametrization to that of the previous exercise. We use this parametrization in the short discussion of the Sklyanin algebra in appendix C.

Appendix C
Sklyanin algebra

There exists an associative algebra $\mathscr{S}$, associated with the eight-vertex model, of considerable interest. It has four generators S_μ ($\mu = 0,\ldots,3$) satisfying the relations

$$\begin{aligned} [S_i, S_0] &= -iJ_{jk}\{S_j, S_k\} \\ [S_i, S_j] &= i\{S_0, S_k\} \end{aligned} \qquad (ijk) = (123) \text{ and cyclic} \tag{C4.1}$$

Properly speaking, the Sklyanin algebra $\mathscr{S}$ is the quotient $\mathscr{F}/\mathscr{I}$, where $\mathscr{F}$ is the free associative algebra generated by S_μ and $\mathscr{I}$ is the ideal defined by the relations (C4.1).

The "structure constants" J_{jk} are given in terms of the elliptic functions α_i appearing in the eight-vertex R-matrix (see exercise B4.2)

$$\mathscr{R}^{8v}(u) = \mathbf{1} \otimes \mathbf{1} + \sum_{i=1}^{3} \alpha_i(u)\, \sigma^i \otimes \sigma^i \tag{C4.2}$$

Explicitly,

$$J_{ij} = \frac{\alpha_i^2 - \alpha_j^2}{\alpha_k^2 - 1} \qquad (ijk) = (123) \text{ and cyclic} \tag{C4.3}$$

The defining relations (C4.1) are obtained from

$$\mathscr{R}(u-v)L_1(u)L_2(v) = L_2(v)L_1(u)\mathscr{R}(u-v) \tag{C4.4}$$

with

$$L_1(u) = L(u) \otimes \mathbf{1}, \qquad L_2(u) = \mathbf{1} \otimes L(u) \tag{C4.5}$$

and

$$L(u) = S_0 + \sum_{i=1}^{3} \alpha_i(u)\, S_i\, \sigma^i \tag{C4.6}$$

a 2×2 matrix in "auxiliary space", the entries of which are elements of the algebra $\mathscr{S}$:

$$L(u) = \begin{pmatrix} S_0 + \alpha_3(u)\, S_3 & \alpha_1(u)\, S_1 - i\alpha_2(u)\, S_2 \\ \alpha_1(u)\, S_1 + i\alpha_2(u)\, S_2 & S_0 - \alpha_3(u)\, S_3 \end{pmatrix} \tag{C4.7}$$

To derive (C4.1) from (C4.4), we need the identities

$$\begin{aligned} \alpha_i(u)\,\alpha_j(v) \;+\; \alpha_j(u-v)\,\alpha_k(u) \;-\; \alpha_i(u-v)\,\alpha_k(v) \\ = \alpha_i(v)\,\alpha_j(u)\,\alpha_k(u-v) \end{aligned} \tag{C4.8}$$

which reduce to the Yang–Baxter equation for the eight-vertex R-matrix (C4.3), and the identifications

$$J_{jk} = \frac{\alpha_j(u-v)\alpha_k(u)\alpha_j(v) - \alpha_k(u-v)\alpha_j(u)\alpha_k(v)}{\alpha_i(u) - \alpha_i(u-v)\alpha_i(v)} \tag{C4.9}$$

We will not repeat here the explicit expressions for $\alpha_i(u)$, equations (B4.26) and (B4.28). It is possible and useful to introduce the constants J_i ($i = 1,2,3$) such that

$$J_{ij} = \frac{-J_i + J_j}{J_k} \qquad (ijk) = (123) \text{ and cyclic} \tag{C4.10}$$

They are given by

$$J_1 = \frac{\theta_2(2\gamma)\theta_2(0)}{\theta_2^2(\gamma)}\,, \quad J_2 = \frac{\theta_3(2\gamma)\theta_3(0)}{\theta_3^2(\gamma)}\,, \quad J_3 = \frac{\theta_4(2\gamma)\theta_4(0)}{\theta_4^2(\gamma)} \tag{C4.11}$$

The algebra $\mathscr{S}$ is well defined even in the trigonometric limit, i.e. for the six-vertex model. The structure constants become in this extreme case

$$J_{12} = -J_{31} = \tan^2\gamma\,, \qquad J_{23} = 0 \tag{C4.12}$$

as can be checked by taking $k \to 0$ in (C4.3) or in (C4.11).

The two-dimensional representation of the Sklyanin algebra $\mathscr{S}$ is simply given by

$$S_0 = \mathbf{1}\,, \qquad S_i = \sigma^i \tag{C4.13}$$

whereas its three-dimensional representation is, in matrix notation, the following:

$$S_0 = \begin{pmatrix} J_3 & 0 & J_1 \\ 0 & J_1 + J_2 - J_3 & 0 \\ J_1 - J_2 & 0 & J_3 \end{pmatrix} \tag{C4.14a}$$

$$S_1 = \sqrt{2J_2J_3}\begin{pmatrix} 0 & 1 & 0 \\ 1 & 0 & 1 \\ 0 & 1 & 0 \end{pmatrix} \tag{C4.14b}$$

$$S_2 = \sqrt{2J_3J_1}\begin{pmatrix} 0 & -i & 0 \\ i & 0 & -i \\ 0 & i & 0 \end{pmatrix} \tag{C4.14c}$$

$$S_3 = 2\sqrt{J_1J_3}\begin{pmatrix} 1 & & \\ & 0 & \\ & & -1 \end{pmatrix} \tag{C4.14d}$$

Looking at the "one-point" transfer matrix for this two-dimensional representation

$L_{ij}(u) \;=$ 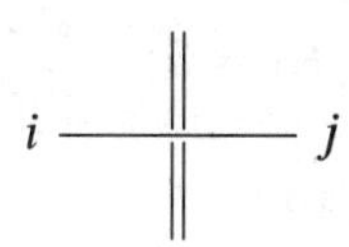 $\qquad i,j \in \{0,1\}$ (C4.15)

we note that the "reference state" $|\Uparrow\rangle$ does not satisfy the condition

$$L_{10}(u)\,|\Uparrow\rangle \;=\; 0 \tag{C4.16}$$

except in the six-vertex limit, $k \to 0$.

The classical version of $\mathscr{S}$ reads as follows:

$$\{S_i, S_0\}_{\mathrm{PB}} = 2J_{jk}S_jS_k\,, \quad \{S_i, S_j\}_{\mathrm{PB}} = -2S_0S_k \tag{C4.17}$$

where the usual correspondence principle

$$[\,,\,] \sim -i\hbar\{\,,\,\}_{\mathrm{PB}}$$

between classical Poisson brackets and quantum commutators has been used.

Exercises

Ex. C4.1 Using Baxter's vertex–face map, let us define the following family of 2×2 matrices:

$$L_{k,\ell}(u) \;=$$

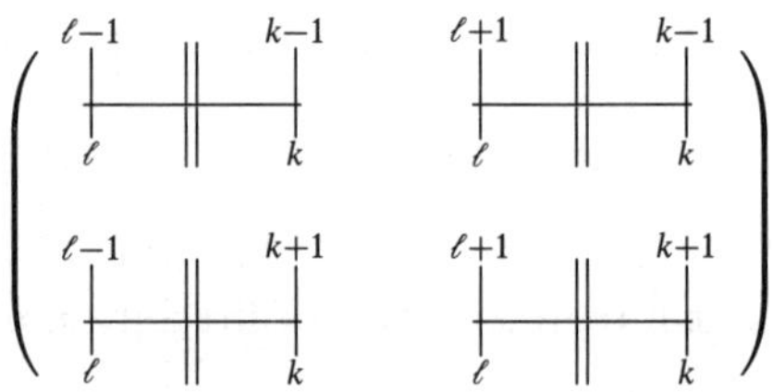

(C4.18)

with k and ℓ integers. Write down the equation equivalent to (C4.4) in this new basis.

Ex. C4.2 Prove that the following elements belong to the center of the Sklyanin algebra $\mathscr{S}$ (C4.1):

$$\begin{aligned} K_0 &= \sum_{\mu=0}^{3} S_\mu^2 \\ K_1 &= S_0^2 + \sum_{i=1}^{3} (1 - J_i)\, S_i^2 \\ D(u) &= S_0^2 - \sum_{i=1}^{3} \alpha_i(u)\alpha_i(u - 2\gamma) S_i^2 \end{aligned} \tag{C4.19}$$

Write $D(u)$ as a quantum determinant.

Ex. C4.3 Prove that the elements

$$K_0 = \sum_{i=1}^{3} S_i^2\,, \quad K_1 = S_0^2 + \sum_{i=1}^{3} J_i S_i^2 \tag{C4.20}$$

are central for the classical algebra (C4.17).

Ex. C4.4 Given the classical central elements (C4.20), define the spaces

$$\Gamma(k_0, k_1) = \left\{ \vec{S} = (S_0, S_1, S_2, S_3) \ / \ K_0(\vec{S}) = k_0, K_1(\vec{S}) = k_1 \right\} \qquad \text{(C4.21)}$$

Prove that

- if $k_1 \in (k_0 J_1, \infty)$, then the space $\Gamma(k_0, k_1)$ is homeomorphic to the sphere S^2;
- if $k_1 \in (k_0 J_3, k_0 J_2)$, then $\Gamma(k_0, k_1)$ is topologically equivalent to the disjoint union of two spheres, which become a torus for $J_2 = J_3$.

5

Face models

5.1 Weights and graphs: the definitions

In this chapter, we will study a class of integrable models different from vertex models, the so-called interaction round a face or (restricted) solid on solid models, face models for short. The lattice variables are assigned to the lattice sites instead of to the edges. The dynamical content of the model is embodied in the Boltzmann weights of the plaquettes, the graphical representation of which is the following (we use the notation B for face Boltzmann weights and reserve W for the vertex Boltzmann weights for extra clarity):

$$B\left(\begin{array}{cc|c} \ell_4 & \ell_3 & \\ \ell_1 & \ell_2 & u \end{array}\right) = \begin{array}{ccc} \ell_4 \times & \text{——} & \times \ell_3 \\ | & & | \\ \ell_1 \times & \text{——} & \times \ell_2 \end{array} \tag{5.1}$$

The data we need in order to define the model are the geometry of the lattice (square and regular unless otherwise stated), the set of lattice variables and the Boltzmann weights of the plaquettes, with arbitrary distribution of lattice variables on the corners.

Integrability of the model is characterized, just as for vertex models, by a set of equations on the Boltzmann weights. These equations are the necessary and sufficient conditions to have a family of commuting transfer matrices or, equivalently, an infinite number of integrals of motion, i.e. conserved quantities.

The characteristic feature of the models that we will work out in this section is that they satisfy some selection rules, meaning that the Boltzmann weights for some configurations of lattice variables vanish. The reason for restricting our attention to such models is that they are connected in a very natural way to rational conformal field theories, the subject of later chapters.

Let us denote by ℓ the lattice variables, which may take values from 1 to n. The selection rules we have just mentioned can be described, quite generally, by an $n \times n$ matrix Λ, the elements of which are all either 0 or 1. If $\Lambda_{\ell\ell'} = 0$, then the lattice variables ℓ and ℓ' cannot be assigned to adjacent sites of the lattice. The matrix Λ characterizes the set of all possible configurations contributing to the partition function of the model. Configurations not derived from Λ necessarily have zero Boltzmann weight.

The matrix Λ is called the incidence matrix of the face model, and we can use it to define a graph whose vertices are associated to the lattice variables

$\ell \in \{1, \ldots, n\}$, with r bonds between the vertices of ℓ and ℓ' iff $\Lambda_{\ell\ell'} = r$. We shall often refer to the model by its graph, much as a semisimple Lie algebra is referred to by its Dynkin diagram. Particularly interesting "graph" face models are those defined on Coxeter graphs. For example, the graph associated to A_5 is

$$\begin{array}{ccccccccc} \bullet & - & \bullet & - & \bullet & - & \bullet & - & \bullet \\ 1 & & 2 & & 3 & & 4 & & 5 \end{array} \tag{5.2}$$

In order to define the partition function of these models, we shall use, once again, the transfer matrix formalism. Denoting by $|\ell\rangle = |\ell_1 \ldots \ell_L\rangle$ the "fixed time" states of the system, the transfer matrix $t_{\vec{\ell}\vec{\ell}'}$ is the amplitude for state $|\ell\rangle$ to become $|\ell'\rangle$ in one time step. Explicitly,

$$t_{\vec{\ell}\vec{\ell}'} = \left\langle \vec{\ell}' \middle| t \middle| \vec{\ell} \right\rangle = B\begin{pmatrix} \ell_1' & \ell_2' \\ \ell_1 & \ell_2 \end{pmatrix} \cdots B\begin{pmatrix} \ell_j' & \ell_{j+1}' \\ \ell_j & \ell_{j+1} \end{pmatrix} \cdots B\begin{pmatrix} \ell_L' & \ell_1' \\ \ell_L & \ell_1 \end{pmatrix} \tag{5.3}$$

where we assume periodic boundary conditions, i.e. $\ell_{L+1} = \ell_1$. In diagrammatic form,

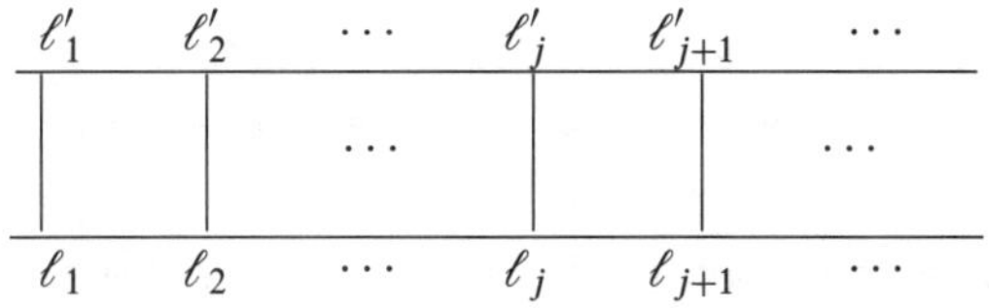

(5.4)

To be precise, the time evolution we are considering here takes us from one row in the square lattice to the next: the transfer matrix $t_{\vec{\ell}\vec{\ell}'}$ is thus known as the "row to row transfer matrix", to be contrasted with the "diagonal transfer matrix", for which time flows in the north-easterly direction.

We wish to find the conditions imposed on the Boltzmann weights by the commutativity of the transfer matrices, i.e. by the integrability condition

$$(t\,t')_{\vec{\ell}\vec{\ell}'} = (t'\,t)_{\vec{\ell}\vec{\ell}'} \tag{5.5}$$

where t' is a new transfer matrix defined by different Boltzmann weights $B'\begin{pmatrix} \ell_i & \ell_j \\ \ell_\ell & \ell_k \end{pmatrix}$. Following exactly the same procedure as for vertex models, we find the face Yang–Baxter equation

$$\begin{aligned} &\sum_{\ell_0} B\left(\begin{matrix} \ell_1 & \ell_0 \\ \ell_2 & \ell_3 \end{matrix}\middle| u\right) B\left(\begin{matrix} \ell_0 & \ell_5 \\ \ell_3 & \ell_4 \end{matrix}\middle| u+v\right) B\left(\begin{matrix} \ell_1 & \ell_6 \\ \ell_0 & \ell_5 \end{matrix}\middle| v\right) \\ &= \sum_{\ell_0} B\left(\begin{matrix} \ell_2 & \ell_0 \\ \ell_3 & \ell_4 \end{matrix}\middle| v\right) B\left(\begin{matrix} \ell_1 & \ell_6 \\ \ell_2 & \ell_0 \end{matrix}\middle| u+v\right) B\left(\begin{matrix} \ell_6 & \ell_5 \\ \ell_0 & \ell_4 \end{matrix}\middle| u\right) \end{aligned} \tag{5.6}$$

A graphical representation of this equation is shown in figure 5.1.

The alert reader will notice that this is not the first time we encounter the Yang–Baxter equation for face models: we have already met it in an altogether

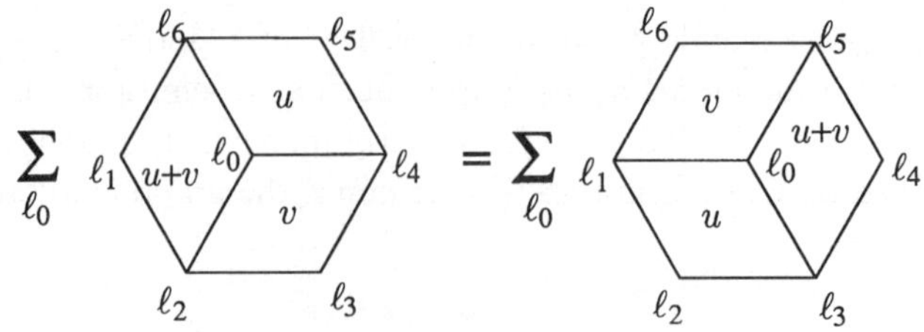

Fig. 5.1 Yang–Baxter equation (5.6) for face models.

different context in chapter 4, where we considered the algebraic Bethe ansatz for the eight-vertex model. Observe that the B coefficients introduced in equation (4.18) to define a basis on which the eight-vertex R-matrix would have a six-vertex texture are precisely the Boltzmann weights of a face model. Moreover, the extra variables $\ell \in \mathbb{Z}$ appearing in equations (4.17) and (4.18) can be nicely interpreted as face variables. For a generic value of the anisotropy γ, the ℓ variables of the eight-vertex model may take any integer value. Thus they are associated with an infinite type A graph

$$\begin{array}{c} \cdots - \bullet - \bullet - \bullet - \bullet - \bullet - \bullet - \cdots \\ -2 \;\; -1 \;\;\; 0 \;\;\; 1 \;\;\; 2 \;\;\; 3 \end{array} \tag{5.7}$$

We may now interpret equation (4.22), which changed the texture of the eight-vertex R-matrix, as defining a map from the eight-vertex model into the face model defined by the unrestricted graph (5.7) (recall equations (4.23)). We have anticipated this map by using the same notation for both the Boltzmann weights of face models and for the coefficients in the clever change of basis of the eight-vertex model which allowed us to solve it.

In the next section, and quite independently of the above discussion, we shall consider a general method for deriving trigonometric solutions for graph face models when the graph is of Coxeter type. The elliptic solutions already found in chapter 4 will be studied in section 5.3.

In chapter 1, we saw that the same Yang–Baxter equation describes the integrability of both factorizable S-matrix theories and vertex models. In the same spirit, we might wonder whether the face Yang–Baxter equation (5.6) allows also an interpretation in terms of scattering matrices of some particle theory. This is indeed the case for a theory with solitons (or kinks) connecting different vacua. Let us denote by $K_{ab}(\theta)$ a kink or soliton, with rapidity θ, which interpolates from the vacuum a at $x = -\infty$ to the vacuum b at $x = +\infty$. If there are N vacua, the pairs (a, b) of vacua which can be connected by a kink define an $N \times N$ incidence matrix Λ_{ab} and thus a graph face model, as we shall discuss in the next section. The elastic scattering of two solitons then defines an S-matrix of face type:

$$K_{ab}(\theta_1) K_{bc}(\theta_2) = \sum_d S\begin{pmatrix} a & d \\ b & c \end{pmatrix}\theta_1 - \theta_2\Big) \; K_{ad}(\theta_2) K_{dc}(\theta_1) \tag{5.8}$$

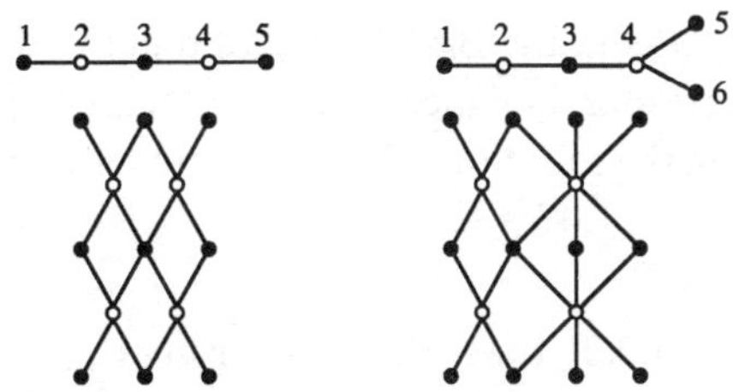

Fig. 5.2 Two examples of Bratelli diagrams.

We leave it as an exercise to check that the factorization equation for this S-matrix is precisely the face Yang–Baxter equation (5.6).

5.2 Trigonometric solutions

This section is devoted to developing a method for finding trigonometric solutions to the face Yang–Baxter equation (5.6) for models defined on a graph.

5.2.1 Bratelli diagrams

Given a graph with incidence matrix Λ, we define its associated Bratelli diagram as follows. First, split the vertices of the graph into two subsets, say black and white (or even and odd), in such a way that like-colored vertices are not connected by a bond. It is not hard to find graphs which do not admit such coloring: for example, any graph with a loop from one vertex to itself. In what follows, we shall consider bi-colored graphs only. Once all the vertices in the graph are consistently colored in black and white, deform the graph so that all the black vertices lie in a row, above the row of all the white vertices. The bonds link vertices in one row to the next. The Bratelli diagram is obtained by repeating this pattern downwards (figure 5.2). We shall refer to the rows in this diagram as levels.

Next, let us define paths on the Bratelli diagram. In order to do this, we add to the diagram the "point at infinity" $(*)$ connected to all points in the first row. A path is a set of points $\{\sigma(i)\}$ on the diagram such that $\sigma(0) = (*)$ and $\Lambda_{\sigma(i)\sigma(i+1)} = 1$, for any level i. Figure 5.3 illustrates this construction.

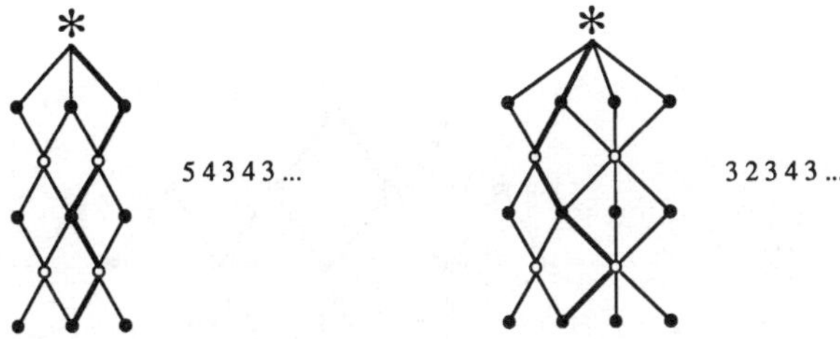

Fig. 5.3 Example of paths in the diagrams of figure 5.2 and the associated states.

It is easy to see that paths in the Bratelli diagram are in direct correspondence with the states of the corresponding face model. In fact, any path is a distribution of site variables satisfying the selection rules of the model encoded in the incidence matrix Λ.

5.2.2 Yang–Baxter operators

We define the face Yang–Baxter operators as

$$X_j(u)\,|\ell_1\ell_2\ldots\ell_j\ldots\rangle = \sum_{\ell'_j} B\left(\begin{matrix}\ell_{j-1} & \ell'_j\\ \ell_j & \ell_{j+1}\end{matrix}\,\middle|\, u\right)|\ell_1\ell_2\ldots\ell_{j-1}\ell'_j\ell_{j+1}\ldots\rangle \tag{5.9}$$

Now, the direction of time is at 45° to the lattice, so these operators act on "diagonal" states. The associated transfer matrix is called the diagonal to diagonal (as opposed to row to row) transfer matrix. The Yang–Baxter operators act on one such "diagonal state" by adding a face weight (a plaquette) on one site, and the identity elsewhere (see figure 5.4) – after X_j acts on a state, it yields a superposition of states with ℓ'_j instead of ℓ_j, weighed by the appropriate Boltzmann weights.

The Yang–Baxter equation for the diagonal to diagonal transfer matrix implies

$$\begin{aligned} X_i(u)X_{i+1}(u+v)X_i(v) &= X_{i+1}(v)X_i(u+v)X_{i+1}(u) \\ X_i(u)X_j(v) &= X_j(v)X_i(u) \qquad |i-j|\geq 2 \end{aligned} \tag{5.10}$$

These Yang–Baxter operators $X_i(u)$ define a non-trivial representation ϕ^{face} of the braid group (2.68):

$$\phi^{\text{face}}(\sigma_i) = \lim_{u\to\infty}\frac{X_i(u)}{\rho(u)} \tag{5.11}$$

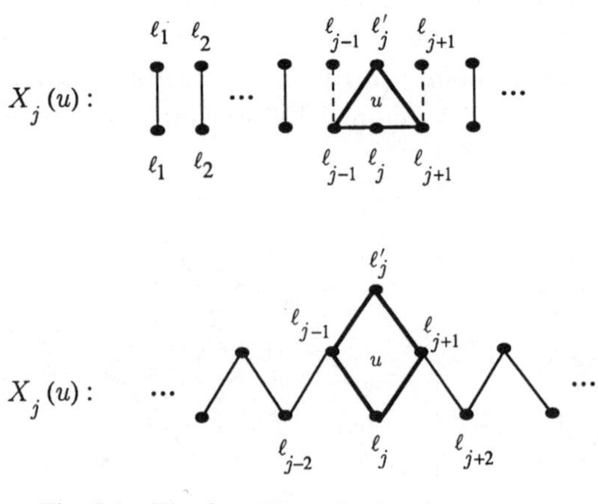

Fig. 5.4 The face Yang–Baxter operator.

where the normalization or unitarity function $\rho(u)$ satisfies

$$\sum_{\ell'_i} B\left(\begin{matrix}\ell_{i-1} & \ell''_i \\ \ell'_i & \ell_{i+1}\end{matrix}\middle| u\right) B\left(\begin{matrix}\ell_{i-1} & \ell'_i \\ \ell_i & \ell_{i+1}\end{matrix}\middle| -u\right) = \rho(u)\rho(-u)\ \delta(\ell_i, \ell''_i) \tag{5.12}$$

which is a necessary condition for the consistency of the representation, namely for having

$$\phi^{\text{face}}\left(\sigma_i \cdot \sigma_i^{-1}\right) = \phi^{\text{face}}(\sigma_i)\, \phi^{\text{face}}(\sigma_i^{-1}) = \mathbf{1}$$

We shall find an explicit expression for the Yang–Baxter operators $X_i(u)$ of face models shortly, but first we need some additional mathematical machinery.

5.2.3 The Temperley–Lieb–Jones algebra

The braid group B_n is closely connected to the Temperley–Lieb–Jones algebra $A_{n,\beta}$, which is a $\mathbb{C}^*$-algebra generated by n elements $\{e_0 = \mathbf{1}, e_1, \ldots, e_{n-1}\}$ satisfying the relations

$$e_i e_{i\pm 1} e_i = e_i \tag{5.13a}$$

$$e_i e_j = e_j e_i \quad ; \quad |j-i| \geq 2 \tag{5.13b}$$

$$e_i^2 = \sqrt{\beta} e_i \tag{5.13c}$$

$$e_i^* = e_i \tag{5.13d}$$

where β is a free parameter. The algebra $A_{n,\beta}$ is also equipped with a Markov trace:

$$\mathrm{tr}\,(x e_n) = \frac{1}{\sqrt{\beta}} \mathrm{tr}\,(x) \tag{5.14}$$

where x is any word of $\{e_i, i < n\}$.

The Temperley–Lieb–Jones algebra $A_{n,\beta}$ is intimately connected with the Hecke algebra $H_n(q)$, the generators of which, g_i $(1 \leq i \leq n-1)$, satisfy:

$$g_i g_{i+1} g_i = g_{i+1} g_i g_{i+1} \tag{5.15a}$$

$$g_i g_j = g_j g_i \ ; \quad |j-i| \geq 2 \tag{5.15b}$$

$$g_i^2 = (q^2 - 1)\, g_i + q^2 \tag{5.15c}$$

The map from Hecke to Temperley–Lieb–Jones is

$$\begin{aligned} \pi : H_n(q) &\to A_{n,\beta} \\ g_i &\mapsto q e_i - \mathbf{1} \end{aligned} \tag{5.16}$$

with the identification

$$\beta = 2 + q^2 + q^{-2} = [2]_q^2 = \left(\frac{q^2 - q^{-2}}{q - q^{-1}}\right)^2 \tag{5.17}$$

or equivalently

$$\sqrt{\beta} = q + q^{-1} \tag{5.18}$$

Indeed, equation (5.13c) is equivalent to (5.15c), and (5.13a) implies for the Hecke generators the equation

$$g_i g_{i\pm 1} g_i + g_i g_{i\pm 1} + g_{i\pm 1} g_i + g_i + g_{i\pm 1} + \mathbf{1} = 0 \tag{5.19}$$

which in turn implies (5.15a). Elements of $H_n(q)$ such as the one on the left-hand side of (5.19) belong to the kernel of π.

Note that the first two (q-independent) relations in (5.15) are just the defining properties of the braid group (2.68), whereas the last one implements a deformation or quantization of the permutation group S_n of n letters:

$$S_n = \lim_{q\to 1} H_n(q) \tag{5.20}$$

In chapter 6, we shall study in more detail the Hecke algebra in the general context of quantum groups.

Motivated by (5.16), we may propose the following natural ansatz for the Yang–Baxter operators $X_i(u)$:

$$X_i(u) = \mathbf{1} + f_\beta(u)\ e_i \tag{5.21}$$

where $f_\beta(u)$ depends on the parameter β in the defining relations for the Temperley–Lieb–Jones algebra (5.13). The conditions imposed by (5.10) reduce to the following equation on the functions $f_\beta(u)$:

$$f_\beta(u) + f_\beta(v) + \sqrt{\beta} f_\beta(u) f_\beta(v) + f_\beta(u) f_\beta(v) f_\beta(u+v) = f_\beta(u+v) \tag{5.22}$$

which have the trigonometric solution

$$f_\beta(u) = \frac{\sin(u)}{\sin(\gamma - u)} \tag{5.23}$$

where the parameter γ is directly related to β (or equivalently, q) by

$$\beta = 4\cos^2\gamma\ , \qquad q = e^{i\gamma} \tag{5.24}$$

Given the ansatz (5.21), a solution to the face Yang–Baxter equation is provided by any representation of the Temperley–Lieb–Jones algebra on the space of face states or, equivalently, representations of the Temperley–Lieb–Jones algebra on the space of paths of the Bratelli diagram associated with the face graph. To find these representations, we will use the general theory of towers of algebras, which are closely linked to type II_1 subfactors (see appendix E).

5.2.4 Towers of algebras associated with a graph

In this section, we shall construct the Temperley–Lieb–Jones algebra $A_{n,\beta}$ in the general framework of towers of algebras.

Consider a multi-matrix algebra $M^{(1)} = \bigoplus_{i=1}^m M^{(1)}_{\mu_i}$, where $M^{(1)}_{\mu_i}$ is the space of all $\mu_i \times \mu_i$ complex matrices. Let $M^{(0)} = \bigoplus_{i=1}^n M^{(0)}_{\nu_i}$ be a multi-matrix subalgebra of $M^{(1)}$ containing the identity. The inclusion $M^{(0)} \subset M^{(1)}$ is defined by an $n \times m$ matrix Λ such that

$$\mu_i = \sum_{j=1}^{n} v_j \Lambda_{ji} \tag{5.25}$$

Through this inclusion, an element $y = (y_1, \ldots, y_n) \in M^{(0)}$ is mapped to a bigger matrix $\pi(y) = (x_1, \ldots, x_m) \in M^{(1)}$, with

$$x_i = \mathrm{diag}\,(y_1, \ldots, y_1, \cdots, y_n, \ldots, y_n) \tag{5.26}$$

and y_j appearing Λ_{ji} times.

A third algebra, $M^{(2)} = \mathrm{End}_{M^{(0)}} M^{(1)}$, is naturally associated with the two algebras $M^{(0)} \subset M^{(1)}$. It consists of the endomorphisms of $M^{(1)}$ viewed as a right $M^{(0)}$ module. Thus, $M^{(2)}$ consists of the endomorphisms $\hat{a} : M^{(1)} \to M^{(1)}$ satisfying $\hat{a}(xy) = \hat{a}(x)y$ for any $x \in M^{(1)}$ and $y \in M^{(0)}$. The space $M^{(2)}$ is itself a multi-matrix algebra, $M^{(2)} = \bigoplus_{j=1}^{n} M^{(2)}_{\rho_j}$, with $\rho_j = \sum_{i=1}^{m} \mu_i \Lambda^T_{ij}$.

The procedure described above is called the fundamental construction. It can be repeated, whereby a fourth algebra $M^{(3)} = \mathrm{End}_{M^{(1)}} M^{(2)}$ is induced naturally from the pair $(M^{(1)}, M^{(2)})$, and so on by recursion. Thus, the original pair of algebras $M^{(0)} \subset M^{(1)}$ generates a semi-infinite tower of algebras

$$M^{(0)} \subset M^{(1)} \subset M^{(2)} \subset M^{(3)} \subset \cdots \tag{5.27}$$

The incidence or inclusion matrix for $M^{(2k)} \subset M^{(2k+1)}$ is Λ, whereas that for $M^{(2k+1)} \subset M^{(2k+2)}$ is its transpose Λ^T.

Given a pair of multi-matrix algebras $M^{(0)} \subset M^{(1)}$ with inclusion matrix Λ, we may associate with them a Bratelli diagram as follows. If $M^{(0)} = \bigoplus_{j=1}^{n} M^{(0)}_{v_j}$ and $M^{(1)} = \bigoplus_{i=1}^{m} M^{(1)}_{\mu_i}$ as above, we draw n black points and m white ones, and connect the jth black point to the ith white point whenever $\Lambda_{ij} \neq 0$. We then iterate, alternating white and black rows.

For example, if $M^{(0)} = \mathbb{C} \oplus M^{(0)}_2 \oplus \mathbb{C}$ and $M^{(1)} = \mathbb{C} \oplus M^{(1)}_3 \oplus M^{(1)}_2 \oplus M^{(1)}_3 \oplus \mathbb{C}$ with inclusion matrix

$$\Lambda = \begin{pmatrix} 1 & 1 & 0 & 0 & 0 \\ 0 & 1 & 1 & 1 & 0 \\ 0 & 0 & 0 & 1 & 1 \end{pmatrix} \tag{5.28}$$

then the Bratelli diagram is obtained by iterating downwards the graph shown in figure 5.5.

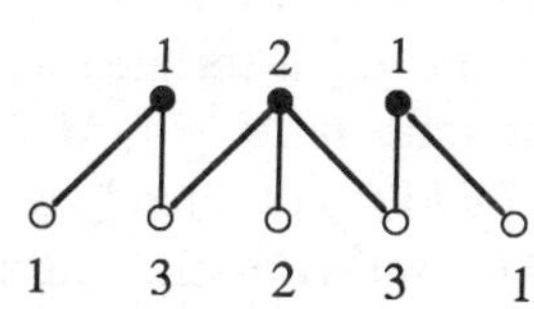

Fig. 5.5 Bratelli diagram associated with the incidence matrix (5.28).

The tower of algebras whose inclusions are described by a Bratelli diagram define a model for the diagram.

Let us consider the pair $M^{(0)} \subset M^{(1)}$ of multi-matrix algebras with inclusion matrix Λ, and let us define traces consistent with the inclusion. We may associate with any m-dimensional vector $(t_1^{(1)}, \ldots, t_m^{(1)})$ a trace in the multi-matrix algebra $M^{(1)}$:

$$\mathrm{tr}^{(1)}(x) = \sum_{i=1}^{m} t_i^{(1)} \mathrm{tr}(x_i) \qquad \forall x \in M^{(1)} \tag{5.29}$$

The trace tr is just the usual matrix trace. Similarly, any n-dimensional vector $(t_1^{(0)}, \ldots, t_n^{(0)})$ defines a trace $\mathrm{tr}^{(0)}$ in $M^{(0)}$. The traces $\mathrm{tr}^{(1)}$ and $\mathrm{tr}^{(0)}$ are consistent with the inclusion Λ iff

$$\sum_{i=1}^{m} \Lambda_{ji} t_i^{(1)} = t_j^{(0)} \tag{5.30}$$

which means simply

$$\mathrm{tr}^{(1)}(\pi(y)) = \mathrm{tr}^{(0)}(y) \qquad \forall y \in M^{(0)} \tag{5.31}$$

where π is the inclusion map (5.26). We shall call a trace $\mathrm{tr}^{(k)}$ faithful if all the components of the vector $t^{(k)}$ are non-zero.

Given a pair of algebras $M^{(0)} \subset M^{(1)}$, equipped with faithful and consistent traces in both algebras, then there exists a unique linear map $E : M^{(1)} \to M^{(0)}$ such that

(i) $\mathrm{tr}^{(0)}(E(x)) = \mathrm{tr}^{(1)}(x) \qquad x \in M^{(1)}$
(ii) $E(y_1 x y_2) = y_1 E(x) y_2 \qquad y_1, y_2 \in M^{(0)},\ x \in M^{(1)}$
(iii) $E(y) = y \qquad y \in M^{(0)}$

The map E is called the conditional expectation from $M^{(1)}$ to $M^{(0)}$. Note that E is the orthogonal projection on $M^{(0)}$ of the scalar product $\langle x, y\rangle = \mathrm{tr}^{(1)}(xy)$ in $M^{(1)}$. Since both the trace in $M^{(1)}$ and its restriction to $M^{(0)}$ are faithful, then $M^{(1)} = M^{(0)} \oplus M^{(0)\perp}$.

Property (*ii*) implies that $E \in \mathrm{End}_{M^{(0)}} M^{(1)}$, and thus E is an element of the algebra $M^{(2)}$ obtained from $M^{(0)}$ and $M^{(1)}$ by the recursive fundamental construction above. Moreover, it is easy to prove that $M^{(2)}$ is generated, as a vector space, by $M^{(1)}$ and by elements of the form xEy, with $x, y \in M^{(1)}$.

Having constructed the conditional expectation $E = E_1$ for the pair $M^{(0)} \subset M^{(1)}$, we would like to find the family of conditional expectations E_k for the whole tower of algebras, and to discover the relationships between these E_k. This is how we will obtain the algebra $A_{n,\beta}$ associated with the tower.

We first need to extend the trace in $M^{(1)}$ to $M^{(2)} \supset M^{(1)}$. We take advantage of the fact that any element of $M^{(2)}$ not already in $M^{(1)}$ can be written as xEy, with $x, y \in M^{(1)}$. A trace in $M^{(1)}$ satisfying the Markov property

$$\mathrm{tr}(xE) = \beta^{-1} \mathrm{tr}(x) \qquad x \in M^{(1)} \tag{5.32}$$

automatically induces a trace in $M^{(2)}$ that propagates to the whole tower. Let us thus assume that the trace $\mathrm{tr}^{(1)}$ satisfies the Markov property (5.32) for some β. Then for any element xEy in $M^{(2)}$ we define $\mathrm{tr}^{(2)}$ using (5.32), and we thus obtain the sought-after conditional expectation $E_2 : M^{(2)} \to M^{(1)}$. By recursion, a family of conditional expectations for the tower of algebras follows, $E_k : M^{(k)} \to M^{(k-1)}$, ($k \in \mathbb{N} - \{0\}$). These conditional expectations are the natural candidates expected to generate $A_{n,\beta}$.

Before studying the algebraic relationships among the conditional expectations of a tower of algebras, we should check under what conditions the traces at hand are indeed Markov, and for what values of β. Let us focus on the multi-matrix algebras $M^{(0)} \subset M^{(1)}$. If their Bratelli diagram is connected, then the existence of a trace with the Markov property is assured, and

$$\beta = ||\Lambda||^2 \tag{5.33}$$

where the norm for a generic matrix $X \in M_{m,n}(\mathbb{R})$, $X : \mathbb{R}^n \to \mathbb{R}^m$, is defined as

$$||X|| = \max\{||X\xi||,\ \xi \in \mathbb{R}^n,\ ||\xi|| \leq 1\} \tag{5.34}$$

The reader may check that, for the incidence matrix Λ of a Coxeter graph (see table 5.1), β is given by

$$\beta = 4\cos^2\frac{\pi}{h} \tag{5.35}$$

where h is the Coxeter number of the graph.

Let us return to the problem of determining the algebraic relations satisfied by the E_k. The crucial statement is the following: if the pair $M^{(0)} \subset M^{(1)}$ is equipped with a Markov trace, then the conditional expectations satisfy

$$\begin{aligned} &E_i^2 = E_i \\ &\beta E_i E_{i\pm 1} E_i = E_i \\ &E_i E_j = E_j E_i \qquad |i = j| \geq 2 \\ &\beta \,\mathrm{tr}\,(x E_i) = \mathrm{tr}\,(x) \qquad x \in M^{(i)} \end{aligned} \tag{5.36}$$

From these relations, it follows that the algebra generated by $\{\mathbf{1}, E_1, E_2, \ldots, E_{n-1}\}$ is isomorphic to $A_{n,\beta}$ defined in (5.13). The condition $E_i^2 = E_i$ is automatic, because the conditional expectations are projectors. The relationship between (5.13) and (5.36) is simply a scaling,

$$E_i = \frac{1}{\sqrt{\beta}}\, e_i \tag{5.37}$$

To prove (5.36), we need to make use of the Markov property of the trace. Consider the first relation with $i = 1$. The element $\beta E_1 E_2 E_1$ belongs to $M^{(3)}$, and thus acts on $M^{(2)}$ as an endomorphism. A generic element in $M^{(2)}$ is of the kind xE_1y ($x, y \in M^{(1)}$). Acting on it with $\beta E_1 E_2 E_1$, we find

$$\beta E_1 E_2 E_1[xE_1y] = \beta E_1 E_2[E_1(x)E_1y] = E_1[E_1(x)\beta E_2(E_1)y] \tag{5.38}$$

To compute $\beta E_2(E_1)$, observe first that, for any $x \in M^{(1)}$,

$$\operatorname{tr}(\beta E_1 x - x) = \beta \operatorname{tr}(E_1 x) - \operatorname{tr}(x) = \beta \operatorname{tr}(x E_1) - \operatorname{tr}(x) = 0 \tag{5.39}$$

where we have used the Markov property of the trace. If we define the scalar product with the help of the trace as above, we obtain the orthogonal decomposition of βE_1:

$$\beta E_1 = (\beta E_1 - \mathbf{1}) + \mathbf{1} \in M^{(1)\perp} \oplus M^{(1)} = M^{(2)} \tag{5.40}$$

Recalling now that E_2 is the orthogonal projection on $M^{(1)}$, we obtain

$$\beta E_2(E_1) = \mathbf{1} \tag{5.41}$$

Finally, from (5.38) and (5.41), we find

$$\beta E_1 E_2 E_1 [x E_1 y] = E_1[E_1(x) y] = E_1(x) E_1(y) = E_1(x E_1 y) \tag{5.42}$$

which concludes the proof of the first condition in (5.36). Performing similar manipulations, the other equalities can be proved without difficulty (exercise 5.13).

We have thus shown that the Temperly–Lieb–Jones algebra may be interpreted as the set of conditional expectations associated with a tower of algebras.

5.2.5 The algebra of face observables

In section 5.2.1, we mapped the states of a graph face model into the space of paths on the associated Bratelli diagram. Physically, the diagram can be interpreted as some sort of discretized spacetime, and the states (paths) as strings living in this spacetime. Now we would like to interpret the observables of the face model as elements of a tower of algebras with incidence matrix given by the face graph. To this end, we introduce Ocneanu's string operators.

A string operator at level n is a pair of paths $(\xi_+, \xi_-)^{(n)}$, both starting at the point at infinity $(*)$ and ending at the same point at level n (figure 5.6). A string operator is simply a closed path in the Bratelli diagram which contains the point at infinity. The string operator $(\xi_+, \xi_-)^{(n)}$ acts on paths η which coincide up to level n with ξ_+ by replacing that piece with ξ_- (see figure 5.7). More formally, any path η can be decomposed into the path $\eta^{n]}$ from $(*)$ to level n and the path $\eta^{[n}$ from level n downwards, $\eta = \eta^{n]} \circ \eta^{[n}$. Then the action of the string operator on a path is given by

$$(\xi_+, \xi_-)^{(n)} \eta = \delta(\eta^{n]}, \xi_+) \quad \xi_- \circ \eta^{[n} \tag{5.43}$$

The delta function of paths is either 1 if the paths coincide or 0 otherwise.

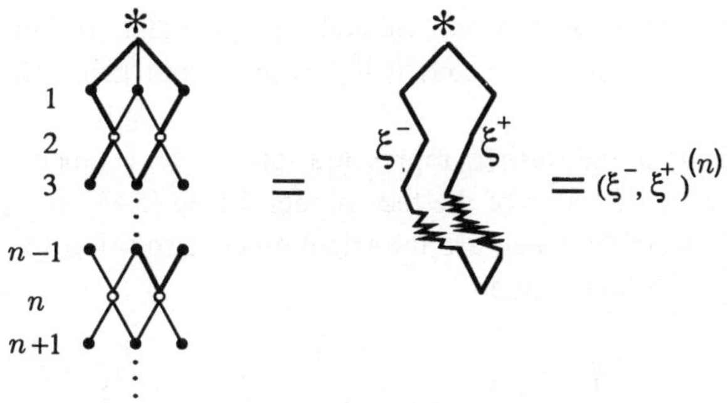

Fig. 5.6 A string operator of level n.

Consider now the operator algebra $A^{(n)}$ generated by all the string operators $(\xi_+, \xi_-)^{(n)}$. The tower of algebras $A^{(1)} \subset A^{(2)} \subset \ldots$ is a model of the Bratelli face diagram. From the inclusion $A^{(1)} \subset A^{(2)}$, we obtain the conditional expectation $E \in A^{(3)}$. Repeating this construction for the other inclusions $A^{(i)} \subset A^{(i+1)}$, we construct all the generators of the Temperely–Lieb–Jones algebra represented as string operators. Using the face variables, we obtain

$$e_j \left|\ell_1 \ldots \ell_j \ldots\right\rangle = \sum_{\ell'_j} \sqrt{\frac{t(\ell_j)t(\ell'_j)}{t(\ell_{j+1})t(\ell_{j-1})}}\ \delta(\ell_{j-1}\ell_{j+1})\ \left|\ell_1 \ldots \ell'_j \ldots\right\rangle \tag{5.44}$$

where the functional $t(\ell_j)$ of the lattice variables ℓ_i is determined by

$$\sum_j \Lambda_{ij} t(\ell_j) = \sqrt{\beta}\ t(\ell_i) \tag{5.45}$$

Equation (5.33) implies that $t(\ell_i)$ defined by (5.45) is the eigenvector of highest eigenvalue of the model's incidence matrix Λ. Given a finite matrix whose elements are all positive or zero, the Perron–Frobenius theorem guarantees that

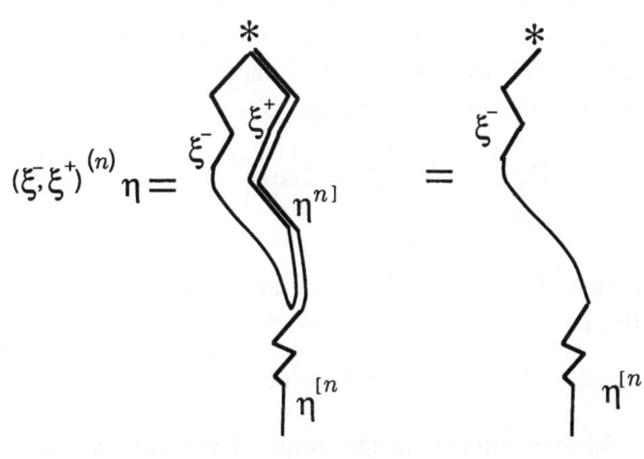

Fig. 5.7 The action (5.43) of a string operator on a path.

the components of its eigenvector of highest eigenvalue are all strictly positive. The difference between (5.30) and (5.45) comes from using the generator $e_i = \sqrt{\beta}\, E_i$.

Relation (5.44) is the desired representation of the Temperley–Lieb–Jones algebra in the space of states of the face model. Using (5.44) and (5.21), we obtain the Boltzmann weights which are the trigonometric solution to the Yang–Baxter equation for graph face models,

$$B\left(\begin{matrix} d & c \\ a & b \end{matrix} \middle| u \right) = \delta_{ac} + \delta_{bd} \frac{\sin u}{\sin(\gamma - u)} \sqrt{\frac{t(a)t(c)}{t(b)t(d)}} \tag{5.46}$$

where $t(x)$ is as in (5.45) and

$$\beta = ||\Lambda||^2 = 4\cos^2\gamma \tag{5.47}$$

Before studying some particular examples, let us summarize how we have derived equation (5.44):

(i) Given the bi-colored graph with incidence matrix Λ, we compute $\beta = ||\Lambda||^2$ and draw the associated Bratelli diagram.
(ii) The face states are interpreted as paths in this diagram.
(iii) The face observables span a model of the Bratelli diagram, i.e. a tower of algebras with inclusion matrix Λ.
(iv) The Temperley–Lieb–Jones algebra $A_{n,\beta}$ is realized as the algebra of conditional expectations.

5.2.6 The trigonometric solution for Coxeter models

We restrict ourselves now to the face graphs of type A, D and E shown in table 5.1. A classical theorem due to Kronecker says that if X is a finite matrix whose elements are integers, then either $||X|| = 2\cos(\pi/r)$ for some integer $r \geq 2$, or else $||X|| > 2$. Moreover, for any integer $r \geq 3$, there exists a $p \times p$ matrix X, all of whose elements are either 0 or 1, such that

$$||X|| = 2\cos\frac{\pi}{r} \tag{5.48}$$

with p the largest integer such that $p \leq r/2$.

For face models defined by Coxeter graphs, the highest eigenvalue of the incidence matrix is

$$\beta = 4\cos^2(\pi/h)$$

where h is the Coxeter number of the graph. The trigonometric solution to the Yang–Baxter equation for all these models is known. In fact, all we need is the

Table 5.1 Coexeter graphs.
Coxeter graphs are the Dynkin diagrams of simply laced semi-simple Lie algebras, with as many dots • as the rank r of the Lie algebra $\mathcal{G}_r$. The Coxeter number of the graph is denoted by h.

Name	Coxeter graph	h
$A_r \quad r \geq 1$		$r+1$
$D_r \quad r \geq 4$		$2r-2$
E_6	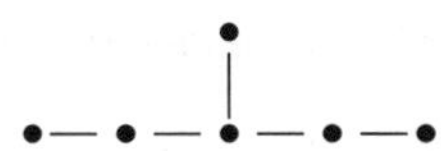	12
E_7	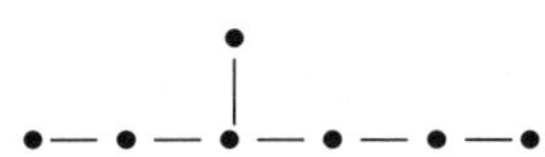	18
E_8		30

appropriate solution to (5.45). For A_r graphs, for instance, it is

$$t^{(r)}(\ell) = \sin\left(\frac{\ell\pi}{r+1}\right) \qquad \ell = 1, \ldots, r \tag{5.49}$$

For the graph A_r, the value of q is $q = \exp i\pi/(r+1)$, i.e. a root of unity, and equation (5.45) becomes

$$t^{(r)}(\ell+1) + t^{(r)}(\ell-1) = \sqrt{\beta}\, t^{(r)}(\ell) \tag{5.50}$$

Note that $t^{(r)}(0) = t^{(r)}(r+1) = 0$. For other cases, we refer the reader to exercise 5.6.

5.3 Elliptic solutions

The elliptic solutions to the Yang–Baxter equation are particularly interesting for various reasons. What concerns us now is that they arise from applying the vertex–face map to elliptic vertex R-matrices such as that of the eight-vertex model. Hence, they provide an alternative way to explore the integrability in the "elliptic regimes"; in order to solve the eight-vertex model (chapter 4) we had to map it into an elliptic face model. On the other hand, elliptic face Boltzmann weights appear also quite naturally in the context of q-deformed conformal field theories, much like the trigonometric face solutions appear in ordinary conformal field theory (chapter 8). The face weights yield the connection matrices of the solutions to the Knizhnik–Zamolodchikov equation, in its usual or q-deformed version (chapter 10).

5.3.1 An example: the Ising model

Let us consider the elliptic solution for the simplest non-trivial case, for which $r = 3$. The lattice variables can thus take the values 1,2,3. We will parametrize the Boltzmann weights as follows:

$$\begin{aligned}
B\left(\begin{matrix} 1 & 2 \\ 2 & 3 \end{matrix}\middle| u\right) &= B\left(\begin{matrix} 3 & 2 \\ 2 & 1 \end{matrix}\middle| u\right) = w_1(u) \\
B\left(\begin{matrix} 1 & 2 \\ 2 & 1 \end{matrix}\middle| u\right) &= B\left(\begin{matrix} 3 & 2 \\ 2 & 3 \end{matrix}\middle| u\right) = w_2(u) \\
B\left(\begin{matrix} 2 & 3 \\ 1 & 2 \end{matrix}\middle| u\right) &= B\left(\begin{matrix} 2 & 1 \\ 3 & 2 \end{matrix}\middle| u\right) = w_3(u) \\
B\left(\begin{matrix} 2 & 1 \\ 1 & 2 \end{matrix}\middle| u\right) &= B\left(\begin{matrix} 2 & 3 \\ 3 & 2 \end{matrix}\middle| u\right) = w_4(u)
\end{aligned} \tag{5.51}$$

Note that these Boltzmann weights satisfy, of course, the reflection symmetries ($a \leftrightarrow c$ and $b \leftrightarrow d$ independently) in the solution (5.46). The Yang–Baxter equation (5.6) reads in this case as

$$\begin{aligned}
w_4(u)w_2(u+v)w_4(v) + w_3(u)w_1(u+v)w_3(v) &= w_2(v)w_4(u+v)w_2(u) \\
w_4(u)w_1(u+v)w_4(v) + w_3(u)w_2(u+v)w_3(v) &= w_1(v)w_4(u+v)w_1(u) \\
w_4(u)w_2(u+v)w_3(v) + w_3(u)w_1(u+v)w_4(v) &= w_2(v)w_3(u+v)w_1(u)
\end{aligned} \tag{5.52}$$

Using the relations (B4.5), we obtain

$$
\begin{aligned}
w_1(u) &= \rho\,\frac{\theta_1(u+\pi/4)}{\theta_1(\pi/4)}\\
w_2(u) &= \rho\,\frac{\theta_1(\pi/4-u)}{\theta_1(\pi/4)}\\
w_3(u) &= \rho\,\frac{\theta_1(u)}{\theta_1(\pi/2)}\\
w_4(u) &= \rho\,\frac{\theta_1(\pi/2-u)}{\theta_1(\pi/2)}
\end{aligned}
\tag{5.53}
$$

where ρ is an overall factor.

In order to interpret this A_3 model as the Ising model, we will proceed as follows. For a model defined by an A_r graph, the allowed configurations are such that the odd and even lattice variables are placed in two different sublattices. An example of an allowed configuration for the A_3 graph model is shown in figure 5.8. The even sublattice contains the single value of lattice variables 2, whereas the odd sublattice contains at each site either 1 or 3. Under the change of variables $\ell_i \to \sigma_i = \ell_i - 2$, the even sublattice will contain only zeroes, and the odd sublattice will display a distribution of "spin variables" ± 1. It is this distribution on the odd sublattice that we shall interpret as a spin configuration for the Ising model. The spin variable configuration derived on the odd sublattice from the configuration of figure 5.8 is shown in figure 5.9.

The partition function for the Ising model is defined as

$$
Z = \sum_{\text{configurations}} \left[\prod_{\substack{\langle i,j\rangle \\ \text{vertical}}} \exp(\beta J_1 \sigma_i \sigma_j) \prod_{\substack{\langle i,j\rangle \\ \text{horizontal}}} \exp(\beta J_2 \sigma_i \sigma_j) \right] \tag{5.54}
$$

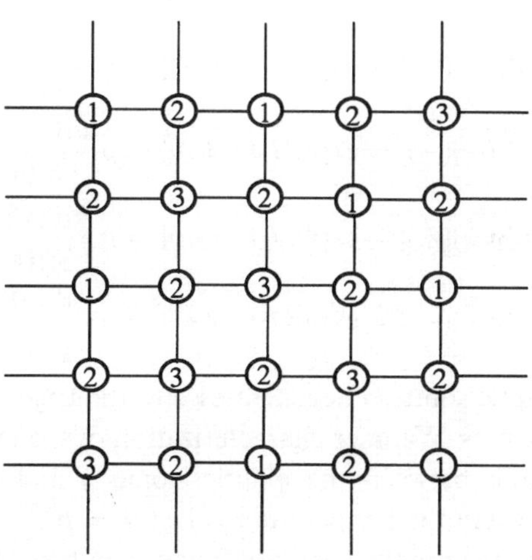

Fig. 5.8 Example of a configuration for the A_3 model.

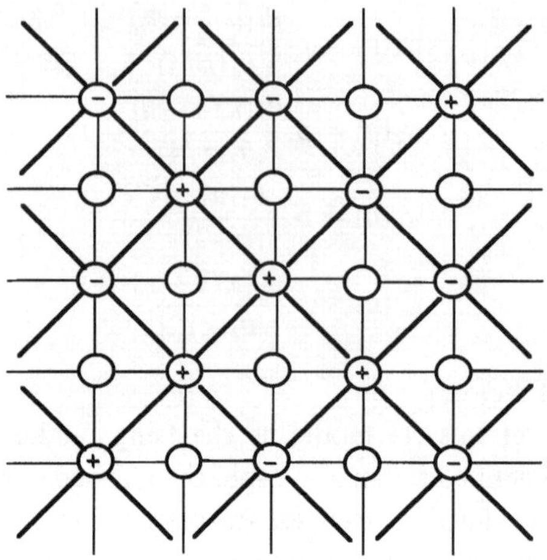

Fig. 5.9 Ising model configurations associated with figure 5.8.

where the sum is over all possible configurations and the notation $\langle i, j\rangle$ stands for nearest neighbor pairs. The constants J_1 and J_2 denote the vertical and horizontal couplings of the spin variables, and are not necessarily equal. The face representation of the Ising model is achieved by the identifications

$$\begin{aligned}\exp(\beta J_1\sigma_i\sigma_j) &= B\left(\begin{matrix}\sigma_i & 0\\ 0 & \sigma_j\end{matrix}\middle| u\right) = B\left(\begin{matrix}\ell_i & 2\\ 2 & \ell_j\end{matrix}\middle| u\right)\\ \exp(\beta J_2\sigma_i\sigma_k) &= B\left(\begin{matrix}0 & \sigma_i\\ \sigma_k & 0\end{matrix}\middle| u\right) = B\left(\begin{matrix}2 & \ell_i\\ \ell_k & 2\end{matrix}\middle| u\right)\end{aligned} \tag{5.55}$$

With the help of the Boltzmann weights (5.51), we find the following possibilities for the factors in (5.54):

$$\begin{aligned}\exp[\beta J_1(+)(+)] &= \exp[\beta J_1(-)(-)] = \rho\,\frac{\theta_1(\pi/4-u)}{\theta_1(\pi/4)}\\ \exp[\beta J_1(+)(-)] &= \exp[\beta J_1(-)(+)] = \rho\,\frac{\theta_1(u+\pi/4)}{\theta_1(\pi/4)}\\ \exp[\beta J_2(+)(+)] &= \exp[\beta J_2(-)(-)] = \rho\,\frac{\theta_1(\pi/2-u)}{\theta_1(\pi/2)}\\ \exp[\beta J_2(+)(-)] &= \exp[\beta J_2(-)(+)] = \rho\,\frac{\theta_1(u)}{\theta_1(\pi/2)}\end{aligned} \tag{5.56}$$

When $p \to 0$, the elliptic solution degenerates into the trigonometric one.

Let us use the Kramers–Wannier characterization of the critical point to make explicit the connection between the elliptic nome p and the departure from criticality $(T - T_c)$, where the temperature is just $T = \beta^{-1}$.

First of all, we shall rewrite the partition function in two different ways, usually referred to as the low and high temperature expansions. Let us consider a square lattice $\mathscr{L}$ of size L lattice units, and, to simplify the notation, let $K_1 = \beta J_1$ and $K_2 = \beta J_2$. Imagine now a generic spin configuration with r_1 (respectively, r_2) unlinked nearest neighbor vertical (respectively, horizontal) pairs. A nearest neighbor pair is linked (unlinked) if the spins at the two sites are equal (opposite). There are thus $(L - r_1)$ linked vertical pairs and $(L - r_2)$ linked horizontal pairs. Simple combinatorics show that the contribution of this configuration to the partition function (5.54) is

$$\exp[K_1(L - 2r_1) + K_2(L - 2r_2)] \tag{5.57}$$

Now, on a two-dimensional surface, the dual to a lattice is obtained by drawing the medians to each original edge. In the simple case at hand of a square lattice on the plane, the dual is a square lattice whose sites are at the center of the original squares. Consider this dual lattice $\mathscr{L}_D$, and draw a thick line on a link of $\mathscr{L}_D$ when adjacent spins are unlike. The previous configuration yields r_1 horizontal thick segments and r_2 vertical ones (see figure 5.10). The thick lines naturally join to form polygonal boundaries which separate the two-dimensional space into spin up and spin down domains. The partition function can be written as a sum over all possible boundary configurations in $\mathscr{L}_D$:

$$Z = \exp[L(K_1 + K_2)] \sum_{\text{polygons} \in \mathscr{L}_D} \exp(-2K_1 r_1 - 2K_2 r_2) \tag{5.58}$$

This is the low temperature representation (K_i large) for the partition function of the Ising model. For each term in this sum, (r_1, r_2) is the number of (horizontal, vertical) polygonal segments in the boundary: two configurations with the same total vertical and horizontal boundary lengths contribute equally.

Let us now find the high temperature representation of the partition function of the Ising model. Then we shall establish the Kramers–Wannier duality by comparison. The high temperature (K_i small) expansion relies on the simple

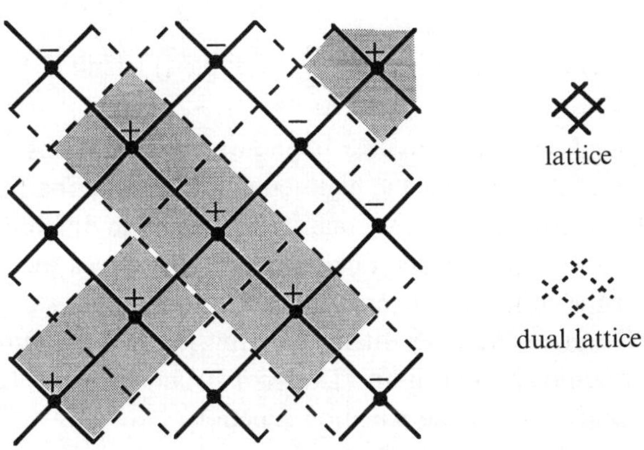

Fig. 5.10

identity

$$\exp K\sigma_i\sigma_j = \cosh K + \sinh K\sigma_i\sigma_j \tag{5.59}$$

which holds for spin variables restricted to $\sigma_i = \pm 1$. The partition function (5.54) is thus

$$Z = (\cosh K_1 \cosh K_2)^L \sum_\sigma \prod_{\substack{\langle i,j\rangle \\ \text{vertical}}} (1 + \tanh K_1\sigma_i\sigma_j) \times \prod_{\substack{\langle m,n\rangle \\ \text{horizontal}}} (1 + \tanh K_2\sigma_m\sigma_n) \tag{5.60}$$

which can be rewritten in a way very similar to (5.58). Imagine first expanding the parentheses in (5.60), so that Z is a single sum of terms. For each nearest neighbor pair of sites, any term in this expansion will contain either a 1 or a tanh. If in the term under consideration the tanh appears, then we draw a thick line on the corresponding edge in $\mathscr{L}$. If, on the contrary, the factor corresponding to a given nearest neighbor pair in the term is 1, no thick line is drawn between the two sites. Such is the procedure for associating a configuration of lines (boundaries, polygons) in $\mathscr{L}$ with every term in the expansion of Z. Each term in the expansion contributes something of the form

$$(\tanh K_1)^{r_1}(\tanh K_2)^{r_2}\sigma_1^{n_1}\sigma_2^{n_2}\cdots \tag{5.61}$$

where r_1 and r_2 are now the total number of vertical and horizontal thick lines, respectively, and n_i is the number of thick lines (segments) with the site i as an end-point. The sum over σ_1, σ_2, ... vanishes unless all the n_i are even, because $\sigma_i = \pm 1$. The result of the sum depends only on the geometry of the lattice; in our simple case, it is

$$\sum_\sigma \sigma_1^{n_1}\sigma_2^{n_2}\cdots\sigma_L^{n_L} = \begin{cases} 2^L & \text{iff } n_i \in \{0,2,4\} \quad \forall i \\ 0 & \text{otherwise} \end{cases} \tag{5.62}$$

where L is the total number of sites on a row. The partition function (5.60) is thus

$$Z = (2\cosh K_1 \cosh K_2)^L \sum_{\text{polygons}\in\mathscr{L}} (\tanh K_1)^{r_1}(\tanh K_2)^{r_2} \tag{5.63}$$

where the sum extends over all possible boundary configurations (polygons for short) in the lattice $\mathscr{L}$. Notice that the high temperature partition function (5.63) resembles strongly the low temperature one (5.58); the main difference is that the sum in (5.58) is over polygons in the dual lattice $\mathscr{L}_D$, whereas the sum in (5.63) is over polygons in the original lattice $\mathscr{L}$.

To obtain the Kramers–Wannier duality relationship, let us compute the free energy F in the thermodynamic limit ($L \to \infty$), from both the low and high temperature representations of the partition function:

$$-\frac{1}{T}F = \lim_{L\to\infty} \frac{1}{L}\log Z \tag{5.64}$$

A simple calculation yields the two results

$$\begin{aligned} -\frac{1}{T}F(K,L) &= K + L + \Psi(e^{-2K_2}, e^{-2K_1}) \\ &= \log(2\cosh K_1 \cosh K_2) + \Psi(\tanh K_1, \tanh K_2) \end{aligned} \tag{5.65}$$

The use of the same function

$$\Psi(a,b) = \lim_{L\to\infty} \frac{1}{L} \log \left[\sum_{\text{polygons}} a^{r_1}\, b^{r_2} \right] \tag{5.66}$$

in the two expressions is only justified in the thermodynamic limit, when the difference between the polygons in $\mathscr{L}$ and those in $\mathscr{L}_D$ becomes irrelevant.

Equation (5.65) indicates the explicit form of the Kramers–Wannier duality transformation. If we define

$$\tanh K_1^* = e^{-2K_2} \ ; \qquad \tanh K_2^* = e^{-2K_1} \tag{5.67}$$

then from (5.65) we find the relationship

$$-\frac{1}{T} F(K_1^*, K_2^*) = -\frac{1}{T} F(K_1, K_2) + \frac{1}{2} \log \left[\sinh 2K_1^* \ \sinh 2K_2^*\right] \tag{5.68}$$

between the high and low temperature regimes; in fact, K_1 and K_2 are large if K_1^* and K_2^* are small, and vice versa. The variables K_i are dual to K_i^*. In equation (5.67), the subscripts 1 and 2 are interchanged because what is vertical on $\mathscr{L}$ is horizontal on $\mathscr{L}_D$, and vice versa.

The critical point is the point in the phase diagram where the free energy is singular; this singularity should be apparent both for $F(K_1, K_2)$ and for $F(K_1^*, K_2^*)$, and therefore, if only a single critical point exists, it must be at the fixed (self-dual) point of the duality transformation (5.67). At the critical point, $F(K_1, K_2) = F(K_1^*, K_2^*)$, so it follows from (5.68) that

$$\sinh 2K_1^* \ \sinh 2K_2^* = 1 \tag{5.69}$$

In terms of the original β, J_1 and J_2, this statement reads as

$$\sinh 2\beta J_1 \ \sinh 2\beta J_2 = 1 \tag{5.70}$$

We are now in a position to understand the physical meaning of the nome p appearing in the elliptic solution to the face Yang–Baxter equation, i.e. in the Boltzmann weights (5.56). Indeed, substituting (5.70) into (5.56), we obtain

$$\left[\frac{\theta_1(\pi/2)}{\theta_1(\pi/4)}\right]^4 = 4 \tag{5.71}$$

Using now the explicit expression for the θ function from appendix B, we find that the nome p vanishes at the critical point $T = T_c$. Hence, p measures the departure from criticality. This result explains also the physical meaning of the trigonometric limit, corresponding to $p = 0$. In fact, the trigonometric solution for the Boltzmann weights obtained using (5.45) and (5.46) describes the critical point of the corresponding face graph model.

5.3.2 Restricted and unrestricted models

Let us study now the solution (4.18) to the eight-vertex model. As we have already seen, it corresponds to the infinite A_∞ graph

$$\begin{array}{ccccccccccccc} \cdots - & \bullet & - & \bullet & - & \bullet & - & \bullet & - & \bullet & - & \bullet & - \cdots \\ & -2 & & -1 & & 0 & & 1 & & 2 & & 3 & \end{array} \tag{5.72}$$

Using the representation (5.44), we may take the trigonometric limit of (4.18) and check that the weights are indeed of the graph face form. We leave it as an exercise for the reader to check that the values of the functions $t(\ell)$ for the unrestricted A_∞ graph thus obtained are

$$t^{(\infty)}(\ell) = \sin(\gamma\ell + \tau) \tag{5.73}$$

In addition to the anisotropy γ, the solution (4.18) already contained the free parameter τ, which also survives the limit. Equation (5.73) allows us to develop some understanding for the parameter τ. We will compare it with the corresponding solution to (5.45) for the restricted A_r graph

$$\begin{array}{ccccccc} \bullet & - & \bullet & - & \cdots & - & \bullet \\ 1 & & 2 & & & & r \end{array} \tag{5.74}$$

which is (5.49):

$$t^{(r)}(\ell) = \sin\left(\frac{\ell\pi}{r+1}\right) \tag{5.75}$$

with $\beta = 4\cos^2 \pi/(r+1)$. The restricted face graph model thus corresponds to

$$\gamma = \frac{\pi}{r+1}, \qquad \tau = 0 \tag{5.76}$$

This value of γ takes us to the special points (4.56) discussed in chapter 4. The conclusion, to be clarified below, is that the special points (4.56) of the unrestricted graph model describe the restricted graph models. As noted before, at these points $q = e^{i\gamma}$ is a root of unity. Moreover, the parameter τ, free in the unrestricted case, must be zero for restricted graphs. Intuitively, we associate the freedom in the choice of τ with the extra translation invariance of the unrestricted graph.

In what precise sense does the eight-vertex model correspond to the restricted A_r face model, when $\gamma = \pi/(r+1)$? For these special values of the anisotropy, the eight-vertex Yang–Baxter equation allows a solution truncated to the lattice variables of the finite graph. In principle, in the face Yang–Baxter equation (5.6), we may always restrict the "external" variables to lie in whatever subset of $\mathbb{Z}$ we choose, but the sum over the "internal" variable ℓ must remain over the whole $\mathbb{Z}$. When the anisotropy is at one of these special values, however, the internal sum truncates; i.e., instead of having a sum over all $\ell \in \mathbb{Z}$, we effectively have a sum from $\ell = 1$ to $\ell = r$ on both sides of the Yang–Baxter equation. This means that, provided we take the external variables in $\{1, \ldots, r\}$, the internal sum is also only over this restricted range: the rest of the sum cancels out. For convenience, we summarize this discussion in table 5.2.

5.4 Fusion for face models

In this section, we shall extend to face models the fusion procedure used in chapter 2 for the construction of higher-spin descendants of the six-vertex model. The models we shall obtain are called (p, q) face models.

Let us start with the unrestricted face model defined on the graph A_∞. This means that the site variables take values in the integers, and the only restriction is that the difference between the lattice variables of two neighboring sites is ± 1. To fix the notation, we shall call this model the $(1, 1)$ face model:

$$B_{1,1}\left(\begin{matrix} a & b \\ c & d \end{matrix}\middle| u\right) = \quad (u) \qquad (5.77)$$

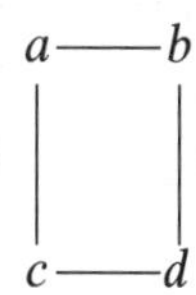

The $(1, 1)$ subscripts refer to the rule $|\ell - \ell'| = 1$ for the variables on both verticallly and horizontally neighboring sites.

Table 5.2 Face models associated with the eight-vertex model.
The face Boltzmann weights are the $B\left(\begin{matrix} \ell_1 & \ell_2 \\ \ell_3 & \ell_4 \end{matrix}\middle| u\right)$ coefficients, and the basis is $X_\ell^\pm(u)$.

Face model	γ	τ	Graph
Unrestricted	generic	generic	$\cdots - \bullet - \bullet - \bullet - \bullet - \cdots$ $-1 \quad 0 \quad 1 \quad 2$
Restricted	$\frac{\pi}{r+1}$	0	$\bullet - \bullet - \cdots - \bullet$ $1 \quad 2 \qquad r$

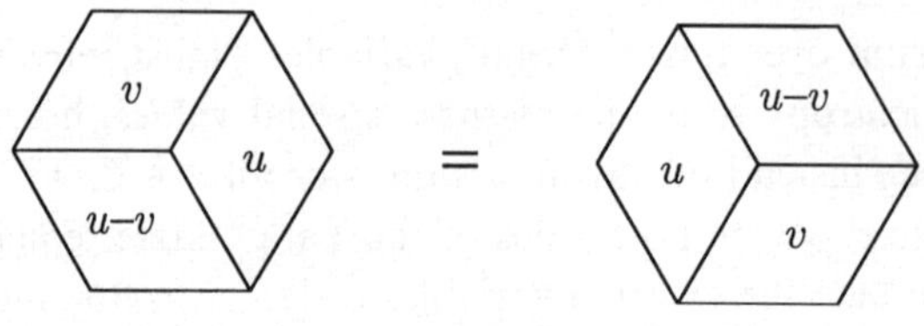

Fig. 5.11 The standard, or (1, 1), face Yang–Baxter equation.

We now construct the (unrestricted) (1,2) model by fusing two (1,1) models. The Boltzmann weights of the fused (1,2) model are simply

$$B_{1,2}\left(\begin{matrix} a & b \\ c & d \end{matrix}\middle| u\right) = \sum_{a',c'} B_{1,1}\left(\begin{matrix} a & a' \\ c & c' \end{matrix}\middle| u+1\right) B_{1,1}\left(\begin{matrix} a' & b \\ c' & d \end{matrix}\middle| u\right)$$

$$= \qquad (5.78)$$

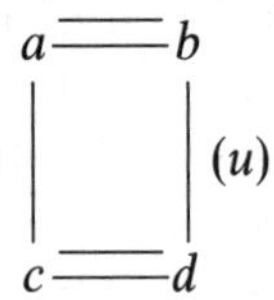

(Note that the rapidity of the first $B_{1,1}$ factor is really $u + \gamma$; for conciseness, we set $\gamma = 1$ in this section.) Whereas adjoining vertical variables must still differ by 1, $|a - c| = |b - d| = 1$, horizontal variables may now differ by −2, 0 or 2:

$$\frac{a - b + 2}{2}, \frac{c - d + 2}{2} \in \{0, 1, 2\} \qquad (5.79)$$

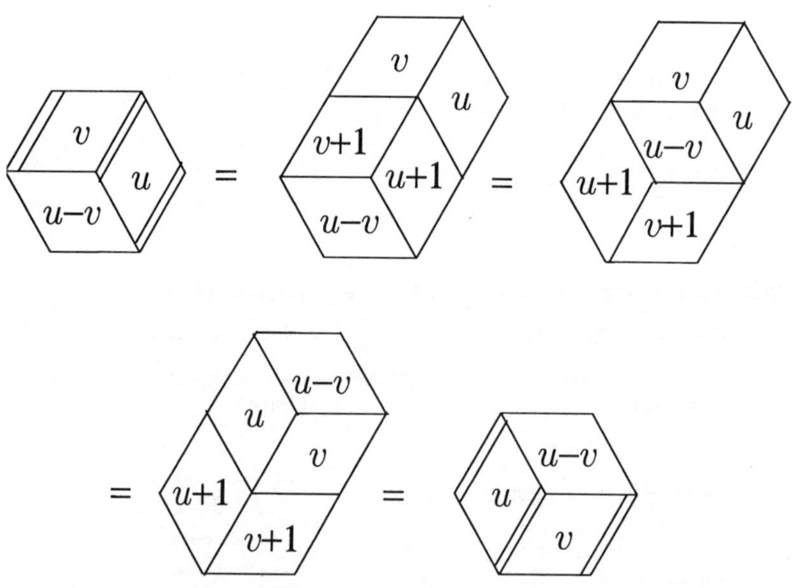

Fig. 5.12. The Yang–Baxter equation for (1, 2) and (1, 1) Boltzmann weights, and its proof using figure 5.11.

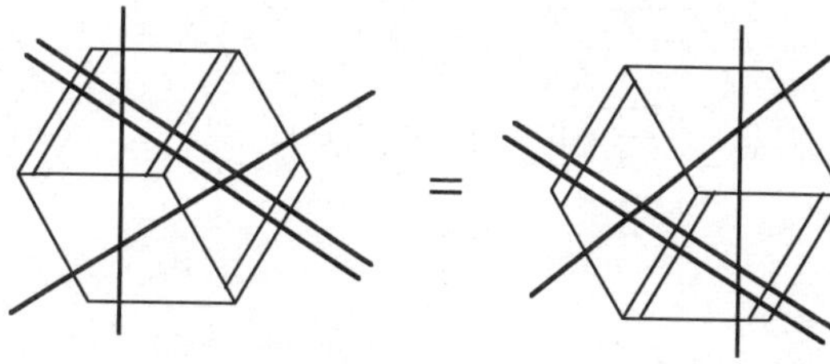

Fig. 5.13 The vertex–face map applied to the Yang–Baxter equation of figure 5.12.

The immediate question to ask is what kind of Yang–Baxter equation do the fused Boltzmann weights $B_{1,2}$ satisfy? It suffices to find it graphically. The original $(1,1)$ face Yang–Baxter equation can be drawn as in figure 5.11. Using this and the definition (5.78), we find the equation shown in figure 5.12.

It is now instructive to use the pictorial version of the vertex–face map to compare figure 5.13 with the first descent equation (2.121).

Using the same rules as above, let us proceed to define the $(2,2)$ Boltzmann weights:

$$B_{2,2}\left(\begin{matrix} a & b \\ c & d \end{matrix}\middle| u\right) = \sum_{a',b'} B_{1,2}\left(\begin{matrix} a & b \\ a' & b' \end{matrix}\middle| u-1\right) B_{1,2}\left(\begin{matrix} a' & b' \\ c & d \end{matrix}\middle| u\right)$$

$$= \tag{5.80}$$

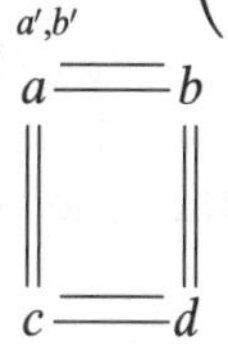

It is easy to check, using the $(1,1)$ and $(1,2)$–$(1,1)$ Yang–Baxter equations above, that $B_{2,2}$ satisfy the face version of the second descent equation (2.125), shown in figure 5.14. The proof of figure 5.14 requires using the $(1,1)$ Yang–Baxter equation four times, as shown in figure 5.15.

With this kind of technology, it is possible to produce a whole bunch of integrable unrestricted (p,q) face models.

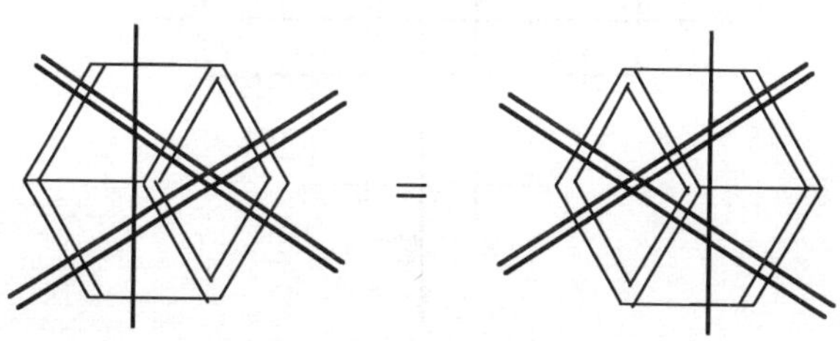

Fig. 5.14. The Yang–Baxter equation satisfied by the Boltzmann weights of the $(2,2)$ and $(1,2)$ fused face models, and its vertex version.

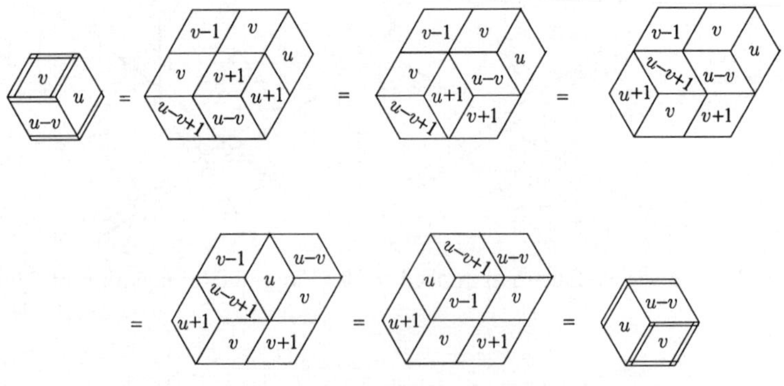

Fig. 5.15 Proof of the Yang–Baxter equation of figure 5.14.

5.5 The corner transfer matrix

Consider a graph face model with graph G defined on a lattice $\mathscr{L}$. Denote by σ_i ($i \in \mathscr{L}$) a lattice field configuration:

$$\sigma : \mathscr{L} \longrightarrow G \tag{5.81}$$

and denote by a, b, c ... the lattice variables. Let (b, c) be an admissible pair of lattice variables, i.e. two lattice variables which are adjacent on the graph G. Impose (b, c) boundary conditions on the lattice shown in figure 5.16.

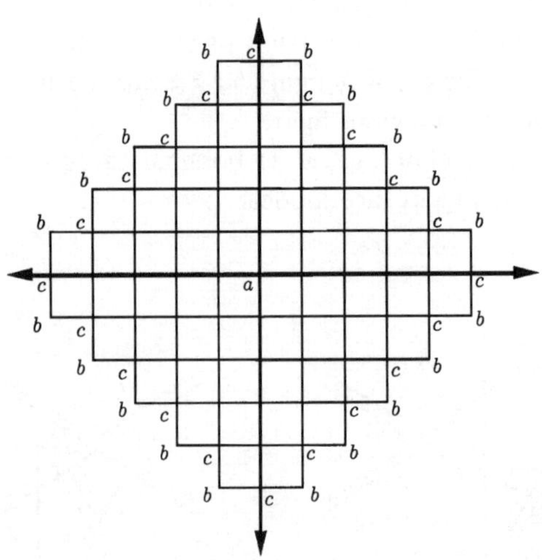

Fig. 5.16. The lattice for the corner transfer matrix computation with lattice variable a at the origin and (b, c) boundary conditions.

Let us concentrate on the lower right quadrant, shown in figure 5.17, and define its corner transfer matrix A as follows:

$$A_{\vec{\sigma},\vec{\sigma}'} = \begin{cases} \prod B\begin{pmatrix} \sigma_i & \sigma_j \\ \sigma_k & \sigma_\ell \end{pmatrix} & \text{if } \sigma_1 = \sigma_1' \\ 0 & \text{if } \sigma_1 \neq \sigma_1' \end{cases} \tag{5.82}$$

The product in (5.82) is over all plaquettes in the lower right quadrant A of figure 5.16. The vectors $\vec{\sigma} = (\sigma_1, \sigma_2, \ldots, \sigma_m)$ and $\vec{\sigma}' = (\sigma_1', \sigma_2', \ldots, \sigma_m')$ are spin configurations on the x and y axes bordering the quadrant A. Similarly, we may define three other corner transfer matrices B, C and D, one for each quadrant.

The partition function can be written in terms of the corner transfer matrices as

$$Z = \operatorname{tr}(ABCD) = \sum_{\vec{\sigma},\vec{\sigma}',\vec{\sigma}'',\vec{\sigma}'''} A_{\vec{\sigma},\vec{\sigma}'} B_{\vec{\sigma}',\vec{\sigma}''} C_{\vec{\sigma}'',\vec{\sigma}'''} D_{\vec{\sigma}''',\vec{\sigma}} \tag{5.83}$$

The corner transfer matrix A can be obtained as a product of the Yang–Baxter operators introduced in section 5.2.2. For example, if the horizontal and vertical states between which A interpolates consisted of only three spins, the corner transfer matrix would read

$$A_{\vec{\sigma},\vec{\sigma}'} = X_3^{(\sigma_3,\sigma_3'')} X_2^{(\sigma_2,\sigma_2')} X_3^{(\sigma_3'',\sigma_3)} \tag{5.84}$$

To solve the model, we must diagonalize the matrix $ABCD$ on the space of states $\vec{\sigma} = (\sigma_1, \sigma_2, \ldots, \sigma_m)$:

$$ABCD\vec{\sigma} = M(\vec{\sigma})\vec{\sigma} \tag{5.85}$$

In terms of the eigenvalues $M(\vec{\sigma})$ of the matrix $ABCD$, the partition function in the thermodynamic limit reads as follows:

$$Z = \lim_{m\to\infty} \sum_{\sigma_1\cdots\sigma_m} M(\sigma_1, \ldots, \sigma_m) \tag{5.86}$$

The interest fact concerning (5.86) is that it simplifies the sum over two-dimensional lattice configurations into a sum over one-dimensional configurations,

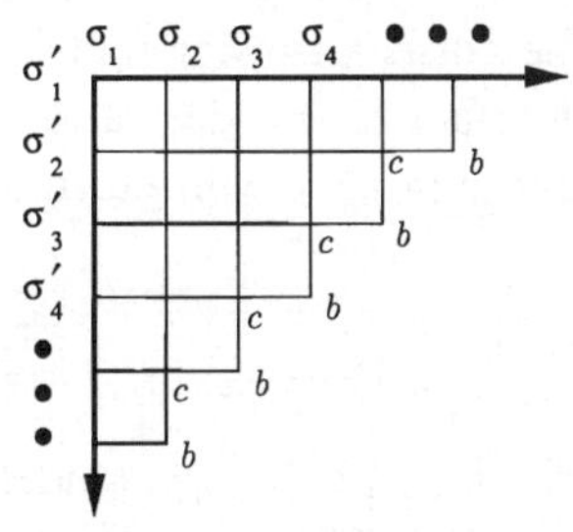

Fig. 5.17 The configurations $\vec{\sigma}$ and $\vec{\sigma}'$ (with $\sigma_1 = \sigma_1'$ at the origin) in the definition (5.82).

which is much more manageable analytically. Indeed, we can associate formally a one-dimensional lattice model to the original two-dimensional lattice model via

$$\exp -\beta E(\sigma_1,\ldots,\sigma_m) = M(\sigma_1,\ldots,\sigma_m) \tag{5.87}$$

The previous discussion applies in the face language, which is the one originally used by Baxter to introduce the concept of the corner transfer matrix. To conclude this section, let us consider the corner transfer matrix from the vertex viewpoint. Picture the dual of figure 5.16: the corner transfer matrix A takes us from the "vertical" state $|\alpha\rangle = |\alpha_1,\ldots,\alpha_n\rangle$ to the "horizontal" state $|\beta\rangle = |\beta_1,\ldots,\beta_n\rangle$. The lattice variables on the edge are frozen by the boundary conditions, and the corner transfer matrix A is just the sum over all possible vertex configurations in the internal lines. Using a synthetic notation, we may write

$$A_{\alpha,\beta}(u) = \langle 0|\, T_1\, |\zeta_1\rangle\, \langle\zeta_1|\, T_2\, |\xi_2\rangle \cdots \langle\delta_{n-1}|\, T_{n-1}\, |\gamma_n\rangle\, \langle\gamma_n|\, T_n\, |\alpha\rangle \tag{5.88}$$

The infinitesimal representation of the corner transfer matrix is $A = 1 - uK$, with $K = \sum_{j=1}^{n} jH_j$, and H_j precisely the local operator appearing in the row to row hamiltonian $H = \sum_{j\in\mathbb{Z}} H_j$. In general, H_j is a local functional. It is quite remarkable that the same local functional H_j appears in K and in H. A noteworthy feature of the corner transfer matrix hamiltonian K is that it is defined on a semi-infinite, not infinite, chain.

It is an obvious choice to interpret K as some sort of boost operator because A causes a rotation by $\pi/2$ in the (x,t) plane of the lattice. Indeed, we might imagine that the operator K, along with the row to row hamiltonian H (generating time translations) and the momentum operator P (generating spatial shifts), span the lattice version of the two-dimensional Poincaré algebra. From the definition of K, we find that

$$[K, t(u)] = \frac{dt(u)}{du} \tag{5.89}$$

whereas the integrability condition $[t(u), t(v)] = 0$ implies that

$$\left[K, H^{(n)}\right] = H^{(n+1)} \quad (n \geq 0) \tag{5.90}$$

where we have used the Taylor expansion of the logarithm of the transfer matrix, $\log t(u) = \sum \frac{u^n}{n!} H^{(n)}$, to identify the hamiltonians, with $H^{(0)} = P$ and $H^{(1)} = H$.

Although relations (5.90) do yield a lattice analog of the Poincaré algebra, the mathematical meaning of the corner transfer matrix is clearer from (5.89): the corner transfer matrix hamiltonian K is a derivation for the spectral parameter. (See also exercises 5.9, 5.10 and 6.14.) We will return to the corner transfer matrix hamiltonians in chapter 10.

Exercises

Ex. 5.1 We say that a matrix X ($X_{ij} \in \{0,1\}$) is indecomposable if its associated graph $\Gamma(X)$ is connected, and that X is equivalent to X' if $\Gamma(X)$ and $\Gamma(X')$ are isomorphic bi-colored graphs. Show that X and X^T are equivalent.

Ex. 5.2 Associated with any finite matrix $X \in M_{m,n}$, the elements of which are all either 0 or 1, we draw a bi-colored graph by first drawing m black vertices and n white ones, and then linking the ith black dot with the jth white dot if $X_{ij} = 1$. For instance, the 6×2 matrix

$$X = \begin{pmatrix} 0 & 1 & 0 & 1 & 1 & 0 \\ 1 & 0 & 1 & 0 & 0 & 1 \end{pmatrix} \tag{5.91}$$

corresponds to the bi-colored graph shown in figure 5.18. Find the Boltzmann weights associated with this graph, and check that they satisfy the face Yang–Baxter equation.

Ex. 5.3 Check that the Yang–Baxter operators given by (5.21) and (5.23) yield a representation (5.11) of the braid group. What is $\rho(u)$?

Ex. 5.4 We have seen that the Temperley–Lieb–Jones algebra $A_{n,\beta} = \langle \mathbf{1}, e_1, \ldots, e_{n-1} \rangle$ is intimately connected to the Hecke algebra, which is a particular realization of the braid group. It is thus natural to wonder about the braid group interpretation of the generators e_i. A pictorial representation for the e_i due to Brauer clarifies this connection. As for the braid group B_n, consider a set of n strands. The identity is represented by

$\mathbf{1}$ =

(5.92)

Now let the Temperley–Lieb–Jones generator e_i be represented by

e_i =

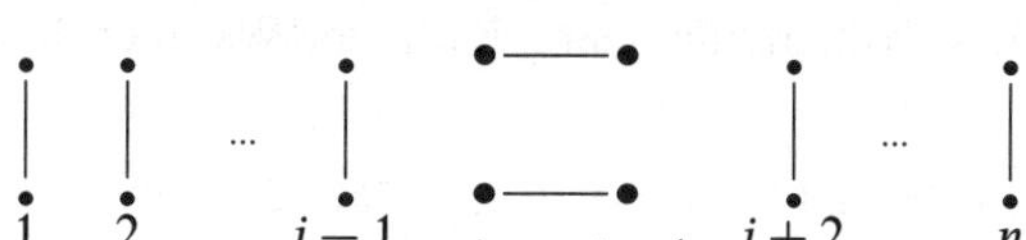

(5.93)

To multiply Temperley–Lieb–Jones generators, their associated graphs must be composed as follows. First, draw the two diagrams with one on top of the other. Secondly, insert an identity $\mathbf{1}$ to link the two diagrams. Finally, disregard all internal dots and eliminate all loops while picking a factor $\sqrt{\beta}$ for each loop. For instance, the diagram for $e_i^2 = \sqrt{\beta}\, e_i$ is shown in figure 5.19. Using this construction, show that $e_i e_{i+1} e_i = e_i$.

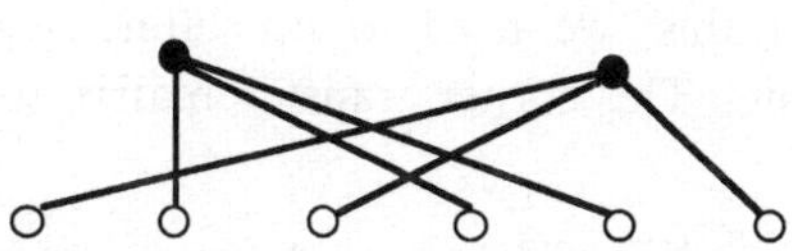

Fig. 5.18 Bi-colored graph associated with the matrix (5.91).

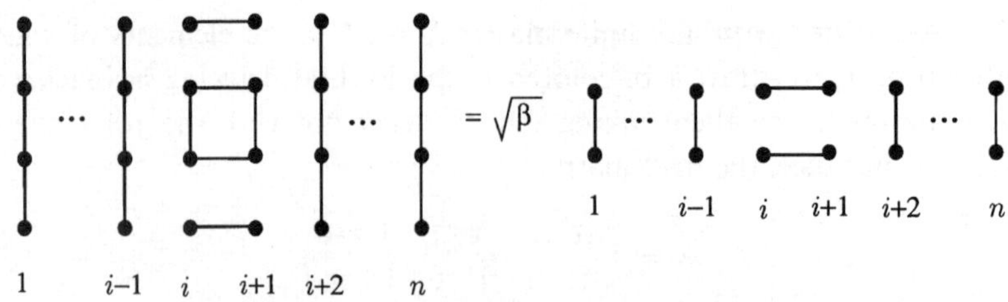

Fig. 5.19 Brauer's representation of $e_i^2 = \sqrt{\beta}\, e_i$.

Ex. 5.5 Verify the truncation of the face Yang–Baxter equation for the special values $\gamma = \pi/r$, $\tau = 0$.

Ex. 5.6 Find the trigonometric solution for D and E graph face models.

Ex. 5.7 A Q-state Potts model is a face model defined by the action

$$S = J \sum_{\langle j,k \rangle} \delta\left(\sigma_j, \sigma_k\right) \tag{5.94}$$

where the variables σ_i take values from 1 to Q. These models are known to experience a second-order phase transition provided $Q \in [0, 4]$. The connection with the A_r face models results from the identification

$$Q = 4\cos^2\left(\frac{\pi}{r+1}\right) \tag{5.95}$$

Following the analysis carried out above for the Ising model, find the free energy of the three-state Potts model using Kramers–Wannier duality.

Ex. 5.8 [A. Kuniba and T. Yajima, *J. Stat. Phys.* **52** (1988) 829.]
Check that the vectors $t(a)$ shown in table 5.3 are trigonometric solutions (5.46) to the Yang–Baxter equation for graph face models. Note that, although the dependence on the lattice variable is elliptic, the dependence on the spectral parameter is still trigonometric. In fact, graphs other than A_n are believed to have no elliptic solutions.

Ex. 5.9 Although the Bethe ansatz gives the energy, spin and momentum of the low lying excitations, it does not exhibit clearly the algebraic structure of the eigenstates. For this, we need to use affine quantum symmetries, in the thermodynamic limit. The corner transfer matrix for the XXZ model is $A = \mathrm{e}^{-u H_{CTM}}$, with

$$H_{CTM} = \sum_{k \geq 1} k \left(\sigma_k^x \sigma_{k+1}^x + \sigma_k^y \sigma_{k+1}^y + \Delta \sigma_k^z \sigma_{k+1}^z\right) \tag{5.96}$$

Table 5.3 Trigonometric solutions (5.46) for some graph face models.

Graph	Lattice variables	$t(a)$
A_r	$\{1,2,\ldots,r\}$	$t(a) = \Theta_1\left(\frac{\pi a}{r+1}, p\right)$
D_{r+1}	$\{0,\bar{0},1,2,\ldots,r-1\}$	$t(a) = \Theta_1\left(\pi\frac{r+a}{2r}, p\right) \quad a \neq \bar{0}$ $t(\bar{0}) = t(0)$
$D^{(1)}_{r+2}$	$\{0,\bar{0},1,2,\ldots,r-1,r,\bar{r}\}$	$t(a) = \Theta_4\left(\pi\frac{a}{r}, p\right) \quad a \neq \bar{0},\bar{r}$ $t(\bar{0}) = t(0) \qquad t(\bar{r}) = t(r)$
$A^{(1)}_r$	$\{0,1,\ldots,r\}$	$t(a) = \Theta_4\left(\pi\frac{a}{r+1}, p\right)$

Show that

$$\begin{aligned}
[H_{CTM}, \Delta^\infty K_i] &= 0 \\
[H_{CTM}, \Delta^\infty E_i] &= \Delta^\infty E_i \\
[H_{CTM}, \Delta^\infty F_i] &= -\Delta^\infty F_i
\end{aligned} \tag{5.97}$$

where K_i, E_i and F_i $(i = 0, 1)$ are the generators of the quantum group $U_q(A_1^{(1)})$. Do not confuse the parameter $\Delta = (q + q^{-1})/2$ with the co-multiplication Δ^∞: $\Delta^\infty E_i$ denotes the action of E_i on the semi-infinite spin chain, and similarly for the other generators.

Ex. 5.10 The transfer matrix $t(u)$ of the XXZ spin chain can be expanded about $u = 0$, as in chapter 2, yielding

$$\begin{aligned}
\log t(0)^{-1}t(u) &= \sum_{\ell\geq 1} u^\ell H_\ell \\
&= u\, t(0)^{-1}\partial_u t(0) + \frac{1}{2}u^2\left[t(0)^{-1}\partial_u^2 t(0) - \left(t(0)^{-1}\partial_u t(0)\right)^2\right] + \cdots
\end{aligned} \tag{5.98}$$

Show that the corner transfer matrix hamiltonian $K = H_{CTM}$ connects all these conserved charges, i.e. $[K, H_\ell] = H_{\ell+1}$.

Ex. 5.11 [G.E. Andrews, R.J. Baxter and P.J. Forrester, *J. Stat. Phys.* **35** (1984) 193.]

The restricted eight-vertex face model has four different regimes, depending on the sign of the Boltzmann weight d and the rapidity u. Prove that in the regions with $u > 0$ and $d > 0$, the corner transfer matrix hamiltonian $H_{\text{CTM}} = \sum jH_j$ is given by $H_j \propto |\sigma_j - \sigma_{j+2}|$.

Ex. 5.12 In section 5.4, we introduced a general class of face models, called (p,q) solid on solid models, by fusing the Boltzmann weights associated with the graph A_∞. Show that the restrictions on the lattice variables ℓ_i for a (p,q) model with non-vanishing Boltzmann weight $B_{p,q}\begin{pmatrix} \ell_1 & \ell_4 \\ \ell_2 & \ell_3 \end{pmatrix}$ are $(\ell_1 - \ell_2 + p)/2$, $(\ell_3 - \ell_4 + p)/2 \in \{0, 1, \ldots, p\}$ and $(\ell_1 - \ell_4 + q)/2$, $(\ell_2 - \ell_3 + q)/2 \in \{0, 1, \ldots, q\}$. Interpret this result graphically.

Ex. 5.13 Prove all the properties of the conditional expectations, equation (5.36), which follow from the fact that $M^{(0)}$ and $M^{(1)} \supset M^{(0)}$ are equipped with a Markov trace.

Appendix D
Knots and integrable models

D5.1 Introduction: the Jones polynomial

The concept of knot theory was introduced in the 19th century when William Thomson (later known as Lord Kelvin) tried to explain the stability and variety of atoms by viewing them as knots. Unfortunately, there are many more topologically inequivalent knots than there are atomic species.

A knot is a smooth embedding of a circle in three-dimensional euclidean space. Two knots are equivalent if one can be deformed continuously into the other, without crossings. A link is the embedding of several disjoint circles in a three-dimensional space, usually $\mathbb{R}^3$ or S^3, so it consists of several knots; in general, the knots in a link cannot be pulled apart. Mathematically, knot theory is an attempt to understand the first homotopy group of three-space without the link L, $\pi_1(\mathbb{R}^3 - L)$. This problem is so difficult to solve, that only bits and pieces of the answer are known, and these are encoded in the link invariants we shall discuss shortly.

To represent a knot, we may start with a ϕ^4 vacuum to vacuum Feynman diagram and then replace every vertex with a non-planar crossing. A diagram with N vertices thus gives rise to 2^N knots, some of which may be equivalent. One such possibility is shown in figure D5.1. It is actually the projections of the links that we draw; however, we call these projections links anyway. Of course, when drawing the projections one should avoid pathological constructions such as placing two crossings at the same point or drawing two segments of a link on top of each other.

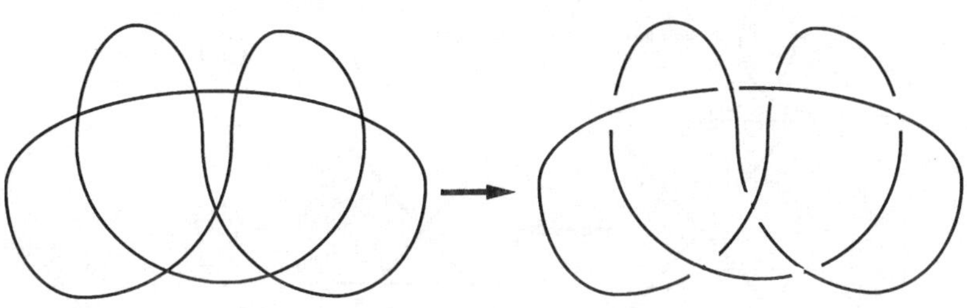

Fig. D5.1

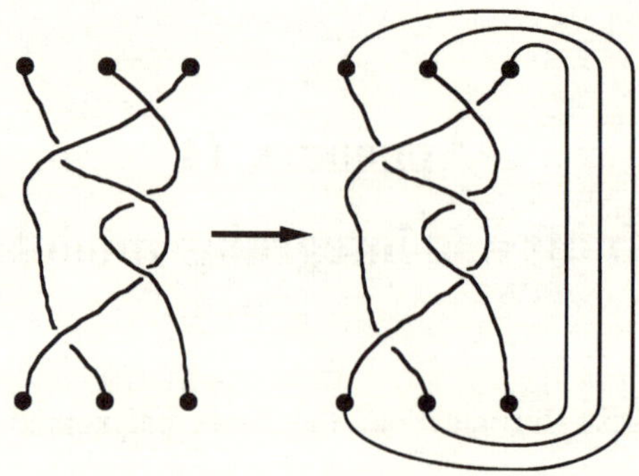

Fig. D5.2 The braid $\alpha = \sigma_1\sigma_2^{-2}\sigma_1\sigma_2^{-1} \in B_3$ gives rise to the link $\hat{\alpha}$.

A more algebraic way to obtain knots is by closing a braid. With each word α in the braid group, we associate a link $\hat{\alpha}$ by joining the upper and lower strands, as shown in figure D5.2. The link $\hat{\alpha}$ is called the closure of α.

The equivalence classes of knots and links can be systematized thanks to the Reidemeister theorem: two knots or links are topologically equivalent if they can be transformed into each other by repeated use of the three moves in figure D5.3. The first move undoes curls, and is, in fact, the trickiest one. The second move encodes one of the defining relations of the braid group, $\sigma_i\sigma_i^{-1} = 1$. Similarly, the third move reflects the Yang–Baxter equation, or, rather, the braid group property $\sigma_i\sigma_{i+1}\sigma_i = \sigma_{i+1}\sigma_i\sigma_{i+1}$.

After having assembled the various knots into equivalence classes according to the Reidemeister moves, we would like to characterize each class by some topological invariant. This problem would be completely solved if we could characterize the first homotopy group of the three-sphere minus the link, $\pi_1(S^3 - K)$. At present, nobody fully understands this group, but we do have a variety of "labels" for equivalence classes of knots. The idea is to assign a polynomial to

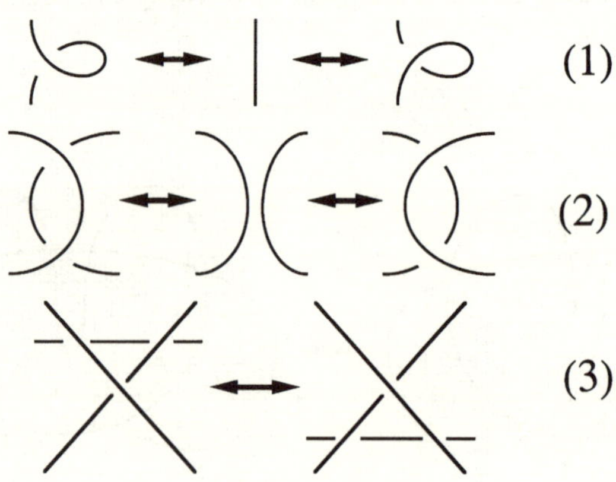

Fig. D5.3

a link in such a way that if two links are equivalent, then their polynomials are the same. With every link we shall associate a Laurent polynomial, which can be computed using the skein relations, which are best explained with an example.

The simplest link polynomial is the Alexander–Conway polynomial $\Delta_K(t)$ in one variable t, normalized such that the polynomial of the unknot is 1:

$$\Delta_{\bigcirc}(t) = 1 \tag{D5.1}$$

Sometimes, the unknot is referred to as the identity link. Although this polynomial was invented by Alexander in 1928, and was very well understood by Seifert in 1934 as the homology of the covering of the knot, only in 1960 did Conway come up with the most useful skein relations, in terms of which the Alexander–Conway polynomial $\Delta_K(t)$ for any link K can be computed by recursion from the normalization $\Delta_{\bigcirc}(t) = 1$. The skein relations encode mathematically the ingenious solution found by Alexander the Great to undo the complicated gordian knot: cut the rope!

The skein relations satisfied by $\Delta_K(t)$ are

$$\Delta_{K^+}(t) - \Delta_{K^-}(t) = t\Delta_{K^0}(t) \tag{D5.2}$$

where the links K^+, K^- and K^0 are related as follows. Given a link K, zoom in on any crossing. Then replace that crossing by its inverse and by the non-crossing, as in figure D5.4. The original link is K^+; the link with the inverse crossing is K^-; and the link with the crossing replaced by the uncrossing is K^0. The operation of deforming a link into its skein relatives is not continuous. Were we dealing with actual ropes or strings, we would have to cut and paste the strands to deform K^+ into K^- and K^0. The polynomials associated with these three links must be related by (D5.2), regardless of what crossing we chose to fiddle around with.

By applying the skein relations (D5.2) a sufficient number of times, any link or knot may be eventually transformed into the unknot. This procedure allows, therefore, for the construction of the polynomial in t and t^{-1}. The magic of the skein relation (D5.2) is that the polynomial obtained for two equivalent knots K and K' is the same, and thus we have indeed found a topological invariant.

It is important to stress, however, that inequivalent links may be assigned the same polynomial. For example, the Alexander–Conway polynomial, defined by the normalization (D5.1) and the skein rule (D5.2), is the same for a link and for its mirror image, even when these two are not topologically equivalent.

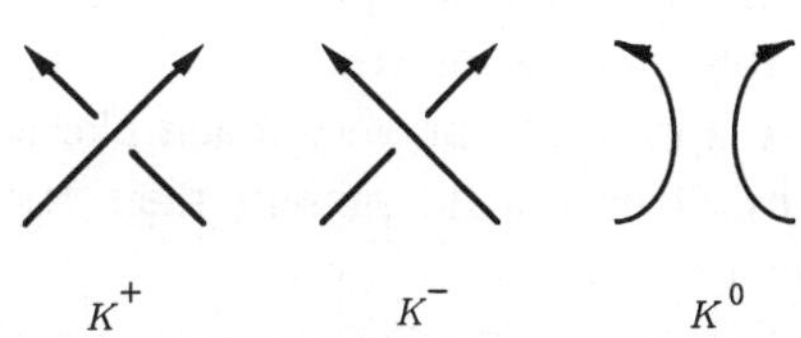

Fig. D5.4

The power of a link polynomial resides in how finely it distinguishes different inequivalent links: the ideal would be a topological index which is different for different equivalence classes.

For more than half a century, the Alexander–Conway polynomial was the only known invariant of links. In 1984, Jones discovered another polynomial, which, unlike the previous one, distinguishes between a link and its mirror image when these are not topologically equivalent. The Jones polynomial $V_K(t)$ is a Laurent polynomial in t (a polynomial in t and t^{-1}), normalized so that $V_{\bigcirc} = 1$, and characterized by the skein rule

$$t^{-1}V_{K^+} - tV_{K^-} = \left(\sqrt{t} - \frac{1}{\sqrt{t}}\right) V_{K^0} \tag{D5.3}$$

Although the Jones polynomial is still not completely understood (see the exercises), it allows many generalizations. It was discovered, for instance, that both the Jones and Alexander–Conway polynomials are particular cases of a more general two-variable Laurent polynomial $P_K(\lambda, t)$ (the HOMFLY polynomial), which is again normalized, $P_{\bigcirc} = 1$, and which satisfies the skein relation

$$\frac{1}{\sqrt{\lambda t}}P_{K^+} - \sqrt{\lambda t}P_{K^-} = \left(\sqrt{t} - \frac{1}{\sqrt{t}}\right) P_{K^0} \tag{D5.4}$$

Comparing (D5.2) and (D5.3) with (D5.4), we see that $V_K(t) = P_K(t, t)$ and $\Delta_K(t') = P_K(t^{-1}, t)$, with $t' = \sqrt{t} - 1/\sqrt{t}$.

D5.2 Markov moves

The skein relations are certainly the simplest way to characterize a link polynomial, but it is perhaps somewhat unclear why the result should be a topological invariant. In cruder terms, which skein relations yield topological invariants and which ones yield nonsense? An algebraic and systematic approach to link invariants exploits the following remarkable result by Markov. Given two braid words $\alpha \in B_n$, $\beta \in B_m$, the links $\hat{\alpha}$ and $\hat{\beta}$ derived from them by closure are topologically equivalent if and only if α and β are related by a finite number of the two Markov moves

$$\begin{array}{lll} \text{Markov I} & \alpha_1 \cdot \alpha_2 \to \alpha_2 \cdot \alpha_1 & \alpha_1, \alpha_2 \in B_n = \langle \mathbf{1}, \sigma_1, \ldots, \sigma_{n-1} \rangle \\ \text{Markov II} & \alpha\sigma_n^{\pm 1} \to \alpha & \alpha \in B_n \end{array} \tag{D5.5}$$

Two links are topologically equivalent if their braid group representations can be transformed into each other through the Markov moves (D5.5), which embody, in an algebraic way, the Reidemeister moves.

The first Markov move appears to be very much like a trace. This idea was pursued by Jones and by Ocneanu: to present their results we shall use the material covered in chapter 5.

Luckily, in our study of graph face models, we have come across the $\mathbb{C}^*$ algebra $A_{n,\beta}$ (the Temperley–Lieb–Jones algebra) with generators $\{\mathbf{1}, e_1, e_2, \ldots, e_{n-1}\}$ and defining relations

$$\begin{aligned} e_i^2 &= \sqrt{\beta}\, e_i \\ e_i e_j &= e_j e_i \qquad |i-j| \geq 2 \\ e_i e_{i\pm 1} e_i &= e_i \end{aligned} \tag{D5.6}$$

The parameter β is determined by the incidence matrix of the model, $\beta = ||\Lambda||^2$. This algebra admits the interpretation as the algebra of conditional expectations for a tower of algebras with inclusion matrix Λ, and it carries a trace functional satisfying the Markov property

$$\operatorname{tr}(x e_n) = \frac{1}{\sqrt{\beta}} \operatorname{tr} x \qquad x \in A_{n-1,\beta} \tag{D5.7}$$

This trace is the natural tool to use when building a functional on braid words which is invariant under the two Markov moves.

For this purpose, we also need the representation of the braid group B_n in the Temperley–Lieb–Jones algebra. We obtain it through the map π defined in (5.16).

The starting point when building an invariant for the link $K = \hat{\alpha}$, obtained by closure from the braid group word $\alpha \in B_n$, is thus the trace $\operatorname{tr}[\pi(\alpha)]$. This functional is, of course, invariant under Markov I, because the trace is cyclic. Comparing Markov II with (D5.7) and normalizing the unknotted circle to 1, the sought-after link invariant turns out to be the Jones polynomial

$$V_K(t) = \left(\sqrt{t} - \frac{1}{\sqrt{t}}\right)^{n-1} \left(\sqrt{t}\right)^{w} \operatorname{tr} \pi(\alpha) \tag{D5.8}$$

where $K = \hat{\alpha}$, $\alpha \in B_n$, $t = q^2$, and w is the net twist or writhe number of K, namely the algebraic sum of the exponents of σ_i in the word α. Formally,

$$\begin{aligned} w(\sigma_i) &= -w(\sigma_i^{-1}) = 1 \\ w(\alpha_1 \alpha_2) &= w(\alpha_1) + w(\alpha_2) \end{aligned} \tag{D5.9}$$

For example, $w(\sigma_1^3) = 3$, $w(\sigma_1 \sigma_2^{-2} \sigma_3) = 0$, etc.

Having found the polynomial through a systematic procedure, we may come back to the skein characterization. The skein rule satisfied by (D5.8) can now be easily understood in terms of the Hecke algebra relation $g_i^2 = (q^2 - 1)g_i + q^2$. In fact, using the map (5.16), we have mapped braid words into the Hecke algebra. Moreover, as is clear from figure D5.5, the three knots participating in the skein relationship can be represented by

$$\begin{aligned} K^0 &= \widehat{AB} \\ K^- &= \widehat{A\sigma_i^{-1}B} \\ K^+ &= \widehat{A\sigma_i B} \end{aligned} \tag{D5.10}$$

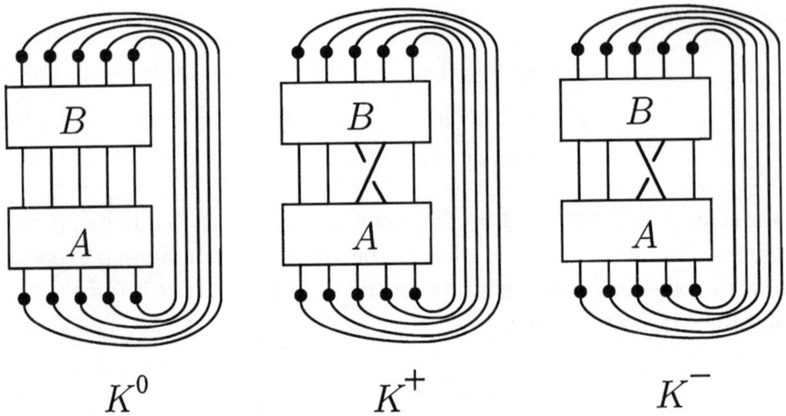

Fig. D5.5. The three links K^+, K^- and K^0 related by the skein rule differ by the insertion of σ_i, σ_i^{-1} or **1** at a given crossing.

Using the map (5.16) and the Hecke relations (5.15), the polynomial (D5.8) is seen to satisfy the skein relations (D5.3).

D5.3 Markov traces for the Hecke algebra

To build the Jones polynomial, we have used the algebra $A_{\beta,n}$. An essential tool used in the construction of the invariant polynomials that we have presented is the Markov property of the trace. Let us now consider in more detail Markov traces on the Hecke algebra.

First of all, notice that the Hecke algebra $H_n(q)$ reduces to the permutation algebra $\mathscr{S}_n$ when $q = 1$. In fact, the irreducible representations of $H_n(q)$ for generic values of q are in one to one correspondence with Young diagrams. We shall define a trace on $H_\infty(q) = \cup_{n=1}^{\infty} H_n(q)$ satisfying

$$\mathrm{tr}\,(xg_n) = z\ \mathrm{tr}\,(x) \qquad x \in H_n \tag{D5.11}$$

The following construction is quite analogous to the standard procedure employed for finding the dimensionality of an irrep of $SU(n)$. Let Y be a Young diagram and let π_Y be the associated irreducible representation. We assign a function $S_Y(q,z)$ to this irrep by multiplying together the entries of table D5.1 which are covered by the Young diagram. In the table, $w = 1 - q + z$. The ith column of the top row is $w - q^i z$, and the ith row of the first column is $q^i - wz$. The other entries are obtained by multiplying in a south-easterly direction by q.

Define now the function $Q_Y(q)$ as follows. A hook of the Young diagram of length m contributes a factor $(1 - q^m)$. The product of these factors for all the hooks is $Q_Y(q)$.

The functional of interest is

$$W_Y(q,z) = \frac{S_Y(q,z)}{Q_Y(q)} \tag{D5.12}$$

Table D5.1 Entries for the numerator $S_Y(q,z)$ of the functional $W_Y(q,z)$, with which a Markov trace is built.
The numerator is obtained by multiplying the entries in this table which overlap with the Young diagram Y.

$S_Y(q,z)$			
$w-z$	$w-qz$	$w-q^2z$	$\cdots$
$qw-z$	$qw-qz$	$qw-q^2z$	$\ddots$
q^2w-z	q^2w-qz	q^2w-q^2z	$\ddots$
$\vdots$	$\ddots$	$\ddots$	$\ddots$

in terms of which the desired trace in $H_\infty(q)$ with the Markov property is

$$\mathrm{tr}\,(x) = \sum_Y W_Y(q,z)\ \mathrm{tr}_Y(x) \tag{D5.13}$$

In this expression, tr_Y stands for the ordinary matrix trace over the representation Y of $x \in H_\infty(q)$, and the sum is over all irreps Y.

The connection between the trace (D5.13) and the Markov trace on $A_{\beta,n}$ used in the construction of the Jones polynomial is provided by the particular realization (5.16) of the Hecke algebra. In fact, the generators g_i satisfy not only the Hecke defining relations, but also the constraint (5.19), which implies that the only relevant Young diagrams are those with at most two rows. This is the first indication of the connection between the Jones polynomial and the quantum group $U_q(s\ell(2))$, to be discussed in chapter 6. In fact, the Jones polynomial is a special case of a family of link polynomials, derived from the HOMFLY polynomial by setting $\lambda = t^{N-1}$ in (D5.4). The case $N = 0$ corresponds to the Alexander–Conway polynomial, whereas the Jones polynomial is obtained when $N = 2$.

Recall that the Hecke algebra was a step in the construction of the Temperley–Lieb–Jones algebra, the elements of which were identified with the face Boltzmann weights. This is why link invariants can be thought of as partition functions for face models. The topological invariance of a link polynomial is, in fact, a reflection of the integrability of the associated face model. The rules for obtaining link polynomials are, essentially, the same as those for finding trigonometric solutions to the face Yang–Baxter equation. The general case with finite spectral parameter is not covered by this construction.

D5.4 The Burau representation

The Alexander–Conway polynomial admits a realization similar to that given above for the Jones polynomial. It is based on a particular representation of the braid group known as the Burau representation, which we now discuss.

Let t be a non-zero complex number and let β_i $(i = 1, \ldots, n-1)$ be the $n \times n$ matrices

$$\beta_i = \begin{array}{c} \\ i \\ i+1 \\ \\ \end{array} \begin{pmatrix} \mathbf{1} & \mathbf{0} & & \mathbf{0} \\ & 1-t & 1 & \\ \mathbf{0} & & & \mathbf{0} \\ & t & 0 & \\ \mathbf{0} & \mathbf{0} & & \mathbf{1} \end{pmatrix} \qquad \text{(columns } i,\ i+1\text{)} \tag{D5.14}$$

It is not hard to see that $\beta_i\beta_{i+1}\beta_i = \beta_{i+1}\beta_i\beta_{i+1}$ and that $\beta_i\beta_j = \beta_j\beta_i$ if $|i-j| \geq 2$. The Burau representation of the braid group generator σ_i is the matrix β_i. The matrices (D5.14) leave invariant the subspace of $\mathbb{C}^n$ whose entries add up to zero. Taking the quotient by this subspace, we obtain the following representation:

$$b_1 = \begin{pmatrix} -t & 0 & \\ -1 & 1 & \mathbf{0} \\ & \mathbf{0} & \mathbf{1} \end{pmatrix} \tag{D5.15a}$$

$$b_i = \begin{array}{c} \\ i-1 \\ i \\ i+1 \\ \\ \end{array} \begin{pmatrix} \mathbf{1} & & \mathbf{0} & & \mathbf{0} \\ & 1 & -t & 0 & \\ \mathbf{0} & 0 & -t & 0 & \mathbf{0} \\ & 0 & -1 & 1 & \\ \mathbf{0} & & \mathbf{0} & & \mathbf{1} \end{pmatrix} \qquad \text{(columns } i-1,\ i,\ i+1\text{)} \qquad 1 < i < n-1 \tag{D5.15b}$$

$$b_{n-1} = \begin{pmatrix} \mathbf{1} & \mathbf{0} & \\ & 1 & -t \\ \mathbf{0} & 0 & -t \end{pmatrix} \tag{D5.15c}$$

Let ψ denote the representation of the braid group B_n by the above matrices, $\psi(\sigma_i) = b_i$. Burau's noteworthy result is the following expression for the Alexander–Conway polynomial:

$$\Delta_{\hat{\alpha}}(t) = \left(\sum_{i=1}^{n-1} t^i \right)^{-1} \det\left[1 - \psi(\alpha)\right] \qquad \alpha \in B_n \tag{D5.16}$$

This determinant hints strongly at the intimate connection between the Alexander–Conway polynomial and fermions.

D5.5 Extended Yang–Baxter systems

In chapter 2, we obtained finite-dimensional representations of the braid group associated with statistical models. It is natural to use these representations to define knot invariants. The braid limit (spectral parameter $u \to \infty$) of the Yang–Baxter operators $X_i(u)$ for vertex models yields the representation

$$\rho(\sigma_i) = \mathbf{1} \otimes \cdots \otimes \mathbf{1} \otimes R \otimes \mathbf{1} \otimes \cdots \otimes \mathbf{1} \in \mathrm{End}\left[\bigotimes^n V\right] \qquad \sigma_i \in B_n \tag{D5.17}$$

where V is a vector space whose dimension equals the number of lattice variables of the model, R interchanges the ith and $(i+1)$th copies of V, and the matrices $\rho(\sigma_i) = R_i$ satisfy

$$\begin{aligned} R_i R_{i+1} R_i &= R_{i+1} R_i R_{i+1} \\ R_i R_j &= R_j R_i \qquad |i-j| \geq 2 \end{aligned} \tag{D5.18}$$

Following Turaev, we define an extended Yang–Baxter system by an R-matrix satisfying (D5.18) and a diagonal endomorphism of V

$$\mu : e_i \mapsto \mu_i e_i \text{ (no sum)} \qquad \{e_i\} \text{ basis of } V \tag{D5.19}$$

such that

$$\begin{aligned} &\left(\mu_i \mu_j - \mu_k \mu_\ell\right) R_{ij}^{k\ell} = 0 \\ &\sum_j R_{ij}^{kj} \mu_j = a\, b\; \delta_i^k \\ &\sum_j \left(R^{-1}\right)_{ij}^{kj} \mu_j = a^{-1}\, b\; \delta_i^k \end{aligned} \tag{D5.20}$$

with non-zero $a, b \in \mathbb{C}$. A trivial example of an extended Yang–Baxter system is provided by $R = \mathbf{1} \otimes \mathbf{1}$, μ any homomorphism, $a = 1$ and $b = \mathrm{tr}\,(\mu) = \sum \mu_i$.

Given an extended Yang–Baxter system, the following functional is invariant under Markov moves:

$$T(\alpha) = a^{-w(\alpha)} b^{-n}\ \mathrm{tr}\left(\rho(\alpha)\mu^{\otimes n}\right) \qquad \alpha \in B_n \tag{D5.21}$$

The writhe number $w(\alpha)$ was introduced in (D5.9) above.

If R has a finite number of different eigenvalues, then it must satisfy an equation of the form

$$\sum_{i=m}^{p} k_i R^i = 0$$

with $k_i \in \mathbb{C}$ fixed. It is not hard to show, using the above definitions, that

$$\sum_{i=m}^{p} k_i a^i T(L_i) = 0 \tag{D5.22}$$

The L_i generalize the deformed links we obtained from a given one by zooming into a crossing. Outside of a small disk, they are all equal, and inside they are as shown in figure D5.6. For the usual skein rule, $m = -1$ and $p = +1$: the R-

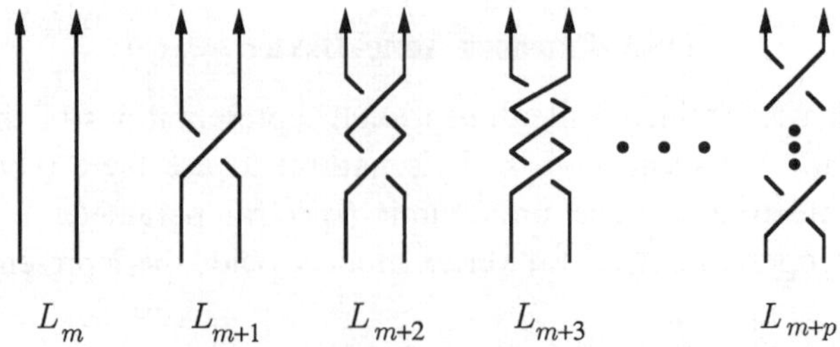

Fig. D5.6

matrices which yield intuitive link invariants have only two different eigenvalues. This is another way of understanding the importance of the Hecke algebra for standard link invariants. Similarly, the Birman–Wenzl algebra is relevant for R's with three different eigenvalues.

Let us note that if $S = (R, \mu, a, b)$ is an extended Yang–Baxter system with link invariant T_S, then $S' = (a^{-1}R, b^{-1}\mu, 1, 1)$ is also an extended Yang–Baxter system, and $T_{S'} = T_S$. Although not always convenient, we could thus restrict ourselves to systems with $a = b = 1$ without loss of generality. Also, letting $S_1 = (-R, -\mu, a, b)$ and $S_2 = (R, \mu, -a, -b)$, then $T_{S_1}(\hat{K}) = T_{S_2}(\hat{K}) = (-1)^n T_S(\hat{K})$ if $K \in B_n$.

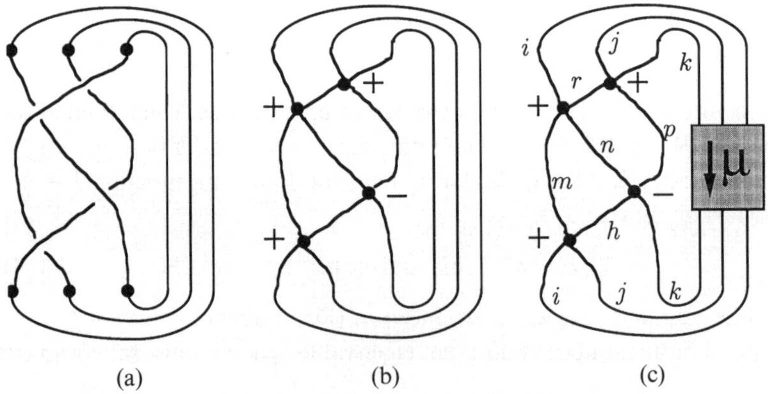

Fig. D5.7

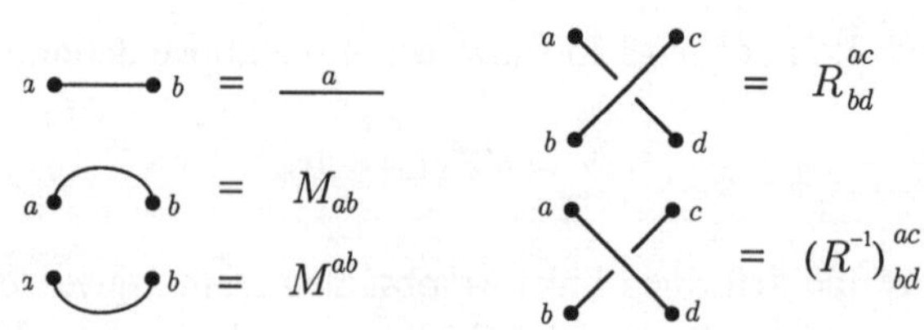

Fig. D5.8

A physical model of (D5.21) follows from the original vertex model. Consider the link formed by closing a braid group element $\alpha \in B_n$, for example the one shown in figure D5.7(a), and its corresponding ϕ^4 diagram, figure D5.7(b). We distinguish between two kinds of ϕ^4 vertices, + and −, depending on whether the crossing comes from a σ_i or a σ_i^{-1}. Assign lattice variables to each line (edge, propagator) of the diagram, as shown in figure D5.7(c). Such an assignment is interpreted as a state configuration on the link. Each state of this type contributes a term

$$\prod_{+\text{ vertices}} R \prod_{-\text{ vertices}} R^{-1} \prod_{\substack{\text{closure}\\ \text{strands}}} \mu \tag{D5.23}$$

to the partition function, i.e. to the link invariant. This contribution can be thought of as consisting of $\rho(\alpha)$ from the braid word and the μ's from the closure identification (through the isomorphism μ) between the in-coming and out-going strands. The trace in (D5.21) is thus the sum over all possible lattice configurations of the terms (D5.23).

For example, the state configuration in figure D5.7(c) contributes

$$R_{ij}^{mh} \left(R^{-1}\right)_{hk}^{np} R_{mn}^{ir} R_{rp}^{jk} \; \mu_i \, \mu_j \, \mu_k \tag{D5.24}$$

A diagrammatic representation of the previous construction is due to Kauffman. This approach calls for colored open strings, where the color index is just a lattice variable. The rules of the game are shown in figure D5.8.

The constraints on the matrices M are determined by simple geometric arguments, as shown in figure D5.9. They read, explicitly, as

$$M_{ab} M^{bc} = \delta_a^c \,, \qquad M_{ia} R_{bd}^{ac} M^{dk} = \left(R^{-1}\right)_{ib}^{ck} \tag{D5.25}$$

We assume that the R-matrices satisfy the Yang–Baxter equation, and thus they yield a braid group representation ρ. The trace in (D5.21) is defined as

$$\operatorname{tr}(\alpha) = \operatorname{Tr}\left((M\, M^T)^{\otimes n} \rho(\alpha) \right) \tag{D5.26}$$

In this pictorial approach, the matrices M can be interpreted quite nicely as the braid limit of scattering matrices for the elastic collision of a particle on a reflecting wall.

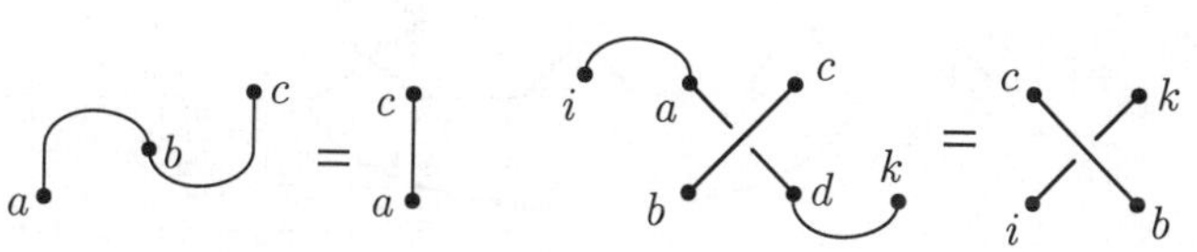

Fig. D5.9

Exercises

Ex. D5.1 Using l'Hospital's rule, find the value of z in (D5.12) such that $\lim_{q\to 1} W_Y(q,z)$ reproduces the usual formula for the dimension of an $SU(N)$ irrep.

Ex. D5.2 Show that the Alexander–Conway polynomial of the knot shown in figure D5.10 is 1, just as it is for the unknot.

Ex. D5.3 Given a Laurent polynomial $P(t)$, is there a knot or link whose Alexander–Conway polynomial is precisely $P(t)$? Show that the conditions

$$\begin{aligned} P(t) &= P(t^{-1}) \\ P(1) &= 1 \end{aligned} \tag{D5.27}$$

are necessary and sufficient.

Ex. D5.4 (Proposed by Vaughn Jones, answer unknown.) Is there a knot K, other than the unknot, the Jones polynomial of which, $V_K(t)$, is 1? Given a polynomial $P(t)$, what are the necessary and sufficient conditions for the existence of a knot with Jones polynomial $P(t)$?

Ex. D5.5 Find the extended Yang–Baxter systems associated with the braid limits of the six- and eight-vertex R-matrices.

Ex. D5.6 Try to include a dependence on the rapidity in the notion of an extended Yang–Baxter system and, therefore, in that of a link invariant.

Fig. D5.10. This knot is constructed by gluing three tangles, containing −3, +5 and +7 windings, respectively.

Appendix E
Spin models

E5.1 Factors and subfactors

In this appendix, we will briefly review spin models and some of the mathematics associated with them, following Jones. The Yang–Baxter equation in its true star-triangle form will appear eventually, but first let us start with the artillery.

Let M be a closed, $*$-closed algebra of bounded operators on a Hilbert space, with an identity $\mathbf{1} \in M$. Then M is called a von Neumann algebra. A factor is a von Neumann algebra M with trivial center, $Z(M) = \mathbb{C}\mathbf{1}$. Factors have been classified by Connes, but we are interested only in II_1 factors, which are infinite-dimensional factors with a trace, i.e. a linear map $\mathrm{tr} : M \to \mathbb{C}$ such that $\mathrm{tr}(ab) = \mathrm{tr}(ba)$.

For example, consider a group G and its unitary representations. Then the set of all intertwiners is a von Neumann algebra, and it is a II_1 factor in some special cases, for example if G is $SL(n, \mathbb{Z})$.

Another example is based on the space $L = L^\infty([0, 1])$. Define the operator $T \in \mathrm{End}(L)$ by $(Tf)(x) = f(x + \lambda \bmod 1)$. The set of formal power series $M = \{a = \sum a_i z^i;\ a_i \in L\}$ is a von Neumann algebra if z is unitary $z^* = z^{-1}$ and a II_1 factor if we require $za = T(a)z^{-1}$. In this case, the trace is given by $\mathrm{tr}\left(\sum a_i z^i\right) = \int_0^1 a_0 dx$.

If M_1 and M_2 are factors, and $M_1 \subset M_2$, then M_1 is a subfactor of M_2 if they have the same identity. Just as for the towers of matrix algebras in section 5.2.4, the basic construction of a third subfactor $M_3 = \mathrm{End}_{M_1}(M_2)$ (viewing M_2 as a right M_1 module) is such that, if M_1 and M_2 are II_1 subfactors, then so is $M_3 \supset M_2 \supset M_1$. The projector $E_{M_1}^{M_2} = E_1 : M_2 \to M_1$ is an element of M_3, and the von Neumann–Jones index of M_2 relative to M_1 is defined as

$$[M_2 : M_1] \equiv [\mathrm{tr}(E_1)]^{-1} \tag{E5.1}$$

This index measures the number of copies of M_1 that can be embedded in M_2, and it is important because its possible values are known:

$$[M_2 : M_1] = \begin{cases} 4\cos^2 \frac{\pi}{n} & n = 3, 4, 5, \ldots \\ \in \mathbb{R} & \text{if} > 4 \end{cases} \tag{E5.2}$$

For example, if $M_1 = \mathbb{C}$ and M_2 are the $n \times n$ complex matrices, then $[M_2 : M_1] = n^2$. Another example uses as the only input a finite group G: for any M_1, if M_2 is the semi-direct product of M_1 and G, then $[M_2 : M_1] = |G|$. In this case, the centralizer of M_1 in M_2 is precisely G.

An example of physical interest is obtained by taking M_2 to be the set of transfer matrices of the six-vertex model, and letting $M_1 = \phi(M_2)$, with $\phi \in \mathrm{End}(M_2)$ the lattice shift operator, $\phi(\sigma_i) = \sigma_{i+1}$. Then $[M_2, \phi(M_2)] = 2+q+q^{-1}$. Similarly, if M_2 is the set of all transfer matrices for A_n face models, then $[M_2, \phi(M_2)] = 4\cos^2 \pi/n$.

A useful theorem implies that the study of II_1 subfactors is reduced to the study of finite-dimensional II_1 subfactors. Indeed, let N_i (respectively P_i) be the centralizer of M_0 in M_i (respectively of M_1 in M_i), with $M_i \subset M_{i+1}$. Then

$$\begin{array}{cccccc} \cdots \subset & N_i \cap M_i & \subset & N_{i+1} \cap M_{i+1} & \subset \cdots \subset & N \\ & \cup & & \cup & & \cup \\ \cdots \subset & P_i \cap M_i & \subset & P_{i+1} \cap P_{i+1} & \subset \cdots \subset & P \end{array} \tag{E5.3}$$

and the theorem says that $(M_0 \subset M_1) \cong (P \subset N)$.

An interesting and quite non-trivial generalization of the towers of algebras is obtained when considering inclusions of subfactors which are not tree-like. Whereas in tree-like inclusion, i.e. towers, there is no room for continuous parameters, as soon as we consider loops there is the possibility of generating moduli for the algebra of paths. The simplest building block for such purposes is a "commuting square", defined by four algebras

$$\begin{array}{ccc} C & \subset & D \\ \cup & & \cup \\ A & \subset & B \end{array} \tag{E5.4}$$

such that D has a trace and the projections are compatible, $E^D_C E^D_B = E^D_A$. This means that the projection $D \to C$ induces a projection $B \to A$ (see figure E5.1).

There are other equivalent ways of stating the commuting square condition, i.e. this projection compatibility:

(i) $E^D_B(C) = A$;
(ii) $E^D_C(B) = A$;
(iii) $A = B \cap C$ and $E^D_B E^D_C = E^D_C E^D_B$;
(iv) $E^D_A(ab) = E^D_A(a)E^D_A(b) \quad \forall a \in A, b \in B$;
(v) B and C are orthogonal modulo A, where the scalar product is defined as $\langle a, b\rangle = \mathrm{tr}(b^* a)$;
(vi) the connection in the space of paths is flat.

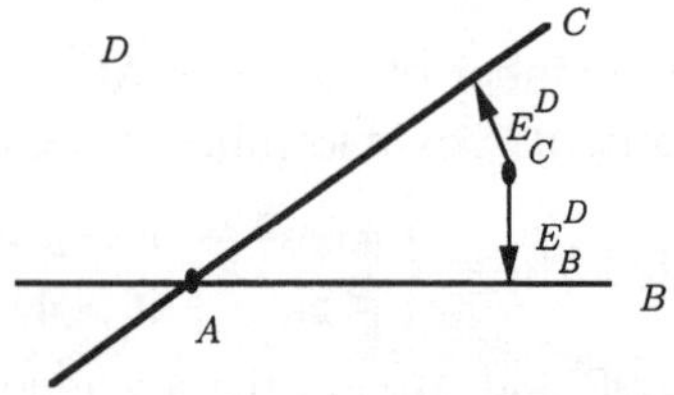

Fig. E5.1. If D is the complex plane, and B and C are two straight lines going through the point A, the commuting square condition requires that the angle between B and C is 90°.

For example, if D denotes the $n \times n$ complex matrices, A is just $\mathbb{C}$ and B denotes the diagonal $n \times n$ matrices, a commuting square is obtained with

$$D \supset C = \{u^* du, d \in B, u \in D, u^* = u^{-1} \text{ and } |u_{ij}| = 1\} \tag{E5.5}$$

Thus, an element of C is the unitary conjugate of a diagonal matrix, where the unitary transformation is such that all its matrix elements have the same absolute value. Such a matrix u is called bi-unitary, so C are the bi-unitary conjugates of diagonal matrices. The standard example of an $n \times n$ bi-unitary matrix is the $\mathbb{Z}_n$ Fourier transform, $u_{ab} = \exp(2\pi iab/n)/\sqrt{n}$.

Quite generally, given the commuting square (E5.4) with C unknown, Ocneanu has found a solution for C of the form $C = uBu^{-1}$, $u \in D$ unitary ($u^* = u^{-1}$). In terms of the algebra of paths implied by (E5.4), any element of D can be viewed as a pair of paths from A to D, one through B and one through C. Thus, u must be a linear combination of pairs of paths, i.e. of squares

$$u = \sum w(a, b, c, d) \text{ Path}(a, b, c, d) \tag{E5.6}$$

where

$$\text{Path}(a, b, c, d) = \begin{array}{ccc} c & \text{—} & d \\ | & & | \\ a & \text{—} & b \end{array} \tag{E5.7}$$

The commuting square condition amounts to requiring bi-unitarity of u, which in the path language is equivalent to crossing symmetry.

Imagine that the commuting square above, with u of the form (E5.6), is extended by the basic construction on the top line and by a slight modification thereof on the bottom line:

$$\begin{array}{ccccc} C & \subset & D & \subset & \text{End}_C(D) \\ \cup & & \cup & & \cup \\ A & \subset & B & \subset & \text{End}_C(B) \end{array} \tag{E5.8}$$

such that each square is commuting. It is not hard to see that an element x of $\text{End}_C(B)$ can also be written as a (different) linear combination of the same paths as u:

$$x = \sum w'(a, b, c, d) \text{ Path}(a, b, c, d) \tag{E5.9}$$

Jones has found that the requirement that x be in the centralizer of C, i.e. that $[x, C] = 0$, is equivalent to the existence of a third set of coefficients $w''(a, b, c, d)$ such that w, w' and w'' satisfy the face Yang–Baxter equation . This result explains why knot theory and the braid group pop up so often around integrability.

E5.2 Spin models

Let us define now a third kind of lattice models, close to graph face models but somewhat simpler. The lattice variables are taken just as in graph face models, and states are built in the same way. The difference lies in the Boltzmann weights. Instead of assigning weights to a plaquette labeled by the four lattice variables at the vertices, we assign the Boltzmann weights to links labeled by the two lattice variables at the ends of the link. We therefore consider an $N \times N$ matrix R, where $R_{ab} = w(a,b)$ are the weights, and the graph is assumed to contain N points.

If, for example, we choose $w(a,b) = 1 - \delta_{a,b}$, then the partition function

$$Z = \sum_{\text{states}} \prod_{\text{edges}} w(\sigma, \sigma') \tag{E5.10}$$

on a lattice $\mathscr{L}$ counts the number of ways of coloring the $\mathscr{L}$ with N colors, such that the color of a site is different from the color of any adjoining site.

To describe spin models formally, it is convenient to introduce two matrices of Boltzmann weights, $\{w_1(a,b)\}$ and $\{w_2(a,b)\}$, in principle independent. Given a state $|\ell_1, \ell_2, \cdots\rangle \in \otimes V$, with $\dim V = N$, we define the action of two operators R_1 and $R_2 \in \otimes \mathrm{End}\ V$ as follows:

$$\begin{aligned} R_1 \,|\ell_1, \ell_2, \cdots\rangle &= \sum_{k=1}^{N} w_1(\ell_1, k)\, |k, \ell_2, \ell_3, \ldots\rangle \\ R_2 \,|\ell_1, \ell_2, \cdots\rangle &= w_2(\ell_1, \ell_2)\, |\ell_1, \ell_2, \ell_3, \ldots\rangle \end{aligned} \tag{E5.11}$$

Thus, R_1 changes the spin variable on the first site and yields a linear combination of states weighted by w_1, whereas R_2 just "measures" the first link (ℓ_1, ℓ_2) with weight w_2. In diagrams, it is convenient to represent w_1 by a vertical line and w_2 by a horizontal one:

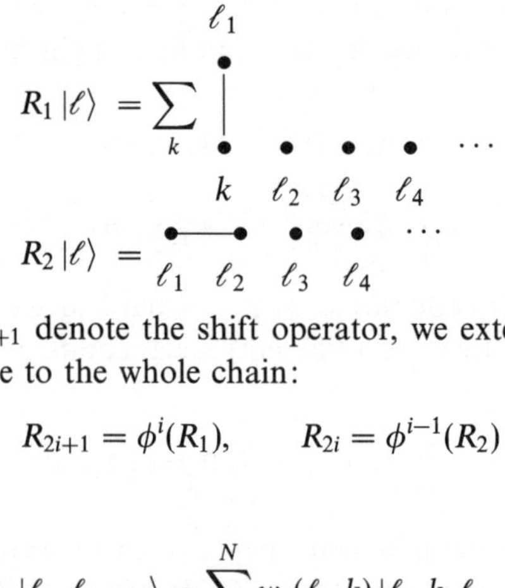

(E5.12)

Letting $\phi : \ell_i \to \ell_{i+1}$ denote the shift operator, we extend the two actions R_1 and R_2 on the first site to the whole chain:

$$R_{2i+1} = \phi^i(R_1), \qquad R_{2i} = \phi^{i-1}(R_2) \tag{E5.13}$$

so that, for instance,

$$\begin{aligned} R_3 \,|\ell_1, \ell_2, \cdots\rangle &= \sum_{k=1}^{N} w_1(\ell_2, k)\, |\ell_1, k, \ell_3, \ldots\rangle \\ R_4 \,|\ell_1, \ell_2, \cdots\rangle &= w_2(\ell_2, \ell_3)\, |\ell_1, \ell_2, \ell_3, \ldots\rangle \end{aligned} \tag{E5.14}$$

The algebra J_n generated by $\{\mathbf{1}, R_1, R_2, \ldots, R_n\}$ is semi-simple if R_1 is real and symmetric, whereby R_{2i+1} is diagonalizable (note that R_{2i} is already diagonal anyway). The interesting part of this statement is that any semi-simple associative algebra over $\mathbb{C}$ is a direct sum of finite-dimensional matrices.

The operator $R_1 R_2 \cdots R_{2n-1}$ acting on $|\ell_1, \ell_2, \ldots, \ell_n\rangle$ is just the diagonal to diagonal transfer matrix of the spin model, namely

$$R_1 R_2 \cdots R_{2n-1} \, |\ell_1, \ell_2, \ldots, \ell_n\rangle = \sum_{k_1,\ldots,k_n} \left(\prod_{i=1}^{n} w_1(\ell_i, k_i) \right) \times \left(\prod_{i=1}^{n-1} w_2(\ell_i, k_{i+1}) \right) |k_1, k_2, \ldots, k_n\rangle \qquad \text{(E5.15)}$$

Using (E5.12), this transfer matrix can be represented schematically as

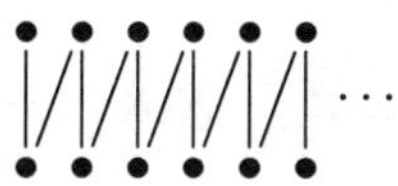

(E5.16)

For example, if we take $w_1(a,b) = 1$ and $w_2(a,b) = \delta_{a,b}$ $(1 \leq a, b \leq N)$, then from the relations

$$\begin{aligned} R_1^2 &= N R_1 \\ R_2^2 &= R_2 \\ R_1 R_2 R_1 &= R_1 \\ R_2 R_1 R_2 &= R_2 \end{aligned} \qquad \text{(E5.17)}$$

we see that the algebra J_n is just the Temperley–Lieb–Jones algebra $A_{\sqrt{N},n}$, with the identifications

$$e_{2i-1} = \frac{1}{\sqrt{N}} R_{2i-1}, \qquad e_{2i} = \sqrt{N} R_{2i} \qquad \text{(E5.18)}$$

The Bratelli diagram for the tower $\mathbf{1} \subset J_1 \subset \cdots \subset J_n \subset J_{n+1} \subset \cdots$ coincides with that of A_3 if $N = 2$, with that of A_5 if $N = 3$, and with the non-truncated Bratelli of A_∞ if $N \geq 4$.

Another example is provided by the choice $w_1(a,b) = \delta_{a,b+1}$, $w_2(a,b) = \epsilon^{2(a-b)}$, where $1 \leq a, b \leq N$ with N odd and $\epsilon = \exp 2\pi i/N$. Now the operators R_i satisfy the algebra

$$\begin{aligned} R_i^N &= 1 \\ R_i R_{i+1} &= \epsilon^2 R_{i+1} R_i \\ R_i R_j &= R_j R_i \qquad |i-j| \geq 2 \end{aligned} \qquad \text{(E5.19)}$$

In order to implement the Reidemeister moves, an interesting choice for R_2 is $w_2(a,b) = w(b,a)^*$, with $w_1(a,b) = w(a,b)$ and $R^* = R^{-1}$, where R denotes the unitary matrix $R_{ab} = w(a,b)$. Effectively, instead of having two kinds of Boltzmann weights for the links, we have only one. Now we can use the unitary matrix R to build a commuting square, as in the previous section. The commuting square condition is fulfilled if and only if R is bi-unitary, i.e. if $|w(a,b)|$ is a constant. We may now extend the commuting square, via the fundamental construction, vertically in the square (E5.4), although, for convenience, we switch the positions of B and C in what follows and obtain

$$\begin{array}{ccccccccc} \Delta & \subset & M & \subset & \langle M, E_1\rangle & \subset & \langle M_1, E_1, E_2\rangle & \subset \cdots \\ \cup & & \cup & & \cup & & \cup & \\ \mathbb{C} & \subset & R\Delta R^{-1} & \subset & \left\langle R\Delta R^{-1}, E_1\right\rangle & \subset & \left\langle R\Delta R^{-1}, E_1, E_2\right\rangle & \subset \cdots \end{array} \qquad \text{(E5.20)}$$

where M denotes the $N \times N$ complex matrices, Δ denotes the diagonal ones, and the projectors E_i are obtained by the fundamental construction. Note that $\left[R\Delta R^{-1}, \Delta\right] = \mathbb{C}$ and $[E_i, \Delta] = 0$.

Next, we wish to find the generators E_i of the Temperley–Lieb–Jones algebra explicitly. Observe first that, if R_1 is unitary and $|w_2(a,b)|$ is a constant $(\forall a, b)$, then

$$\rho = \lim_{n\to\infty} R_1 R_2 \cdots R_n R_n^{-1} \cdots R_2^{-1} R_1^{-1} \qquad \text{(E5.21)}$$

is an endomorphism of operators on the semi-infinite lattice. If $R_1 = R$ is bi-unitary and $R_2 = R_1^{-1}$ as above, then $\rho(E_i) = E_{i+1}$, so conjugation by the transfer matrix ρ acts in the diagram (E5.20) downwards and to the right. The endomorphism ρ is not the shift operator $\phi : E_i \to E_{i+2}$, although $\rho^2 = \phi$.

Now, in the tower $\Delta \subset M \subset \langle M, E_1\rangle \subset \langle M_1, E_1, E_2\rangle \subset \cdots$, we may represent each new Temperley–Lieb–Jones generator as one more index, so that if M denotes the $N \times N$ matrices, then $\langle M, E_1\rangle$ denotes the $(N+1)\times(N+1)$ matrices, and generally $M_n = \langle M, E_1, \ldots, E_n\rangle$ denotes the $(N+n)\times(N+n)$ matrices. If $X \in M_n$, requiring $[X, \Delta] = 0$ means that X, as a tensor, is diagonal in the first component, i.e. $X_{a_1 a_2 \cdots a_n}^{b_1 b_2 \cdots b_n} = \delta_{a_1}^{b_1} Y_{a_2 \cdots a_n}^{b_2 \cdots b_n}$. Conjugation by the transfer matrix yields the equation $RXR^{-1} = Y$, i.e. $RX = YR$, shown in figure E5.2(a) .

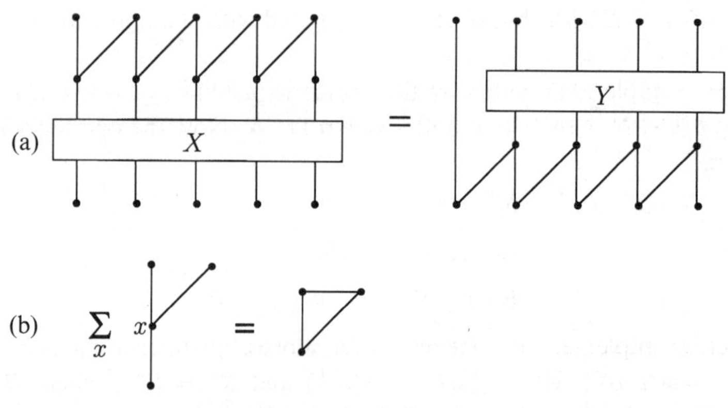

Fig. E5.2. (a) The equation which follows from the requirement that the commutant of the diagonal matrices in an algebra of the tower vanishes. (b) The star-triangle relation follows from requiring that the commutant of diagonal operators Δ in $\left\langle R\Delta R^{-1}, E_1\right\rangle$ vanish.

The equation $RX = YR$ is a generalization of the star-triangle relation: note that there is an implicit sum over an internal index in RX, a sum that is absent in the term YR. This can be seen explicitly by finding the commutant of Δ in $\langle R\Delta R^{-1}, E_1\rangle$ and requiring it to be zero. This yields, in fact, the star-triangle relation

$$\sum_x w_1(x,c)\, w_2(b,x)\, w_1(a,x) \;=\; w_2(c,b)\, w_1(a,c)\, w_2(a,b) \tag{E5.22}$$

shown in figure E5.2(b), which is another form of the Yang–Baxter equation. This is the original guise in which the Yang–Baxter equation was discovered by Onsager in his pioneering solution of the two-dimensional Ising model. The name of star-triangle equation derives from an ancient duality equation for coils in electromechanics; we shall use (E5.22) in our discussion of the chiral Potts model in chapter 7.

Exercises

Ex. E5.1 Draw the Bratelli diagrams for the Temperley–Lieb–Jones algebras $\mathbf{1} \subset A_{\sqrt{N},1} \subset \cdots \subset A_{\sqrt{N},n} \cdots$ in (E5.18) for $N \geq 4$, and assign to each site the number of paths from the $\mathbf{1}$ site to that site. Show that the dimension of $A_{\sqrt{N},n}$, which is the sum of the squares of the numbers on the Bratelli diagram at level n, is given by the Catalan number $C_n = \frac{1}{n+1}\binom{2n}{n}$.

Ex. E5.2 From the matrices R_i in (E5.19), define the operators $S_i = \sum_{j=0}^{N-1} \epsilon^{2j} R_i^j$ and show that they form a representation of the braid group.

Ex. E5.3 Given a subfactor $N \subset M$, build a tower of inclusions

$$N \subset M \subset M_1 = \langle M, E_1\rangle \subset M_2 = \langle M_1, E_2\rangle \subset \cdots \tag{E5.23}$$

with the help of the fundamental construction. Letting X' denote the commutant or centralizer of X, show that

$$\begin{array}{ccc} N' \cap M & \subset & N' \cap M_1 \\ \cup & & \cup \\ M' \cap M & \subset & M' \cap M_1 \end{array} \tag{E5.24}$$

is a commuting square.

Ex. E5.4 Show that if D denotes all complex $N \times N$ matrices, A denotes the complex numbers, B denotes the diagonal $N \times N$ matrices and C is spanned by $\{\mathbf{1}, u, u^2, \ldots, u^{N-1}\}$ with $u_{ab} = \delta_{a,1}\delta_{b,N} + \delta_{a+1,b}$, the diagram (E5.4) is a commuting square. Check all (seven) equivalent conditions in section E5.1.

6
Quantum groups: mathematical review

We provide in this chapter a brief review of mathematical definitions and results concerning quantum groups, both finite and affine.

6.1 Hopf algebras

Let $(\mathscr{A}, m, \imath)$ be an algebra whose multiplication $m : \mathscr{A} \otimes \mathscr{A} \to \mathscr{A}$ is associative, i.e.

$$[m(m \otimes \mathbf{1})](a \otimes b \otimes c) = [m(\mathbf{1} \otimes m)](a \otimes b \otimes c) \qquad \forall a, b, c \in \mathscr{A} \tag{6.1}$$

We shall often write $ab = m(a \otimes b)$, $\forall a, b \in \mathscr{A}$. If $a \in \mathscr{A}$ and $\lambda \in \mathbb{C}$, to make formal sense of $\lambda a \in \mathscr{A}$, we need the unit map $\imath : \mathbb{C} \to \mathscr{A}$ of $\mathscr{A}$, which is intimately linked to the identity $\mathbf{1} \in \mathscr{A}$:

$$\imath : \lambda \in \mathbb{C} \mapsto \lambda \mathbf{1} \in \mathscr{A} \tag{6.2}$$

The unit $\imath$ and the multiplication m are compatible in the sense that

$$m(a \otimes \imath(\lambda)) = a\lambda = \lambda a = m(\imath(\lambda) \otimes a) \qquad \forall a \in \mathscr{A} \quad \forall \lambda \in \mathbb{C} \tag{6.3}$$

Let us consider now a co-multiplication or co-product $\Delta : \mathscr{A} \to \mathscr{A} \otimes \mathscr{A}$, which should be "co-associative":

$$(\Delta \otimes \mathbf{1})(\Delta(a)) = (\mathbf{1} \otimes \Delta)(\Delta(a)) \qquad \forall a \in \mathscr{A} \tag{6.4}$$

Mathematicians like to depict such properties by commutative diagrams; the image of any element under any of the various paths should be the same. In this case, the appropriate diagram is

$$\begin{array}{ccccc} \mathscr{A} & \xrightarrow{\Delta} & \mathscr{A} \otimes \mathscr{A} & \xrightarrow{\Delta \otimes \mathbf{1}} & \mathscr{A} \otimes \mathscr{A} \otimes \mathscr{A} \\ \| & & & & \| \\ \mathscr{A} & \xrightarrow{\Delta} & \mathscr{A} \otimes \mathscr{A} & \xrightarrow{\mathbf{1} \otimes \Delta} & \mathscr{A} \otimes \mathscr{A} \otimes \mathscr{A} \end{array} \tag{6.5}$$

In short-hand, we write the co-associativity condition as

$$(\mathbf{1} \otimes \Delta)\Delta = (\Delta \otimes \mathbf{1})\Delta \tag{6.6}$$

We also need a co-unit map $\epsilon : \mathscr{A} \to \mathbb{C}$ to define a co-algebra $(\mathscr{A}, \Delta, \epsilon)$; it must satisfy

$$(\mathbf{1} \otimes \epsilon)\Delta = (\epsilon \otimes \mathbf{1})\Delta = \mathbf{1} \tag{6.7}$$

which in diagrams reads as follows:

$$\begin{array}{ccc} \mathscr{A} & \xrightarrow{\mathbf{1}} & \mathscr{A} \\ \downarrow{\scriptstyle\Delta} & & \| \\ \mathscr{A}\otimes\mathscr{A} & \xrightarrow{\mathbf{1}\otimes\epsilon} & \mathscr{A}\otimes\mathbb{C} \end{array} \quad \text{and} \quad \begin{array}{ccc} \mathscr{A} & \xrightarrow{\mathbf{1}} & \mathscr{A} \\ \downarrow{\scriptstyle\Delta} & & \| \\ \mathscr{A}\otimes\mathscr{A} & \xrightarrow{\epsilon\otimes\mathbf{1}} & \mathbb{C}\otimes\mathscr{A} \end{array} \tag{6.8}$$

Now, a structure $(\mathscr{A}, m, \imath, \Delta, \epsilon)$, i.e. a simultaneous algebra and co-algebra, is called a bi-algebra if the co-multiplication Δ and the co-unit ϵ are consistent with the multiplication m, i.e. if they are homomorphisms:

$$\epsilon(ab) = \epsilon(a)\epsilon(b), \qquad \Delta(ab) = \Delta(a)\Delta(b) \tag{6.9}$$

Actually, the unit $\imath$ and co-unit ϵ must also be compatible:

$$\imath(\epsilon(a)) = \epsilon(a)\mathbf{1} \qquad (\forall a \in \mathscr{A}) \tag{6.10}$$

A Hopf algebra is a bi-algebra enjoying an antipode $\gamma : \mathscr{A} \to \mathscr{A}$, which is an antihomomorphism

$$\gamma(ab) = \gamma(b)\gamma(a) \tag{6.11}$$

satisfying the following condition:

$$m(\gamma \otimes \mathbf{1})\Delta(a) = m(\mathbf{1} \otimes \gamma)\Delta(a) = \epsilon(a)\mathbf{1} \qquad \forall a \in \mathscr{A} \tag{6.12}$$

This condition involves all the ingredients of the bi-algebra structure, as can be seen from its diagrammatic version:

$$\begin{array}{ccccc} \mathscr{A} & \xrightarrow{\Delta} & \mathscr{A}\otimes\mathscr{A} & \xrightarrow{\mathbf{1}\otimes\gamma} & \mathscr{A}\otimes\mathscr{A} \\ \| & & & & \downarrow{\scriptstyle m} \\ \mathscr{A} & \xrightarrow{\epsilon} & \mathbb{C} & \xrightarrow{\imath} & \mathscr{A} \end{array} \tag{6.13}$$

A similar diagram also commutes with $\gamma \otimes \mathbf{1}$ instead of $\mathbf{1} \otimes \gamma$.

If the multiplication m is commutative (respectively, not commutative) , we call the algebra commutative (respectively, non-commutative). Just like the multiplication, the co-multiplication may or may not be commutative. The Hopf algebra is accordingly co-commutative or non-co-commutative. Of primary interest are those Hopf algebras which are neither commutative nor co-commutative. If the Hopf algebra is commutative, then the following diagram holds:

$$\begin{array}{ccc} \mathscr{A}\otimes\mathscr{A} & \xrightarrow{m} & \mathscr{A} \\ \downarrow{\scriptstyle\sigma} & & \| \\ \mathscr{A}\otimes\mathscr{A} & \xrightarrow{m} & \mathscr{A} \end{array} \tag{6.14}$$

We have introduced the permutation map

$$\begin{aligned} \sigma : \mathscr{A}\otimes\mathscr{A} &\to \mathscr{A}\otimes\mathscr{A} \\ a \otimes b &\mapsto b \otimes a \end{aligned} \tag{6.15}$$

which merely interchanges the order of the operands. Thus, commutativity means

$$ab \equiv m(a \otimes b) = m(\sigma(a \otimes b)) \equiv m(b \otimes a) \equiv ba \qquad \forall a, b \in \mathscr{A} \tag{6.16}$$

On the other hand, if the algebra is co-commutative, then the diagram is

$$\begin{array}{ccc} \mathscr{A} & \xrightarrow{\Delta} & \mathscr{A} \otimes \mathscr{A} \\ \Big\| & & \Big\downarrow \sigma \\ \mathscr{A} & \xrightarrow{\Delta} & \mathscr{A} \otimes \mathscr{A} \end{array} \tag{6.17}$$

or equivalently

$$\Delta(a) = \sigma \cdot \Delta(a) \equiv \Delta'(a) \qquad \forall a \in \mathscr{A} \tag{6.18}$$

To end this set of basic definitions, note that the commutative diagrams employed above are quite useful: if we reverse the arrows in a diagram expressing a property of the algebra, we obtain the analogous diagram for the co-algebra property, and vice versa. Arrow reversal interchanges m with Δ and ι with ϵ, but leaves γ alone. The antipode must thus be viewed on an equal footing as an inverse and a co-inverse, and its existence is highly non-trivial: Hopf algebras are much more interesting than mere bi-algebras.

6.2 Quasi-triangular Hopf algebras

Given a co-multiplication Δ, it is not hard to check that the operation $\Delta' = \sigma \circ \Delta \in \text{End}(\mathscr{A} \otimes \mathscr{A})$ is also a co-multiplication, with modified antipode $\gamma'(a) = [\gamma(a)]^{-1}$, $(\forall a \in \mathscr{A})$. A given Hopf algebra $\mathscr{A}$ is called quasi-triangular if there exists a universal $\mathscr{R}$-matrix $\mathscr{R} \in \mathscr{A} \otimes \mathscr{A}$ such that

$$\Delta'(a) = \mathscr{R}\Delta(a)\mathscr{R}^{-1} \qquad \forall a \in \mathscr{A} \tag{6.19}$$

and

$$\begin{aligned} (\mathbf{1} \otimes \Delta)\mathscr{R} &= \mathscr{R}_{13}\mathscr{R}_{12} = \sum_{i,j} A_i A_j \otimes B_j \otimes B_i \\ (\Delta \otimes \mathbf{1})\mathscr{R} &= \mathscr{R}_{13}\mathscr{R}_{23} = \sum_{i,j} A_i \otimes A_j \otimes B_i B_j \end{aligned} \tag{6.20}$$

and

$$(\gamma \otimes \mathbf{1})\mathscr{R} = (\mathbf{1} \otimes \gamma^{-1})\mathscr{R} = \mathscr{R}^{-1} \tag{6.21}$$

where $\mathscr{R}$ is called the universal $\mathscr{R}$-matrix. We write $\mathscr{R} = \sum_i A_i \otimes B_i$ and let

$$\begin{aligned} \mathscr{R}_{12} &= \sum_i A_i \otimes B_i \otimes \mathbf{1} \\ \mathscr{R}_{13} &= \sum_i A_i \otimes \mathbf{1} \otimes B_i \\ \mathscr{R}_{23} &= \sum_i \mathbf{1} \otimes A_i \otimes B_i \end{aligned} \tag{6.22}$$

Essentially, quasi-triangularity means that the co-multiplication Δ and its "transpose" Δ' are related linearly. In some sense, it establishes an equivalence between two different ways of "adding things up".

A co-commutative algebra is trivially quasi-triangular, with $\mathscr{R} = \mathbf{1} \otimes \mathbf{1}$. A Hopf algebra is called triangular if $\mathscr{R}_{12}\mathscr{R}_{21} = \mathbf{1} \otimes \mathbf{1}$, where $\mathscr{R}_{21} = \sum B_i \otimes A_i$.

A non-co-commutative quasi-triangular Hopf algebra is called a quantum group.

The interest in quasi-triangular Hopf algebras is that they produce solutions to the Yang–Baxter equation naturally. Indeed, we may derive from (6.19) and (6.20) the Yang–Baxter equation in a "universal form" without spectral parameter:

$$\mathscr{R}_{12}\mathscr{R}_{13}\mathscr{R}_{23} = \mathscr{R}_{23}\mathscr{R}_{13}\mathscr{R}_{12} \tag{6.23}$$

The proof goes like this:

$$\begin{aligned}
[(\sigma \circ \Delta) \otimes \mathbf{1}]\, \mathscr{R} &= \sum_i \Delta'(A_i) \otimes B_i \\
&= \sum_i \mathscr{R}_{12}\Delta(A_i)\mathscr{R}_{12}^{-1} \otimes B_i \\
&= \mathscr{R}_{12} \left(\sum_i \Delta(A_i) \otimes B_i \right) \mathscr{R}_{12}^{-1} \\
&= \mathscr{R}_{12}\, [(\Delta \otimes \mathbf{1})\mathscr{R}]\, \mathscr{R}_{12}^{-1} \\
&= \mathscr{R}_{12}\mathscr{R}_{13}\mathscr{R}_{23}\mathscr{R}_{12}^{-1}
\end{aligned} \tag{6.24}$$

On the other hand,

$$\begin{aligned}
[(\sigma \circ \Delta) \otimes \mathbf{1}]\, \mathscr{R} &= \sigma_{12}(\Delta \otimes \mathbf{1})\mathscr{R} \\
&= \sigma_{12}(\mathscr{R}_{13}\mathscr{R}_{23}) \\
&= \mathscr{R}_{23}\mathscr{R}_{13}
\end{aligned} \tag{6.25}$$

and thus (6.23) follows.

6.3 Drinfeld's quantum double

We shall now give the instructions for building a quasi-triangular Hopf algebra starting from any Hopf algebra. Actually, we shall start from two Hopf algebras that are dual to each other, to be defined below in a very intuitive sense. Other than duality, the fundamental ingredient in Drinfeld's quantum double construction is the notion of normal ordering, quite familiar from quantum mechanics.

Let us start from two Hopf algebras $\mathscr{A}$ and $\mathscr{A}^*$, isomorphic as vector spaces. We say that they are dual to each other if the product and co-product of $\mathscr{A}$ coincide, respectively, with the co-product and product of $\mathscr{A}^*$. Formally, letting e_a be the basis vectors for $\mathscr{A}$ and e^a those for $\mathscr{A}^*$, we require that

$$\begin{aligned}
e_a e_b &= m_{ab}^{c}\, e_c \qquad & \Delta(e^a) &= m_{cb}^{a}\, e^b \otimes e^c \\
e^a e^b &= \mu_c^{ab}\, e^c \qquad & \Delta(e_a) &= \mu_a^{bc} e_b \otimes e_c
\end{aligned} \tag{6.26}$$

where the "matrices" m and μ are such that both $\mathscr{A}$ and $\mathscr{A}^*$ are co-algebras. The permutation in the subscripts of the definition of the co-multiplication of $\mathscr{A}^*$ should be stressed.

If the antipode for $\mathscr{A}$ is

$$\gamma(e_a) = \gamma_a^b \, e_b \qquad e_a \text{ basis of } \mathscr{A} \tag{6.27}$$

then the antipode for $\mathscr{A}^*$ is obtained from a simple matrix inversion:

$$\gamma(e^a) = \left(\gamma^{-1}\right)_b^a \, e^b \qquad e^a \text{ basis of } \mathscr{A}^* \tag{6.28}$$

From a physical point of view, we think of $\mathscr{A}$ as being made up of creation operators, whereas $\mathscr{A}^*$ contains the annihilation operators. To assemble the two algebras into Drinfeld's double $\mathscr{D}(\mathscr{A}, \mathscr{A}^*) \subset \mathscr{A} \otimes \mathscr{A}^*$, we must come up with a notion of normal ordering. Following tradition, we shall write the annihilation operators to the right, so a basis of $\mathscr{D}(\mathscr{A}, \mathscr{A}^*)$ is simply $\{e_a e^b\}$.

The co-product in $\mathscr{D}(\mathscr{A}, \mathscr{A}^*)$ follows without any difficulty from the definitions above:

$$\Delta(e_a e^b) = \mu_a^{cd} \, m_{vu}^b \, e_c e^u \otimes e_d e^v \tag{6.29}$$

The product in $\mathscr{D}(\mathscr{A}, \mathscr{A}^*)$, however, calls for some rearranging of the basis elements into normal form, both before and after the definitions (6.26) can be applied. The normal ordering prescription which we use to rewrite things like $e^b e_a$ in the double's basis $e_c e^d$ must be such that the co-product (6.29) acts like an algebra homomorphism, i.e. such that the product and the co-product in $\mathscr{D}(\mathscr{A}, \mathscr{A}^*)$ are compatible. Following Drinfeld, the normal ordering we use is

$$e^a e_b = m_{kd}^x \, m_{xu}^a \, \mu_b^{vy} \, \mu_y^{ck} \, \left(\gamma^{-1}\right)_v^u \, e_c e^d \tag{6.30}$$

where a sum over all repeated indices (Einstein's convention) is understood.

A suggested diagrammatic way for writing this is

$$\begin{array}{ccccccc}
e_c \longrightarrow & \mu_y^{ck} & \xrightarrow{\;y\;} & \mu_b^{vy} & \longrightarrow e_b \\
& \uparrow k & & \uparrow v & \\
& R & & S & \\
& \downarrow k & & \downarrow u & \\
e^d \longrightarrow & m_{kd}^x & \xrightarrow{\;x\;} & m_{xu}^a & \longrightarrow e^a
\end{array} \tag{6.31}$$

Of course, in this diagram, we start with e_c on top of e^d (the pictorial representation for $\mathscr{A}$ to the left of $\mathscr{A}^*$) and still end up with e_b on top of e^a. The diagram is not quite correct, but it does yield a clue or two about what this normal ordering is doing. The crucial features of the diagram are the vertical arrows, labeled R and S. The action of S is not mysterious; using the antipode of $\mathscr{A}^*$, it lowers the index v and raises the index u for smooth index matching. On the other hand, whatever goes on at the point R, what results is a pair of basis elements dual to each other. This is the technical soul of the double construction: what is happening at the point R can be isolated and understood as a universal $\mathscr{R}$-matrix for the Hopf algebra $\mathscr{D}(\mathscr{A},\mathscr{A}^*)$.

Indeed, if we write

$$\mathscr{R} = \sum_a e_a \otimes \mathbf{1} \otimes \mathbf{1} \otimes e^a \in \mathscr{D}(\mathscr{A},\mathscr{A}^*) \otimes \mathscr{D}(\mathscr{A},\mathscr{A}^*) \tag{6.32}$$

then, using the normal ordering recipe (6.30), it is a little tedious but straightforward to prove that

$$\mathscr{R}\Delta(x) = \Delta'(x)\mathscr{R} \qquad \forall x \in \mathscr{D}(\mathscr{A},\mathscr{A}^*) \tag{6.33}$$

where the transposed co-product in the double is just, reading from (6.29) above,

$$\Delta'(e_a e^b) = \mu_a^{cd}\, m_{vu}^b\, e_d e^v \otimes e_c e^u \tag{6.34}$$

A better diagram than (6.31) would have the upper and lower lines crossing at R, to emphasize that the normal ordering prescription embodies the "braiding" by an $\mathscr{R}$-matrix.

To complete the characterization of the quantum double $\mathscr{D}(\mathscr{A},\mathscr{A}^*)$, we still need its antipode. It follows directly from the normal ordering (6.30) and the fact (6.11) that the antipode is an antihomomorphism. Explicitly,

$$\gamma(e_a^b) = \mu_a^{kx}\, \mu_x^{cv}\, m_{ud}^y\, m_{yk}^b\, \gamma(e_c)\, \gamma(e^f) \tag{6.35}$$

We postpone to an exercise the complete verification that $\mathscr{D}(\mathscr{A},\mathscr{A}^*)$ is indeed a quasi-triangular Hopf algebra, i.e. a quantum group.

To summarize this important construction, let us repeat that the two essential ingredients are duality and normal ordering. In practice, to build a quantum group all we need is two dual Hopf algebras, dual to each other. We shall turn next to an example.

6.4 The quantum group $U_q(s\ell(2))$

6.4.1 Quantum double construction

Drinfeld used the quantum double to "deform" any classical Lie algebra into a quantum group. The initial Hopf algebra $\mathscr{A}$ is the universal enveloping algebra of the Borel subalgebra of a Lie algebra, i.e. the algebra of all polynomials in the generators associated with positive or Cartan roots.

Let us study in some detail the simplest case of $SU(2)$; or more correctly $s\ell(2)$: the distinction has to do with the $*$-automorphisms of the Lie algebra A_1. Before using Drinfeld's double construction to deform $s\ell(2)$, let us consider its Hopf algebra structure. We shall see that $U(s\ell(2))$, the universal enveloping algebra of $s\ell(2)$, is a co-commutative Hopf algebra.

The classical Lie algebra is generated by X_+, X_- and H, with non-vanishing commutators

$$[X_+, X_-] = H$$
$$[H, X_\pm] = \pm 2X_\pm \tag{6.36}$$

Note that $X_\pm$ are the usual raising and lowering operators of angular momentum, whereas H is twice the usual J_z operator. The co-multiplication of the angular momentum operators is well known from elementary quantum mechanics. The action of an angular momentum operator J on a pair of particles is, in fact, the action of the co-multiplication of the operator $\Delta(J)$ on the tensor product of two one-particle Hilbert spaces. The rule $J = J_1 + J_2$ means that the co-multiplication is trivial:

$$\Delta(J) = J \otimes \mathbf{1} + \mathbf{1} \otimes J \qquad J = X_\pm, H \tag{6.37}$$

The antipode and co-unit of the Hopf algebra follow from the inverse operation in the Lie group; they are given by

$$\gamma(X_\pm) = -X_\pm \qquad \gamma(H) = -H \qquad \gamma(\mathbf{1}) = \mathbf{1} \tag{6.38}$$

$$\epsilon(X_\pm) = \epsilon(H) = 0 \qquad \epsilon(\mathbf{1}) = 1 \tag{6.39}$$

Note that, if we write an element of the Lie group as $g = \exp i[\theta_+ X_+ + \theta_- X_- + \theta_3 H]$, then $\gamma(g) = g^{-1}$ and $\epsilon(g) = 1$.

The Borel subalgebra B_+ of $s\ell(2)$ is generated by $\mathbf{1}$, X_+ and H, and the algebra of all polynomials in X_+ and H, denoted by $U(B_+)$, is itself a co-commutative Hopf algebra. The quantum group $U_q(s\ell(2))$ is obtained by applying the double construction to $\mathscr{A} = U(B_+)$, as follows.

First of all, we need a basis of $U(B_+)$. A convenient choice is the set of symmetric polynomials in the two generators H and X_+. A basis of $\mathscr{A} = U(B_+)$ is thus $\mathbf{1}$, H, X_+, H^2, X_+^2, $\{HX_+\} = (HX_+ + X_+H)/2$, H^3, X_+^3, $\{H^2X_+\} = (H^2X_+ + HX_+H + X_+H^2)/3$, …, and so on. If we actually wanted to label these basis elements, we could use the notation $e_{\alpha,\beta} = \{H^\alpha X_+^\beta\}$ $(m, n \in \mathbb{N})$, where the brackets indicate full normalized symmetrization.

The multiplication in $U(B_+)$ is derived from the basic relations

$$\mathbf{1}\,H = H\,\mathbf{1} = H = e_{10}$$
$$\mathbf{1}\,X_+ = X_+\mathbf{1} = X_+ = e_{01}$$
$$HX_+ = \{HX_+\} + X_+ = e_{11} + e_{01} \tag{6.40}$$
$$X_+H = \{HX_+\} - X_+ = e_{11} - e_{01}$$

The reader should write out $H^n X_+ \in U(B_+)$ in terms of the basis elements $e_{\alpha\beta}$ to prove that the modest $X_+ = e_{01}$ appears in the expansion, for any n. This simple fact is crucial in understanding the quantum double, as we shall see shortly.

The next step is to find the algebra $\mathscr{A}^*$, isomorphic to $\mathscr{A}$ as vector spaces, such that $\mathscr{A}$ and $\mathscr{A}^*$ are dual to each other. From the (known) multiplication of $\mathscr{A} = U(B_+)$, we should be able to derive the co-multiplication of $\mathscr{A}^*$.

In fact, $\mathscr{A}^*$ is identified with the universal enveloping algebra $U(B_-)$ (in obvious notation, the lower Borel subalgebra B_- of $s\ell(2)$ is generated by $\mathbf{1}$, H and X_-), and we choose the following duality map:

$$\begin{aligned} * \quad : \quad & \mathscr{A} \to \mathscr{A}^* \\ & a \mapsto a^* \\ & ab \mapsto (ab)^* = b^* a^* \\ & \mathbf{1} \mapsto \mathbf{1} \\ & H \mapsto H^* = hH \\ & X_+ \mapsto (X_+)^* = hX_- \end{aligned} \tag{6.41}$$

We have introduced an arbitrary (but fixed) constant $h \in \mathbb{C}$. According to this duality map, the dual basis to $\{e_{\alpha,\beta}\}$ is $\{e^{\beta,\alpha} = h^{\alpha+\beta} X_-^{\beta} H^{\alpha}\}$.

The co-multiplication in $\mathscr{A}^* = U(B_-)$ follows from its definition (6.26), duality and the multiplication rules (6.40). All we have to do to find the co-product of $e^{\beta\alpha}$ is to see when $e_{\alpha\beta}$ appears in a product $e_{\gamma\delta}e_{\gamma'\delta'}$. From the multiplication rules, it is clear that the only way to obtain $\mathbf{1}$ in the product of two basis elements of $\mathscr{A}$ is to multiply $\mathbf{1}$ by $\mathbf{1}$. Thus,

$$\Delta(\mathbf{1}) = \mathbf{1} \otimes \mathbf{1} \tag{6.42}$$

Similarly, there are only two ways to obtain $H = e_{10}$: either you multiply H by $\mathbf{1}$, or $\mathbf{1}$ by H; accordingly, the co-product of $H^* = hH$ is merely

$$\Delta(H) = H \otimes \mathbf{1} + \mathbf{1} \otimes H \tag{6.43}$$

Now, for $X_+ = e_{01}$ the situation is trickier. There are many ways of getting X_+ in the decomposition into symmetric polynomials of the product of two symmetric polynomials. As noted above, in fact, X_+ appears in the decomposition of all products of the form $H^n X_+$ and $X_+ H^n$. This means that the co-product of X_+ is an infinite power series. The "quantization constant" h in (6.41) was introduced as a means of weighing the various terms differently to assure good behavior of this formal series. The diligent reader will have no problems in verifying that the co-product of X_+ is

$$\Delta(X_+) = X_+ \otimes \mathbf{1} + \mathbf{1} \otimes X_+ + \sum_{n \geq 1} \frac{h^n}{n!} (X_+ \otimes H^n + (-1)^n H^n \otimes X_+) \tag{6.44}$$

Defining

$$q = \mathrm{e}^{2h} \tag{6.45}$$

we may write (6.44) as

$$\Delta X_+ = X_+ \otimes q^{\frac{H}{2}} + q^{-\frac{H}{2}} \otimes X_+ \tag{6.46}$$

The "classical" limit $h \to 0$ yields the trivial co-multiplication $\Delta X_+ = X_+ \otimes \mathbf{1} + \mathbf{1} \otimes X_+$.

Next, we need a product rule for the dual elements H and X_-, from which we shall obtain a co-product for H (which should be the same as above) and X_-. Repeating the trick given above, i.e. using as basis of $\mathscr{A}^* = U(B_-)$ the symmetric polynomials in H and X_-, we set

$$\begin{aligned} H\,X_- &= \{HX_-\} - X_- \\ X_-\,H &= \{HX_-\} + X_- \\ \mathbf{1}\,X_- &= X_-\,\mathbf{1} = X_- \end{aligned} \tag{6.47}$$

in agreement with the B_- commutation relation

$$[H, X_-] = -2X_- \tag{6.48}$$

By the same construction as given above, we check (6.43) and find

$$\begin{aligned} \Delta X_- &= X_- \otimes \mathbf{1} + \mathbf{1} \otimes X_- + \sum_{n \geq 1} \frac{h^n}{n!}(X_- \otimes H^n + (-1)^n H^n \otimes X_-) \\ &= X_- \otimes q^{\frac{H}{2}} + q^{-\frac{H}{2}} \otimes X_- \end{aligned} \tag{6.49}$$

Note that it is crucial to retain the order of indices in (6.26) to obtain (6.46) and (6.49).

We are now ready to define the quantum double $\mathscr{D}(U(B_+), U(B_-)) = U_q(s\ell(2))$, by simply imposing the normal ordering (6.30) on the structure derived so far. The outcome is that one of the three defining commutation relations for $s\ell(2)$ is q-deformed into

$$[X_+, X_-] = \frac{q^H - q^{-H}}{q - q^{-1}} \tag{6.50}$$

Let us point out that the root of q-deformations is to be found in the curious q-numbers

$$[n] = [n]_q = \frac{q^n - q^{-n}}{q - q^{-1}} \xrightarrow[q \to 1]{} n \tag{6.51}$$

which preserve many of the properties of ordinary numbers. The classical Lie algebra is recovered in the limit $q \to 1$ (i.e. $h \to 0$).

The deformation parameter even appears in the antipode, though the co-unit does not change from the classical one:

$$\gamma(X_\pm) = -q^{\pm 1} X_\pm \qquad \gamma(H) = -H \qquad \gamma(\mathbf{1}) = \mathbf{1} \tag{6.52}$$

$$\epsilon(X_\pm) = \epsilon(H) = 0 \qquad \epsilon(\mathbf{1}) = 1 \tag{6.53}$$

The universal $\mathscr{R}$-matrix is of the form $\mathscr{R} = \sum_i e_i \otimes e^i$, where $\{e_i\}$ is a basis of $U_q(B_+)$ and $\{e^i\}$ is a basis of $\left(U_q(B_-)\right)$. Explicitly,

$$\mathscr{R} = q^{\frac{H \otimes H}{2}} \sum_{n=0}^{\infty} \frac{\left(1-q^{-2}\right)^n}{[n]_q!} q^{\frac{n(1-n)}{2}} q^{\frac{nH}{2}} X_+^n \otimes q^{\frac{-nH}{2}} X_-^n \tag{6.54}$$

with

$$\begin{aligned} &[0]_q! = 1 \\ &[n]_q! = [n]_q[n-1]_q \cdots [1]_q \qquad n \in \mathbb{N} - \{0\} \end{aligned} \tag{6.55}$$

The q-phases are doubtlessly very important, but the crux of the structure is contained in $\mathscr{R} \sim \sum X_+^n \otimes X_-^n$, which reflects the basic trick that $X_\mp$ are dual to $X_\pm$ (with the suitable factors of h). We always assume that formal series such as the one in (6.55) are well defined, if necessary by completing the universal enveloping algebra appropriately.

The universal $\mathscr{R}$-matrix (6.54) of $U_q(s\ell(2))$ satisfies

$$\mathscr{R}\Delta(x) = \Delta'(x)\mathscr{R} \qquad \forall x \in U_q(s\ell(2)) \tag{6.56a}$$

$$(\Delta \otimes \mathbf{1})\mathscr{R} = \mathscr{R}_{13}\mathscr{R}_{23} \tag{6.56b}$$

$$(\mathbf{1} \otimes \Delta)\mathscr{R} = \mathscr{R}_{13}\mathscr{R}_{12} \tag{6.56c}$$

$$(\gamma \otimes \mathbf{1})\mathscr{R} = (\mathbf{1} \otimes \gamma^{-1})\mathscr{R} = \mathscr{R}^{-1} \tag{6.56d}$$

$$(\epsilon \otimes \mathbf{1})\mathscr{R} = (\mathbf{1} \otimes \epsilon)\mathscr{R} = \mathbf{1} \tag{6.56e}$$

which imply the Yang–Baxter equation for $\mathscr{R}$ (in the braid limit):

$$\mathscr{R}_{12}\mathscr{R}_{13}\mathscr{R}_{23} = \mathscr{R}_{23}\mathscr{R}_{13}\mathscr{R}_{12} \tag{6.57}$$

The quadratic Casimir is

$$C = X_-X_+ + \left[\frac{H+1}{2}\right]_q^2 = X_-X_+ + \left(\frac{q^{\frac{H+1}{2}} - q^{-\frac{H+1}{2}}}{q-q^{-1}}\right)^2 \tag{6.58}$$

It is interesting to observe from the commutation relations for $U_q(s\ell(2))$

$$\begin{aligned} &[X_+, X_-] = [H]_q = \frac{q^H - q^{-H}}{q-q^{-1}} \\ &\left[H, X_\pm\right] = \pm 2X_\pm \end{aligned} \tag{6.59}$$

and the q-deformed co-product

$$\begin{aligned} &\Delta X_\pm = X_\pm \otimes q^{\frac{H}{2}} + q^{-\frac{H}{2}} \otimes X_\pm \\ &\Delta H = H \otimes \mathbf{1} + \mathbf{1} \otimes H \end{aligned} \tag{6.60}$$

that the generators $E = X^+ q^{\frac{H}{2}}$, $F = X^- q^{-\frac{H}{2}}$ and $K = q^H$ form a subalgebra of the whole Hopf algebra $U_q(s\ell(2))$. We shall denote it by $U_q'(s\ell(2))$. We prefer to distinguish between them because the universal $\mathscr{R}$-matrix (6.54) really belongs to $U_q(s\ell(2)) \otimes U_q(s\ell(2))$, and not to $U_q'(s\ell(2)) \otimes U_q'(s\ell(2))$, since $q^{H \otimes \frac{H}{2}}$ is by no means equal to $q^{\frac{H}{2}} \otimes q^{\frac{H}{2}}$. This subtlety is not very relevant for generic q, but acquires importance when q is a root of unity.

6.4.2 Irreducible representations

Let us consider the case when the formal parameter q is not a root of unity, i.e. $q \neq e^{2\pi is/p}$, $(s,p) = 1$. When $q^p = 1$ (with $p \in \mathbb{N}$) , the situation is much more intricate, and will be dealt with separately in chapter 7. For generic q, the finite-dimensional irreducible representations of $U_q(s\ell(2))$ are essentially the same as for the classical or undeformed $U(s\ell(2)) = U_{q=1}(s\ell(2))$. They are labeled by an integer or half-odd j, with the dimension of the representation being $2j+1$. If V^j denotes the $(2j+1)$-dimensional space spanned by $|j,m\rangle$, with $m \in \{-j, -j+1, \dots, j-1, j\}$, then

$$\begin{aligned} X_\pm \, |j,m\rangle &= \sqrt{[j \mp m]_q [j \pm m + 1]_q} \, |j, m \pm 1\rangle \\ H \, |j,m\rangle &= 2m \, |j,m\rangle \\ C \, |j,m\rangle &= \left[\frac{2j+1}{2} \right]_q^2 |j,m\rangle \end{aligned} \tag{6.61}$$

where C is the q-deformed quadratic Casimir (6.58).

We have chosen a basis in which X_+ is the transpose of X_- (in formulae, $X_+ = X_-^T$); this can be done for any value of $q \in \mathbb{C}$. In addition, we may impose a hermiticity condition on the representation, of the form $(X_+)^\dagger = \theta X_-$, where $\theta = \pm 1$ is called the spin parity. This imposes certain restrictions relating j and q:

$$X_+^\dagger = \theta X_- \iff \theta [j-m]_q [j+m+1]_q > 0 \quad \forall m \in \{-j, \dots, j\} \tag{6.62}$$

If $q \in \mathbb{R}$, then (6.62) always holds with $\theta = 1$. However, if q is a pure phase, then certain values of j are not allowed. For the time being, we shall not consider this case any further.

Letting $\rho^{(j)} : U_q(s\ell(2)) \to \mathrm{End}\,(V^j)$ denote the representation j, we define

$$\begin{aligned} &\Delta^{(j_1,j_2)}(a) = \rho^{(j_1)} \otimes \rho^{(j_2)} \Delta(a) \qquad \forall a \in U_q(s\ell(2)) \\ &\rho^{(j_1)} \otimes \rho^{(j_2)} \mathcal{R} = \mathcal{R}^{(j_1,j_2)} \\ &R^{(j_1,j_2)} = P^{(j_1,j_2)} \mathcal{R}^{(j_1,j_2)} \end{aligned} \tag{6.63}$$

where the permutation map $P^{(j_1,j_2)} : V^{j_1} \otimes V^{j_2} \to V^{j_2} \otimes V^{j_1}$ is introduced for later convenience. The representation $\mathcal{R}^{(j_1,j_2)}$ of the universal $\mathcal{R}$-matrix differs by this P from the braiding matrix R^{j_1,j_2} (see section 2.5.1). The Yang–Baxter relation

(6.57) reads now as

$$\mathscr{R}_{12}^{(j_1,j_2)}\mathscr{R}_{13}^{(j_1,j_3)}\mathscr{R}_{23}^{(j_2,j_3)} = \mathscr{R}_{23}^{(j_2,j_3)}\mathscr{R}_{13}^{(j_1,j_3)}\mathscr{R}_{12}^{(j_1,j_2)} \tag{6.64}$$

with the matrices acting on $V^{j_1} \otimes V^{j_2} \otimes V^{j_3}$. In terms of the permuted R-matrices, the Yang–Baxter equation is

$$\begin{aligned}&\left(\mathbf{1} \otimes R^{(j_1,j_2)}\right)\left(R^{(j_1,j_3)} \otimes \mathbf{1}\right)\left(\mathbf{1} \otimes R^{(j_2,j_3)}\right)\\ &\qquad = \left(R^{(j_2,j_3)} \otimes \mathbf{1}\right)\left(\mathbf{1} \otimes R^{(j_1,j_3)}\right)\left(R^{(j_1,j_2)} \otimes \mathbf{1}\right)\end{aligned} \tag{6.65}$$

The quasi-triangular relation $\Delta'(a) = R\Delta(a)R^{-1}$ implies

$$R^{(j_1,j_2)}\Delta^{(j_1,j_2)}(a) = \Delta^{(j_2,j_1)}(a)R^{(j_1,j_2)} \tag{6.66}$$

In a diagram, the importance of the permutation map P in the definition of R is apparent:

$$\begin{array}{ccc} V^{j_1} \otimes V^{j_2} & \xrightarrow{\Delta^{j_1,j_2}(a)} & V^{j_1} \otimes V^{j_2} \\ \Big\downarrow R^{j_1,j_2} & & \Big\downarrow R^{j_1,j_2} \\ V^{j_2} \otimes V^{j_1} & \xrightarrow{\Delta^{j_2,j_1}(a)} & V^{j_2} \otimes V^{j_1} \end{array} \qquad \forall a \in \mathscr{A} \tag{6.67}$$

As an example, in the fundamental irrep ($j = \frac{1}{2}$) the explicit expression for $\mathscr{R}^{\frac{1}{2}\frac{1}{2}}$ is

$$\mathscr{R}^{\frac{1}{2}\frac{1}{2}} = \begin{array}{c} \\ ++ \\ +- \\ -+ \\ -- \end{array} \begin{array}{c} \begin{array}{cccc} ++ & +- & -+ & -- \end{array} \\ \begin{pmatrix} q & 0 & 0 & 0 \\ 0 & 1 & q-q^{-1} & 0 \\ 0 & 0 & 1 & 0 \\ 0 & 0 & 0 & q \end{pmatrix} \end{array} \tag{6.68}$$

which we have encountered in equation (2.82) as the braid limit of the matrix of Boltzmann weights for the six-vertex model.

To simplify the formulae, it is useful to introduce a graphical representation for the R-matrix, as shown in figure 6.1. In this notation, the Yang–Baxter equation (6.64) is obviously true, corresponding to the braid relation $\sigma_i\sigma_{i+1}\sigma_i = \sigma_{i+1}\sigma_i\sigma_{i+1}$ (figure 6.2).

Acting on each space V^j, we can define the matrix

$$\left(w^j\right)_{mm'} = q^{-c_j}(-1)^{j-m}\delta_{m+m',0}q^{-m} \tag{6.69}$$

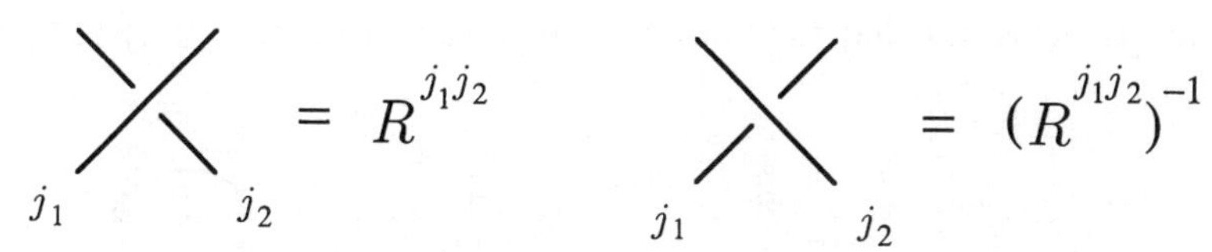

Fig. 6.1

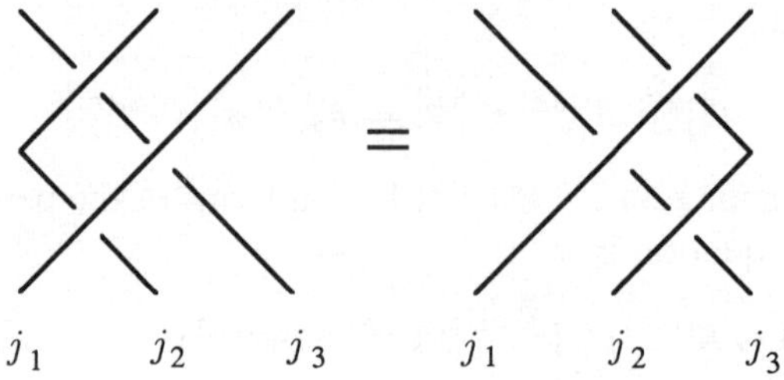

Fig. 6.2

where $c_j = j(j+1)$ is the classical Casimir. With the help of the automorphism

$$\tau(X^{\pm}) = X^{\mp}, \qquad \tau(H) = H \tag{6.70}$$

it is easy to verify that

$$waw^{-1} = \tau\gamma(a) \qquad \forall a \in U_q(s\ell(2)) \tag{6.71}$$

and that

$$(\tau \otimes \mathbf{1})\mathscr{R}^{-1} = (w \otimes \mathbf{1})\mathscr{R}(w^{-1} \otimes \mathbf{1}) \tag{6.72}$$

This last equation, particularized to two finite-dimensional representations j_1 and j_2, reads as

$$\left[\left(\mathscr{R}^{(j_1,j_2)}\right)^{-1}\right]^{T_1} = w_1 \mathscr{R}^{(j_1,j_2)} w_1^{-1} \tag{6.73}$$

where T_1 means transposition in the first space of $V^{j_1} \otimes V^{j_2}$, and $w_1 = w^{j_1} \otimes \mathbf{1}$. The existence of w in $U_q(s\ell(2))$ provides us with an example of the general structure of ribbon Hopf algebras, a subclass of Hopf algebras which are generally interesting when they are also quantum groups. The matrix w^j is essential in the derivation of link invariants from quantum groups, as we will discuss in the next section. We can begin to see this in the graphical representation for w^j shown in figure 6.3, which the reader is invited to compare with the matrices M introduced in appendix D. With the help of the identifications in figure 6.3, equation (6.73) is seen to be the exact analog of equation (D5.25).

Let us come back to the finite-dimensional irreducible representations of $U_q(s\ell(2))$. Once we have representations, we may consider taking their tensor product. It turns out that the tensor product of two irreps is fully reducible

Fig. 6.3

into a linear combination of irreps, just as in ordinary $SU(2)$:

$$V^{j_1} \otimes V^{j_2} = \sum_{j=|j_1-j_2|}^{j_1+j_2} V^j \tag{6.74}$$

For the nth tensor product, we obtain

$$V^{j_1} \otimes V^{j_2} \otimes \cdots \otimes V^{j_n} = \bigoplus_j \{ W^{(j_1,\ldots,j_n;j)} \otimes V^j \} \tag{6.75}$$

where the dimension of $W^{(j_1,\ldots,j_n;j)}$ is determined by the multiplicity of V^j in $V^{j_1} \otimes V^{j_2} \otimes \cdots \otimes V^{j_n}$. An appropriate graphical representation for the basis of the vector space $W^{(j_1,\ldots,j_n;j)}$ is the "conformal block"

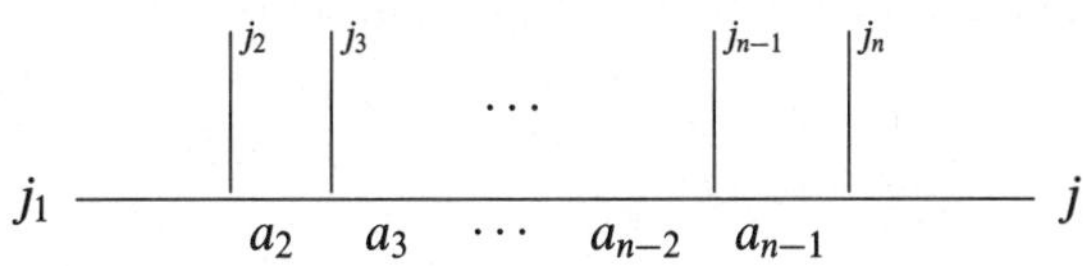

(6.76)

where the different channels $(a_2, a_3, \ldots, a_{n-1})$ are those allowed by the decomposition rules (6.74).

We may extend to $U_q(s\ell(2))$ the standard group theoretic $3j$ and $6j$ symbols. The quantum $3j$ symbols or quantum Clebsch–Gordan coefficients arise from the projectors

$$\begin{aligned} K_j^{j_1 j_2} : \quad & V^j \to V^{j_1} \otimes V^{j_2} \\ & e_m^j \mapsto \sum_{m_1,m_2} \begin{bmatrix} j_1 & j_2 & j \\ m_1 & m_2 & m \end{bmatrix}_q e_{m_1}^{j_1} \otimes e_{m_2}^{j_2} \end{aligned} \tag{6.77}$$

where e_m^j is a basis in V^j. For generic values of q, this map can be inverted to yield

$$\begin{aligned} V^{j_1} \otimes V^{j_2} & \longrightarrow \bigoplus_{j=|j_1-j_2|}^{j_1+j_2} V^j \\ e_{m_1}^{j_1} \otimes e_{m_2}^{j_2} & \mapsto \sum_j \begin{bmatrix} j_1 & j_2 & j \\ m_1 & m_2 & m \end{bmatrix}_q e_m^j \end{aligned} \tag{6.78}$$

The quantum Clebsch–Gordan coefficients $\begin{bmatrix} j_1 & j_2 & j \\ m_1 & m_2 & m \end{bmatrix}_q$ appearing in both formulae above are the same provided all the bases are chosen as in (6.61). Note that such a basis is not necessarily orthonormal, although orthonormality is assured if q is real. The quantum Clebsch–Gordan coefficients are computed as in the classical case. For instance, from the decomposition $\left|\frac{1}{2}\right\rangle \otimes \left|\frac{1}{2}\right\rangle = |0\rangle + |1\rangle$, we find (with $|+\rangle = |\frac{1}{2}, +\frac{1}{2}\rangle$ and $|-\rangle = |\frac{1}{2}, -\frac{1}{2}\rangle$) the following "Clebsches":

$$\begin{aligned}
|1,1\rangle &= |+\rangle \otimes |+\rangle \\
|1,0\rangle &= \frac{1}{\sqrt{[2]_q}} \left(\sqrt{q}\,|-\rangle \otimes |+\rangle + \frac{1}{\sqrt{q}}\,|+\rangle \otimes |-\rangle \right) \\
|1,-1\rangle &= |-\rangle \otimes |-\rangle \\
|0,0\rangle &= \frac{1}{\sqrt{[2]_q}} \left(\frac{1}{\sqrt{q}}\,|-\rangle \otimes |+\rangle - \sqrt{q}\,|+\rangle \otimes |-\rangle \right)
\end{aligned} \tag{6.79}$$

Note that, if q is real, then $\langle 0,0|1,0\rangle = 0$, but if q is a phase ($q^\dagger = q^{-1}$), then $\langle 0,0|1,0\rangle \neq 0$ and $\langle 1,0|1,0\rangle \neq 1$. In both cases, nevertheless, the state $|0,0\rangle$ is annihilated by X_+, X_- and H, and hence it yields a spin zero irrep indeed (see exercise 6.1).

Explicitly, the quantum Clebsch–Gordan coefficients are given in general by

$$\begin{aligned}
\begin{bmatrix} j_1 & j_2 & j \\ m_1 & m_2 & m \end{bmatrix}_q &= \delta^m_{m_1+m_2}\, \Delta(j_1, j_2, j) \\
&\times q^{[(j_1+j_2-j)(j_1+j_2+j+1)+j_1 m_2 - j_2 m_1]/2} \\
&\times \sqrt{[j_1+m_1]![j_1-m_1]![j_2+m_2]!} \\
&\times \sqrt{[j_2-m_2]![j+m]![j-m]![2j+1]} \\
&\times \sum_{r\geq 0} \left[\frac{(-1)^r q^{-r(j_1+j_2+j+1)}}{[r]![j_1+j_2-j-r]![j_1-m_1-r]!} \right. \\
&\left. \times \frac{1}{[j_2+m_2-r]![j-j_2+m_1+r]![j-j_1+r-m_2]!} \right]
\end{aligned} \tag{6.80}$$

where

$$\Delta(a,b,c) = \sqrt{\frac{[-a+b+c]![a-b+c]![a+b-c]!}{[a+b+c+1]!}} \tag{6.81}$$

and we adopt the convention $[0]! = 1$. Let us point out the following symmetry properties of the quantum Clebsch–Gordan coefficients:

$$\begin{aligned}
\begin{bmatrix} j_1 & j_2 & j \\ m_1 & m_2 & m \end{bmatrix}_q &= (-1)^{j_1+j_2-j} \begin{bmatrix} j_2 & j_1 & j \\ m_2 & m_1 & m \end{bmatrix}_{q^{-1}} \\
&= (-1)^{j_1-m_1} q^{-m_1} (-1)^{j_1+j-j_2} \\
&\times \sqrt{\frac{[2j+1]}{[2j_2+1]}} \begin{bmatrix} j_1 & j & j_2 \\ -m_1 & m & m_2 \end{bmatrix}_q
\end{aligned} \tag{6.82}$$

An interesting particularization of (6.80) is

$$\begin{bmatrix} j_1 & j_2 & 0 \\ m_1 & m_2 & 0 \end{bmatrix}_q = \frac{w^j_{m_1 m_2}}{\sqrt{[2j+1]}}\, q^{c_j} \tag{6.83}$$

With the help of figure 6.3, and using a suitable graphical representation for the quantum Clebsch–Gordan coefficients, we obtain the curious relation shown in figure 6.4.

The quasi-triangularity condition (6.66) implies that the following diagram commutes:

$$\begin{array}{ccc} V^j & \xrightarrow{K_j^{j_1 j_2}} & V^{j_1} \otimes V^{j_2} \\ \Big\downarrow \rho(j_1, j_2; j) & & \Big\downarrow R^{(j_1, j_2)} \\ V^j & \xrightarrow{K_j^{j_2 j_1}} & V^{j_2} \otimes V^{j_1} \end{array} \tag{6.84}$$

where $\rho(j_1, j_2; j)$ is a c-number. From the explicit expression (6.80), we obtain

$$\rho(j_1, j_2; j) = (-1)^{j_1+j_2-j} q^{c_j - c_{j_1} - c_{j_2}} \tag{6.85}$$

the graphical representation of which, as well as its particularization to $j = 0$, are shown in figure 6.5. This reminds us of the first Reidemeister move (see appendix D).

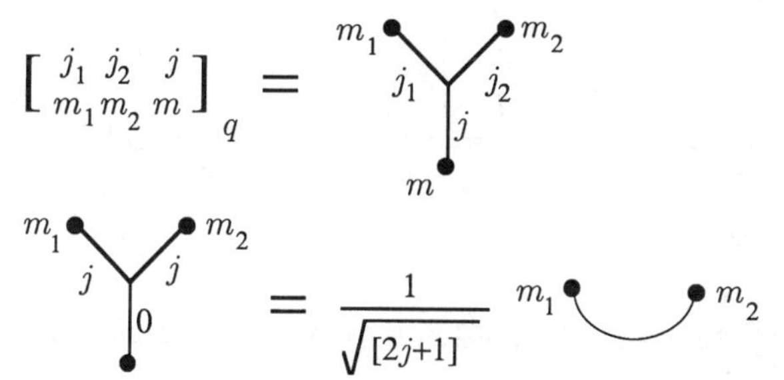

Fig. 6.4. Diagrammatic representation of a Clebsch–Gordan coefficient and of relation (6.83).

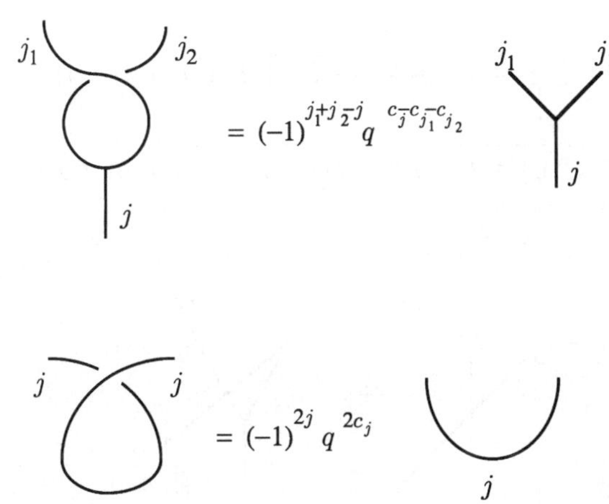

Fig. 6.5

From equation (6.56b), we find

$$\left(K_j^{j_1 j_1}\right)_{12} \mathscr{R}^{(j,j_3)} = \mathscr{R}_{13}^{(j_1,j_3)} \mathscr{R}_{23}^{(j_2,j_3)} K_j^{j_1 j_2} \tag{6.86}$$

which admits the diagrammatic representation shown in figure 6.6.

The quantum $6j$ symbols reflect the associativity of the tensor product of irreps

$$\left(V^{j_1} \otimes V^{j_2}\right) \otimes V^{j_3} \xrightarrow{6j} V^{j_1} \otimes \left(V^{j_2} \otimes V^{j_3}\right) \tag{6.87}$$

Associativity itself implies *a priori* the pentagonal equation for the $6j$ symbols:

$$\begin{array}{ccc}
\left(\left(V^{j_1} \otimes V^{j_2}\right) \otimes V^{j_3}\right) \otimes V^{j_4} & \longrightarrow & \left(V^{j_1} \otimes \left(V^{j_2} \otimes V^{j_3}\right)\right) \otimes V^{j_4} \\
\downarrow & & \downarrow \\
\left(V^{j_1} \otimes V^{j_2}\right) \otimes \left(V^{j_3} \otimes V^{j_4}\right) & & V^{j_1} \otimes \left(\left(V^{j_2} \otimes V^{j_3}\right) \otimes V^{j_4}\right) \\
& \searrow & \downarrow \\
& & V^{j_1} \otimes \left(V^{j_2} \otimes \left(V^{j_3} \otimes V^{j_4}\right)\right)
\end{array} \tag{6.88}$$

The quantum $6j$ symbol appearing in

$$\left(V^{j_1} \otimes V^{j_2}\right)_{j_{12}} \otimes V^{j_3} \xrightarrow{\left\{\begin{matrix} j_1 & j_2 & j_{12} \\ j_3 & j & j_{23} \end{matrix}\right\}_q} V^{j_1} \otimes \left(V^{j_2} \otimes V^{j_3}\right)_{j_{23}} \tag{6.89}$$

is given explicitly by

$$\begin{aligned}
\left\{\begin{matrix} j_1 & j_2 & j_{12} \\ j_3 & j & j_{23} \end{matrix}\right\}_q &= \Delta(j_1, j_2, j_{12})\Delta(j_3, j, j_{12})\Delta(j_1, j, j_{23})\Delta(j_2, j_3, j_{23}) \\
&\times \sum_{z \geq 0} \left[\frac{(-1)^z [z+1]!}{[z - j_1 - j_2 - j_{12}]![z - j_3 - j - j_{12}]![z - j_1 - j - j_3]!} \right. \\
&\qquad \times \frac{1}{[z - j_2 - j_3 - j_{23}]![j_1 + j_2 + j_3 + j - z]!} \\
&\qquad \left. \times \frac{1}{[j_1 + j_3 + j_{12} + j_{23} - z]![j_2 + j + j_{12} + j_{23} - z]!} \right]
\end{aligned} \tag{6.90}$$

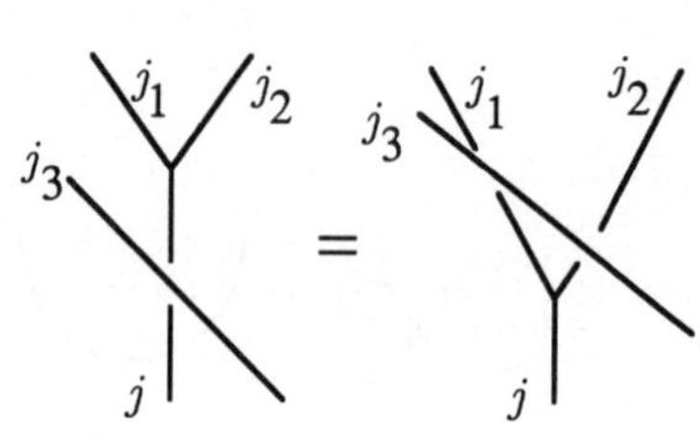

Fig. 6.6

where the sum runs only over those $z \in \mathbb{N}$ yielding non-negative arguments in the square brackets. In diagrams, equation (6.89) reads as

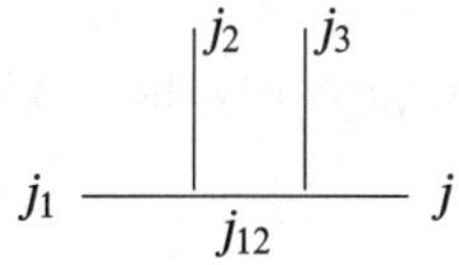

$$= \sum_{j_{23}} \begin{Bmatrix} j_1 & j_2 & j_{12} \\ j_3 & j & j_{23} \end{Bmatrix}_q \tag{6.91}$$

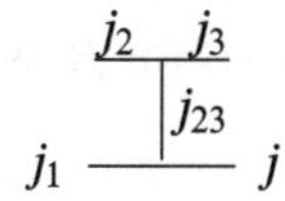

which will allow us to interpret the quantum $6j$ symbols as formal analogs of the fusion matrices to be discussed in the context of rational conformal field theories in chapter 9.

Quantum $6j$ symbols satisfy the following symmetry properties:

$$\begin{aligned} \begin{Bmatrix} j_1 & j_2 & j_{12} \\ j_3 & j & j_{23} \end{Bmatrix}_q &= \begin{Bmatrix} j_2 & j_1 & j_{12} \\ j & j_3 & j_{23} \end{Bmatrix}_q \\ &= \begin{Bmatrix} j_1 & j_{12} & j_2 \\ j_3 & j_{23} & j \end{Bmatrix}_q = \begin{Bmatrix} j_1 & j_{23} & j \\ j_3 & j_{12} & j_2 \end{Bmatrix}_q \end{aligned} \tag{6.92}$$

They can be derived from (6.90), and can be obtained as the symmetries of the tetrahedral representation of the quantum $6j$ symbols, shown in figure 6.7.

6.5 Centralizer and Hecke algebra

From the quasi-triangularity relation (6.66), it follows that

$$\Delta^{(j_1,j_2)}(a) = \left(R^{(j_1,j_2)}\right)^{-1} \Delta^{(j_2,j_1)}(a) R^{(j_1,j_2)} \qquad \forall a \in U_q(s\ell(2)) \tag{6.93}$$

If we fix the representation $j_1 = j_2 = j$, then the R-matrix $R^{j,j}$ commutes with the whole quantum group:

$$\left[\Delta^{(j,j)}(a), R^{(j,j)}\right] = 0 \qquad \forall a \in \mathscr{A} = U_q(s\ell(2)) \tag{6.94}$$

Consider now an n-fold tensor product $\left(V^j\right)^{\otimes n}$, which can be thought of as the Hilbert space for n identical particles in one dimension. The co-multiplication extends linearly, and thus we may define the action of the quantum group generators on $\left(V^j\right)^{\otimes n}$. From equation (6.94), it follows that the endomorphisms of $\left(V^j\right)^{\otimes n}$ of the form

$$g_i = \mathbf{1} \otimes \mathbf{1} \otimes \cdots \otimes \mathbf{1} \otimes qR^{(j,j)} \otimes \mathbf{1} \otimes \cdots \otimes \mathbf{1} \qquad (i = 1, \ldots, n-1) \tag{6.95}$$

$$\begin{Bmatrix} j_1 & j_2 & j_{12} \\ j_3 & j & j_{23} \end{Bmatrix}_q =$$

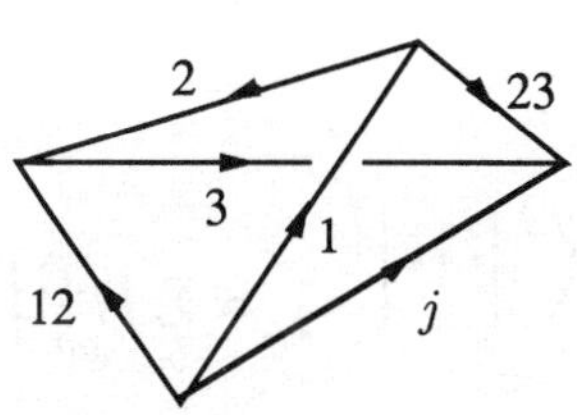

Fig. 6.7 Tetrahedral representation of the quantum $6j$ symbols.

(which act as the identity everywhere except on the ith and $(i+1)$th copies of V^j, where they braid with the help of the R-matrix) commute with the quantum group. The factor q is just convenient for normalizations. From the Yang–Baxter equation (6.64) for R, we deduce that the operators g_i satisfy the conditions

$$\begin{aligned} g_i g_{i+1} g_i &= g_{i+1} g_i g_{i+1} \\ g_i g_k &= g_k g_i \qquad |i-k| \geq 2 \end{aligned} \tag{6.96}$$

and thus they represent the braid group B_n.

In certain cases, the elements g_i generate the whole centralizer $C^j_{q,n}$. This happens, for instance, if $j = \frac{1}{2}$; then they fulfil also

$$g_i^2 = (q^2 - 1) g_i + q^2 \mathbf{1} \tag{6.97}$$

and thus $C^{\frac{1}{2}}_{q,n}$ is isomorphic to the Hecke algebra $H_n(q)$, a q-deformation of the permutation group $\mathscr{S}_n$. The Hecke condition (6.97) follows from the fact that $R^{(\frac{1}{2},\frac{1}{2})}$ has only two distinct eigenvalues, namely -1 and q^2.

The decomposition rules (6.75) allow us to induce the action of $C^j_{q,n}$ on the space $W^{j,\ell}_n = W^{j,\dots,j;\ell}$:

$$\begin{array}{ccc} \overbrace{V^j \otimes \cdots \otimes V^j}^{n \text{ times}} & = & \bigoplus_\ell \left(W^{j,\ell}_n \otimes V^\ell \right) \\ \Big\downarrow g_i & & \Big\downarrow \rho^j_{(\ell)}(g_i) \\ \underbrace{V^j \otimes \cdots \otimes V^j}_{n \text{ times}} & = & \bigoplus_\ell \left(W^{j,\ell}_n \otimes V^\ell \right) \end{array} \tag{6.98}$$

This induced action of $C^j_{q,n}$ defines different representations $\rho^j_{(\ell)}$ of $C^j_{q,n}$ on the spaces $W^{j,\ell}_n$. They are labeled by the representation ℓ, and their dimension is given by the multiplicity of V^ℓ in $\otimes^n V^j$. Using the graphical representation (6.76) for the basis elements in $W^{j,\ell}_n$, we can write the matrix realization of $\rho^j_{(\ell)}(g_i)$ as in figure 6.8. Note that we use the same notation for the matrix representation of $\rho^j_{(\ell)}(g_i)$ as we will use for the braiding matrices of conformal field theory to stress the similarities between both concepts.

We can obtain the explicit form of the B matrices in terms of quantum $6j$

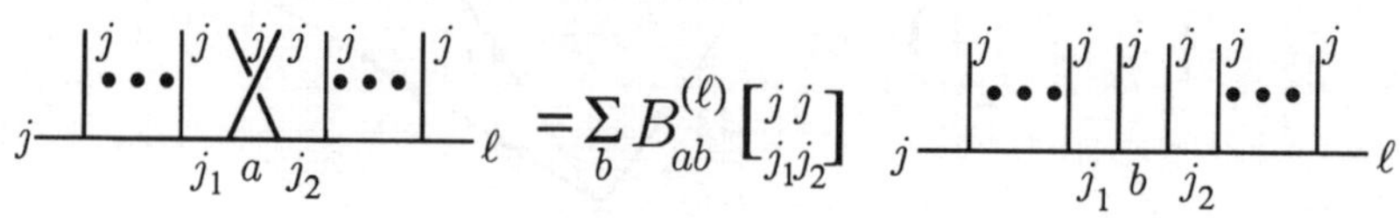

Fig. 6.8

symbols. Indeed, from figure 6.5, or rather, from its inverse, we have that

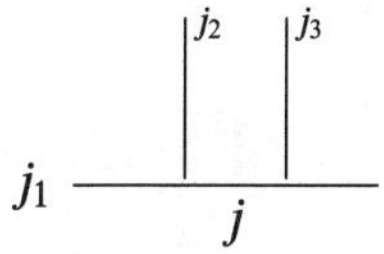

$$= \alpha \times$$
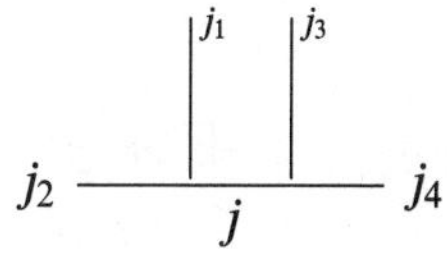

(6.99)

where $\alpha = (-1)^{j_1+j_2-j} q^{-c_j+c_{j_1}+c_{j_2}}$, whereas from the definition (6.91) of quantum $6j$ symbols we obtain

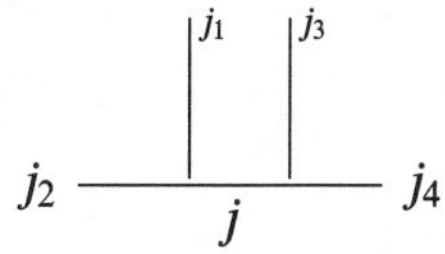

$$= \sum_{j'} \left\{ \begin{matrix} j_2 & j_1 & j \\ j_3 & j_4 & j' \end{matrix} \right\}_q$$
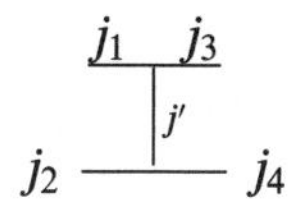

(6.100)

With the help of figure 6.6 and figure 6.5,

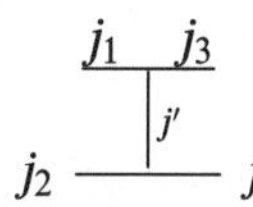

$$= \beta \times$$
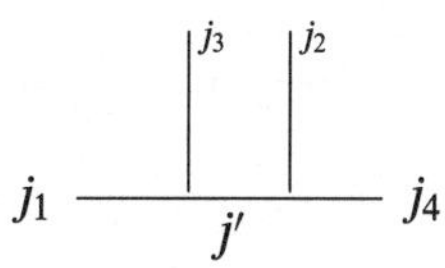

(6.101)

where $\beta = (-1)^{j_4-j_2-j'} q^{c_{j_4}-c_{j_2}-c_{j'}}$. Combining the three steps above, we obtain

$$B_{jj'} \begin{bmatrix} j_2 & j_3 \\ j_1 & j_4 \end{bmatrix} = (-1)^{j_1+j_4-j-j'} q^{c_{j_1}+c_{j_4}-c_j-c_{j'}} \left\{ \begin{matrix} j_2 & j_1 & j \\ j_3 & j_4 & j' \end{matrix} \right\}_q \tag{6.102}$$

For this braiding matrix, we may use the diagrammatic representations in figure 6.9.

6.5.1 *Representations of $H_n(q)$*

Let us comment on the special case $j = \frac{1}{2}$ when the spaces $W_n^{j,\ell}$ define representations of the Hecke algebra $H_n(q)$. These are naturally labeled by ℓ. As mentioned in appendix D, for generic q they are in one to one correspondence with the irreps of $\mathscr{S}_n$, the permutation group, and they can be characterized by Young tableaux. Let us consider some examples of irreducible representations of $H_n(q)$, for small n. We show their "conformal block" representation and their dimension:

$$B_{ab} \left[{}^{j\ j}_{j_1 j_2} \right] =$$
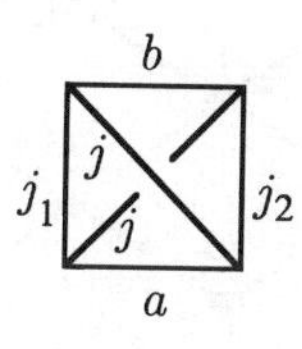

$$=$$
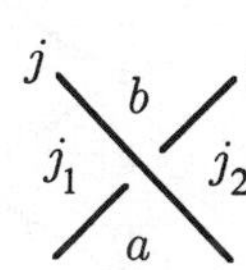

Fig. 6.9

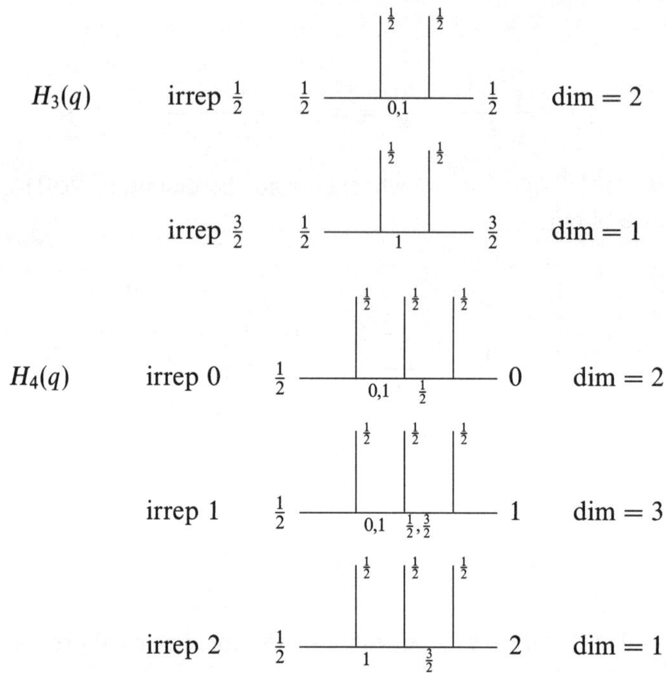

Note that when n is even we only obtain integer spins. Similarly, when n is odd only half-odd spins appear. These representations can be ordered in a Bratelli diagram, as shown in figure 6.10.

Note, though, that the tower in figure 6.10 is not like those found from the fundamental construction discussed in section 5.2.4. Here, when the level increases, the algebra becomes larger and larger. Though we shall not treat the case of q a root of unity until chapter 7, let us advance some of the peculiarities of the representations of $H_n(e^{\pi i/p})$ found by Wenzl. Letting $k = p - 2$, unitary and irreducible representations of $H_n(e^{\pi i/p})$ are those with $(2, k)$ Young diagrams, i.e. Young diagrams with two rows and $\lambda_1 - \lambda_2 \leq k$, where λ_i is the number of boxes in the ith row. In our framework, with $H_n(q)$ arising as the centralizer of

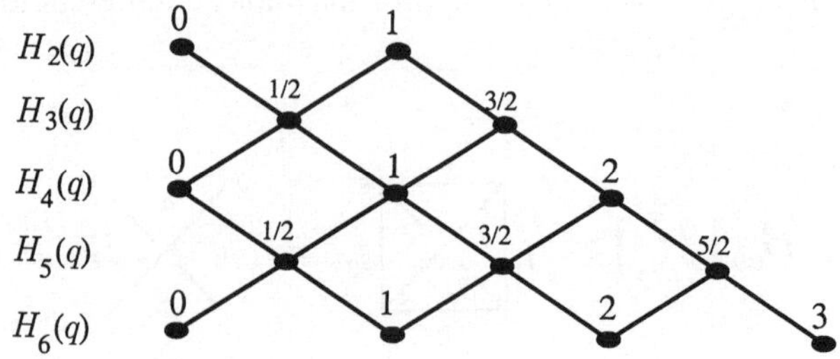

Fig. 6.10

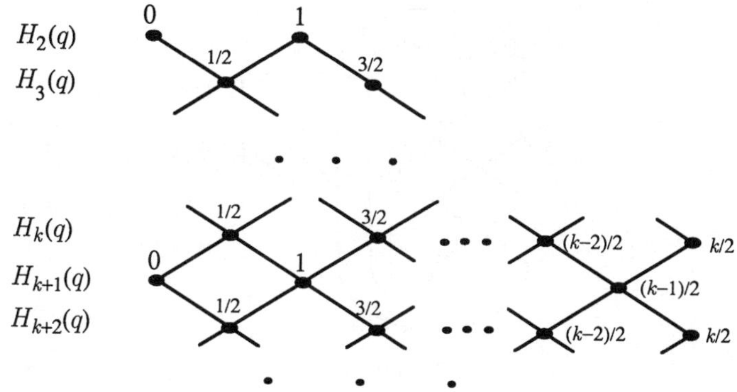

Fig. 6.11. Truncation of the Bratelli tower of figure 6.9 after level $k-1$. Shown here is the case with k odd.

$U_q(s\ell(2))$, if $q = \exp[\pi i/(k+2)]$ then the unitary irreps are those corresponding to spaces $W_n^{\frac{1}{2},\ell}$ where the Young tableau of the irrep ℓ is of the kind $(2,k)$. Recall that an irrep j of $SU(2)$ has a Young diagram with at most two rows and $\lambda_1 - \lambda_2 = 2j$. Therefore, the unitary irreps of $H_n(\mathrm{e}^{\pi i/(k+2)})$ are those with spin $j \le k/2$. If we restrict the Bratelli diagram of figure 6.10 to unitary irreps, then the tower stabilizes after level $(k-1)$: from there on, it is isomorphic to a tower obtained by the fundamental construction and bi-colored graph A_{k+1} (see figure 6.11). We shall use some of these results later on.

6.6 Link invariants from quantum groups

This section can be considered as an update on appendix D. As we have shown in section 6.5, we can find a representation of the braid group in the centralizers $C_q^{j,n}$ of $U_q(s\ell(2))$. Following the discourse in appendix D, to define a link invariant we should first find a Markov trace in $C_q^{j,n}$. For $j = \frac{1}{2}$, this defines a Markov trace in $H_n(q)$. Due to the existence in $U_q(s\ell(2))$ of the ribbon element w of (6.69), and of Kauffman's relation in figure D5.9, it is possible to use Turaev's concept of an extended Yang–Baxter system (section D5.4) to proceed. Associated with the R-matrix $R^{(j,j)}$, with j an arbitrary irrep of $U_q(s\ell(2))$, we should find an isomorphism $\mu^{(j)} : V^j \to V^j$ satisfying the conditions (D5.20). From figure D5.8, the obvious candidate is

$$\mu^{(j)} = \rho^j\left(q^H\right) : V^j \longrightarrow V^j \tag{6.103}$$

The conditions (D5.20) are satisfied with

$$a = (-1)^{2j} q^{2c_j}\,, \qquad b = [2j+1] \tag{6.104}$$

Therefore, the link invariant associated with the irrep j of $U_q(s\ell(2))$ is

$$P(\hat{\alpha}) = a^{-w(\alpha)} \frac{1}{[2j+1]^n} \operatorname{tr}\left(q^H \otimes \cdots \otimes q^H \cdot \pi^{(j)}(\alpha)\right) \tag{6.105}$$

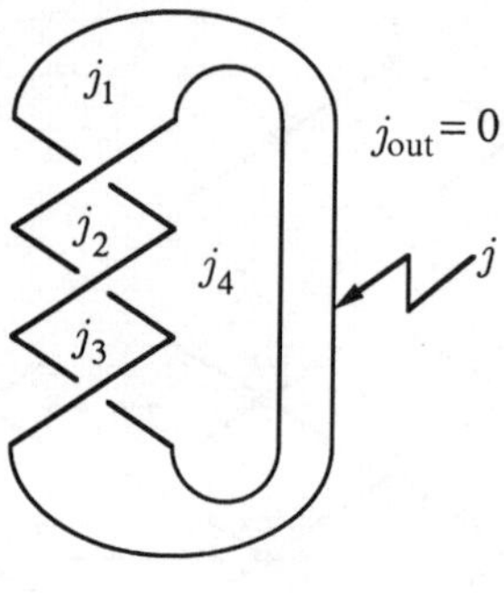

Fig. 6.12

where α is a word in B_n, $\hat{\alpha}$ is the link obtained through its closure, and $\pi^{(j)}(\alpha)$ is its representation in the centralizer $C^j_{q,n}$ from (6.95),

$$\pi^{(j)}(\sigma_i) = \mathbf{1} \otimes \cdots \otimes \mathbf{1} \otimes R^{(j,j)}_{i,i+1} \otimes \mathbf{1} \otimes \cdots \otimes \mathbf{1} \tag{6.106}$$

For $j = \frac{1}{2}$, the polynomial in q obtained by (6.105) is, in fact, the Jones polynomial.

To bring this in line with the results of appendix D, in particular with equation (D5.13), we may use the decomposition rules (6.75) to write the trace in (6.105) as

$$\mathrm{tr}\,(\hat{\alpha}) = \sum_{\ell} [2\ell + 1]\ \mathrm{tr}_{\,W^{j,\ell}_n} \left(\rho^j_{(\ell)}(\alpha)\right) \tag{6.107}$$

where the representation $\rho^j_{(\ell)}$ is defined on $W^{j,\ell}_n$, via (6.98). Note that only representations of $H_n(q)$ with Young diagrams of at most two rows contribute to the Jones polynomial. In appendix D, this was explained in terms of the relation between the Temperley–Lieb–Jones algebra $A_{\beta,n}$ and the Hecke algebra $H_n(q)$. Here, we have derived the same result from the structure of the representations of $U_q(s\ell(2))$.

To conclude this brief discussion on link polynomials, we present a graphical way of computing the link polynomial (6.105). Imagine we are given a link with a chosen orientation for each strand (figure 6.12). We choose a representation j of $U_q(s\ell(2))$, which we associate to the strands. Then irreps are assigned to the various two-dimensional domains, into which the projection of the knot breaks up the plane, as follows:

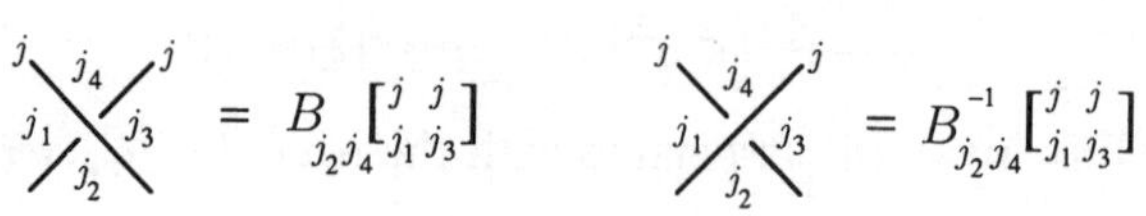

Fig. 6.13

(i) the exterior of the knot is assigned the representation $j_{\text{out}} = 0$;
(ii) each inner domain is assigned a representation j_i such that, if j_i and j_k are on adjoining domains, $j_2 \in j \times j_1$. Such an assignment of representations is called a "state configuration" for the knot.

To compute $P(\hat{\alpha})$, we sum over all possible state configurations. The contribution from each configuration is determined by the rules shown in figure 6.13. For example, for the trefoil of figure 6.12 we obtain

$$P(\widehat{\sigma_1^3}) = \sum_{j_4=0}^{2j} [2j+1] \left(B_{jj}^{-1} \begin{bmatrix} j & j \\ 0 & j_4 \end{bmatrix} \right)^3 \tag{6.108}$$

where we have used rules (*i*) and (*ii*); the last one forces $j_1 = j_2 = j_3 = j$ and $j_4 \in j \otimes j$.

6.7 The quantum group $U_q(\mathcal{G})$

In this section, we present a discussion of the invention of quantum semi-simple algebras, due to Drinfeld and Jimbo, generalizing the construction of $U_q(s\ell(2))$ given above.

Let $\mathcal{G}$ be a semi-simple Lie algebra, $A = (a_{ij})$ $(i, j = 1, \ldots, n = \text{rank}\, \mathcal{G})$ its Cartan matrix and $D = (D_i)$ the vector or diagonal matrix such that $D_i a_{ij} = a_{ij} D_j$.

The quantum algebra $U_q(\mathcal{G})$ is defined as the algebra of formal power series in q with generators e_i, f_i, k_i $(i = 1, \ldots, n = \text{rank}\, \mathcal{G})$ subject to the following relations:

$$\begin{aligned} k_i k_j &= k_j k_i \\ k_i e_j &= q_i^{a_{ij}} e_j k_i \\ k_i f_j &= q_i^{-a_{ij}} f_j k_i \\ e_i f_j - f_j e_i &= \delta_{ij} \frac{k_i - k_i^{-1}}{q_i - q_i^{-1}} \end{aligned} \tag{6.109}$$

$$\sum_{\ell=0}^{1-a_{ij}} (-1)^\ell \begin{bmatrix} 1 - a_{ij} \\ \ell \end{bmatrix}_{q_i} e_i^{1-a_{ij}-\ell} e_j e_i^\ell = 0 \quad i \neq j \tag{6.110a}$$

$$\sum_{\ell=0}^{1-a_{ij}} (-1)^\ell \begin{bmatrix} 1 - a_{ij} \\ \ell \end{bmatrix}_{q_i} f_i^{1-a_{ij}-\ell} f_j f_i^\ell = 0 \quad i \neq j \tag{6.110b}$$

We have used the notations

$$q_i = q^{D_i}, \qquad [x]_{q_i} = \frac{q_i^x - q_i^{-x}}{q_i - q_i^{-1}} \tag{6.111}$$

The equations (6.110) are called the quantum Serre relations. For the simplest case of $\mathscr{G} = A_1$, there are no Serre relations, and equations (6.109) coincide with (2.113).

The generators e_i, f_i and k_i, where the index i ranges over the positive simple roots, constitute the Chevalley basis of the algebra. Supplemented by the Serre relations, it is equivalent to the Cartan basis (one raising operator for each positive root). In the quantum case, the Chevalley basis is much more convenient than the Cartan basis, due to the profusion of q-factors.

The co-multiplication of $U_q(\mathscr{G})$ is given by

$$\begin{aligned} \Delta(k_i) &= k_i \otimes k_i \\ \Delta(e_i) &= e_i \otimes k_i + \mathbf{1} \otimes e_i \\ \Delta(f_i) &= f_i \otimes \mathbf{1} + k_i^{-1} \otimes f_i \end{aligned} \tag{6.112}$$

and the antipode is given by

$$\gamma(e_i) = -e_i k_i^{-1} \quad , \quad \gamma(f_i) = -k_i f_i \quad , \quad \gamma(k_i) = k_i^{-1} \tag{6.113}$$

To write the universal $\mathscr{R}$-matrix of $U_q(\mathscr{G})$, we need the "logarithm" of k_i,

$$H_i = \frac{1}{2hD_i} \log\left(k_i^2\right) \tag{6.114}$$

where we have set $q = \exp h$. Then

$$\mathscr{R} = \exp\left[h \sum_{i,j=1}^{n} \left(B^{-1}\right)_{ij} H_i \otimes H_j\right] \left[1 + \sum_{i=1}^{n} \left(1 - q_i^{-2}\right) e_i \otimes f_i + \cdots\right] \tag{6.115}$$

where $B_{ij} = D_i a_{ij}$ is the symmetrized Cartan matrix. This $\mathscr{R}$-matrix follows from Drinfeld's quantum double construction.

6.8 *R*-matrices: an incomplete catalog

The R-matrix in the fundamental of $U_q(s\ell(n))$, $R_n = P\mathscr{R}^{(\mathbf{n},\mathbf{n})}$, is

$$R_n = q \sum_{i=1}^{n} e_{ii} \otimes e_{ii} + \sum_{i \neq j} e_{ij} \otimes e_{ji} + \left(q - q^{-1}\right) \sum_{i<j} e_{jj} \otimes e_{ii} \tag{6.116}$$

where e_{ij} is an $n \times n$ matrix, the only non-zero entry of which is the (i, j)th one. In matrix notation,

$$(R_n)_{ij}^{k\ell} = \begin{cases} q & \text{if } i = j = k = \ell \\ 1 & \text{if } i = \ell \neq k = j \\ q - q^{-1} & \text{if } i = k < \ell = j \\ 0 & \text{otherwise} \end{cases} \tag{6.117}$$

It is not hard to verify that

$$(R_n)^{-1} = q^{-1} \sum_{i=1}^{n} e_{ii} \otimes e_{ii} + \sum_{i \neq j} e_{ij} \otimes e_{ji} + \left(q^{-1} - q\right) \sum_{i>j} e_{jj} \otimes e_{ii} \tag{6.118}$$

and thus

$$R_n - (R_n)^{-1} = \left(q - q^{-1}\right) \mathbf{1}_{V \otimes V} \tag{6.119}$$

where V is the $\mathbf{n}$-dimensional space on which $U_q(s\ell(n))$ is represented.

Due to equation (6.119), we expect the link invariant derived from these R-matrices to satisfy the customary skein rules, relating the uncrossing, the crossing over and the crossing under (respectively, $\mathbf{1}$, R and R^{-1}). In the notation of appendix D, the extended Yang–Baxter system associated with the R-matrix $R_n = R^{(\mathbf{n},\mathbf{n})}$ calls for the isomorphism $\mu = \operatorname{diag}(\mu_i)$, $\mu_i = (-q)^{2i-n-1}$, and the values $a = (-1)^{n+1}q^n$, $b = 1$.

For completeness, we will also write down the R-matrices in the fundamental representations corresponding to $U_q(so(2n+1))$, $U_q(sp(2n))$, $U_q(so(2n))$ and $U_q(s\ell(n+1)/\mathbb{Z}_2)$, respectively. To write down Jimbo's compact and general R-matrix for all these cases, we require some notation. We set

$$(m, v) = \begin{cases} (2n+1, -1) & \text{for } B_n \\ (2n, 1) & \text{for } C_n \\ (2n, -1) & \text{for } D_n \\ (n+1, -1) & \text{for } A_n/\mathbb{Z}_2 \end{cases} \tag{6.120}$$

The indices denoted with small latin letters run from 1 to m. We let

$$i' = m + 1 - i \tag{6.121}$$

and

$$\bar{\imath} = \begin{cases} i - v/2 & \text{if } 1 \le i < (m+1)/2 \\ i & \text{if } i = (m+1)/2, \text{ with } m \text{ odd} \\ i + v/2 & \text{if } (m+1)/2 < i \le m \end{cases} \tag{6.122}$$

Finally, we set

$$s(i) = \begin{cases} 1 & \text{if } 1 \le i \le (m+1)/2 \\ -v & \text{if } (m+1)/2 \le i \le m \end{cases} \tag{6.123}$$

The R-matrix is then

$$\begin{aligned} R_{m,v} = & -q \sum_{\substack{i \\ i \neq i'}} e_{ii} \otimes e_{ii} + \sum_{\substack{i \\ i = i'}} e_{ii} \otimes e_{ii} + \sum_{\substack{i,j \\ i \neq j,j'}} e_{ij} \otimes e_{ji} \\ & -q \sum_{\substack{i \\ i \neq i'}} e_{ii'} \otimes e_{i'i} - \left(q - q^{-1}\right) \sum_{i<j} e_{ii} \otimes e_{jj} \\ & + \left(q - q^{-1}\right) \sum_{i<j} s(i)s(j)(-q)^{(\bar{\imath}-\bar{\jmath})} e_{ij'} \otimes e_{i'j} \end{aligned} \tag{6.124}$$

A curious property of these R-matrices is the fact that their inverses take a very simple form:

$$\left(R_{m,v}^{-1}\right)_{ij}^{k\ell} = \phi\left(\left(R_{m,v}\right)_{i'j'}^{k'\ell'}\right) \tag{6.125}$$

where ϕ is the automorphism sending q to q^{-1}.

It is not hard to check that the extended Yang–Baxter system $S_{m,v}$ associated with the R-matrix $R_{m,v}$ has $\mu_i = (-q)^{\bar{i}-(m+1)}$, $a = (-q)^{(m+v)}$ and $b = 1$. It defines an invariant $T_{m,v}$ for oriented links through equation (D5.21), from which an invariant for unoriented links follows:

$$\overline{T}_{m,v}(\hat{\alpha}) = \left((-1)^m v q^{m+v}\right)^{w(\alpha)} T_{m,v}(\hat{\alpha}) \tag{6.126}$$

This link invariant satisfies the following rather surprising skein relation:

$$\begin{aligned} 0 =& \overline{T}_{m,v}(L_+) + v\overline{T}_{m,v}(L_-) \\ &+ \left((vq) + (-vq)^{-1}\right)\left[\overline{T}_{m,v}(L_0) + v\overline{T}_{m,v}(L_\infty)\right] \end{aligned} \tag{6.127}$$

where the deformed links in the expression differ at a crossing, as shown in figure 6.14.

6.9 Classical Yang–Baxter equation

At the time of writing, there is no classification of the solutions to the Yang–Baxter equation. Nevertheless, Belavin and Drinfeld have established a classification of the solutions to the simpler classical Yang–Baxter equation, which we now review. The basic ingredient is a deformation parameter q with a value very close to its classical value $q = 1$:

$$q = e^h = 1 + h + h^2/2 + \cdots \tag{6.128}$$

A solution $\mathscr{R}(u, q)$ to the Yang–Baxter equation is said to be quasi-classical if it admits an expansion in h about the identity (up to an overall irrelevant constant)

$$\mathscr{R}(u, q) = 1 + h\,r(u) + O(h^2) \tag{6.129}$$

The matrix $r(u)$ is called the classical r-matrix, and it satisfies the classical Yang–Baxter equation

$$[r_{12}(u), r_{13}(u+v)] + [r_{12}(u), r_{23}(v)] + [r_{13}(u+v), r_{23}(v)] = 0 \tag{6.130}$$

which is just the $O(h^2)$ term of the perturbative expansion in powers of h of the full (quantum) Yang–Baxter equation.

Let $\mathscr{G}$ be a simple Lie algebra with basis I_μ. We may expand $r(u) \in \mathrm{End}(\mathscr{G} \otimes \mathscr{G})$ as

$$r(u) = \sum_{\mu,v} r_{\mu v}(u) I_\mu \otimes I_v \tag{6.131}$$

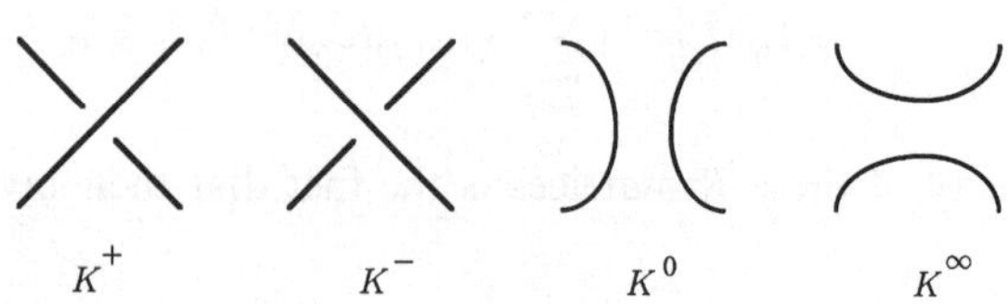

Fig. 6.14. The four local surgery operations entering the skein rule (6.127) for unoriented links.

If $\det r_{\mu\nu} \neq 0$, then $r(u)$ extends to a meromorphic function on $\mathbb{C}$, and all of its poles are simple. Moreover, the set Γ of poles is discrete and closed under the addition (in $\mathbb{C}$). The rank of Γ is defined as the dimension of the vector space (over $\mathbb{R}$) spanned by its elements. Clearly, it can only be 0, 1 or 2. Accordingly, the functions $r_{\mu\nu}$ are either rational, trigonometric or elliptic. Elliptic solutions to the classical Yang–Baxter equation exist only for the A_n series, i.e. only for $\mathscr{G} = SU(N)$.

The classical trigonometric solutions are obtained as follows. Let X_α denote the normalized non-diagonal generators of the Lie algebra $\mathscr{G}$, $(X_\alpha, X_{-\alpha}) = 1$. Let

$$\begin{aligned} t &= \sum_\mu I_\mu \otimes I_\mu \\ r &= \sum_{\alpha>0} (X_\alpha \otimes X_{-\alpha} - X_{-\alpha} \otimes X_\alpha) \end{aligned} \tag{6.132}$$

Then the trigonometric solution to the classical Yang–Baxter equation is

$$r(u) = r - t + \frac{2t}{1 - e^u} \tag{6.133}$$

6.10 Affine quantum groups

To extend the quantum deformation of semi-simple Lie algebras presented above to the quantum deformation of affine Lie algebras, we first review some basic facts about Cartan matrices. Let $A = (a_{ij})$ be an $n \times n$ matrix with integer entries, satisfying the following conditions:

(i) A is indecomposable, i.e. it is not of the form $\begin{pmatrix} A_1 & 0 \\ 0 & A_2 \end{pmatrix}$;
(ii) $a_{ij} \leq 0$ for $i \neq j$;
(iii) $a_{ij} = 0 \Leftrightarrow a_{ji} = 0$.

We also normalize A such that $a_{ii} = 2$.

Such a matrix A is said to be of affine type if the following two conditions hold, for v any n-component vector:

$$Av \geq 0 \Rightarrow Av = 0 \qquad \text{and} \qquad \dim\{u; Au = 0\} = 1 \tag{6.134}$$

It turns out that affine matrices satisfy $\det A = 0$, $a_{ij}a_{ji} \leq 4$ (no summation), and are all classified. There always exists a diagonal matrix D such that $B = D \cdot A$ is symmetric. The matrix B is called the symmetrized Cartan matrix, and it is used to define a symmetric bi-linear degenerate form on the affine Lie algebra. For finite Lie algebras, the bi-linear form is not degenerate.

The Dynkin diagrams of affine Cartan matrices are shown in tables 6.1 and 6.2. The algebras with superscript (1) are called untwisted; the others are called twisted. Dynkin diagrams of an untwisted Kac–Moody algebra $\mathscr{G}_\ell^{(1)}$ of rank

Table 6.1 Dynkin diagrams of untwisted affine $\mathscr{G}_\ell^{(1)}$.
The dots • represent the simple roots α_i. The numbers v_i under each simple root characterize the eigenvector $v = v_i\alpha_i$ of eigenvalue zero, $Av = 0$.

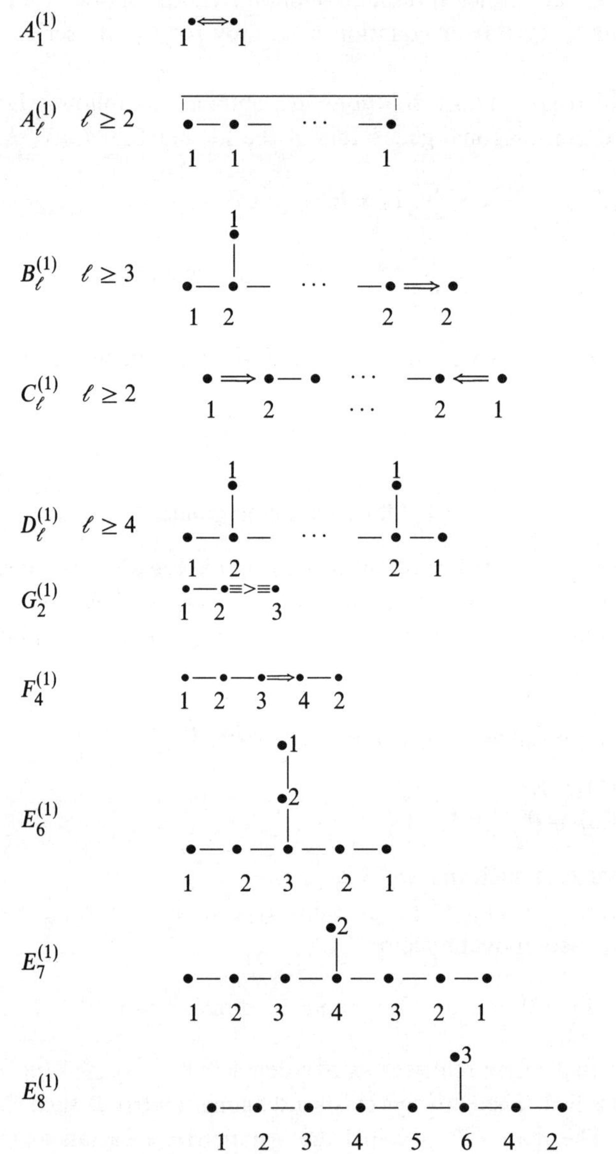

Table 6.2 Dynkin diagrams of twisted affine Lie algebras $\mathscr{G}_\ell^{(x)}$, $x > 1$. Notation as in table 6.1.

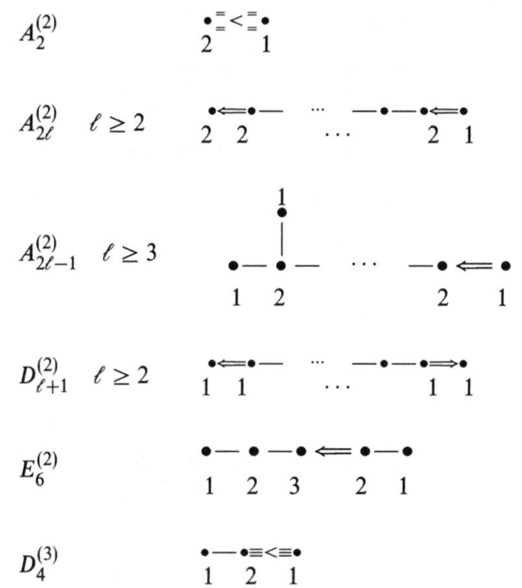

$\ell + 1$ have $\ell + 1$ vertices. The Dynkin diagrams of finite type (table 6.3) are subdiagrams of those of affine type, and their Cartan matrix satisfies $\det A > 0$.

From the diagrams in tables 6.1 and 6.3 we see that, given a semi-simple Lie algebra $\mathscr{G}_\ell$ of rank ℓ with an $\ell \times \ell$ Cartan matrix A ($\det A > 0$), and its untwisted affine extension $\mathscr{G}_\ell^{(1)} = \widehat{\mathscr{G}}_\ell$ with an $(\ell+1) \times (\ell+1)$ Cartan matrix $\hat{A}$ ($\det \widehat{A} = 0$), we may recover A from $\hat{A}$ by removing from $\hat{A}$ its zeroth row and zeroth column.

Quite naturally, we define the quantum algebra $U_q(\widehat{\mathscr{G}}_\ell)$ to have $3(\ell+1)$ generators e_i, f_i and k_i $(i = 0, 1, \ldots, \ell)$ satisfying the same relations (6.109), where a_{ij} are the entries of the affine Cartan matrix of $\mathscr{G}^{(1)}$.

Similarly, the full affine quantum group $U_q(\widetilde{\mathscr{G}}_\ell)$ is obtained by adding to $U_q(\widehat{\mathscr{G}}_\ell)$ the grading operator or derivation d satisfying the relations

$$[d, e_i] = \delta_{i,0} e_i\,, \qquad [d, f_i] = -\delta_{i,0}\, f_i\,, \qquad [d, k_i] = 0 \tag{6.135}$$

and enjoying the co-multiplication $\Delta(d) = d \otimes \mathbf{1} + \mathbf{1} \otimes d$.

Before delving into a more detailed study of the affine quantum group $U_q(\widetilde{\mathscr{G}}_\ell)$, it is worth recalling some basic results about the undeformed extended algebra $\hat{\mathscr{G}}_\ell$ associated with the affine Cartan matrix $\mathscr{G}_\ell^{(1)}$.

Table 6.3 Dynkin diagrams of finite Cartan matrices $\mathcal{G}_\ell$.
All diagrams have ℓ vertices •, representing the simple roots α_i.

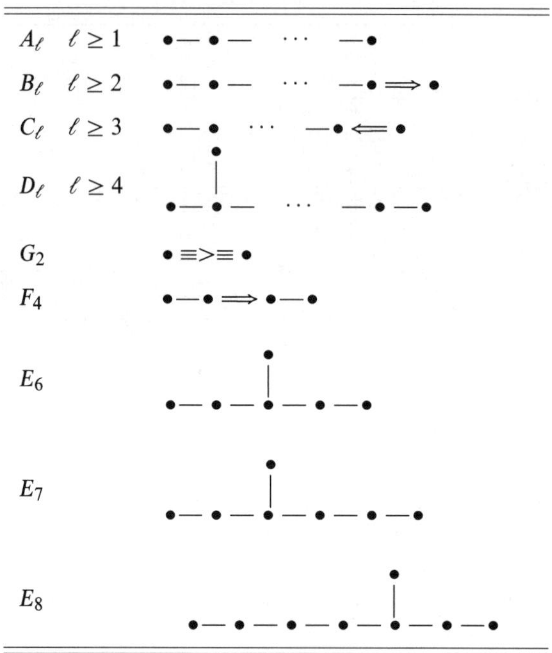

Table 6.4
The links among the roots • in the Dynkin diagrams of tables 6.1, 6.2 and 6.3 represent the non-zero off-diagonal entries of the Cartan matrix according to this dictionary. Note that the labels i, j indicate here the root α_i, whereas the labels in tables 6.1 and 6.2 refer to something else.

$i\bullet \quad \bullet j$	$a_{ij} = a_{ji} = 0$
$i\bullet - \bullet j$	$a_{ij} = a_{ji} = -1$
$i\bullet \Longrightarrow \bullet j$	$a_{ij} = -2,\ a_{ji} = -1$
$i\bullet \Longleftrightarrow \bullet j$	$a_{ij} = a_{ji} = -2$
$i\bullet \equiv\!>\!\equiv \bullet j$	$a_{ij} = -3,\ a_{ji} = -1$
$i\bullet \overset{=}{\underset{=}{}} > \overset{=}{\underset{=}{}} \bullet j$	$a_{ij} = -4,\ a_{ji} = -1$

Given a Lie algebra $\mathcal{G}_\ell$ with an $\ell \times \ell$ Cartan matrix of finite type $\mathcal{G}_\ell$, we define the loop algebra

$$L(\mathcal{G}) = \mathcal{G} \otimes \mathbb{C}\left[t, t^{-1}\right] \tag{6.136}$$

(where $\mathbb{C}[t,t^{-1}]$ is the algebra of Laurent polynomials in t, i.e. the algebra of polynomials in t and t^{-1}) using the following Lie bracket:

$$[x \otimes P, y \otimes Q] = [x,y] \otimes PQ \qquad P,Q \in \mathbb{C}\left[t,t^{-1}\right], \quad x,y \in \mathcal{G}_\ell \tag{6.137}$$

From the non-degenerate invariant symmetric bi-linear form (,) of $\mathcal{G}_\ell$ we induce one such form on $L(\mathcal{G}_\ell)$, namely $(x \otimes P, y \otimes Q) = (x,y)PQ$.

A two-cocycle on $L(\mathcal{G}_\ell)$ is defined by

$$\psi(a,b) = \text{Res}\left(\frac{da}{dt}, b\right) \qquad a,b \in L(\mathcal{G}_\ell) \tag{6.138}$$

where the residue of a Laurent series is Res $\left(\sum_{i\in\mathbb{Z}} a_i t^i\right) = a_{-1}$. The two-cocycle satisfies

$$\begin{aligned} &\psi(a,b) = -\psi(b,a) \\ &\psi([a,b],c) + \psi([b,c],a) + \psi([c,a],b) = 0 \end{aligned} \tag{6.139}$$

Given the cocycle ψ, we may define $\hat{L}(\mathcal{G}_\ell)$ as the central extension of $L(\mathcal{G}_\ell)$ by ψ. In formulae, we set $\hat{L}(\mathcal{G}_\ell) = L(\mathcal{G}_\ell) \oplus \mathbb{C}c$, where c is a new central element, $\left[c, \hat{L}(\mathcal{G}_\ell)\right] = 0$, and the new bracket is defined in terms of the old one as $[\ ,\ ] + \psi(\ ,\ )c$. Finally, $\tilde{L}(\mathcal{G}_\ell)$ is defined by adding to $\hat{L}(\mathcal{G}_\ell)$ the derivation d, which acts on $\hat{L}(\mathcal{G}_\ell)$ as $t\,\frac{d}{dt}$, and kills the central element c.

The algebra $\hat{L}(\mathcal{G}_\ell)$ is the one associated with the affine Cartan matrix $\mathcal{G}_\ell^{(1)}$. Just as a finite-dimensional semi-simple Lie algebra $\mathcal{G}_\ell$ is associated with a Cartan matrix of finite type, so an infinite-dimensional loop algebra $\hat{L}(\mathcal{G}_\ell)$ is associated with a Cartan matrix of untwisted affine type.

In agreement with the usual notation employed by physicists, let us denote by

$$J_n^x = x \otimes t^n \qquad (n \in \mathbb{Z}, x \in \mathcal{G}_\ell) \tag{6.140}$$

the generators of $L(\mathcal{G}_\ell)$. Using the modified bracket of $\hat{L}(\mathcal{G}_\ell)$ and the definition (6.138) of the cocycle, we find the Kac–Moody algebra commutators to be:

$$\left[J_n^x, J_m^y\right] = J_{n+m}^{[x,y]} + n\,\delta_{m+n,0}\,(x,y)\,c \tag{6.141}$$

From (6.140), it is easy to find representations of the algebra (6.141) starting from finite-dimensional irreducible representations of $\mathcal{G}_\ell$. Let V denote a finite-dimensional irrep of $\mathcal{G}_\ell$. We define a representation $V(t) = V \otimes \mathbb{C}\left[t,t^{-1}\right]$ of (6.141) quite naturally as

$$\rho^V\left(J_n^x\right) = \rho^V(x) \otimes t^n \tag{6.142}$$

It is a trivial exercise to check that $\rho^V(c) = 0$. These representations $V(t)$ are thus referred to as being of level zero.

To end these brief comments on Kac–Mody algebras, we should also mention their most typical representations, namely the highest weight modules. They are characterized by a "highest weight" vector $|v\rangle$ such that

$$\begin{aligned} J_n^x \,|v\rangle &= 0 \qquad n > 0 \\ c\,|v\rangle &= k\,|v\rangle \end{aligned} \tag{6.143}$$

where k is called the level of the representation. If $k \in \mathbb{Z}$, the highest weight module is called unitary, and we fix conventions so that $k \in \mathbb{N}$.

As an example, consider the affine extension of $s\ell(2)$, i.e. $A_1^{(1)}$. A highest weight module of spin j and level k is characterized by a highest weight vector $|j,k\rangle$ such that

$$\begin{aligned} J_n^x \,|j,k\rangle &= 0 \qquad n > 0 \quad x \in \{+,-,z\} \\ J_0^z \,|j,k\rangle &= j\,|j,k\rangle \\ J_0^+ \,|j,k\rangle &= 0 \end{aligned} \tag{6.144}$$

These representations will be called $V_{j,k}$. Note the difference between $V_{j,k}$ and the level zero $V_j(t)$ considered above.

We may now return to the quantum affine extension $U_q(\tilde{\mathscr{G}}_\ell)$. Its universal $\mathscr{R}$-matrix was found, by Drinfeld, using the quantum double construction. Here, we introduce a universal $\mathscr{R}$-matrix for $U_q(\tilde{\mathscr{G}}_\ell)$ depending explicitly on a formal parameter which plays the role of a spectral parameter.

Let us start by considering the automorphism

$$\begin{aligned} D_t : U_q(\tilde{\mathscr{G}}_\ell) \otimes \mathbb{C}\left[t,t^{-1}\right] &\longrightarrow U_q(\tilde{\mathscr{G}}_\ell) \otimes \mathbb{C}\left[t,t^{-1}\right] \\ D_t(e_i) &= t^{\delta_{i,0}} e_i \\ D_t(f_i) &= t^{-\delta_{i,0}} f_i \\ D_t(k_i) &= t^{\delta_{i,0}} k_i \\ D_t(d) &= d \end{aligned} \tag{6.145}$$

and a new set of co-products:

$$\Delta_t(a) = (D_t \otimes \mathbf{1})\,\Delta(a)\,, \qquad \Delta_t'(a) = (D_t \otimes \mathbf{1})\,\Delta'(a) \tag{6.146}$$

Letting $\mathscr{R}$ denote the universal $\mathscr{R}$-matrix of $U_q(\tilde{\mathscr{G}}_\ell)$, we define

$$\mathscr{R}(t) = (D_t \otimes \mathbf{1})\,\mathscr{R} \tag{6.147}$$

which satisfies

$$\begin{aligned} \mathscr{R}(t)\Delta_t(a) &= \Delta_t'(a)\mathscr{R}(t) \\ (\Delta_z \otimes \mathbf{1})\,\mathscr{R}(u) &= \mathscr{R}_{13}(zu)\mathscr{R}_{23}(u) \\ (\mathbf{1} \otimes \Delta_u)\,\mathscr{R}(zu) &= \mathscr{R}_{13}(z)\mathscr{R}_{12}(zu) \end{aligned} \tag{6.148}$$

These expressions should be compared with (6.20). The matrix $\mathcal{R}(t)$ is given by

$$\begin{aligned}\mathcal{R}(t) = \exp h \left((c \otimes d + d \otimes c) + \sum_{i,j=1}^{\ell} \left(B^{-1}\right)_{ij} H_i \otimes H_j \right) \\ \times \left(\mathbf{1} + \sum_{i=1}^{\ell} 2 \sinh (hD_i)\ e_i \otimes f_i + 2t \sinh h\, e_0 \otimes f_0 + \cdots \right)\end{aligned} \tag{6.149}$$

where $B = D \cdot A$ is the symmetrized Cartan matrix, and the quantum central extension is $c = H_0 + H_\theta$, with θ the maximal root of $\mathcal{G}_\ell$.

The $\mathcal{R}$-matrix $\mathcal{R}(t)$ satisfies the following Yang–Baxter equation:

$$\mathcal{R}_{12}(t)\mathcal{R}_{13}(tw)\mathcal{R}_{23}(w) = \mathcal{R}_{23}(w)\mathcal{R}_{13}(tw)\mathcal{R}_{12}(t) \tag{6.150}$$

Making a change of variables $t = \exp u$ and $w = \exp v$, we recover the well-known Yang–Baxter equation with the difference property.

We can remove from the universal $\mathcal{R}$-matrix (6.149) for $U_q(\tilde{\mathcal{G}}_\ell)$ the term with an explicit dependence on the central charge and the derivation, defining the $\mathcal{R}$-matrix

$$\overline{\mathcal{R}}(t) = \exp\left(-h(c \otimes d + d \otimes c)\right) \mathcal{R}(t) \tag{6.151}$$

which satisfies a more involved Yang–Baxter equation than (6.150), namely

$$\overline{\mathcal{R}}_{12}(t)\overline{\mathcal{R}}_{13}(twq^{c_2})\overline{\mathcal{R}}_{23}(w) = \overline{\mathcal{R}}_{23}(w)\overline{\mathcal{R}}_{13}(twq^{-c_2})\overline{\mathcal{R}}_{12}(t) \tag{6.152}$$

with the notation $c_2 = \mathbf{1} \otimes c \otimes \mathbf{1}$. For finite-dimensional irreps, the central charge is zero, and we recover the Yang–Baxter equation (6.150), with $\overline{\mathcal{R}}(t)$ instead of $\mathcal{R}(t)$. Given two zero-level finite-dimensional irreducible representations V and W, we define

$$\overline{\mathcal{R}}^{VW}(t) = \left[\rho^V \otimes \rho^W\right] \overline{\mathcal{R}}(t) \tag{6.153}$$

Note that finite-dimensional irreps are representations of $U_q(\hat{\mathcal{G}}_\ell)$, but are not really representations of $U_q(\tilde{\mathcal{G}}_\ell)$. This is why we need to define the $\mathcal{R}$-matrix (6.151), where we effectively eliminate (or rather, gauge away) the derivation d of $U_q(\tilde{\mathcal{G}}_\ell)$.

Given a finite-dimensional irrep V of $U_q(\mathcal{G}_\ell^{(1)})$, we define a new irrep $V(t) = V \otimes \mathbb{C}\left[t, t^{-1}\right]$ as follows:

$$\rho^{V(t)}(a) = \rho^V\left(D_t a\right) \qquad a \in U_q(\mathcal{G}_\ell^{(1)}) \tag{6.154}$$

In this way, an irrep of $U_q(\mathcal{G}_\ell^{(1)})$ is characterized by the affine parameter t. This construction is essentially the same as that introduced in equations (2.113) in order to affinize the irreps of $SU(2)$.

For these representations, it is possible to prove that $\overline{\mathcal{R}}^{VW}(t)$ is the intertwiner for the representations $V(tz)$ and $W(z)$. Indeed, using (6.148) and the definition (6.151), it is easy to check that

$$\overline{\mathcal{R}}(t)\Delta_t(a) = \left(D_{q^c}^{-1} \otimes D_{q^c}^{-1}\right) \Delta_t'(a)\overline{\mathcal{R}}(t) \tag{6.155}$$

with the prefactor on the right-hand side coming from

$$\exp(-h(c \otimes d + d \otimes c))\Delta_t'(a) = \left[D_{q^c}^{-1} \otimes D_{q^c}^{-1}\right] \Delta_t'(a)\, \mathrm{e}^{-h(c\otimes d+d\otimes c)} \tag{6.156}$$

The prefactor is just one for finite-dimensional irreps, because $c = 0$. Using (6.146) and (6.154), we may write

$$\rho^{V(z)} \otimes \rho^{W(z)} [\Delta_t(a)] = \rho^{V(zt)} \otimes \rho^{W(z)} [\Delta(a)] \tag{6.157}$$

and finally, from (6.155), we obtain

$$\overline{\mathscr{R}}(t)\rho^{V(zt)} \otimes \rho^{W(z)} [\Delta(a)] = \rho^{V(zt)} \otimes \rho^{W(z)} \left[\Delta'(a)\right] \overline{\mathscr{R}}(t) \tag{6.158}$$

In the context of the six-vertex model (chapter 2), we have worked out the particular case $\mathscr{G}_\ell = A_1$, and $V = W$ the spin-$\frac{1}{2}$ irrep, whereas above we have sketched out the general formalism.

6.11 Quasi-Hopf algebras

The structure of Hopf algebras that we have studied so far admits an interesting generalization. In this section, we review the basic definitions of quasi-Hopf algebras, and postpone until chapters 10 and 11 their physical applications. The reader will not suffer if this section is left until later.

In a Hopf algebra, the co-product Δ satisfies the co-associativity axiom (6.4), which implies, of course, the strict associativity of the tensor product of its representations, namely

$$(\Delta \otimes \mathbf{1})\Delta = (\mathbf{1} \otimes \Delta)\Delta \quad \Rightarrow \quad (\pi_1 \otimes \pi_2) \otimes \pi_3 = \pi_1 \otimes (\pi_2 \otimes \pi_3) \tag{6.159}$$

If, moreover, the Hopf algebra is co-commutative, $\Delta = \Delta'$, then the tensor product of two representations is independent of their order:

$$\Delta = \Delta' \quad \Rightarrow \quad \pi_1 \otimes \pi_2 = \pi_2 \otimes \pi_1 \tag{6.160}$$

To construct quantum groups, we relax condition (6.160) and require only that the two tensor products be equivalent, not equal:

$$\Delta' = \mathscr{R}\Delta\mathscr{R}^{-1} \quad \Rightarrow \quad \pi_1 \otimes \pi_2 \approx \pi_2 \otimes \pi_1 \tag{6.161}$$

It is precisely the matrix $\mathscr{R}_{12} = (\pi_1 \otimes \pi_2)\mathscr{R}$ which establishes this equivalence. From this perspective, it is natural to relax the co-associativity (6.159) into an equivalence instead of an equality. This step takes us to quasi-Hopf algebras.

Instead of (6.4), we require

$$(\Delta \otimes \mathbf{1})\Delta(a) = \phi(\mathbf{1} \otimes \Delta)\Delta(a)\phi^{-1} \qquad (\forall a \in \mathscr{A}) \tag{6.162}$$

where $\phi \in \mathscr{A} \otimes \mathscr{A} \otimes \mathscr{A}$ measures the lack of strict associativity. Any such associativity functor satisfies a pentagonal equation, which is simply the commutativity of the diagram (6.88). Explicitly, ϕ must satisfy the following equation:

$$(\mathbf{1} \otimes \mathbf{1} \otimes \Delta)(\phi)\, (\Delta \otimes \mathbf{1} \otimes \mathbf{1})(\phi) = (\mathbf{1} \otimes \phi)\, (\mathbf{1} \otimes \Delta \otimes \mathbf{1})(\phi)\, (\phi \otimes \mathbf{1}) \tag{6.163}$$

A quasi-Hopf algebra is a Hopf algebra with an element $\phi \in \mathscr{A}^{\otimes 3}$ satisfying (6.163). Its quasi-triangularity can also be implemented, requiring the following conditions, which can be derived from consistency between commutative and associative operations:

$$\begin{aligned} &\Delta'(a) = \mathscr{R}\,\Delta(a)\,\mathscr{R}^{-1} \qquad (\forall a \in \mathscr{A}) \\ &(\Delta \otimes \mathbf{1})(\mathscr{R}) = \phi_{312}\mathscr{R}_{13}\phi_{132}^{-1}\mathscr{R}_{23}\phi \\ &(\mathbf{1} \otimes \Delta)(\mathscr{R}) = \phi_{231}^{-1}\mathscr{R}_{13}\phi_{213}\mathscr{R}_{12}\phi^{-1} \end{aligned} \tag{6.164}$$

The Yang–Baxter equation for quasi-triangular quasi-Hopf algebras (sometimes called quasi-quantum groups for short) takes a somewhat unfamiliar form:

$$\mathscr{R}_{12}\phi_{312}\mathscr{R}_{13}\phi_{132}^{-1}\mathscr{R}_{23}\phi_{123} = \phi_{321}\mathscr{R}_{23}\phi_{231}^{-1}\mathscr{R}_{13}\phi_{213}\mathscr{R}_{12} \tag{6.165}$$

This equation can be understood step by step. The left-hand side arises, schematically, from the transformations

$$(12)3 \stackrel{\phi_{123}}{\rightarrow} 1(23) \stackrel{\mathscr{R}_{23}}{\rightarrow} 1(32) \stackrel{\phi_{132}^{-1}}{\rightarrow} (13)2 \stackrel{\mathscr{R}_{13}}{\rightarrow} (31)2 \stackrel{\phi_{312}}{\rightarrow} 3(12) \stackrel{\mathscr{R}_{12}}{\rightarrow} 3(21) \tag{6.166}$$

whereas the right-hand side comes from

$$(12)3 \stackrel{\mathscr{R}_{12}}{\rightarrow} (21)3 \stackrel{\phi_{213}}{\rightarrow} 2(13) \stackrel{\mathscr{R}_{13}}{\rightarrow} 2(31) \stackrel{\phi_{231}^{-1}}{\rightarrow} (23)1 \stackrel{\mathscr{R}_{23}}{\rightarrow} (32)1 \stackrel{\phi_{321}}{\rightarrow} 3(21) \tag{6.167}$$

There exist gauge or "twist" transformations among these algebras. Indeed, let $(\mathscr{A}, \Delta, \mathscr{R}, \phi)$ be a quasi-triangular quasi-Hopf algebra (we leave the multiplication, unit, antipode and co-unit for the reader, who should work it out if this is the first contact with quasi-quantum groups). Then a new quasi-triangular quasi-Hopf algebra $(\mathscr{A}, \tilde{\Delta}, \tilde{\mathscr{R}}, \tilde{\phi})$ can be constructed from it with the help of any invertible element $\Omega \in \mathscr{A} \otimes \mathscr{A}$, called a twist operator:

$$\begin{aligned} &\tilde{\Delta}(a) = \Omega\,\Delta(a)\,\Omega^{-1} \\ &\tilde{\phi} = \Omega_{23}(\mathbf{1} \otimes \Delta)(\Omega)\,\phi(\Delta \otimes \mathbf{1})\left(\Omega^{-1}\right)\,\Omega_{12}^{-1} \\ &\tilde{\mathscr{R}} = \Omega_{21}\,\mathscr{R}\,\Omega_{12}^{-1} \end{aligned} \tag{6.168}$$

An example of a quasi-triangular quasi-Hopf algebra is proposed in exercise 6.18.

Exercises

Ex. 6.1 Study the relationship between the lack of orthogonality $\langle 0,0|1,0\rangle \neq 0$ in the decomposition of $\frac{1}{2} \times \frac{1}{2}$ (section 6.4.2) when q is a phase with the lack of hermiticity of the quantum group invariant hamiltonian (3.131). (See also exercise 3.16.)

Ex. 6.2 The Borel–Weil–Bott construction for $U_q(s\ell(2))$. Consider the space of polynomials in two variables u and v. The spin-j irrep of $SU(2)$ is given by the homogeneous polynomials of degree $2j$ with basis

$$|j,m\rangle = \frac{u^{j+m}v^{j-m}}{\sqrt{(j+m)!(j-m)!}} \qquad m=-j,\ldots,j \tag{6.169}$$

In this space, the generators of $s\ell(2)$ are represented by

$$X^+ = u\frac{\partial}{\partial v}\ , \quad X^- = v\frac{\partial}{\partial u}\ , \quad H = u\frac{\partial}{\partial u} - v\frac{\partial}{\partial v} \tag{6.170}$$

The metric in the space of polynomials is

$$\langle f|g\rangle = f\left(\frac{d}{du},\frac{d}{dv}\right) g(u,v)\bigg|_{u=v=0} \tag{6.171}$$

Define now the q-derivatives or q-difference operators $D_u^{(q)}$ by their action on functions, namely

$$D_u^{(q)} f(u) = \frac{f(qu) - f(q^{-1}u)}{(q-q^{-1})\,u} \tag{6.172}$$

and similarly for $D_v^{(q)}$. Prove that the operators

$$X^+ = uD_v^{(q)}\ , \quad X^- = vD_u^{(q)}\ , \quad H = u\frac{\partial}{\partial u} - v\frac{\partial}{\partial v} \tag{6.173}$$

constitute a faithful representation of the generators of $U_q(s\ell(2))$. The metric (6.171) is now replaced by

$$\langle f|g\rangle_q = f\left(D_u^{(q)}, D_v^{(q)}\right) g(u,v)\bigg|_{u=v=0} \tag{6.174}$$

Using now the q-derivatives (6.172), define a q-generalization of the harmonic oscillator, and find its eigenvalues and eigenstates. In the limit $q \to 1$, you should recover the usual spectrum $(n+\frac{1}{2})$.

Ex. 6.3 Consider a group G and the algebra $\mathscr{A} = \mathrm{Fun}\,(G)$ of functions on the group, which has a natural Hopf structure. The group G itself is contained in $\mathscr{A}$, because it is isomorphic to the subset of constant functions. Accordingly, the group multiplication extends uniquely to a multiplication m in the algebra, and the co-multiplication is induced by the group multiplication:

$$\Delta f(g_1 \otimes g_2) = f(m(g_1 \otimes g_2)) = f(g_1 g_2) \qquad \forall f \in \mathrm{Fun}\,(G) \ \ \forall g_1, g_2 \in G \tag{6.175}$$

Similarly, the co-unit and antipode derive from the group identity e and inverse:

$$\epsilon(f) = f(e)\ \ ,\ \ [\gamma(f)](g) = f(g^{-1}) \qquad \forall f \in \mathrm{Fun}\,(G) \ \ \forall g \in G \tag{6.176}$$

Show that if G is abelian, then $\mathscr{A}$ is a co-commutative Hopf algebra. Under what conditions is $\mathscr{A}$ a quantum group?

Ex. 6.4 Given the bi-algebra $(\mathscr{A}, m, \imath, \Delta, \epsilon)$, show that $(\mathscr{A} \otimes \mathscr{A}, m^{\otimes}, \imath^{\otimes}, \Delta^{\otimes}, \epsilon^{\otimes})$ is a bi-algebra with $\imath^{\otimes} = \imath \otimes \imath$, $\epsilon^{\otimes} = \epsilon \otimes \epsilon$, $m^{\otimes} = (m \otimes m) \circ (\mathbf{1} \otimes \sigma \otimes \mathbf{1})$, and $\Delta^{\otimes} = (\mathbf{1} \otimes \sigma \otimes \mathbf{1}) \circ (\Delta \otimes \Delta)$. Note the importance of the permutation map $\sigma : a \otimes b \mapsto b \otimes a$. If $\mathscr{A}$ was a Hopf algebra, i.e. if it also had an antipode γ, what would $\gamma^{\otimes}$ be in order for $\mathscr{A} \otimes \mathscr{A}$ to be a Hopf algebra?

Ex. 6.5 A familiar example of a deformed co-multiplication is given by the lorentzian deformation of the galilean group. Classically, the composition of velocities is described by $\Delta^G(\vec{v}) = \vec{v} \otimes \mathbf{1} + \mathbf{1} \otimes \vec{v}$, whereas the relativistic deformation of this co-product is $\Delta^L_c(\vec{v}) = (\vec{v} \otimes \mathbf{1} + \mathbf{1} \otimes \vec{v})/(1 - c^{-2} \vec{v} \otimes \vec{v})$. Clearly, $\lim_{c \to \infty} \Delta^L_c = \Delta^G$. Work out the full Hopf algebra for both cases. Verify, in particular, that $\epsilon(\vec{v}) = 0$ and $\gamma(\vec{v}) = -\vec{v}$.

Ex. 6.6 Show that the quasi-triangularity conditions on a Hopf algebra, equations (6.19), (6.20) and (6.21), imply that

$$\left.\begin{array}{l}(\Delta \otimes \mathbf{1})\mathscr{R}(a) = \mathscr{R}_{13}\mathscr{R}_{23}(\Delta \otimes \mathbf{1})(a) \\ (\mathbf{1} \otimes \Delta)\mathscr{R}(a) = \mathscr{R}_{13}\mathscr{R}_{12}(\mathbf{1} \otimes \Delta)(a)\end{array}\right. \quad \forall a \in \mathscr{A} \otimes \mathscr{A} \tag{6.177}$$

Letting each line represent a copy of $\mathscr{A}$, and representing $\mathscr{R}$ and Δ graphically, as in figure 6.15, equations (6.177) and (6.23) can be represented graphically, as in figure 6.16. *Caveat*: these diagrams hold for the algebra, independently of representations. Throughout the book, in all other diagrams we always work with representations of the quantum group.

Ex. 6.7 Verify that Drinfeld's quantum double does yield a quasi-triangular Hopf algebra. Other than the intertwiner condition (6.33), check that the $\mathscr{R}$-matrix (6.32), the co-product (6.30) and the antipode (6.35) of the double satisfy conditions (6.20) and (6.21).

Ex. 6.8 Given a Hopf algebra $\mathscr{A}$, its naive dual $\mathscr{A}^{\circ} = \mathrm{End}(\mathscr{A}, \mathbb{C})$ is also a Hopf algebra. Indeed, the multiplication m° and co-multiplication Δ° of $\mathscr{A}^{\circ}$ are just

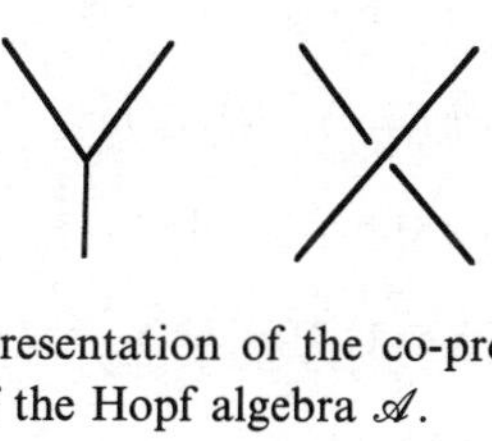

Fig. 6.15. Diagrammatic representation of the co-product and the universal $\mathscr{R}$-matrix. Each line represents a copy of the Hopf algebra $\mathscr{A}$.

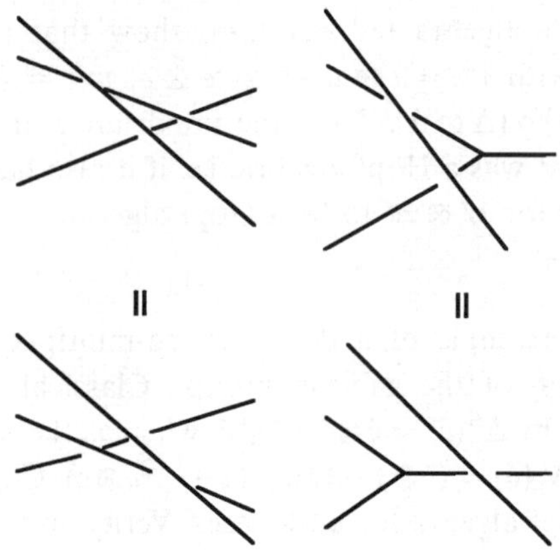

Fig. 6.16. Diagrammatic representation of equations (6.23) and (6.177), consequences of quasi-triangularity.

the co-multiplication Δ and multiplication m of $\mathscr{A}$:

$$\begin{aligned} (\Delta^\circ \tilde{a})\,(b \otimes c) &= \tilde{a}(m(b \otimes c)) = \tilde{a}(bc) \qquad && \forall a, b, c \in \mathscr{A} \\ m^\circ(\tilde{a} \otimes \tilde{b})(c) &= \tilde{a} \otimes \tilde{b}(\Delta(c)) && \forall \tilde{a}, \tilde{b}, \tilde{c} \in \mathscr{A}^\circ \end{aligned} \tag{6.178}$$

The useful dual $\mathscr{A}^*$ to A, used in section 6.3, differs from A° in that its co-multiplication Δ^* is given by $\Delta^*(\tilde{a})(b \otimes c) = \tilde{a}(cb)$. Define the co-unit ϵ^* and antipode γ^* of A^*, and check that A^* is a Hopf algebra. Define $\mathscr{R} = \sum e_a \otimes e^a \in \mathscr{A} \otimes \mathscr{A}^*$, and show that

$$(\Delta \otimes \mathbf{1})\mathscr{R} = \mathscr{R}_{23}\mathscr{R}_{13}\,, \quad (\mathbf{1} \otimes \Delta^*)\mathscr{R} = \mathscr{R}_{12}\mathscr{R}_{13}\,, \quad (\mathbf{1} \otimes \gamma^*)\mathscr{R} = \mathscr{R}^{-1} \tag{6.179}$$

Show that the co-multiplication of the quantum double $\mathscr{D}(\mathscr{A}, \mathscr{A}^*) \subset \mathscr{A} \otimes \mathscr{A}^*$ can be written as

$$\Delta(x \otimes y) = \mathscr{R}_{23}\Delta_{13}(x)\Delta^\circ_{24}(y)\mathscr{R}^{-1}_{23} \qquad \forall x, y \in \mathscr{D}(\mathscr{A}) \tag{6.180}$$

where the subscripts in the co-multiplications indicate on which of the four factors in $\mathscr{D} \otimes \mathscr{D}$ they act. Define also the transposed co-product Δ' in $\mathscr{D}$, and check that the universal $\mathscr{R}$-matrix $\mathscr{R}_{\mathscr{D}} = \sum e_a \otimes \mathbf{1} \otimes \mathbf{1} \otimes e^a$ intertwines between Δ and Δ'. Pay careful attention to the order of the various algebras in the tensor products.

Ex. 6.9 [S.L. Woronowicz, *Lett. Math. Phys.* **23** (1991) 143.]
Let us denote the co-product of an element $b \in \mathscr{A}$ by

$$\Delta(b) = \sum b^{(1)} \otimes b^{(2)} \tag{6.181}$$

Co-associativity is implicit in this short-hand:

$$(\mathbf{1} \otimes \Delta)\,\Delta(b) = (\Delta \otimes \mathbf{1})\,\Delta(b) = \sum b^{(1)} \otimes b^{(2)} \otimes b^{(3)} \tag{6.182}$$

- What is the relationship between the symbols $b^{(1)}$ in (6.181) and (6.182)? And $b^{(2)}$?
- Show that the linear operator $Q : \mathscr{A} \to \mathscr{A} \otimes \mathscr{A}$, with $Q(b) = \sum b^{(2)} \otimes \gamma\left(b^{(1)}\right) b^{(3)}$, satisfies the identity $(Q \otimes \mathbf{1})\, Q(b) = (\mathbf{1} \otimes \Delta)\, Q(b)$.
- Consider now the linear operator $T : \mathscr{A} \otimes \mathscr{A} \to \mathscr{A} \otimes \mathscr{A}$, with $T(a \otimes b) = (\mathbf{1} \otimes a)\, Q(b)$, and show that it satisfies a Yang–Baxter equation

$$(T \otimes \mathbf{1})(\mathbf{1} \otimes T)(T \otimes \mathbf{1}) = (\mathbf{1} \otimes T)(T \otimes \mathbf{1})(\mathbf{1} \otimes T) \tag{6.183}$$

[Hint: note first that if $\mathscr{A}$ is co-commutative, then $T(a \otimes b) = b \otimes a$. Show that T is invertible, with $T^{-1}(a \otimes b) = \sum b\gamma^{-1}\left(a^{(3)}\right) a^{(1)} \otimes a^{(2)}$. Show also that the action of T on a co-product is trivial; more generally, $T\,[(a \otimes b)\Delta(c)] = T(a \otimes b)\Delta(c)$.]
- Similarly, define $T'(a \otimes b) = \sum b^{(1)} \otimes \gamma\left(b^{(2)}\right) ab^{(3)}$, and prove that, if $\mathscr{A}$ is commutative, then T' is just the transposition, that T' is invertible, and that T' satisfies a Yang–Baxter equation just like (6.183).

Ex. 6.10 To appreciate better the close similarity between loop algebras and their associated semi-simple Lie algebras, prove that the representation (6.142) corresponds, in the Chevalley basis of the extended algebra $\hat{\mathscr{G}}_\ell$, to the requirement

$$\rho^V(e_0) = t\rho^V(x_{-\theta}), \quad \rho^V(f_0) = t^{-1}\rho^V(x_\theta) \tag{6.184}$$

where x_θ and $x_{-\theta}$ are the elements of $\mathscr{G}$ associated with the maximal root θ and its negative, respectively.

Ex. 6.11 [D. Bernard, 'Hidden yangians in two-dimensional massive current algebra', *Commun. Math. Phys.* **137** (1991) 191.]

In two spacetime dimensions, if a local current $J_\mu = J_\mu^{(0)} = A_\mu$ is conserved ($\partial^\mu J_\mu^{(0)} = 0$), then there always exists a pseudo-scalar potential $\mathscr{X}^{(0)}$ for it: $J_\mu^{(0)} = \epsilon_{\mu\nu}\partial^\nu \mathscr{X}^{(0)}$. The algebra satisfied by the charges $Q_0^a = \int A_0^a$ just reproduces that of the gauge fields A_μ^a, $\left[Q_0^a, Q_0^b\right] = if^{abc}Q_0^c$. These charges are local, in the sense that $\Delta Q_0^a = Q_0^a \otimes \mathbf{1} + \mathbf{1} \otimes Q_0^a$. Now define, for $n \geq 1$,

$$J_\mu^{(n)} = \left(\partial_\mu + A_\mu\right) \mathscr{X}^{(n-1)} = \epsilon_{\mu\nu}\partial^\nu \mathscr{X}^{(n)} \tag{6.185}$$

Clearly, $\partial^\mu J_\mu^{(n)} = 0$ for all $n \geq 0$. Let us require Δ to be a homomorphism. The yangian is the bi-algebra of the charges $Q_n = \int J_0^{(n)}$ associated with these infinitely many conserved currents. Check that $\left[Q_0^a, Q_1^b\right] = if^{abc}Q_1^c$. Note that, although $J_\mu^{(0)}$ is local, $\mathscr{X}^{(0)}$ is not, and neither are $J_\mu^{(n)}$ for $n \geq 1$. Check, for instance, that

$$\Delta Q_1^a = Q_1^a \otimes \mathbf{1} + \mathbf{1} \otimes Q_1^a + \frac{1}{2} f^{abc} Q_0^c \otimes Q_0^b \tag{6.186}$$

Find also $f^{d[ab}\left[Q_1^{c]}, Q_1^d\right]$. Try to find the general commutators $[Q_m, Q_n]$ and co-products ΔQ_m.

Ex. 6.12 Extend the definition of the yangian charges Q_n of the previous exercise to $n < 0$. The inverse operator of the derivative is, in this context, the Wilson integral, so that, for starters, $Q_{-1} = P \exp \int_0^x J_\mu^{(0)} dx^\mu$. Find $[Q_0, Q_{-1}]$ and ΔQ_{-1}.

Ex. 6.13 [J. Lukierski, A. Nowicki and H. Ruegg, *Phys. Lett.* **B293** (1992) 344.]
Since the Poincaré algebra (the semi-direct product of the Lorentz algebra and translations) is the limit $R \to \infty$ of anti-de-Sitter $SO(3,2)$, we can find a quantum deformation of Poincaré starting from $U_q(sp(4)) = U_q(C_2)$. In the limit $R \to \infty$, it is necessary to let $q \to 1$, keeping $R \log q = 2/M$ constant. If q is real, the mass M is real and positive. Set $M = 1$, i.e. measure all energies and momenta in units of the new mass scale M_R, and check that the deformed Poincaré algebra satisfied by the rotations, boosts and momenta is

$$\begin{aligned}
&[P_\mu, P_\nu] = 0\,, \quad [J_i, P_0] = 0\,, \quad [P_0, L_k] = -iP_k \\
&[J_i, P_j] = i\epsilon_{ijk} P_k\,, \quad [J_i, J_j] = i\epsilon_{ijk} J_k\,, \quad [J_i, L_j] = i\epsilon_{ijk} L_k \\
&[P_j, L_k] = -\frac{i}{2}\delta_{jk} \sinh 2P_0 \\
&[L_i, L_j] = -i\epsilon_{ijk}\left(J_k \cosh 2P_0 - P_k \vec{P}\cdot\vec{J}\right)
\end{aligned} \tag{6.187}$$

which reduces to the classical Poincaré algebra when $M \to \infty$. Verify that the co-products of the rotations and energy are trivial, whereas

$$\begin{aligned}
\Delta P_i &= P_i \otimes e^{P_0} + e^{-P_0} \otimes P_i \\
\Delta L_i &= L_i \otimes e^{P_0} + e^{-P_0} \otimes L_i + \epsilon_{ijk}\left(P_j \otimes J_k e^{P_0} + e^{-P_0} J_j \otimes P_k\right)
\end{aligned} \tag{6.188}$$

and find also the antipodes of all the generators. Check that the two Casimirs of this algebra are

$$\begin{aligned}
m^2 &= \sinh^2 P_0 - \vec{P}^2 \\
m^2\left(1 - m^2\right) s(s+1) &= \left(\cosh 2P_0 - \vec{P}^2\right) W_0^2 - \vec{W}^2
\end{aligned} \tag{6.189}$$

where the deformed Pauli–Lubanski vector is given by $W_0 = \vec{P}\cdot\vec{J}$, $W_i = \frac{1}{2} J_i \sinh 2P_0 + \epsilon_{ijk} P_j L_k$.

Ex. 6.14 Path representation of affine quantum groups [O. Foda and T. Miwa, 'Corner transfer matrix and quantum affine algebras, *Int. J. Mod. Phys.* **A7** (1992) 279].
Let us consider $U_q(\widehat{s\ell}(2))$. Let V be the spin-$\frac{1}{2}$ finite-dimensional irrep defined by (2.114):

$$\begin{aligned}
F_0(+) = E_1(+) &= 0 & F_0(-) = E_1(-) &= (+) \\
E_0(+) = F_1(+) &= (-) & E_0(-) = F_1(-) &= 0 \\
K_1(+) = K_0^{-1}(+) &= q(+) & K_1(-) = K_0^{-1}(-) &= q^{-1}(-)
\end{aligned} \tag{6.190}$$

Let $\mathscr{W}$ denote the semi-infinite tensor product $\bigotimes_{k=1}^{\infty} V_k$, with $V_k = V$. In $\mathscr{W}$ we define the space $\mathscr{P}$ of paths $|p\rangle$ by fixing the asymptotic behavior of these paths:

$$|p\rangle \in \mathscr{P} \iff \text{for } k >> 0 \qquad p(k) = \begin{cases} + & k \text{ even} \\ - & k \text{ odd} \end{cases} \tag{6.191}$$

where $p(k)$ denotes the state ($+$ or $-$) at the kth position of the path $|p\rangle$. The path $|0\rangle = (-+-+-+-\cdots)$ will be referred to as the vacuum state. Paths are states in $\mathscr{W}$ differing from the "bare" vacuum $|0\rangle$ by a finite amount. To characterize paths, we introduce the following:

- the length $\ell(p)$ of a path $|p\rangle$ is $\ell(p) = \max\{n; p(n) \neq 0(n)\}$;
- the depth $d(p)$ of a path $|p\rangle$ is $d(p) = \min\{n| \ \exists \ i_1, \ldots, i_n$ such that $|p\rangle = Q_{i_1} \cdots Q_{i_n} |0\rangle\}$, where Q_i acts on the path variables at positions i and $i+1$ as follows:

$$Q_i(+-) = (-+)\,, \quad Q_i(-+) = (+-)\,, \quad Q_i(++) = Q_i(--) = 0 \tag{6.192}$$

- the spin $s(p)$ of a path $|p\rangle$ is $S\,|p\rangle = \sum_{k=1}^{\infty} S_k\,|p\rangle = s(p)\,|p\rangle$, with $S_k(+) = \frac{1}{2}(+)$ and $S_k(-) = -\frac{1}{2}(-)$.

We may now define an infinite-dimensional irrep of $U_q(\widehat{s\ell}(2))$ on $\mathscr{P}$. For this, we use the infinite co-product

$$\begin{aligned} \Delta^{(\infty)}(E_i) &= \sum_{j=0}^{\infty} \overbrace{K_i \otimes \cdots \otimes K_i}^{j \text{ times}} \otimes E_i \otimes \mathbf{1} \otimes \mathbf{1} \otimes \cdots \\ \Delta^{(\infty)}(F_i) &= \sum_{j=0}^{\infty} \overbrace{\mathbf{1} \otimes \cdots \otimes \mathbf{1}}^{j \text{ times}} \otimes F_i \otimes K_i^{-1} \otimes K_i^{-1} \otimes \cdots \\ \Delta^{(\infty)}(K_i) &= K_i \otimes K_i \otimes K_i^{-1} \otimes \cdots \end{aligned} \tag{6.193}$$

To simplify the notation, we use X instead of $\Delta^{(\infty)}(X)$ in what follows.

From (6.190), we observe that $K_0K_1 = 1$ on V. Since the central charge is $c = H_0 + H_1$, we deduce that finite-dimensional representations have zero central extension. This is not the case, however, for infinite-dimensional irreps. To obtain an irrep of level 1, for instance, we impose on $\Delta^{(\infty)}(K_0)\Delta^{(\infty)}(K_1)$ the condition $K_0K_1 = q$, which reads more explicitly as

$$K_0\,|p\rangle = q^{1-2s(p)}\,|p\rangle\ , \qquad K_1\,|p\rangle = q^{2s(p)}\,|p\rangle \tag{6.194}$$

The formal action of the non-diagonal generators on $\mathscr{P}$ is given by

$$\begin{aligned} E_0\,|p\rangle &= \sum_{k=1}^{\infty} q^{-2s_k(p)} \sigma_k^{(E_0)}(p) \\ F_0\,|p\rangle &= \sum_{k=1}^{\infty} q^{2s(p)-2s_k(p)} \sigma_k^{(E_1)}(p) \end{aligned}$$

$$E_1 |p\rangle = \sum_{k=1}^{\infty} q^{2s_k(p)} \sigma_k^{(E_1)}(p) \tag{6.195}$$

$$F_1 |p\rangle = \sum_{k=1}^{\infty} q^{2s_k(p)-2s(p)} \sigma_k^{(E_0)}(p)$$

where

$$\sigma_k^{(E_0)}(p) = \begin{cases} 0 & \text{if } p(k) = (-) \\ p(j) & \text{if } j \neq k \text{ and } p(k) = (+) \\ (-) & \text{if } p(k) = (+) \end{cases}$$
$$\sigma_k^{(E_1)}(p) = \begin{cases} 0 & \text{if } p(k) = (+) \\ p(j) & \text{if } j \neq k \text{ and } p(k) = (-) \\ (+) & \text{if } p(k) = (-) \end{cases} \tag{6.196}$$

All these expressions are to be understood as formal power series in q; we shall consider the case $|q| \to 0$, which physically means very low temperature.

In the $q \to 0$ limit, prove that $|0\rangle$ is a highest weight vector, and solve to order q^2 the highest weight vector equations for the path representation.

Ex. 6.15 Consider the XXZ hamiltonian $H = \sum_{j\in\mathbb{Z}} \sigma_j^x \sigma_{j+1}^x + \sigma_j^y \sigma_{j+1}^y + \Delta \sigma_j^z \sigma_{j+1}^z$, with $\Delta = \left(q + q^{-1}\right)/2$.

- Prove that H is invariant under $U_q(\widehat{s\ell}(2))$, i.e. that

$$\left[H, \Delta^{(\infty)}(E_i)\right] = \left[H, \Delta^{(\infty)}(F_i)\right] = \left[H, \Delta^{(\infty)}(K_i)\right] = 0 \tag{6.197}$$

- Using the corner transfer matrix of exercise 5.9, prove in the limit $q \to 0^-$ (i.e. $\Delta \to -\infty$) to order q^2 that the corner transfer matrix ground state is the highest weight vector $|0\rangle$ of the path representation in the previous exercise.
- Prove that the operator

$$d = -S^z_{\text{total}} + \frac{q}{1-q^2} \sum_{j\in\mathbb{Z}} j \left(\sigma_j^x \sigma_{j+1}^x + \sigma_j^y \sigma_{j+1}^y + \Delta \sigma_j^z \sigma_{j+1}^z\right) \tag{6.198}$$

  satisfies (6.135). [Hint: work out the previous exercise.]
- Find a relationship between H_{CTM}, d, and the shift operator $T : \vec{\sigma}_j \to \vec{\sigma}_{j+1}$.

Ex. 6.16 [M. Kashiwara, 'Crystal bases', *C. R. Acad. Sci.* **311** (1990) 277; K.C. Misra and T. Miwa, 'Crystal base for the basic representation of $U_q(\widehat{s\ell}(n))$', *Commun. Math. Phys.* **134** (1990) 79.]

Note that the limit $q \to 0$ of (6.79) is well defined, namely

$$|1,1\rangle = |+\rangle \otimes |+\rangle$$
$$|1,0\rangle = |+\rangle \otimes |-\rangle \qquad |0,0\rangle = |-\rangle \otimes |+\rangle \tag{6.199}$$
$$|1,-1\rangle = |-\rangle \otimes |-\rangle$$

A general property of the limit $q \to 0$ of quantum groups is that the tensor product of two representations is decomposed into monomials only, i.e. all Clebsch–Gordan coefficients are either 0 or 1. Find the new rescaled basis in which this is true for regular representations of $U_q(s\ell(2))$. The group theoretical combinatorics becomes much simpler when $q \to 0$. Since this limit is smooth, it is extremely useful for computing branching rules for the general $q \neq 0$ case. Evaluate the quantum $6j$ symbols (6.90) in the limit $q \to 0$.

Ex. 6.17 Discuss how equation (6.155) can be taken as an indication that, when we intertwine by means of $\overline{\mathscr{R}}(t)$ a finite-dimensional irrep $V(t)$ with an infinite-dimensional one W, we effectively cause a shift in the spectral parameter $t \to q^{-c(W)}t$.

Ex. 6.18 [R. Dijkgraaf, V. Pasquier and P. Roche, in *International Colloquium on Modern Quantum Field Theory*, Tata Institute for Fundamental Research, Bombay (1990).]

Let G be a finite group, and consider the Hopf algebra Fun(G) of complex functions on G. Apply Drinfeld's double construction to Fun(G), and show that the double $\mathscr{D}(G)$ is a quasi-triangular quasi-Hopf algebra with ϕ an element of the co-homology group $H^3(G, U(1))$. Construct the representation theory of this algebra using that of the group G, and show its relationship to the orbifold rational conformal field theories of R. Dijkgraaf, C. Vafa, E. Verlinde and H. Verlinde, *Commun. Math. Phys.* **123** (1989) 485.

7
Integrable models at roots of unity

In this chapter, we analyze in some detail the quantum group $U_q(s\ell(2))$ with $q^\ell = 1$, and consider some of its physical applications. Interest in the special situation arising when q is a root of unity lies in the richness of the finite-dimensional irreducible representations resulting from an enlargement of the center. Some of these irreps are labeled by continuous parameters, and their intertwiners, when they exist, provide solutions to the Yang–Baxter equation with a spectral parameter in the spectrum of the enlarged center. These new solutions to the Yang–Baxter equation can be obtained by the use of the same procedure that was presented in chapter 2 to find higher-spin descendants of the six-vertex model (spin $\frac{1}{2}$), but using now finite-dimensional irreps which transform non-trivially under the enlarged center. The most famous among these new descendants is the chiral Potts model, which we will review briefly. Associated with the new irreps when q is a root of unity, are new spin chain hamiltonians; these are defined, and their solutions via the Bethe ansatz are sketched.

7.1 Mathematical preliminaries

7.1.1 The center of $U_q(s\ell(2))$

We consider the Hopf algebra $U_q(s\ell(2))$ with $q^\ell = 1$ ($\ell \in \mathbb{N}$, $\ell \geq 3$) and work in the Weyl–Chevalley basis $\{E, F, K\}$:

$$\begin{aligned} [E,F] &= EF - FE = \frac{K - K^{-1}}{q - q^{-1}} \\ KEK^{-1} &= q^2 E \\ KFK^{-1} &= q^{-2} F \\ KK^{-1} &= K^{-1}K = \mathbf{1} \end{aligned} \tag{7.1}$$

The co-multiplication, antipode and co-unit are the standard ones (we quote Δ' instead of Δ in this chapter to simplify comparison with the literature):

$$\begin{aligned} \Delta'(E) &= E \otimes \mathbf{1} + K \otimes E \\ \Delta'(F) &= F \otimes K^{-1} + \mathbf{1} \otimes F \\ \Delta'(K) &= K \otimes K \end{aligned} \tag{7.2}$$

$$\gamma(E) = -K^{-1}E\,, \quad \gamma(F) = -FK\,, \quad \gamma(K) = K^{-1} \tag{7.3}$$

$$\epsilon(E) = \epsilon(F) = 0\,, \qquad \epsilon(K) = \epsilon(\mathbf{1}) = 1 \tag{7.4}$$

To distinguish between the deformation parameter q and the value of a primitive root of unity, we let $\epsilon^\ell = 1$ ($\ell \in \mathbb{N} - \{0, 1, 2\}$) and consider the limit $q \to \epsilon$.

The main difference between $U_q(s\ell(2))$ and $U_\epsilon(s\ell(2))$ is the structure of the center, which in the latter case is larger. Indeed, in addition to the quadratic Casimir, it contains the generators $E^{\ell'}$, $F^{\ell'}$ and $K^{\ell'}$, where

$$\ell' = \begin{cases} \ell & \text{if } \ell \text{ is odd} \\ \ell/2 & \text{if } \ell \text{ is even} \end{cases} \tag{7.5}$$

In order to show this, it is convenient to use the following relation, which results directly from (7.1):

$$[E^m, F^s] = \sum_{j=1}^{\min(m,s)} \begin{bmatrix} m \\ j \end{bmatrix} \begin{bmatrix} s \\ j \end{bmatrix} [j]!\; F^{s-j} \prod_{r=j-m-s+1}^{2j-m-s} [K\,;r]\; E^{m-j} \tag{7.6}$$

with the short-hand notation

$$[K\,;r] = \frac{Kq^r - K^{-1}q^{-r}}{q - q^{-1}} \tag{7.7}$$

and the q-binomials

$$\begin{bmatrix} n \\ j \end{bmatrix} = \frac{[n][n-1]\cdots[n-j+1]}{[j][j-1]\cdots[2][1]} = \frac{[n]!}{[n-j]![j]!}\,, \quad \begin{bmatrix} n \\ 0 \end{bmatrix} = 1 \tag{7.8}$$

It is an easy exercise to check that $E^{\ell'}$, $F^{\ell'}$ and $K^{\ell'}$ generate a Hopf subalgebra $Z \subset U_\epsilon(s\ell(2))$ contained in the center Z_ϵ of $U_\epsilon(s\ell(2))$. The center Z_ϵ is generated by, in addition to Z, the Casimir

$$C = \frac{Kq + K^{-1}q^{-1}}{\left(q - q^{-1}\right)^2} + FE \tag{7.9}$$

7.1.2 Finite-dimensional irreps

In chapter 6, we have defined representations of the quantum group $U_q(s\ell(2))$ on which the central Hopf subalgebra Z acts trivially, i.e. the actions of F^ℓ and E^ℓ vanish:

$$\begin{aligned} \pi_j^{\pm}(K)e_i &= \pm q^{2j-2i}e_i \\ \pi_j^{\pm}(F)e_i &= [i+1]e_{i+1} \\ \pi_j^{\pm}(E)e_i &= \pm[2j+1-i]e_{i-1} \end{aligned} \tag{7.10}$$

with $0 \le i \le 2j \in \mathbb{N}$. When q is an ℓth root of unity, the representations (7.10) are irreducible with non-zero q-dimension only if $0 \le j \le (\ell' - 2)/2$. We will call these irreps regular, standard or non-generic.

By Schur's lemma, we may characterize the irreps of $U_\epsilon(s\ell(2))$ with the eigenvalues of all the elements in the center Z_ϵ. Given an irreducible representation π, then

$$\pi(u) = \mathscr{X}_\pi(u)\mathbf{1} \qquad u \in Z_\epsilon \tag{7.11}$$

with $\mathscr{X}_\pi(u)$ ordinary c-numbers. The representations (7.10) above correspond to the particular values

$$\begin{aligned} \mathscr{X}_\pi(E^{\ell'}) &= \mathscr{X}_\pi(F^{\ell'}) = 0 \\ \mathscr{X}_\pi(K^{\ell'}) &= \pm 1 \end{aligned} \tag{7.12}$$

We shall now consider more general representations using

$$\mathscr{X}_\pi(E^{\ell'}) = x\,, \quad \mathscr{X}_\pi(F^{\ell'}) = y\,, \quad \mathscr{X}_\pi(K^{\ell'}) = z = \lambda^{\ell'} \tag{7.13}$$

Cyclic, semi-cyclic and nilpotent irreducible representations are spanned by $\{e_i, 0 \le i \le \ell' - 1\}$, with

$$\begin{aligned} &\pi_{x,y,z}(K)\, e_i = \lambda\, \epsilon^{-2i}\, e_i && \forall i \\ &\pi_{x,y,z}(F)\, e_i = e_{i+1} && i \neq \ell' - 1 \\ &\pi_{x,y,z}(F)\, e_{\ell'-1} = y\, e_0 && \\ &\pi_{x,y,z}(E)\, e_0 = x\, e_{\ell'-1} && \\ &\pi_{x,y,z}(E)\, e_i = \left(\frac{\left(\lambda \epsilon^{1-i} - \lambda^{-1}\epsilon^{i-1}\right)\left(\epsilon^i - \epsilon^{-i}\right)}{\left(\epsilon - \epsilon^{-1}\right)^2} + x\, y \right) e_{i-1} && 1 \neq 0 \end{aligned} \tag{7.14}$$

The ℓ'-dimensional representations defined by (7.14) are irreducible for arbitrary $x, y \in \mathbb{C}$ and $z \in \mathbb{C}^* = \mathbb{C} - \{0\}$. The value $c = \pi_{x,y,z}(C)$ of the Casimir (7.9) is easily computed; it depends on (x, y, z) as follows:

$$\prod_{i=0}^{\ell'-1} \left(c - \frac{\lambda \epsilon \epsilon'^{i} + \lambda^{-1} \epsilon \epsilon'^{-i}}{\left(\epsilon' - \epsilon'^{-1}\right)^2} \right) = x\, y \tag{7.15}$$

In this expression,

$$\epsilon' = \begin{cases} \epsilon & \text{if } \ell \text{ is odd} \\ \epsilon^2 & \text{if } \ell \text{ is even} \end{cases} \tag{7.16}$$

It will be important during the rest of this chapter to map the irreps of $U_\epsilon(s\ell(2))$ into the three-dimensional manifold of co-ordinates $(c, x, y, z) \in \mathbb{C}^3 \otimes \mathbb{C}^*$ subject to (7.15). Every point of this manifold represents an irreducible representation.

Note that all the irreps have dimension $\le \ell'$. We call classical, regular, standard or non-generic irreps those which are a continuous deformation of their finite-dimensional classical analogs, and which exist for arbitrary q. The regular, standard, classical or non-generic representations (7.10) all have dimension $< \ell'$. Among the irreps with dimension equal to ℓ', we have the cyclic, periodic, or generic irreps, for which $x \neq 0$ and $y \neq 0$; we shall consider these representations in section 7.6.5 in connection with the chiral Potts model. Another class of irreps of dimension ℓ' comprises those with $x = y = 0$, but with λ a generic complex value; we will call these irreps nilpotent or diagonal. An intermediate case between the cyclic and the nilpotent irreps is given by the semi-cylic or semi-periodic irreps with $x = 0 \neq y$ (see table 7.1). All cyclic, semi-cyclic and nilpotent irreps exist only for q a root of unity and, in particular, for $q \neq 1$: they have no analog in the representation theory of classical Lie algebras. Nilpotent irreps with $\lambda = \pm q^{-1}$ (and thus $z = \pm 1$), in particular, have no classical $q \to 1$ limit, but deserve a special mention because they cannot be obtained by simply setting $x = y = 0$ in the cyclic case.

7.1.3 The co-adjoint action

An important concept in the study of Lie algebras is that of derivation. Given a generic Lie algebra g, a derivation D is a map $D : g \to g$ satisfying Leibnitz's chain rule $D(xy) = D(x)y + xD(y)$. Inner derivations of g are those such that there exists an $x \in g$ with $D = \mathrm{ad}_x$, where the adjoint action of x is defined by $\mathrm{ad}_x(y) = [x, y] = xy - yx$, for all $y \in g$. If the Lie algebra is semi-simple, then all derivations are inner.

We consider odd ℓ only, for simplicity (to deal with even ℓ, it suffices to replace ℓ in all the formulae below by $\ell' = \ell/2$), and introduce the convenient notation

$$A^{(\ell)} = A^{\ell}/[\ell]! \tag{7.17}$$

We shall be interested in the following inner derivations of $U_\epsilon(s\ell(2))$:

$$\begin{aligned} \mathbf{e}(u) &= \lim_{q\to\epsilon} \left[E^{(\ell)}, u\right] \\ \mathbf{f}(u) &= \lim_{q\to\epsilon} \left[F^{(\ell)}, u\right] \qquad \forall u \in U_\epsilon(s\ell(2)) \\ \mathbf{k}(u) &= \lim_{q\to\epsilon} \left[K^{(\ell)}, u\right] \end{aligned} \tag{7.18}$$

The commutators above should first be computed for arbitrary q, after which the limit $q \to \epsilon$ can be taken.

Table 7.1 Irreps of $U_q(s\ell(2))$ when $q^{\ell} = 1$.
As in the text, $\ell' = \ell$ if ℓ is odd and $\ell' = \ell/2$ if ℓ is even. The last column reminds us whether the irreps have a highest/lowest weight vector.

Irrep π	$\mathscr{X}_\pi\left(E^{\ell'}\right)$	$\mathscr{X}_\pi\left(F^{\ell'}\right)$	$\mathscr{X}_\pi\left(K^{\ell'}\right)$	Dimension	h.w./l.w.
Cyclic	$x \neq 0$	$y \neq 0$	$z \neq \pm 1$	ℓ'	no/no
Semi-cyclic	0	$y \neq 0$	$z \neq \pm 1$	ℓ'	yes/no
Nilpotent	0	0	generic	ℓ'	yes/yes
Classical	0	0	± 1	$< \ell'$	yes/yes

To get a feeling for the action of the inner derivations **e**, **f** and **k**, we display some explicit calculations. Let us evaluate, for instance, $\mathbf{e}(F)$. Letting $\eta = (\epsilon - \epsilon^{-1})$, and using the fact that

$$[\ell - 1]!\Big|_{q=\epsilon} = \frac{\ell}{(\epsilon - \epsilon^{-1})^{\ell-1}} \tag{7.19}$$

we find

$$\begin{aligned} \mathbf{e}(F) &= \frac{1}{[\ell]!}\left[E^{\ell}, F\right]_{q\to\epsilon} \\ &= \frac{\eta^{\ell-2}}{\ell}\left(K\epsilon^{1-\ell} - K^{-1}\epsilon^{\ell-1}\right) E^{\ell-1} \end{aligned} \tag{7.20}$$

A more instructive example is

$$\begin{aligned} \mathbf{e}(K) &= \frac{1}{[\ell]!}\left[E^{\ell}, K\right]_{q\to\epsilon} = (-1)\,\frac{q^{2\ell} - \epsilon^{-2\ell}}{[\ell]!}\,E^{\ell}\,K\Big|_{q\to\epsilon} \\ &= -\eta^{\ell-1}\,\frac{q - q^{-1}}{\ell}\,E^{\ell}\,K\Big|_{q\to\epsilon} = -\frac{\eta^{\ell}}{\ell}\,E^{\ell}\,K \end{aligned} \tag{7.21}$$

Note the trick of setting some of the q's to ϵ in the first step, and using the fact that $\epsilon^{\ell} = 1$. Obviously, we have also

$$\mathbf{e}(E) = 0 \tag{7.22}$$

From equations (7.20), (7.21) and (7.22) we may immediately deduce how **f** acts on the generators. All we need to recall is that inner derivations transform under an automorphism ϕ of the algebra as

$$\mathrm{ad}_{\phi(x)}(y) = \phi\,\mathrm{ad}_x\,\phi^{-1}(y) \qquad \forall y \in U_{\epsilon}(s\ell(2)) \tag{7.23}$$

With the help of the automorphism

$$\phi(E) = F\,, \quad \phi(F) = E\,, \quad \phi(K) = K\,, \quad \phi(q) = q^{-1} \tag{7.24}$$

we find

$$\mathbf{f}(u) = \phi\,\mathbf{e}\left(\phi^{-1}(u)\right) \qquad \forall y \in U_{\epsilon}(s\ell(2)) \tag{7.25}$$

The action of the **k** derivation can be computed as in (7.21). We obtain

$$\mathbf{k}(E) = \frac{1}{[\ell]!}\left[K^{\ell}, E\right]\Big|_{q\to\epsilon} = \frac{q^{2\ell} - \epsilon^{-2\ell}}{[\ell]!}\,K^{\ell}\,E\Big|_{q\to\epsilon} = \frac{\eta^{\ell}}{\ell}\,K^{\ell}E \tag{7.26}$$

The expression for $\mathbf{k}(F)$ follows from this via the automorphism (7.24), whereas $\mathbf{k}(K) = 0$.

The set of derivations $\{\mathbf{e}, \mathbf{f}, \mathbf{k}\}$ generate an ordinary Lie algebra $\hat{\mathfrak{g}}$ which preserves the subalgebra Z, meaning that $\hat{\mathfrak{g}} Z \subset Z$. We would like to exponentiate $\hat{\mathfrak{g}}$ in such a way that the Lie group elements define automorphisms of the Hopf algebra $U_\epsilon(s\ell(2))$. De Concini, Kac and Procesi have shown that the elements of the Lie group $\hat{G} = \exp(\hat{g})$ are automorphisms, not of $U_\epsilon(s\ell(2))$, but of $U_\epsilon(s\ell(2)) \otimes_Z \hat{Z}$, where by $\hat{Z}$ we denote the formal power series in x, y, z and z^{-1}. To illustrate this, we consider

$$(\exp t\mathbf{e})F = F - \left(\frac{e^{-t\eta^\ell x/\ell} - 1}{\eta^2 x} \epsilon K + \frac{e^{t\eta^\ell x/\ell} - 1}{\eta^2 x} \epsilon^{-1} K^{-1} \right) E^{\ell-1}$$

$$(\exp t\mathbf{e})K = e^{-t\eta^\ell x/\ell} K \tag{7.27}$$

Given any automorphism $a \in \hat{G}$, we may use it to "twist" a representation as follows:

$$\pi^a(u) = \pi(a(u)) \tag{7.28}$$

If π is irreducible, then π^a is also irreducible, and by Schur's lemma an action of the automorphism a is induced in the space of eigenvalues which characterize the spectrum of the center:

$$\begin{aligned} \pi &\rightarrow \pi^a \\ \{\mathscr{X}_\pi(u)\} &\rightarrow \{\mathscr{X}_{\pi^a}(u)\} \qquad u \in Z \end{aligned} \tag{7.29}$$

We thus find an action of $\hat{G}$ on the manifold labeling all the irreps of $U_\epsilon(s\ell(2))$. We shall call this manifold Spec Z_ϵ. The action of $\hat{G}$ on Spec Z_ϵ is called the co-adjoint action. Concerning the co-adjoint action of $\hat{G}$ on Spec Z_ϵ, let us note some of the main results obtained by De Concini, Kac and Procesi.

Given any point in Spec Z_ϵ, its orbit with respect to $\hat{G}$ is the set of all the points in Spec Z_ϵ which can be reached by acting with $\hat{G}$ on the selected point. It is easy to check, using (7.27), that any orbit contains at least one point corresponding to a semi-cyclic representation, i.e. a representation π for which $\mathscr{X}_\pi(E^\ell) = 0$.

The fixed points can also be easily obtained from (7.27): they correspond to the finite-dimensional irreducible representations (7.10) with $0 \le j \le (\ell-2)/2$. Among the fixed points, some are singular points of the three-dimensional manifold (7.15). For these points, the map induced by Schur's lemma from the space of representations into Spec Z_ϵ is two to one. This reflects the duality symmetry $(m,n) \leftrightarrow (p-m, p'-n)$ we will encounter in the study of (p,p') conformal minimal models, which we will use to truncate the fusion algebra (see also section 11.5.2).

To finish this subsection, let us consider more closely the mathematical meaning of the group of automorphisms $\hat{G}$. The Gauss decomposition of the classical group $SL(2)$ is given by $a = YZX \in SL(2)$, with

$$Y = \begin{pmatrix} 1 & 0 \\ y & 1 \end{pmatrix}, \quad Z = \begin{pmatrix} z & 0 \\ 0 & z^{-1} \end{pmatrix}, \quad X = \begin{pmatrix} 1 & x \\ 0 & 1 \end{pmatrix} \tag{7.30}$$

It is convenient to introduce new parameters (x', y', z'), which are related to (x, y, z) as follows:

$$x' = (-\eta)^\ell, \qquad y' = \eta^\ell y z^{-1} x, \qquad z' = z \tag{7.31}$$

Using these new parameters, we define the map

$$\begin{aligned} \pi : \text{Spec } Z &\rightarrow SL(2) \\ (x', y', z') &\mapsto Y'Z'X' = \begin{pmatrix} z' & z'x' \\ z'y' & z'^{-1} + z'x'y' \end{pmatrix} \end{aligned} \tag{7.32}$$

We may now compare, with the help of the map π, the action of the derivations in $\hat{\mathfrak{g}}$ with that of the inner derivations in $SL(2)$, in its Chevalley basis. Let us carry out the exercise for the derivation $\mathbf{e}$:

$$\mathbf{e} \begin{pmatrix} z' & z'x' \\ z'y' & z'^{-1} + z'x'y' \end{pmatrix} = \begin{pmatrix} -z'^2 x' & 0 \\ z'^2 - z'^2 x'y' - 1 & z'^2 x' \end{pmatrix} \tag{7.33}$$

which can be written more attractively as

$$\mathbf{e} \begin{pmatrix} z' & z'x' \\ z'y' & z'^{-1} + z'x'y' \end{pmatrix} = \left[\begin{pmatrix} 1 & 0 \\ z' & 1 \end{pmatrix}, \begin{pmatrix} z' & z'x' \\ z'y' & z'^{-1} + z'x'y' \end{pmatrix} \right] \tag{7.34}$$

Therefore,

$$\mathbf{e}(\cdot) = [z'f, \cdot] \tag{7.35}$$

This relation connects the derivations in $\hat{\mathfrak{g}}$ with the Chevalley generators of $SL(2)$.

The map π defined in (7.32) allows us to visualize part of the above results from a different, and quite intriguing, perspective. Using π, we may put in one to one relation orbits of $\hat{G}$ with conjugacy classes in $SL(2)$. Moreover, the set of fixed points of Spec Z_ϵ correspond to the central elements of the classical group $SL(2)$.

7.1.4 Intertwiners

The next problem we shall encounter in the study of the representation theory at roots of unity are the conditions under which two irreps $\xi_1 = (c_1, x_1, y_1, z_1)$ and $\xi_2 = (c_2, x_2, y_2, z_2)$ of $U_\epsilon(s\ell(2)) = U_\epsilon$ admit an intertwiner, i.e. a matrix

$$R(\xi_1, \xi_2) : \rho_{\xi_1}(U_\epsilon) \otimes \rho_{\xi_2}(U_\epsilon) \longrightarrow \rho_{\xi_2}(U_\epsilon) \otimes \rho_{\xi_1}(U_\epsilon) \tag{7.36}$$

such that ($\forall a \in U_\epsilon$)

$$R^{r_1'' r_2''}_{r_1' r_2'}(\xi_1, \xi_2) \Delta'_{\xi_1, \xi_2}(a)^{r_1' r_2'}_{r_1 r_2} = \Delta'_{\xi_2, \xi_1}(a)^{r_1'' r_2''}_{r_1' r_2'} R^{r_1' r_2'}_{r_1 r_2}(\xi_2, \xi_1) \tag{7.37}$$

This was previously discussed for the non-generic representations (7.10) in chapter 6: the intertwiner is given in terms of the universal $\mathcal{R}$-matrix of $U_\epsilon = U_\epsilon(s\ell(2))$. Naively, we could extend this result to cyclic representations by simply setting $R = P(\rho_{\xi_1} \otimes \rho_{\xi_2})\mathcal{R}$. This naive procedure would automatically produce intertwiners for arbitrary pairs of representations (ξ_1, ξ_2), which is the wrong result. In fact, let us assume that there exists an intertwiner $R(\xi_1, \xi_2)$, and consider (7.37) with $a \in Z_\epsilon$. Quite independently of $\mathcal{R}$, we must have

$$\Delta'_{\xi_1 \xi_2}(a) = \Delta'_{\xi_2 \xi_1}(a) \qquad \forall a \in Z_\epsilon \tag{7.38}$$

which severely constrains the points in Spec Z_ϵ between which an intertwiner exists. With the help of the basic identity

$$\left.\begin{array}{r} AB = wBA \\ w^N = 1 \end{array}\right\} \Longrightarrow (A+B)^N = A^N + B^N \tag{7.39}$$

we can prove that

$$\begin{aligned} \Delta'\left(E^{\ell'}\right) &= E^{\ell'} \otimes \mathbf{1} + K^{\ell'} \otimes E^{\ell'} \\ \Delta'\left(F^{\ell'}\right) &= F^{\ell'} \otimes K^{-\ell'} + \mathbf{1} \otimes F^{\ell'} \\ \Delta'\left(K^{\ell'}\right) &= K^{\ell'} \otimes K^{\ell'} \end{aligned} \tag{7.40}$$

This equation means that $E^{\ell'}$, $F^{\ell'}$ and $K^{\ell'}$ form a Hopf subalgebra of $U_\epsilon(s\ell(2))$.

Equation (7.38) defines submanifolds of Spec Z_ϵ labeled by two parameters, let us say γ_E and γ_F:

$$\gamma_E = \frac{1-z}{x}, \qquad \gamma_F = \frac{z^{-1}-1}{y} \tag{7.41}$$

Thus, an intertwiner between ξ_1 and ξ_2 may exist only if they share the same values of (γ_E, γ_F). In general, the condition (7.38) defines submanifolds as algebraic curves of Fermat type. The necessary condition for the existence of an intertwiner between two irreps is that the associated points in Spec Z_ϵ live on the same submanifold.

The actual computation of intertwiners between two irreps on the same spectral variety deserves a whole section to itself: the next one.

7.2 A family of *R*-matrices

In this technical and rather tedious section, we will derive in detail the intertwiner $R_{\xi_1\xi_2}$ matrix (7.37) for the tensor product $\xi_1 \otimes \xi_2$ of two generic irreducible representations with a highest weight vector, i.e. of two nilpotent or two semi-cyclic irreps ξ_1 and ξ_2 of $U_q(A_1^{(1)})$. From (7.41), we note that, for nilpotent representations, $\gamma_E^{-1} = \gamma_F^{-1} = 0$, so no further conditions for the existence of an intertwining *R*-matrix are required in this case, although a spectral condition does have to be satisfied for semi-cyclic irreps among which an intertwining *R*-matrix exists.

7.2.1 Highest weight intertwiner

As already discussed in chapters 2 and 6, any irreducible representation of the algebra $U_q(A_1)$ can be promoted to an irrep of the affine $U_q(A_1^{(1)})$ via the identifications

$$\begin{aligned} E_0 &= \mathrm{e}^u F & E_1 &= \mathrm{e}^u E \\ F_0 &= \mathrm{e}^{-u} E & F_1 &= \mathrm{e}^{-u} F \\ K_0 &= K^{-1} & K_1 &= K \end{aligned} \tag{7.42}$$

where E_i, F_i and K_i are the generators of $U_q(A_1^{(1)})$; E, F and K are the generators of $U_q(A_1)$; and e^u is the affine parameter related to the rapidity u.

From (7.37) for $U_q(A_1^{(1)})$, we obtain the following four equations:

$$\begin{aligned} R_{12}\left(F_1 \otimes \mathbf{1} + e^{-u} K_1^{-1} \otimes F_2\right) &= \left(e^{-u} F_2 \otimes \mathbf{1} + K_2^{-1} \otimes F_1\right) R_{12} \\ R_{12}\left(F_1 \otimes K_2^{-1} + e^{u} \mathbf{1} \otimes F_2\right) &= \left(e^{u} F_2 \otimes K_1^{-1} + \mathbf{1} \otimes F_1\right) R_{12} \\ R_{12}\left(E_1 \otimes \mathbf{1} + e^{-u} K_1 \otimes E_2\right) &= \left(e^{-u} E_2 \otimes \mathbf{1} + K_2 \otimes E_1\right) R_{12} \\ R_{12}\left(E_1 \otimes K_2 + e^{u} \mathbf{1} \otimes E_2\right) &= \left(e^{u} E_2 \otimes K_1 + \mathbf{1} \otimes E_1\right) R_{12} \end{aligned} \tag{7.43}$$

where the subscripts 1 and 2 of the various operators refer to the representations ξ_1 and ξ_2, respectively. We have introduced the short-hand notation

$$e^{u_1 - u_2} = e^u \tag{7.44}$$

For the time being, these representations may be considered to be regular, nilpotent or semi-cyclic. The method we shall use to find R applies to all of them, and relies only on the fact that they are highest weight representations (it does not apply to cyclic irreps).

The main principle of the method is to transform (7.43) into recurrence equations relating "lowest weight entries" of the R-matrix to "highest weight entries", and then to use the fact that R, when acting on the tensor product of the two highest weight vectors $e_0 \otimes e_0$, yields simply $e_0 \otimes e_0$ up to a constant. We fix this normalization to be 1, namely

$$R\left(e_0 \otimes e_0\right) = e_0 \otimes e_0 \tag{7.45}$$

It is easy to derive from (7.44) the following recursion formulae:

$$\begin{aligned} R_{12}\left(F_1\right) &= \left[A R_{12} - e^{-u} B R_{12} K_1^{-1}\right] \frac{1}{\mathbf{1} \otimes \mathbf{1} - e^{-2u} K_1^{-1} K_2^{-1}} \\ R_{12}\left(F_2\right) &= \left[B R_{12} - e^{-u} A R_{12} K_2^{-1}\right] \frac{1}{\mathbf{1} \otimes \mathbf{1} - e^{-2u} K_1^{-1} K_2^{-1}} \end{aligned} \tag{7.46}$$

with

$$A = e^{-u} F_2 \otimes \mathbf{1} + K_2^{-1} \otimes F_1\,, \quad B = F_2 \otimes K_1^{-1} + e^{-u} \mathbf{1} \otimes F_1 \tag{7.47}$$

Similar equations hold also for F replaced by E.

Iterating equations (7.46) and noting that A and B commute, we obtain the general formula

$$\begin{aligned} R_{12}\left(F_1^{r_1} \otimes F_2^{r_2}\right) &= \sum_{s_1=0}^{r_1} \sum_{s_2=0}^{r_2} (-1)^{s_1+s_2} \epsilon^{s_1(s_1-1)+s_2(s_2-1)}\, e^{-u(s_1+s_2)} \\ &\times \left[\begin{bmatrix} r_1 \\ s_1 \end{bmatrix}\right] \left[\begin{bmatrix} r_2 \\ s_2 \end{bmatrix}\right] A^{r_1-s_1+s_2} B^{s_1+r_2-s_2}\, R_{12}\left(K_1^{-s_1} \otimes K_2^{-s_2}\right) \\ &\times \frac{1}{\prod_{n=0}^{r_1+r_2-1}\left(1 - \epsilon^{2n} e^{-2u} K_1^{-1} \otimes K_2^{-1}\right)} \end{aligned} \tag{7.48}$$

where we have used explicitly that $K^{-1}F = \epsilon^2 F K^{-1}$ and

$$[|r|] = \frac{1-\epsilon^{2r}}{1-\epsilon^2}, \qquad \left[\left|\begin{matrix} r \\ s \end{matrix}\right|\right] = \frac{[|r|]!}{[|s|]!\,[|r-s|]!} \tag{7.49}$$

Applying now the equality (7.48) to the highest weight vector $e_0 \otimes e_0$, and recalling that $Ke_0 = \lambda e_0$, we obtain

$$\begin{aligned} R_{12}\left(F_1^{r_1} \otimes F_2^{r_2}\right) e_0 \otimes e_0 = & \sum_{s_1=0}^{r_1} \left[\left|\begin{matrix} r_1 \\ s_1 \end{matrix}\right|\right] \left(\lambda_1^{-1}\,\mathrm{e}^{-u}\right)^{s_1} A^{r_1-s_1} B^{s_1} \\ & \times \sum_{s_2=0}^{r_2} \left[\left|\begin{matrix} r_2 \\ s_2 \end{matrix}\right|\right] \left(\lambda_2^{-1}\,\mathrm{e}^{-u}\right)^{s_2} B^{r_2-s_2} A^{s_2} \\ & \times \frac{1}{\prod_{n=0}^{r_1+r_2-1}\left(1-\epsilon^{2n}\,\mathrm{e}^{-2u}\lambda_1^{-1}\lambda_2^{-1}\right)} \quad e_0 \otimes e_0 \end{aligned} \tag{7.50}$$

Further use of the Gauss binomial formula

$$\sum_{v=0}^{n} (-1)^v \left[\left|\begin{matrix} n \\ v \end{matrix}\right|\right] \; z^v \epsilon^{v(v-1)} = \prod_{v=0}^{n-1}\left(1 - z\,\epsilon^{2v}\right) \tag{7.51}$$

allows us to transform (7.50) into

$$\begin{aligned} R_{12}\left(F_1^{r_1} \otimes F_2^{r_2}\right) e_0 \otimes e_0 = & \frac{1}{\prod_{n=0}^{r_1+r_2-1}\left(1-\epsilon^{2n}\,\mathrm{e}^{-2u}\lambda_1^{-1}\lambda_2^{-1}\right)} \\ \times \prod_{n_1=0}^{r_1-1}\left(A - \epsilon^{2n_1}\lambda_1^{-1}\,\mathrm{e}^{-u} B\right) & \prod_{n_2=0}^{r_2-1}\left(B - \epsilon^{2n_2}\lambda_2^{-1}\,\mathrm{e}^{-u} A\right) e_0 \otimes e_0 \end{aligned} \tag{7.52}$$

Next, we need the following:

$$\begin{aligned} \prod_{n_1=0}^{r_1-1}\left(A - \epsilon^{2n_1}\lambda_1^{-1}\,\mathrm{e}^{-u}B\right) = & \sum_{n_1=0}^{r_1} \left[\left|\begin{matrix} r_1 \\ n_1 \end{matrix}\right|\right] \left(\mathrm{e}^{-u}F_2\right)^{r_1-n_1} \\ & \times \left(K_2^{-1}, \mathrm{e}^{-2u}\lambda_1^{-1}\right)^{\lfloor n_1} \left(1, \lambda_1^{-1}K_1^{-1}\right)^{\lfloor r_1-n_1} F_1^{n_1} \\ \prod_{n_2=0}^{r_2-1}\left(B - \epsilon^{2n_2}\lambda_2^{-1}\,\mathrm{e}^{-u}A\right) = & \sum_{n_2=0}^{r_2} \left[\left|\begin{matrix} r_2 \\ n_2 \end{matrix}\right|\right] \left(1, \lambda_2^{-1}K_2^{-1}\right)^{\lfloor n_2} \\ & \times F_2^{r_2-n_2}\left(\mathrm{e}^{-u}F_1\right)^{n_2}\left(K_1^{-1}, \mathrm{e}^{-2u}\lambda_2^{-1}\right)^{\lfloor r_2-n_2} \end{aligned} \tag{7.53}$$

where we have introduced the symbol

$$(a,b)^{\lfloor r} = \prod_{n=0}^{r-1}\left(a - \epsilon^{2n} b\right) \tag{7.54}$$

Using (7.53) in (7.52), we obtain

$$
\begin{aligned}
R_{12}\left(F_1^{r_1} \otimes F_2^{r_2}\right) e_0 \otimes e_0 = & \frac{1}{\prod_{n=0}^{r_1+r_2-1}\left(1-\epsilon^{2n}\, \mathrm{e}^{-2u} \lambda_1^{-1} \lambda_2^{-1}\right)} \\
& \times \sum_{n_1=0}^{r_1} \sum_{n_2=0}^{r_2} \left[\begin{vmatrix} r_1 \\ n_1 \end{vmatrix}\right] \left[\begin{vmatrix} r_2 \\ n_2 \end{vmatrix}\right] \mathrm{e}^{-u(r_1-n_1+n_2)} \\
& \times \left(\epsilon^{2(r_2-n_2)} \lambda_2^{-1}, \mathrm{e}^{-2u} \lambda_1^{-1}\right)\lfloor_{n_1} \left(1, \lambda_2^{-2} \epsilon^{2(r_2-n_2)}\right)\lfloor_{n_2} \\
& \times \left(1, \lambda_1^{-2} \epsilon^{2(n_1+n_2)}\right)\lfloor_{r_1-n_1} \left(\lambda_1^{-1}, \mathrm{e}^{-2u} \lambda_2^{-1}\right)\lfloor_{r_2-n_2} \\
& F_2^{r_1+r_2-n_1-n_2} e_0 \otimes F_1^{n_1+n_2} e_0
\end{aligned} \tag{7.55}
$$

This is the desired result, since any vector in $V_{\xi_1} \otimes V_{\xi_2}$ can be expressed as $F_1^{r_1} e_0 \otimes F_2^{r_2} e_0$. The action of R_{12} maps it into a vector in $V_{\xi_2} \otimes V_{\xi_1}$. Since in the derivation of (7.55) we have not used the fact that ϵ is a root of unity, the procedure presented above also yields an explicit expression for the R-matrix intertwining between regular representations of higher spin. Such R-matrices were shown in chapter 2 to descend from the intertwiner $\mathscr{R}^{(6v)}$ between two spin-$\frac{1}{2}$ irreps. Equation (7.55) yields more generally the intertwiner between two highest weight representations of $U_q(A_1^{(1)})$, and is therefore valid, not only for regular representations, but for nilpotent and semi-cyclic ones as well.

7.2.2 *The nilpotent R-matrix*

For later convenience, let us introduce now the following basis $\{e_r\}$ for nilpotent representations, which differs slightly from (7.14):

$$
\begin{aligned}
K e_r &= \lambda \epsilon^{-2r} e_r \\
E\ e_r &= d_{r-1}(\lambda) e_{r-1} \\
F\ e_r &= d_r(\lambda) e_{r+1}
\end{aligned} \tag{7.56}
$$

where

$$
d_j^2(\lambda) = [j+1] \frac{\lambda \epsilon^{-j} - \lambda^{-1} \epsilon^j}{\epsilon - \epsilon^{-1}} \tag{7.57}
$$

with the usual notation

$$
[x] = \frac{\epsilon^x - \epsilon^{-x}}{\epsilon - \epsilon^{-1}} \tag{7.58}
$$

If we set $\lambda = \epsilon^{2S}$, with $2S$ a positive integer, we recover the regular representations of spin S.

In the basis (7.56), the R-matrix

$$
R\left(e_{r_1}^{(\lambda_1,u_1)} \otimes e_{r_2}^{(\lambda_2,u_2)}\right) = \sum_{r_2',r_1'} R_{r_1,r_2}^{r_2',r_1'}(\lambda_1,u_1;\lambda_2,u_2) e_{r_2'}^{(\lambda_2,u_2)} \otimes e_{r_1'}^{(\lambda_1,u_1)} \tag{7.59}
$$

is given by (we use the convention that $\prod_a^b = 0$ if $b < a$)

$$
\begin{aligned}
R_{r_1,r_2}^{r_1+r_2-r,r}(\lambda_1,u_1;\lambda_2,u_2) \;=\; & \frac{1}{\prod_{n=0}^{r_1+r_2-1}\left(1-\epsilon^{2n}\,\mathrm{e}^{-2u}\lambda_1^{-1}\lambda_2^{-1}\right)} \\
& \times \frac{\prod_{j=0}^{r-1} d_j(\lambda_1)}{\prod_{j=0}^{r_1-1} d_j(\lambda_1)} \;\frac{\prod_{j=0}^{r_1+r_2-r-1} d_j(\lambda_2)}{\prod_{j=0}^{r_2-1} d_j(\lambda_2)} \\
& \times \sum_{\substack{n_1=0\\ n_1+n_2=r}}^{r_1}\sum_{n_2=0}^{r_2} \left[\left|\begin{matrix} r_1\\ n_1\end{matrix}\right|\right]\left[\left|\begin{matrix} r_2\\ n_2\end{matrix}\right|\right]\mathrm{e}^{-u(r_1-n_1+n_2)} \\
& \times \left(1,\epsilon^{2(n_1+n_2)}\lambda_1^{-2}\right)^{\lfloor r_1-n_1}\left(1,\epsilon^{2(r_2-n_2)}\lambda_2^{-2}\right)^{\lfloor n_2} \\
& \times \left(\lambda_1^{-1},\,\mathrm{e}^{-2u}\lambda_2^{-1}\right)^{\lfloor r_2-n_2}\left(\epsilon^{2(r_2-n_2)}\lambda_2^{-1},\,\mathrm{e}^{-2u}\lambda_1^{-1}\right)^{\lfloor n_1}
\end{aligned}
\tag{7.60}
$$

Recall that $u = u_1 - u_2$.

Taking into account that

$$
\begin{aligned}
\left(1,\epsilon^{2(n_1+n_2)}\lambda_1^{-2}\right)^{\lfloor r_1-n_1} \;=\; & \frac{1}{\lambda_1^{r_1-n_1}}\left(\prod_{j=0}^{r_1-n_1-j}\epsilon^{j+n_1+n_2}\right)\left(\epsilon-\epsilon^{-1}\right)^{r_1-n_1} \\
& \times \frac{\prod_{j=0}^{r_1-n_1-j} d^2_{j+n_1+n_2}(\lambda_1)}{\prod_{j=0}^{r_1-n_1-j}[j+1+n_1+n_2]}
\end{aligned}
\tag{7.61}
$$

and also that

$$
[|x|] = [x]\,\epsilon^{x-1}\,, \qquad \left[\left|\begin{matrix} x\\ y\end{matrix}\right|\right] = \begin{bmatrix} x\\ y\end{bmatrix}\epsilon^{xy-y^2}
\tag{7.62}
$$

we finally end up, after some rearranging of the d_j factors, with the following expression:

$$
\begin{aligned}
R_{r_1,r_2}^{r_1+r_2-r,r}(\lambda_1,\lambda_2|u) \;=\; & \frac{\epsilon^{(r_1+r_2-r)r-r_1r_2}}{\prod_{n=0}^{r_1+r_2-1}\left(\mathrm{e}^{u}\lambda_1\lambda_2\epsilon^{-n}-\mathrm{e}^{-u}\epsilon^{n}\right)} \\
& \times \sum_{\substack{n_1=0\\ n_1+n_2=r}}^{r_1}\sum_{n_2=0}^{r_2} \begin{bmatrix} r_1\\ n_1\end{bmatrix}\begin{bmatrix} r_2\\ n_2\end{bmatrix}\frac{[r]!\,[r_2-n_2]!}{[r_1+n_2]!\,[r_2]!}\left(\epsilon-\epsilon^{-1}\right)^{r_1-n_1+n_2} \\
& \times \prod_{j=r_1}^{r_1+n_2-1} d_j(\lambda_1)\prod_{j=n_1+n_2}^{r_1+n_2-1} d_j(\lambda_1)\prod_{j=r_2-n_2}^{r_2-1} d_j(\lambda_2)\prod_{j=r_2-n_2}^{r_1+r_2-r-1} d_j(\lambda_2) \\
& \times \lambda_1^{n_2}\lambda_2^{r_1-n_1}\prod_{j=0}^{r_2-n_2-1}\left(\mathrm{e}^{u}\lambda_2\epsilon^{-j}-\mathrm{e}^{-u}\lambda_1\epsilon^{j}\right) \\
& \times \prod_{j=0}^{n_1-1}\left(\mathrm{e}^{u}\lambda_1\epsilon^{-j+r_2-n_2}-\mathrm{e}^{-u}\lambda_2\epsilon^{j+n_2-r_2}\right)
\end{aligned}
\tag{7.63}
$$

The surprising thing is that an explicit expression for this R-matrix can be found at all, rather than its apparent complexity.

The R-matrix (7.63) satisfies the Yang–Baxter equation

$$\begin{aligned}(\mathbf{1}\otimes R(\xi_1,\xi_2))\ (R(\xi_1,\xi_3)\otimes\mathbf{1})\,(\mathbf{1}\otimes R(\xi_2,\xi_3))\\ =(R(\xi_2,\xi_3)\otimes\mathbf{1})\,(\mathbf{1}\otimes R(\xi_1,\xi_3))\,(R(\xi_1,\xi_2)\otimes\mathbf{1})\end{aligned} \tag{7.64}$$

where ξ denotes the pair of parameters (λ, u) characterizing the affine nilpotent representation.

In addition to the Yang–Baxter equation (7.64), the R-matrix (7.63) enjoys also the following properties.

(i) Normalization:

$$R(\xi,\xi)=\mathbf{1}\otimes\mathbf{1} \tag{7.65}$$

(ii) Unitarity:

$$R(\xi_1,\xi_2)R(\xi_2,\xi_1)=\mathbf{1}\otimes\mathbf{1} \tag{7.66}$$

(iii) Reflection symmetry:

$$PR(\lambda_1,\lambda_2|u)P = R(\lambda_2,\lambda_1|u) \tag{7.67}$$

where P is the permutation operator. In components,

$$R^{r_1',r_2'}_{r_1,r_2}(\lambda_1,\lambda_2|u) = R^{r_2',r_1'}_{r_2,r_1}(\lambda_2,\lambda_1|u) \tag{7.68}$$

(iv) Hermiticity: if λ_1 and λ_2 are pure phases, u is real, and also

$$E^\dagger = \theta F \tag{7.69}$$

for both irreps, then

$$R^\dagger(\lambda_1,\lambda_2|u) = R(\lambda_2,\lambda_1|-u) \tag{7.70}$$

We will return to this property when discussing the hermiticity of the spin chain hamiltonians derived from the R-matrix (7.63).

7.3 Nilpotent hamiltonians

Given the R-matrix (7.63), we may construct the transfer matrix

$$t_{\lambda_\circ}(\lambda|u) = \mathrm{tr}_a\left(\mathcal{R}_{aL}(\lambda,\lambda_\circ|u)\cdots\mathcal{R}_{a1}(\lambda,\lambda_\circ|u)\right) \tag{7.71}$$

where $\mathcal{R}_{ai}$ denotes the operator $\mathcal{R} = PR$ acting on the space $V_a\otimes V_i$. The auxiliary space V_a is the representation space of the nilpotent irrep (λ, u), whereas the quantum spaces V_i ($1\le i\le L =$ length of the lattice) carry the irrep $(\lambda_\circ, 0)$. Of course, for nilpotent representations it is true that dim V_a=dim $V_i = \ell'$, so quantum and auxiliary spaces have the same dimension, which in turn implies that the hamiltonian

$$H(\lambda_\circ) = i\frac{\partial}{\partial u} \log\, t_{\lambda_\circ}(\lambda|u)\Big|_{\substack{\lambda=\lambda_\circ \\ u=0}} \tag{7.72}$$

is a local operator.

The Yang–Baxter operator (7.64) guarantees that the transfer matrices commute:

$$\left[t_{\lambda_\circ}(\lambda|u)\,,\; t_{\lambda_\circ}(\lambda'|u')\right] = 0 \tag{7.73}$$

Since, in addition to the usual rapidity, the values of λ and λ' may differ, we should expect "more" conserved quantities than in a system based on regular representations. Indeed, in addition to the hamiltonian (7.72) obtained by varying u, we may vary λ in the vicinity of $\lambda_\circ$ to find another local hamiltonian

$$Q(\lambda_\circ) = 2\lambda\frac{\partial}{\partial \lambda} \log\, t_{\lambda_\circ}(\lambda|u)\Big|_{\substack{\lambda=\lambda_\circ \\ u=0}} \tag{7.74}$$

such that

$$[H(\lambda_\circ), Q(\lambda_\circ)] = 0 \tag{7.75}$$

The existence of a second local hamiltonian in addition to the usual one is perhaps the most salient feature of integrable models based on nilpotent irreducible representations of the affine quantum group.

Let us study some explicit examples of these hamiltonians. When $\epsilon^4 = 1$, the nilpotent irreps are two-dimensional, and the model is somewhat trivial, since we find that

$$H(\lambda_\circ) = \frac{i}{\lambda_\circ - \lambda_\circ^{-1}} \sum_{j=1}^{L} \left(\sigma_j^x \sigma_{j+1}^x + \sigma_j^y \sigma_{j+1}^y + \frac{\lambda_\circ + \lambda_\circ^{-1}}{2} \sigma_j^z \right) \tag{7.76}$$

is precisely the XX hamiltonian (i.e. the XXZ hamiltonian with $\Delta = 0$ or $q = i$) with an external magnetic field depending on $\lambda_\circ$. The second hamiltonian is, in this case,

$$Q(\lambda_\circ) = \frac{-i}{\lambda_\circ - \lambda_\circ^{-1}} \sum_{j=1}^{L} \left(\sigma_j^x \sigma_{j+1}^y - \sigma_j^y \sigma_{j+1}^x \right) \tag{7.77}$$

Combining $H(\lambda_\circ)$ and $Q(\lambda_\circ)$, we obtain a model which, after a Jordan–Wigner transformation, is equivalent to a free fermion model with complex hopping parameter and non-vanishing chemical potential.

When $\epsilon^3 = 1$, things become more interesting. Up to an overall factor, the hamiltonian is now

$$
\begin{aligned}
H(\lambda_\circ) =& \frac{1}{2}\sum_{j=1}^{L} \left(\lambda_\circ\epsilon + \lambda_\circ^{-1}\epsilon^{-1}\right)\left(S_j^x S_{j+1}^x + S_j^y S_{j+1}^y\right) \\
& - S_j^z S_{j+1}^z - 2\left(S_j^x S_{j+1}^x + S_j^y S_{j+1}^y\right)^2 \\
& + \left(\lambda_\circ\epsilon + \lambda_\circ^{-1}\epsilon^{-1} - 2i\frac{w(\lambda_\circ)}{\epsilon - \epsilon^{-1}}\right) \\
& \times \left[\left(S_j^x S_{j+1}^x + S_j^y S_{j+1}^y\right) S_j^z S_{j+1}^z + \Longleftrightarrow\right] \\
& - \frac{\lambda_\circ\epsilon - \lambda_\circ^{-1}\epsilon^{-1}}{\epsilon - \epsilon^{-1}}\left(S_j^x S_{j+1}^x + S_j^y S_{j+1}^y\right)\left(S_j^z + S_{j+1}^z\right) \\
& + \left(S_j^z\right)^2\left(S_{j+1}^z\right)^2 - 3\left[\left(S_j^z\right)^2 + \left(S_{j+1}^z\right)^2\right] \\
& + \frac{\lambda_\circ^2\epsilon^{-1} - \lambda_\circ^{-2}\epsilon}{\epsilon - \epsilon^{-1}}\left(S_j^z + S_{j+1}^z\right)
\end{aligned}
\tag{7.78}
$$

where $\vec{S}$ are the spin-1 matrices of classical $SU(2)$ and

$$
w(\lambda_\circ) = \sqrt{\left(\lambda_\circ - \lambda_\circ^{-1}\right)\left(\lambda_\circ\epsilon^{-1} - \lambda_\circ^{-1}\epsilon\right)}
\tag{7.79}
$$

Note that the parameter $\lambda_\circ$ appears in (7.78) not only through an external magnetic field, but also in other terms which do not have such direct interpretation.

It is perhaps worth comparing this hamiltonian with the anisotropic spin-1 XXZ hamiltonian constructed by Fateev and Zamolodchikov:

$$
\begin{aligned}
H^{FZ}(q) =& \sum_{j=1}^{L} S_j^x S_{j+1}^x + S_j^y S_{j+1}^y + \frac{q^2 + q^{-2}}{2} S_j^z S_{j+1}^z \\
& - \left(S_j^x S_{j+1}^x + S_j^y S_{j+1}^y\right)^2 - \frac{q^2 + q^{-2}}{2}\left(S_j^z\right)^2\left(S_{j+1}^z\right)^2 \\
& + \left(1 - q - q^{-1}\right)\left[\left(S_j^x S_{j+1}^x + S_j^y S_{j+1}^y\right) S_j^z S_{j+1}^z + \Longleftrightarrow\right] \\
& + \frac{q^2 + q^{-2} - 2}{2}\left[\left(S_j^z\right)^2 + \left(S_{j+1}^z\right)^2\right]
\end{aligned}
\tag{7.80}
$$

The parameter q controls the anisotropy of this one-parameter family of integrable hamiltonians, which contain in particular (for $q = \pm 1$) the isotropic hamiltonian of Babudjian and Takhtadjan. Comparing (7.78) with (7.80), we see that

$$
H^{FZ}(q = \epsilon, \epsilon^3 = 1) = H(\lambda_\circ = \epsilon^2)
\tag{7.81}
$$

which means (figure 7.1) that the integrable deformation in the λ direction intersects at one point the ordinary q-deformation of the spin-1 hamiltonian.

Let us analyze what happens to the nilpotent hamiltonian (7.78) when $\lambda = \pm 1, \pm\epsilon$. At these points of Spec Z_ϵ, the three-dimensional nilpotent representation

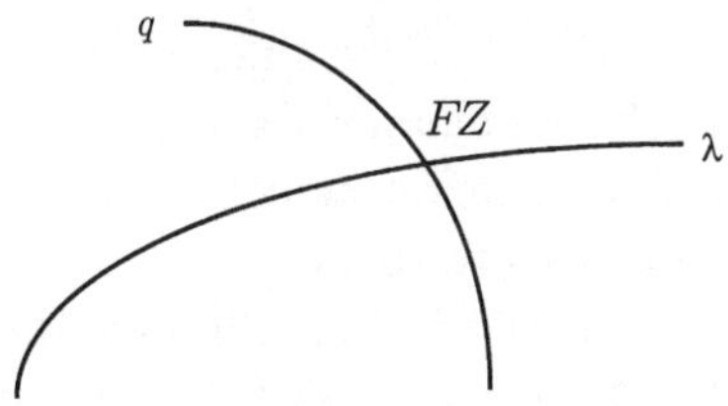

Fig. 7.1. The independent parameters λ and q characterize different families of integrable hamiltonians which intersect at the Fateev–Zamolodchikov hamiltonian with anisotropy $\Delta = (q^2 + q^{-2})/2 = -\frac{1}{2}$.

becomes reducible, and it decomposes into the identity and a doublet, both regular representations. The hamiltonian at these orbifold points has, in fact, an $SU(2) \times U(1)$ symmetry which is absent for generic values of λ. To show this more explicitly, we introduce the following 3×3 matrices:

$$\sigma^+ = \begin{pmatrix} 0 & 0 & 0 \\ 0 & 0 & 1 \\ 0 & 0 & 0 \end{pmatrix} \quad \tau^+ = \begin{pmatrix} 0 & 1 & 0 \\ 0 & 0 & 0 \\ 0 & 0 & 0 \end{pmatrix} \quad \rho^+ = \begin{pmatrix} 0 & 0 & 1 \\ 0 & 0 & 0 \\ 0 & 0 & 0 \end{pmatrix}$$
$$\sigma^- = \begin{pmatrix} 0 & 0 & 0 \\ 0 & 0 & 0 \\ 0 & 1 & 0 \end{pmatrix} \quad \tau^- = \begin{pmatrix} 0 & 0 & 0 \\ 1 & 0 & 0 \\ 0 & 0 & 0 \end{pmatrix} \quad \rho^- = \begin{pmatrix} 0 & 0 & 0 \\ 0 & 0 & 0 \\ 1 & 0 & 0 \end{pmatrix} \tag{7.82}$$
$$\sigma^0 = \begin{pmatrix} 0 & 0 & 0 \\ 0 & 1 & 0 \\ 0 & 0 & 1 \end{pmatrix} \quad \tau^0 = \begin{pmatrix} 1 & 0 & 0 \\ 0 & 1 & 0 \\ 0 & 0 & 0 \end{pmatrix} \quad \rho^0 = \begin{pmatrix} 1 & 0 & 0 \\ 0 & 0 & 0 \\ 0 & 0 & 1 \end{pmatrix}$$

in terms of which the hamiltonian at the orbifold points looks like the sum of two XX hamiltonians:

$$H(\lambda_\circ = 1) = -\sum_{j=1}^{L} \left(\tau_j^+ \tau_{j+1}^- + \tau_j^- \tau_{j+1}^+ + \rho_j^+ \rho_{j+1}^- + \rho_j^- \rho_{j+1}^+ - 2\sigma_j^0 \right)$$
$$H(\lambda_\circ = \epsilon) = H(\lambda_\circ = 1, \text{ with } \sigma \Rightarrow \tau) \tag{7.83}$$

When $\lambda = 1$, the state $e_0 = \begin{pmatrix} 1 \\ 0 \\ 0 \end{pmatrix}$ is a singlet, whereas $e_1 = \begin{pmatrix} 0 \\ 1 \\ 0 \end{pmatrix}$ and $e_2 = \begin{pmatrix} 0 \\ 0 \\ 1 \end{pmatrix}$ form a doublet of the $SU(2)$ generated by $\sum \sigma_j^+$, $\sum \sigma_j^-$ and $\sum \sigma_j^z$. The $U(1)$ is generated by $\sum \sigma_j^0$, which commutes with the whole hamiltonian $H(\lambda_\circ = 1)$, and can thus be viewed as a chemical potential distinguishing between the singlet e_0 and the doublet (e_1, e_2).

The above model (7.83), which is a limiting case of the nilpotent hamiltonian with $\epsilon^3 = 1$, can be shown to be equivalent to a Hubbard model with infinite Coulomb repulsion and a non-zero chemical potential. The repulsion forbids two electrons to occupy the same site, so there are only three possible states for each site (empty, spin up, spin down).

The above discussion shows that the integrable hamiltonians derived from nilpotent representations share common features with various known models, such as higher-spin models or fermion models. Their behavior, especially at the orbifold points, deserves further investigation.

Having dealt at length with the physical interpretation of the cases $\ell' = 2$ and $\ell' = 3$, let us note that, for the general case $\ell' \geq 2$, a very compact expression for the hamiltonian following from the R-matrix (7.63) can be found. It is given by

$$H(\lambda) = -\sum_{j=1}^{L}\sum_{n=1}^{\ell'-1}\left\{\frac{1}{\sin\frac{\pi n}{\ell}}\left[\left(\Sigma_j(\lambda)\Sigma_{j+1}^\dagger(\lambda)\right)^n + \left(\Sigma_j^\dagger(\lambda)\Sigma_{j+1}(\lambda)\right)^n\right]\right.$$
$$\left. + \mu_j(\lambda) + \mu_{j+1}(\lambda)\right\} \tag{7.84}$$

with the diagonal matrices $\Sigma(\lambda)$ and $\mu(\lambda)$ acting as

$$\begin{aligned}
\Sigma(\lambda)\, e_r &= \sqrt{\frac{\epsilon^r - \epsilon^{-r}}{\epsilon^{r-1}\lambda^{-1} - \epsilon^{-r+1}\lambda}}\, e_r \\
\mu(\lambda)\, e_0 &= 0 \\
\mu(\lambda)\, e_r &= -i\sum_{j=0}^{r-1}\frac{\lambda^{-1}\epsilon^j + \lambda\epsilon^{-j}}{\lambda^{-1}\epsilon^j - \lambda\epsilon^{-j}}\, e_r
\end{aligned} \tag{7.85}$$

We shall discuss in section 7.6 the chiral Potts hamiltonian, which has a very similar structure.

7.4 Bethe ansatz

The models discussed in the previous section are the descendants of the six-vertex model through nilpotent irreps. The discussion of the algebraic Bethe ansatz in chapter 2 generalizes immediately to these hamiltonians, which we now present.

We start by introducing the monodromy matrix $T_{\lambda_\circ}(u)$ which acts on the same quantum space as the transfer matrix $t_{\lambda_\circ}(\lambda|u)$, namely $\mathscr{H} = \otimes^L V^{\lambda_\circ}$, but whose auxiliary space is simply the regular two-dimensional irrep $V^{\frac{1}{2}}$:

$$T_{\lambda_\circ}(u) = \mathscr{R}_{aL}^{(\frac{1}{2},\lambda_\circ)}(u)\cdots\mathscr{R}_{a1}^{(\frac{1}{2},\lambda_\circ)}(u) \tag{7.86}$$

where $\mathscr{R}^{(\frac{1}{2},\lambda_\circ)}(u)$ is the intertwiner for $V^{\frac{1}{2}} \otimes V^{\lambda_\circ}$. The main property of the monodromy matrix is that it satisfies the quadratic relations

$$\mathscr{R}_{12}^{(\frac{1}{2},\frac{1}{2})}(u-v)T_{\lambda_\circ}(u)_1 \otimes T_{\lambda_\circ}(v)_2 = T_{\lambda_\circ}(v)_2 \otimes T_{\lambda_\circ}(u)_1\mathscr{R}_{12}^{(\frac{1}{2},\frac{1}{2})}(u-v) \tag{7.87}$$

where $\mathscr{R}^{(\frac{1}{2},\frac{1}{2})}$ is the six-vertex $\mathscr{R}$-matrix, and, as usual, $T_1 = T\otimes\mathbf{1}$ and $T_2 = \mathbf{1}\otimes T$. From (7.87), it follows that the entries of

$$T_{\lambda_\circ}(u) = \begin{pmatrix} A_{\lambda_\circ}(u) & B_{\lambda_\circ}(u) \\ C_{\lambda_\circ}(u) & D_{\lambda_\circ}(u) \end{pmatrix} \tag{7.88}$$

satisfy the familiar relations (2.40). These entries can be computed from the definition (7.86) and the explicit expression for $\mathscr{R}_{an}^{(\frac{1}{2},\lambda_\circ)}(u) = L_n^{\lambda_\circ}(u)$:

$$L_n^{\lambda_\circ}(u) = \frac{1}{e^u\sqrt{\epsilon\lambda} - e^{-u}\frac{1}{\sqrt{\epsilon\lambda}}} \times \begin{pmatrix} e^u\sqrt{\epsilon}K_n^{\frac{1}{2}} - e^{-u}\frac{1}{\sqrt{\epsilon}}K_n^{-\frac{1}{2}} & \frac{\epsilon-\epsilon^{-1}}{\sqrt{\epsilon}}F_nK_n^{\frac{1}{2}} \\ \frac{\epsilon-\epsilon^{-1}}{\sqrt{\epsilon}}E_nK_n^{-\frac{1}{2}} & e^u\sqrt{\epsilon}K_n^{-\frac{1}{2}} - e^{-u}\frac{1}{\sqrt{\epsilon}}K_n^{\frac{1}{2}} \end{pmatrix} \tag{7.89}$$

where E_n, F_n and K_n ($n = 1, \ldots, L$) act on the space $V_n^{\lambda_\circ}$. Since nilpotent irreps are highest weight vector representations, we may take, as the Bethe reference state $|\Omega_\circ\rangle$ the highest weight of the tensor product, namely $|\Omega_\circ\rangle = \overbrace{e_0 \otimes \cdots \otimes e_0}^{L \text{ times}}$, and then it is clear that $C_{\lambda_\circ}|\Omega_\circ\rangle = 0$. From (7.87), it follows also that

$$B_{\lambda_\circ}(u)B_{\lambda_\circ}(v) = B_{\lambda_\circ}(v)B_{\lambda_\circ}(u) \tag{7.90}$$

Hence, to diagonalize the transfer matrix (7.71), we may use the standard algebraic Bethe ansatz states

$$|\Psi_M\rangle = \prod_{i=1}^{M} B_{\lambda_\circ}(u_i)\,|\Omega_\circ\rangle \tag{7.91}$$

The number M of spin waves is a conserved quantum number, and therefore we may diagonalize the transfer matrix in each sector independently, as we did for the XXZ spin-$\frac{1}{2}$ case.

We need the commutation relations between the diagonal entries of the monodromy matrix $T_{\lambda_\circ}(\lambda|u)$ and the off-diagonal $B_{\lambda_\circ}(u)$ operators. They can be obtained from the relation

$$\mathscr{R}_{12}^{(\lambda,\frac{1}{2})}(u-v)T_{\lambda_\circ}(\lambda|u)_1 \otimes T_{\lambda_\circ}\left(\frac{1}{2}|v\right)_2 = T_{\lambda_\circ}\left(\frac{1}{2}|v\right)_2 \otimes T_{\lambda_\circ}(\lambda|u)_1\mathscr{R}_{12}^{(\lambda,\frac{1}{2})}(u-v) \tag{7.92}$$

where $T_{\lambda_\circ}(\frac{1}{2}|v)$ is simply $T_{\lambda_\circ}(v)$. In components, this equation reads as follows:

$$\begin{aligned} &\left(\mathscr{R}^{(\lambda,\frac{1}{2})}(u-v)\right)_{r_2,s_2}^{r_1,s_1}\left(T_{\lambda_\circ}(\lambda|u)\right)_{r_3}^{r_2}\left(T_{\lambda_\circ}\left(\frac{1}{2}|v\right)\right)_{s_3}^{s_2} \\ &\quad = \left(T_{\lambda_\circ}\left(\frac{1}{2}|v\right)\right)_{s_2}^{s_1}\left(T_{\lambda_\circ}(\lambda|u)\right)_{r_2}^{r_1}\left(\mathscr{R}_{12}^{(\lambda,\frac{1}{2})}(u-v)\right)_{r_3,s_3}^{r_2,s_2} \end{aligned} \tag{7.93}$$

Taking into account that

$$\left(\mathscr{R}^{(\lambda_1,\lambda_2)}(u)\right)_{r_1,r_2}^{r'_1,r'_2} = \left(\mathscr{R}^{(\lambda_2,\lambda_1)}(u)\right)_{r_2,r_1}^{r'_2,r'_1} \tag{7.94}$$

and choosing $s_1 = 0$, $r_1 = r$, $s_3 = 1$ and $r_3 = r$, we obtain

$$\begin{aligned} &A(v)T_{r+1}^{r}(u)\mathscr{R}_{1,r}^{0,r+1}(u-v) + B(v)T_r^r(u)\mathscr{R}_{1,r}^{1,r}(u-v) \\ &\quad = \mathscr{R}_{0,r}^{0,r}(u-v)T_r^r(u)B(v) + \mathscr{R}_{1,r-1}^{0,r}(u-v)T_r^{r-1}(u)D(v) \end{aligned} \tag{7.95}$$

where $\mathscr{R} = \mathscr{R}^{(\frac{1}{2},\lambda)}$ and $T_r^s(u) = \left(T_{\lambda_\circ}(\lambda|u)\right)_r^s$ $(r, s = 0, 1, \ldots, \dim V^{\lambda_\circ} - 1)$. Similarly, choosing $s_3 = 0$ and $r_3 = r + 1$ yields

$$A(v)T_{r+1}^r(u)\mathscr{R}_{0,r+1}^{0,r+1}(u-v) + B(v)T_r^r(u)\mathscr{R}_{0,r+1}^{1,r}(u-v) \\ = \mathscr{R}_{0,r}^{0,r}(u-v)T_{r+1}^r(u)A(v) + \mathscr{R}_{1,r-1}^{0,r}(u-v)T_{r+1}^{r-1}(u)C(v) \tag{7.96}$$

Eliminating from these equations the term $A(v)T_{r+1}^r$, we find

$$T_r^r(u)\, B(v) = \mathscr{L}_r^\lambda(u-v)B(v)T_r^r(u) + \frac{\mathscr{R}_{0,r}^{0,r}(u-v)}{\mathscr{R}_{0,r+1}^{0,r+1}(u-v)}T_{r+1}^r(u)A(v) \\ + \frac{\mathscr{R}_{1,r-1}^{0,r}(u-v)}{\mathscr{R}_{0,r+1}^{0,r+1}(u-v)}T_{r+1}^{r-1}(u)C(v) - \frac{\mathscr{R}_{1,r-1}^{0,r}(u-v)}{\mathscr{R}_{1,r}^{0,r+1}(u-v)}T_r^{r-1}(u)D(v) \tag{7.97}$$

where

$$\mathscr{L}_r^\lambda(u) = \frac{\mathscr{R}_{1,r}^{1,r}(u)\mathscr{R}_{0,r+1}^{0,r+1}(u) - \mathscr{R}_{0,r+1}^{1,r}(u)\mathscr{R}_{1,r}^{0,r+1}(u)}{\mathscr{R}_{0,r}^{0,r}(u)\mathscr{R}_{0,r+1}^{0,r+1}(u)} \tag{7.98}$$

The required $\mathscr{R}$-matrix entries can be obtained from equation (7.63) with $\lambda_1 = \epsilon$ (spin $\frac{1}{2}$) and $\lambda_2 = \lambda$. Explicitly, we find

$$\begin{aligned}
\mathscr{R}_{0,r}^{0,r}(u) &= \frac{e^u\epsilon^{-r}\sqrt{\epsilon\lambda} - e^{-u}\epsilon^r/\sqrt{\epsilon\lambda}}{e^u\sqrt{\epsilon\lambda} - e^{-u}/\sqrt{\epsilon\lambda}} \\
\mathscr{R}_{0,r}^{1,r-1}(u) &= \frac{\left(\epsilon - \epsilon^{-1}\right)\epsilon^r}{\sqrt{\epsilon\lambda d_{r-1}(\lambda)}\left(e^u\sqrt{\epsilon\lambda} - e^{-u}/\sqrt{\epsilon\lambda}\right)} \\
\mathscr{R}_{1,r}^{0,r+1}(u) &= \frac{\left(\epsilon - \epsilon^{-1}\right)\epsilon^{-r}\sqrt{\lambda}}{\sqrt{\epsilon d_r(\lambda)}\left(e^u\sqrt{\epsilon\lambda} - e^{-u}/\sqrt{\epsilon\lambda}\right)} \\
\mathscr{R}_{1,r}^{1,r}(u) &= \frac{e^u\epsilon^r\sqrt{\epsilon/\lambda} - e^{-u}\epsilon^{-r}\sqrt{\lambda/\epsilon}}{e^u\sqrt{\epsilon\lambda} - e^{-u}/\sqrt{\epsilon\lambda}}
\end{aligned} \tag{7.99}$$

It is not hard to check that these relations are equivalent to those in (7.89).

With the help of (7.99), we continue our computations to find

$$\mathscr{L}_r^\lambda(u) = \frac{e^u\sqrt{\epsilon\lambda} - e^{-u}/\sqrt{\epsilon\lambda}}{e^u\epsilon^{-r}\sqrt{\epsilon\lambda} - e^{-u}\epsilon^r/\sqrt{\epsilon\lambda}} \\ \times \frac{e^u/\sqrt{\epsilon\lambda} - e^{-u}\sqrt{\epsilon\lambda}}{e^u\epsilon^{-r}\sqrt{\lambda/\epsilon} - e^{-u}\epsilon^r\sqrt{\epsilon/\lambda}} \tag{7.100}$$

Setting $\lambda = \epsilon$ in this equation, we find $\mathscr{L}$ for the regular spin-$\frac{1}{2}$ irrep:

$$\begin{aligned}
\mathscr{L}_0^{s=\frac{1}{2}}(u) &= \mathscr{L}_0^{\lambda=\epsilon}(u) = \frac{e^u\epsilon^{-1} - e^{-u}\epsilon}{e^u - e^{-u}} = \frac{\sinh(u - i\gamma)}{\sinh u} \\
\mathscr{L}_1^{s=\frac{1}{2}}(u) &= \mathscr{L}_1^{\lambda=\epsilon}(u) = \frac{e^u\epsilon - e^{-u}\epsilon^{-1}}{e^u - e^{-u}} = \frac{\sinh(u + i\gamma)}{\sinh u}
\end{aligned} \tag{7.101}$$

We are now in a position to compute the eigenvalues of the transfer matrix $t_{\lambda_\circ}(\lambda|u) = \text{tr}\, T_{\lambda_\circ}(\lambda|u)$:

$$\begin{aligned}
t_{\lambda_\circ}(\lambda|u) \prod_{i=1}^{M} B_{\lambda_\circ}(u_i)\,|\Omega_\circ\rangle &= \sum_r T_r^r(u) \prod_{i=1}^{M} B_{\lambda_\circ}(u_i)\,|\Omega_\circ\rangle \\
&= \sum_r \prod_{i=1}^{M} B_{\lambda_\circ}(u_i) \left(\prod_{j=1}^{M} \mathscr{L}_r^{\lambda}(u-u_j) \right) T_r^r(u)\,|\Omega_\circ\rangle + \text{u.t.} \\
&= \Lambda\left(\lambda, \lambda_\circ | u, \{u_i\}\right) |\Psi_M\rangle + \text{unwanted terms}
\end{aligned} \tag{7.102}$$

with

$$\Lambda\left(\lambda, \lambda_\circ | u, \{u_i\}\right) = \sum_r \left(\mathscr{R}_{r,0}^{r,0}(\lambda, \lambda_\circ|u) \right)^L \prod_{i=1}^{M} \mathscr{L}_r^{\lambda}(u-u_i) \tag{7.103}$$

We have used the fact that

$$T_r^r(u)\,|\Omega_\circ\rangle = \left(\mathscr{R}_{r,0}^{r,0}(\lambda, \lambda_\circ|u) \right)^L |\Omega_\circ\rangle \tag{7.104}$$

and we have chosen the normalization $\mathscr{R}_{00}^{00} = 1$.

Of course, for generic values of the rapidities u_i, the unwanted terms in (7.102) will be present, and the state $|\Psi_M\rangle$ will not be an eigenstate of the monodromy matrix. To find the Bethe equation fixing the rapidities u_i, we use the trick of canceling the poles in the eigenvalues Λ as we vary u. We can already do this for the eigenvalue of the transfer matrix tr $T_{\lambda_\circ}(\frac{1}{2}|u)$, which is obtained from (7.103) by setting $\lambda = \epsilon$, and using equations (7.101):

$$\begin{aligned}
\Lambda(\lambda = \epsilon, \lambda_\circ | u, \{u_i\}) &= \left(\mathscr{R}_{00}^{00}(\lambda = \epsilon, \lambda_\circ|u)\right)^L \prod_{k=1}^{M} \mathscr{L}_0^{\lambda=\epsilon}(u-u_k) \\
&\quad + \left(\mathscr{R}_{10}^{10}(\lambda = \epsilon, \lambda_\circ|u)\right)^L \prod_{k=1}^{M} \mathscr{L}_1^{\lambda=\epsilon}(u-u_k) \\
&= \prod_{k=1}^{M} \frac{\sinh(u-u_k-i\gamma)}{\sinh(u-u_k)} \\
&\quad + \left(\frac{e^{u}\epsilon - e^{-u}\lambda_\circ}{e^{u}\epsilon\lambda_\circ - e^{-u}} \right)^L \prod_{k=1}^{M} \frac{\sinh(u-u_k+i\gamma)}{\sinh(u-u_k)}
\end{aligned} \tag{7.105}$$

If u approaches u_j (for some given j), $\Lambda(u)$ may develop a pole. This is unacceptable, because the eigenvalue should be analytic in u. Hence, the residue of $\Lambda(u)$ at $u = u_j$ should vanish. This condition yields the Bethe ansatz equations for the rapidities u_j:

$$\left(\frac{e^{u_j}\epsilon\lambda_\circ - e^{-u_j}}{e^{u_j}\epsilon - e^{-u_j}\lambda_\circ}\right)^L = \prod_{\substack{k=1\\k\neq j}}^{M}\frac{\sinh(u_j-u_k+i\gamma)}{\sinh(u_j-u_k-i\gamma)} \tag{7.106}$$

Letting

$$\epsilon = e^{i\gamma}, \qquad \lambda_\circ = \epsilon^{2S} = e^{2i\gamma S} \tag{7.107}$$

we may rewrite (7.106) in a more pleasing way:

$$\left(\frac{\sinh\left(u_j+i\frac{\gamma}{2}+i\gamma S\right)}{\sinh\left(u_j+i\frac{\gamma}{2}-i\gamma S\right)}\right)^L = \prod_{\substack{k=1\\k\neq j}}^{M}\frac{\sinh(u_j-u_k+i\gamma)}{\sinh(u_j-u_k-i\gamma)} \tag{7.108}$$

To improve the look of this result even more, we change variables:

$$u_j + i\frac{\gamma}{2} = \frac{\gamma}{2}\lambda_j \tag{7.109}$$

These rapidities λ_j should not be confused with the quantum group labels λ. The final version of the Bethe equations then reads as

$$\left(\frac{\sinh\frac{\gamma}{2}\left(\lambda_j+2iS\right)}{\sinh\frac{\gamma}{2}\left(\lambda_j-2iS\right)}\right)^L = \prod_{\substack{k=1\\k\neq j}}^{M}\frac{\sinh\frac{\gamma}{2}(\lambda_j-\lambda_k+2i)}{\sinh\frac{\gamma}{2}(\lambda_j-\lambda_k-2i)} \tag{7.110}$$

After all this work, it is rewarding to see that if $2S$ is an integer, this equation reproduces the Bethe ansatz equation (3.71) of the spin-S anisotropic XXZ model. This is not at all surprising, since throughout the derivation of (7.110) we have not used explicitly the fact that λ (or equivalently S) may take any complex value when ϵ is a root of unity. Thus, we observe again that nilpotent models constitute an integrable deformation of higher-spin anisotropic XXZ models, when the anisotropy takes some quantized values. Having fixed the anisotropy by the requirement that $\epsilon^\ell = 1$, we are then free to "move the spin", which becomes a continuous parameter of the model.

It can be shown that the unwanted terms in (7.102) cancel provided the Bethe ansatz equation (7.110) holds. In this case, therefore, we diagonalize simultaneously $t_{\lambda_\circ}(\lambda|u)$ and $t_{\lambda_\circ}(\frac{1}{2}|u)$ for all values of λ and u. From the definition (7.72) of the hamiltonian $H(\lambda_\circ)$ and the eigenvalue (7.103), we may obtain the energy of the Bethe states (7.91):

$$\begin{aligned} E_M &= i\frac{\partial}{\partial u}\log\Lambda(\lambda,\lambda_\circ|u,\{u_i\})\Big|_{\substack{\lambda=\lambda_\circ\\u=0}} = i\frac{\partial}{\partial u}\sum_{j=1}^{M}\log\mathscr{L}_0^{\lambda_\circ}(u-u_j)\Big|_{u=0} \\ &= -\sum_{j=1}^{M}\frac{\sin 2\gamma S}{\sinh\left[\frac{\gamma}{2}(\lambda_j+2iS)\right]\sinh\left[\frac{\gamma}{2}(\lambda_j-2iS)\right]} \end{aligned} \tag{7.111}$$

Similarly, the eigenvalue of $Q(\lambda_\circ)$ on the state $|\Psi_M\rangle$ is

$$\begin{aligned}\langle\Psi_M|\,Q(\lambda_\circ)\,|\Psi_M\rangle &= 2\lambda\frac{\partial}{\partial\lambda}\log\Lambda(\lambda,\lambda_\circ|u,\{u_i\})\Big|_{\substack{\lambda=\lambda_\circ\\ u=0}} \\ &= 2\lambda\frac{\partial}{\partial\lambda}\sum_{j=1}^{M}\log\mathscr{L}_0^{\lambda}(u_j)\Big|_{\lambda=\lambda_\circ} \\ &= -\sum_{j=1}^{M}\frac{\sinh\gamma\lambda_j}{\sinh\left[\frac{\gamma}{2}(\lambda_j+2iS)\right]\sinh\left[\frac{\gamma}{2}(\lambda_j-2iS)\right]}\end{aligned} \tag{7.112}$$

and the momentum is

$$\begin{aligned}\langle\Psi_M|\,P\,|\Psi_M\rangle &= -i\log\Lambda(\lambda_\circ,\lambda_\circ|u=0,\{u_i\}) \\ &= -i\sum_{j=1}^{M}\log\left[\frac{\sinh\frac{\gamma}{2}(\lambda_j+2iS)}{\sinh\frac{\gamma}{2}(\lambda_j-2iS)}\right]\end{aligned} \tag{7.113}$$

Formulae (7.111) and (7.113) coincide with those for the spin-S XXZ model if we specify S to be an integer or half-odd value.

The statistical model we have defined starting from nilpotent irreps of $SU(2)_q$ depends on the anisotropy ϵ, which is an ℓth root of unity, and on the "spin" S or $\lambda = \epsilon^{2S}$. Not all values of S give sensible results, however. For a good statistical mechanical interpretation of the two-dimensional model, we are interested in finding a parameter region where all the Boltzmann weights are positive. For a reasonable spin chain picture, on the other hand, unitarity requires that the hamiltonian (7.72) be hermitian. From the hermiticity conditions (7.69) and (7.70), it can be shown that

$$H^\dagger(S) = H(S) \qquad \Longleftrightarrow \qquad E^\dagger = \theta_S F \tag{7.114}$$

which in turn is equivalent to

$$\theta_S \sin\gamma(2S-k)\sin\gamma(k+1) > 0 \qquad \forall\ell\in\{0,1,\ldots,\ell'-2\} \tag{7.115}$$

The parameter θ_S is called the spin parity.

These conditions restrict the value of S to some intervals. If $\epsilon = \exp(2\pi i/\ell)$, then there are two disjoint hermiticity regions of allowed spins S, each associated with a given spin parity. We need to distinguish the cases when ℓ is even:

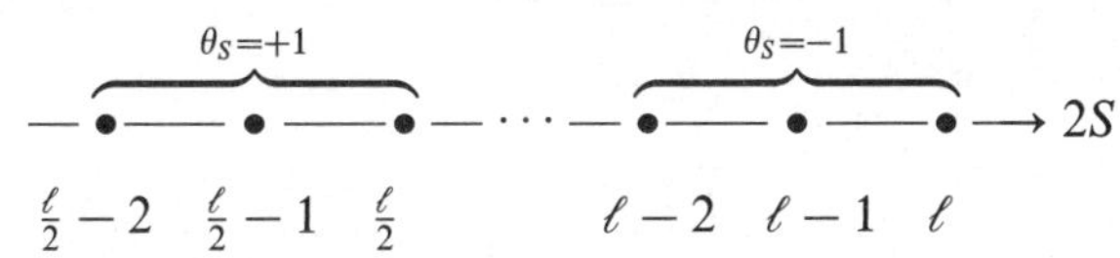

(7.116)

and when ℓ is odd:

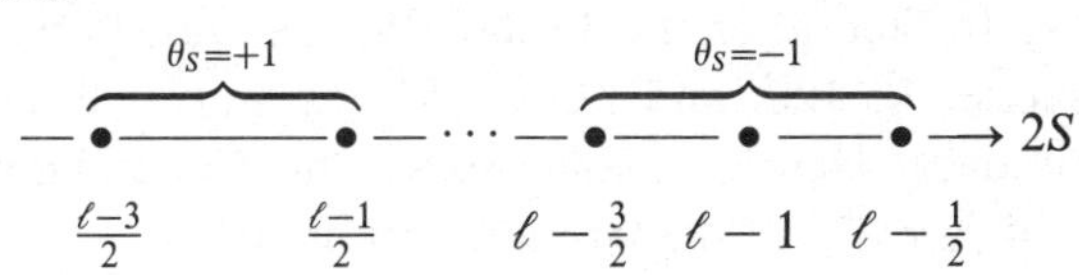

(7.117)

The hermiticity regions for real S shown above contain, in three of the four cases, a mid-point corresponding to a higher-spin anisotropic XXZ hamiltonian. In all four cases, and whenever $2S$ is an integer, the boundary points of the hermiticity regions correspond to "orbifold" hamiltonians, like (7.83) for $\ell = 3$.

7.5 The limit $\ell \to \infty$

We would like to analyze in this section the limit $\ell \to \infty$ or, equivalently, $\epsilon = \exp(2\pi i/\ell) \to 1$ of the nilpotents models introduced above. When taking this limit, we shall keep the labels λ characterizing nilpotent representations finite and fixed. As $\ell \to \infty$, the dimension of the nilpotent representations becomes infinite, as do the intertwining R-matrices.

Let us first consider the limit of the R-matrices. To simplify matters, we set $\lambda_1 = \lambda_2 = \lambda$ in (7.63) and then take $\epsilon \to 1$. The result is the following infinite-dimensional R-matrix:

$$R^{r_1+r_2-n,n}_{r_1,r_2}(\lambda|u) = \sqrt{\frac{\binom{r_1+r_2}{r_1}}{\binom{r_1+r_2}{n}}}\left(e^u\lambda - e^{-u}\lambda^{-1}\right)^{-(r_1+r_2)} \tag{7.118}$$

$$\times \sum_{\substack{n_1=0 \\ n_1+n_2=n}}^{r_1}\sum_{n_2=0}^{r_2} \binom{r_1}{n_1}\binom{r_2}{n_2}\left(\lambda-\lambda^{-1}\right)^{r_1-n_1+n_2}\left(e^u - e^{-u}\right)^{r_2-n_2+n}$$

Since this R-matrix satisfies a Yang–Baxter relation with spectral parameters, we can derive from it an integrable vertex model. From the one-dimensional point of view, the hamiltonian is

$$\begin{aligned} H &= i\frac{\partial}{\partial u}\log t_\lambda^{\ell=\infty}(u) \\ &= \frac{2i}{\lambda-\lambda^{-1}}\sum_{j=1}^{L}\left(a_j a_{j+1}^\dagger + a_j^\dagger a_{j+1} - \left(\lambda+\lambda^{-1}\right)a_j^\dagger a_j\right) \end{aligned} \tag{7.119}$$

where a and $a^\dagger$ are the customary harmonic oscillator operators satisfying

$$\left[a, a^\dagger\right] = 1 \tag{7.120}$$

Similarly, the second hamiltonian Q is given by

$$Q(\lambda) \sim \sum_{j=1}^{L}\left(a_j a_{j+1}^\dagger - a_j^\dagger a_{j+1}\right) \tag{7.121}$$

Expressions (7.119) and (7.121) resemble strongly the expressions (7.76) and (7.77) for the case $\ell = 4$. In fact, after the Jordan–Wigner transformation, (7.76) and (7.77) become precisely (7.120) and (7.121) with a and $a^\dagger$ fermionic operators, instead of bosonic ones. Hence, in some sense, the $\ell = \infty$ limit is the bosonic version of the $\ell = 4$ case, i.e. of the free fermion model.

7.5.1 Quantum harmonic oscillators

To exploit this bosonic interpretation further, let us study the braid group limit of the R-matrix (7.118):

$$\begin{aligned}\left(R^{(\lambda,\lambda)}\right)_{r_1,r_2}^{r_1+r_2-n,n} &= \lim_{u\to+\infty} e^{u(r_1-n)} R(\lambda|u)_{r_1,r_2}^{r_1+r_2-n,n} \\ &= \sqrt{\binom{r_1}{n}\binom{r_1+r_2-n}{r_2}} \left(\lambda-\lambda^{-1}\right)^{r_1-n} \lambda^{-(r_1+r_2)}\end{aligned} \tag{7.122}$$

This R-matrix can be obtained, alternatively, as the intertwiner R-matrix for representations of a quantum deformation of the harmonic oscillator algebra $U_q(h_4)$.

The algebra $U_q(h_4)$ has four generators, $\mathcal{N}$, a, $a^\dagger$ and $\mathcal{M}$, subject to the relations

$$\left[\mathcal{N}, a^\dagger\right] = a^\dagger\,, \quad [\mathcal{N}, a] = -a\,, \quad \left[a, a^\dagger\right] = \frac{q^{\mathcal{M}} - q^{-\mathcal{M}}}{q - q^{-1}} \tag{7.123}$$

and $\mathcal{M}$ central, with co-product

$$\begin{aligned}\Delta(x) &= x \otimes q^{\mathcal{M}/2} + q^{-\mathcal{M}/2} \otimes x \qquad x = a, a^\dagger \\ \Delta(y) &= y \otimes \mathbf{1} + \mathbf{1} \otimes y \qquad y = \mathcal{M}, \mathcal{N}\end{aligned} \tag{7.124}$$

antipode

$$\gamma(x) = -x \qquad x = a, a^\dagger, \mathcal{N}, \mathcal{M} \tag{7.125}$$

and co-unit

$$\epsilon(x) = 0 \qquad x = a, a^\dagger, \mathcal{N}, \mathcal{M} \tag{7.126}$$

In fact, $U_q(h_4)$ is a quasi-triangular ribbon Hopf algebra with universal $\mathcal{R}$-matrix

$$\mathcal{R} = q^{-(\mathcal{M}\otimes\mathcal{N}+\mathcal{N}\otimes\mathcal{M})} \exp\left[\left(q-q^{-1}\right)\left(q^{\mathcal{M}/2}\otimes q^{-\mathcal{M}/2}\right) a \otimes a^\dagger\right] \tag{7.127}$$

It has two Casimirs,

$$C_1 = \frac{q^{\mathcal{M}} - q^{-\mathcal{M}}}{q - q^{-1}} \mathcal{N} - a^\dagger a\,, \quad C_2 = \mathcal{M}^2 \tag{7.128}$$

An irrep of $U_q(h_4)$ is characterized by the eigenvalues (m, n) of $\mathcal{M}$ and $\mathcal{N}$ on the highest weight vector $|0\rangle$. If $m \neq 0$, an irrep of $U_q(h_4)$ is spanned by the orthonormal basis $\{|r\rangle\}_{r=0}^{\infty}$:

$$\begin{aligned}a\,|r\rangle &= \sqrt{[m]\,r}\,|r-1\rangle \\ a^\dagger\,|r\rangle &= \sqrt{[m]\,(r+1)}\,|r+1\rangle \\ \mathcal{M}\,|r\rangle &= m\,|r\rangle \qquad \mathcal{N}\,|r\rangle = (r+n)\,|r\rangle\end{aligned} \tag{7.129}$$

where, as usual, $[x] = \left(q^x - q^{-x}\right) / \left(q - q^{-1}\right)$. By evaluating the universal $\mathcal{R}$-matrix (7.127) in the tensor product $(m_1, n_1) \otimes (m_2, n_2)$, we find

$$R^{(m_1,n_1)\otimes(m_2,n_2)} = q^{m_1 n_2 + m_2 n_1} P \mathcal{R}^{(m_1,n_1)\otimes(m_2,n_2)} \tag{7.130}$$

whose matrix elements in the basis (7.129) are

$$\left(R^{(m_1,m_2)}\right)_{r_1 r_2}^{r'_1 r'_2} = \delta_{r_1+r_2}^{r'_1+r'_2} \sqrt{\binom{r_1}{r'_2}\binom{r'_1}{r_2} \{(q^{m_1}-q^{-m_1})(q^{m_2}-q^{-m_2})\}^{r_1-r'_2}} \times q^{(m_1-m_2)(r_1-r'_2)/2} q^{-m_1 r'_1 - m_2 r'_2} \tag{7.131}$$

This R-matrix satisfies the braid group relation

$$R_i^{(m_2,m_3)} R_{i+1}^{(m_1,m_3)} R_i^{(m_1,m_2)} = R_{i+1}^{(m_1,m_2)} R_i^{(m_1,m_3)} R_{i+1}^{(m_2,m_3)} \tag{7.132}$$

Setting $m_1 = m_2 = m$ in (7.131) and comparing it with $R^{(\lambda,\lambda)}$ in (7.122), we see that the expressions coincide provided $\lambda = q^m$. Similarly, on comparing the representation (7.56) of the generators E, F and K of $U_\epsilon(A_1)$ with the q-deformed harmonic oscillator algebra (7.129), we find the following identification in the limit $\epsilon \to 1$:

$$\sqrt{\epsilon - \epsilon^{-1}} \begin{pmatrix} E \\ F \end{pmatrix} \sim \sqrt{q - q^{-1}} \begin{pmatrix} a \\ a^\dagger \end{pmatrix}, \qquad K \sim q^{\mathcal{M}} \epsilon^{-2\mathcal{N}} \tag{7.133}$$

It is thus clear that, in the limit $\ell \to \infty$, nilpotent representations of $U_\epsilon(s\ell(2))$ become representations of the quantum harmonic oscillator algebra.

7.5.2 Link invariants

The R-matrix (7.122) or, equivalently (7.131), can be used to construct an invariant of links and knots, forming an extended Yang–Baxter system as described in appendix D. It is given by

$$Z_{\hat{\alpha}}(q^m) = q^{m(-w(\alpha)+N-1)} \mathrm{tr}'\left(\pi_m(\alpha)\right) \tag{7.134}$$

where $\hat{\alpha}$ is the closure of a word α of the braid group B_N, $w(\alpha)$ is its writhe number, $\pi_e(\alpha)$ is the representation of α with the matrices $R^{(m,m)}$ (7.131), and tr′ is a trace where we sum over all indices except the first one. This slight modification of the usual Turaev definition of the link invariant associated with an extended Yang–Baxter system is required in order to obtain a finite result.

It is a conjecture that $Z_{\hat{\alpha}}(q^m)$ is the inverse of the Alexander–Conway polynomial $\Delta_{\hat{\alpha}}(t)$:

$$Z_{\hat{\alpha}}(q^m) = \frac{1}{\Delta_{\hat{\alpha}}(t = q^m)} \tag{7.135}$$

7.6 The chiral Potts model

In order to understand some features of the transition between commensurate and incommensurate phases, an asymmetric or chiral clock model was invented as a natural generalization of the Ising model: the spin or face variables at each site are allowed to take N values, which are uniformly spread out around a circle, just like the hours or minutes marks on a clock's face. The discrete group $\mathbb{Z}_N$ is therefore a manifest symmetry, just as $\mathbb{Z}_2$ is a symmetry of the Ising model. Chirality arises from an asymmetry in the Boltzmann weights of the spin model, according to their orientation. It turns out that the chiral Potts model is integrable. Moreover, its spectral variables belong (for $N \geq 3$) to an algebraic curve of genus higher than one. Recall that the spectral variety of all models we have studied so far was of genus zero (the sphere) except for the eight-vertex model of chapter 4, where it was genus one (a torus). Let us start by fixing notation and enjoying a general overview of the various features of the model.

To define the model, we start with a rectangular two-dimensional lattice, and assign lattice variables $\sigma = 0, 1, 2, \ldots, N-1 \bmod N$ to the lattice sites. The variable $\sigma_{i,j} \in \mathbb{Z}_N$ is called the spin at site (i, j).

The interaction energy for the whole lattice, in a configuration with spins $\sigma_{i,j}$, is given by

$$E = -\sum_{j,k} \sum_{n=1}^{N-1} \left[E_n^H \, \omega^{n(\sigma_{j,k} - \sigma_{j,k+1})} + E_n^V \, \omega^{n(\sigma_{j,k} - \sigma_{j+1,k})} \right] \tag{7.136}$$

where

$$\omega = \mathrm{e}^{2\pi i/N} \tag{7.137}$$

and E_n^H (respectively, E_n^V) is the horizontal (respectively, vertical) coupling constant. For the time being, we shall not worry whether the energy (7.137) or these coupling constants are real. When $N = 2$, we recover expression (5.54) for the Ising model. If all the couplings are equal, i.e. if $E_n^H = E_m^V = J$, the model (7.136) becomes the N-state Potts model, with energy (recall exercise 5.7)

$$E = -JN \sum_{\langle j,k \rangle} \left(\delta\left(\sigma_j, \sigma_k\right) - \frac{1}{N} \right) \tag{7.138}$$

It is convenient to use the following Boltzmann weights, rather than the horizontal and vertical coupling constants E_n^H and E_n^V $(n = 1, \ldots, N-1)$:

$$\begin{aligned} W(\sigma - \sigma') = W^H(\sigma - \sigma') &= \exp \beta \sum_{n=1}^{N-1} E_n^H \omega^{n(\sigma - \sigma')} \\ \overline{W}(\sigma - \sigma') = W^V(\sigma - \sigma') &= \exp \beta \sum_{n=1}^{N-1} E_n^V \omega^{n(\sigma - \sigma')} \end{aligned} \tag{7.139}$$

They are periodic:

$$W(x) = W(x + N)\,, \qquad \overline{W}(x) = \overline{W}(x + N) \tag{7.140}$$

but, in general, $W(x) \neq W(-x)$, whence the chirality. To indicate clearly that we deal with $W(\sigma - \sigma')$ and not with $W(\sigma' - \sigma)$, we draw an arrow from σ to σ':

$$W(\sigma - \sigma') = \overset{\sigma}{\bullet} \longrightarrow \overset{\sigma'}{\bullet}\,, \qquad \overline{W}(\sigma - \sigma') = \begin{matrix} \sigma' & \bullet \\ & \uparrow \\ \sigma & \bullet \end{matrix} \tag{7.141}$$

With the help of these Boltzmann weights, we may construct two discrete evolution operators in the horizontal and vertical directions. Let us call them S and T. If we agree that time flows upwards in (7.141), then these operators are

$$\langle\sigma|\otimes\langle\sigma'|\, S\,|\sigma\rangle\otimes|\sigma'\rangle = W(\sigma-\sigma')\,, \quad \langle\sigma'|\, T\,|\sigma\rangle = \overline{W}(\sigma-\sigma') \tag{7.142}$$

Using the periodicity (7.140), the lattice evolution operators can be recast very concisely as follows:

$$S = \sum_{n=1}^{N} \widehat{W}(n)\, \left(Z\otimes Z^{\dagger}\right)^{n}\,, \qquad T = \sum_{n=1}^{N} \overline{W}(n)\, X^{n} \tag{7.143}$$

Here, $\widehat{W}$ is the (discrete) Fourier transform of W, namely

$$\widehat{W}(n) = \frac{1}{N}\sum_{m=1}^{N} \omega^{-nm}\, W(m) \tag{7.144}$$

Also, the $N\times N$ matrices X and Z act on the basis $\{|\sigma\rangle\}_{\sigma=0}^{N-1}$ as

$$X\,|\sigma\rangle = |\sigma-1\rangle \quad , \qquad Z\,|\sigma\rangle = \omega^{\sigma}\,|\sigma\rangle \tag{7.145}$$

or, explicitly,

$$X = \begin{pmatrix} 0 & 1 & & & & \\ 0 & 0 & 1 & & 0 & \\ & & 0 & \ddots & & \\ \vdots & & & \ddots & 1 & \\ 0 & & & & 0 & 1 \\ 1 & 0 & & \cdots & & 0 \end{pmatrix}, \quad Z = \begin{pmatrix} 1 & & & & \\ & \omega & & 0 & \\ & & \omega^{2} & & \\ & 0 & & \ddots & \\ & & & & \omega^{N-1} \end{pmatrix} \tag{7.146}$$

They satisfy the relations

$$X^{N} = Z^{N} = 1\,, \qquad XZ = \omega ZX \tag{7.147}$$

which show that there is no preferred vector in the basis $\{|\sigma\rangle\}$, since everything is completely cyclic modulo N.

The representations for all the models studied so far (XXZ relatives) have had a highest weight vector. However, now we find a cyclic representation space without any highest weight or lowest weight vectors.

To conclude these introductory remarks, let us note the spin chain hamiltonian respecting $\mathbb{Z}_N$ invariance:

$$H = -\sum_{j=1}^{L}\sum_{n=1}^{N-1}\left[\overline{\alpha}_n\,(X_j)^{n} + \alpha_n\left(Z_j Z_{j+1}^{\dagger}\right)^{n}\right] \tag{7.148}$$

where X_j and Z_j are the operators (7.145) acting on the jth site, and α_n and $\overline{\alpha}_n$ are coupling constants (we shall see below how they are related to the Boltzmann weights $W(n)$ and $\overline{W}(n)$ of the chiral Potts model). A general property of the hamiltonian (7.148) is that it commutes with the $\mathbb{Z}_N$ charge operator

$$e^{2\pi i\widehat{Q}/N} = \prod_{j=1}^{L} X_j \tag{7.149}$$

The spectrum of H thus splits into N sectors with definite charge, labeled by $Q = 0, 1, \ldots, N-1$. The hamiltonian H is not, in general, parity invariant, but it does respect translational invariance. Therefore, the total momentum is a good quantum number. We shall see that, because the rapidity lives on a higher genus curve, and therefore does not enjoy the difference property, the hamiltonian for the chiral Potts model contains significantly less information than the Boltzmann weights. The question we shall address next is, what are the conditions on the Boltzmann weights $W(n)$ and $\overline{W}(n)$ for the two-dimensional model (7.136) to be exactly integrable?

7.6.1 Star-triangle relations

The chiral Potts model can be nicely formulated as a particular kind of face model, although the most economical presentation is that of a spin model (appendix E). Let us start with the more familiar face picture. Besides the $\mathbb{Z}_N$ spins σ_i, it is convenient to introduce one more lattice variable, neutral with respect to $\mathbb{Z}_N$ transformations, which we shall call $*$. The relevant graph for the face version of the model is an N-pointed star, with the neutral variable $*$ at its center linked to the $\mathbb{Z}_N$ variables, as shown in figure 7.2. Any allowed configuration on a square lattice consists of two clearly distinguishable sectors (figure 7.3): an odd sublattice, where all the variables take the neutral value $*$, and an even sublattice, where all the variables range over $\mathbb{Z}_N$ (the even or odd labeling for the two sublattices is arbitrary). This is very reminiscent of what we found in section 5.3.1 for the Ising model. Indeed, when $N = 2$, we recover the A_3 graph associated with the Ising model.

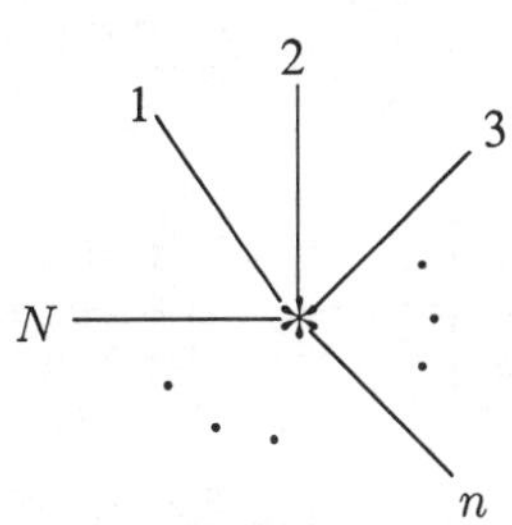

Fig. 7.2 Graph defining the $\mathbb{Z}_N$ chiral Potts model.

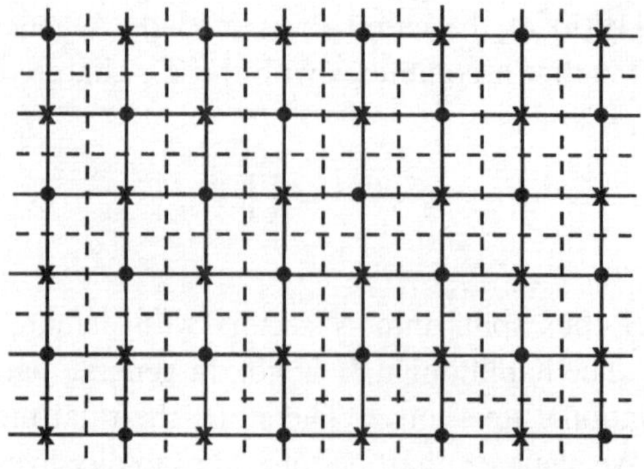

Fig. 7.3. The lattice on which the chiral Potts model is defined is shown by the solid lines. Solid dots represent the sites on the even sublattice ($\mathbb{Z}_N$ variables), and crosses denote those on the odd sublattice (neutral variable). The dual lattice is depicted by the dashed lines.

Consider now the dual lattice, and assign "rapidity variables" $p, q, \ldots$ to them; they are so far unspecified, and possibly multi-dimensional. Each plaquette is crossed by two rapidities p and q. Two of its corners are occupied by the neutral element $*$, and the other two are occupied by $\mathbb{Z}_N$ variables σ and σ'. The split into two sublattices explains the existence of two different kinds of Boltzmann weights: two different kinds of plaquettes arise according to whether the neutral variables are in the upper left and lower right corners, or in the opposite two. We may readily identify these two kinds of Boltzmann weights with those of equation (7.141) after a 45° rotation, up to a rapidity-dependent factor (see figure 7.4):

$$B_{p,q}\begin{pmatrix} \sigma & * \\ * & \sigma' \end{pmatrix} = f_{pq} W_{pq}(\sigma - \sigma')$$
$$B_{p,q}\begin{pmatrix} * & \sigma' \\ \sigma & * \end{pmatrix} = \overline{W}_{pq}(\sigma - \sigma') \tag{7.150}$$

After the identification (7.150), the Yang–Baxter equation for W and $\overline{W}$ follows from its general expression (5.6) for face models; it takes the form of a

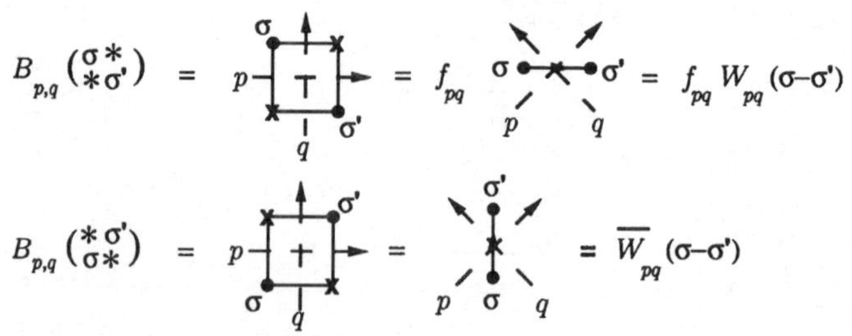

Fig. 7.4. Graphical representation of the Boltzmann weights (7.150) for the chiral Potts model.

star-triangle equation (recall appendix E):

$$\sum_{\sigma_\circ=1}^{N} \overline{W}_{qr}(\sigma_2-\sigma_\circ)\, W_{pr}(\sigma_1-\sigma_\circ)\, \overline{W}_{pq}(\sigma_\circ-\sigma_3) = R_{pqr}\, W_{pq}(\sigma_1-\sigma_2)\, \overline{W}_{pr}(\sigma_2-\sigma_3) W_{qr}(\sigma_1-\sigma_3) \tag{7.151}$$

where the variables σ_i range over $\mathbb{Z}_N$, and

$$R_{pqr} = \frac{f_{pq} f_{qr}}{f_{pr}} \tag{7.152}$$

The integrability of the chiral Potts model is encoded in this hexagon equation, shown in figure 7.5. The crucial dependence of the Boltzmann weights W and $\overline{W}$ on the rapidity variables p and q is emphasized with the help of the rapidity lines on the dual to the hexagon.

The solution to the Yang–Baxter equation in figure 7.5 can be parametrized in a variety of ways. In the original work by Au-Yang and collaborators, each rapidity was shown to consist of four complex numbers $p = (a_p, b_p, c_p, d_p)$ and $q = (a_q, b_q, c_q, d_q)$, and the weights are given by (recall $\omega = e^{2\pi i/N}$)

$$\begin{aligned} \frac{W_{pq}(n)}{W_{pq}(0)} &= \prod_{j=1}^{n} \frac{d_p b_q - a_p c_q \omega^j}{b_p d_q - c_p a_q \omega^j} \\ \frac{\overline{W}_{pq}(n)}{\overline{W}_{pq}(0)} &= \prod_{j=1}^{n} \frac{\omega a_p d_q - d_p a_q \omega^j}{c_p b_q - b_p c_q \omega^j} \end{aligned} \tag{7.153}$$

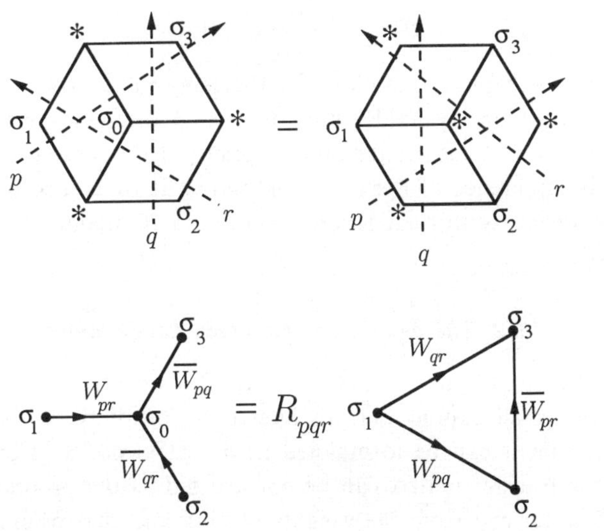

Fig. 7.5. The Yang–Baxter equation for the chiral Potts model in the face and spin versions.

For later use, let us note also the explicit form of the Fourier transform of $W_{pq}(n)$:

$$\frac{\widehat{W}_{pq}(n)}{\widehat{W}_{pq}(0)} = \prod_{j-1}^{n} \frac{b_p d_q - d_p b_q \omega^{j-1}}{c_p a_q - a_p c_q \omega^j} \tag{7.154}$$

The proportionality factor f_{pq} is independent of σ:

$$f_{pq} = \left[\prod_{m=1}^{N} \left(\sum_{k=1}^{N} \omega^{mk} \frac{\overline{W}_{pq}(k)}{W_{pq}(m)} \right) \right]^{1/N} \tag{7.155}$$

Integrability, i.e. the Yang–Baxter equation (7.151), requires all the rapidity variables $x = (a_x, b_x, c_x, d_x)$ to exist on a manifold Γ_κ obtained as the intersection of two Fermat curves:

$$x \in \Gamma_\kappa \ : \quad \begin{cases} a_x^N + \kappa' b_x^N = \kappa\, d_x^N \\ \kappa' a_x^N + b_x^N = \kappa\, c_x^N \\ \kappa^2 + \kappa'^2 = 1 \end{cases} \tag{7.156}$$

For completeness, note that the following are also true:

$$\kappa\, a_p^N + \kappa' c_p^N = d_p^N \,, \qquad \kappa\, b_p^N + \kappa' d_p^N = c_p^N \tag{7.157}$$

The parameter κ is free, and the curve Γ_κ plays the role of the integrability manifold. In fact, in the chiral Potts model, the standard uniformization parameter u is replaced by the two rapidity variables p and q living on Γ_κ. Surprisingly, the genus of Γ_κ is

$$g(\Gamma_\kappa) = N^3 - 2N^2 + 1 \tag{7.158}$$

so the genus of the rapidity manifold for the Ising model ($N = 2$) is just one, and the difference property holds; but for $N = 3$ the genus increases to ten! This result implies that the Boltzmann weights (7.153) cannot be rewritten as functions of the difference of the two rapidities. The difference property is lost because there are no conformal Killing vectors on Riemann surfaces of genus greater than one.

7.6.2 The associated spin chain hamiltonian

Let us derive the hamiltonian (7.148) from the Boltzmann weights (7.153). For this purpose, we shall exploit first an interesting property of the chiral Potts model, namely that it can be formulated as a vertex model. The (diagonal to diagonal) vertex transfer matrix can be constructed as the product of two face transfer matrices formed from the weights (7.150), and the vertex R-matrix can then be read off as the product of four face Boltzmann weights, two W's and two $\overline{W}$'s (see figure 7.6).

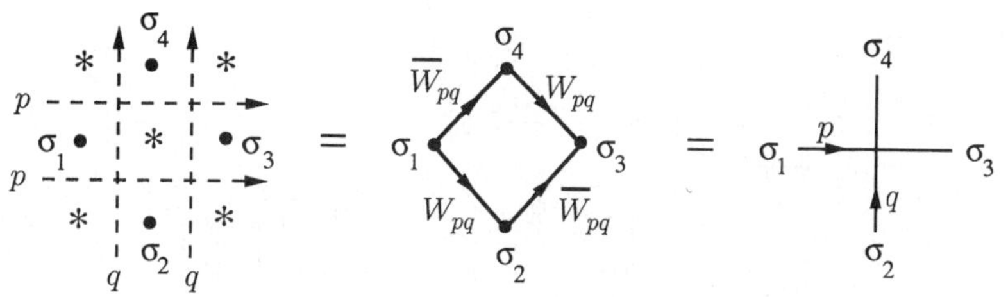

Fig. 7.6 The map from the face to the vertex versions of the chiral Potts model.

For simplicity, we assume that the rapidities carried by the dual lattice take only two values: p in the horizontal direction and q in the vertical direction. Physically, this implies some sort of homogeneity. We find

$$R^{\sigma_4\sigma_3}_{\sigma_1\sigma_2}(p,q) \;=\; W_{pq}(\sigma_1-\sigma_2)\,\overline{W}_{pq}(\sigma_2-\sigma_3)\,W_{pq}(\sigma_4-\sigma_3)\,\overline{W}_{pq}(\sigma_1-\sigma_4) \quad (7.159)$$

In terms of the operators S and T introduced in equation (7.142), we may write this R-matrix as

$$R(p,q) = S_{pq}\left(T_{pq}\otimes T_{pq}\right)S_{pq} \quad (7.160)$$

where S_{pq} and T_{pq} are given by (7.143) with W and $\overline{W}$ replaced by the weights W_{pq} and $\overline{W}_{pq}$ in (7.153). The R-matrix (7.159) satisfies the vertex version of the Yang–Baxter equation (without the difference property). Moreover, if we normalize the weights with

$$W_{pq}(0) = \overline{W}_{pq}(0) = 1 \quad (7.161)$$

then it becomes the identity when the two rapidities are equal:

$$R(p,q)\Big|_{p=q} = \mathbf{1}\otimes\mathbf{1} \quad (7.162)$$

This last property is useful for evaluating the spin chain hamiltonian $H = \sum_{j=1}^{L} h_{j,j+1}$ associated with the chiral Potts model; it suffices to take the logarithmic derivative of R at $p=q$:

$$h(q) \sim \delta_p \log R(p,q)\Big|_{p=q} \quad (7.163)$$

From (7.160) and (7.143), it follows immediately that the two-body hamiltonian $h(q)$ has the structure anticipated in equation (7.148). The coefficients α_n (respectively, $\overline{\alpha}_n$) are proportional to the variation $\delta_p\widehat{W}_{pq}(n)$ (respectively, $\delta_p\overline{W}_{pq}(n)$) at $p=q$. Using the parametrization (7.153) and the spectral conditions (7.156), we obtain

$$\delta_p \widehat{W}_{pq}(n)\Big|_{p=q} = \frac{i}{2}\left(\frac{\delta_p a_p}{a_p} - \frac{\delta_p d_p}{d_p}\right)\Big|_{p=q} \frac{\omega^{n/2}}{\kappa' \sin\frac{\pi n}{N}} \frac{a_q^N}{b_q^N}\left(\frac{a_q c_q}{b_q d_q}\right)^{n-N}$$
$$\delta_p \overline{W}_{pq}(n)\Big|_{p=q} = \frac{i}{2}\left(\frac{\delta_p a_p}{a_p} - \frac{\delta_p d_p}{d_p}\right)\Big|_{p=q} \frac{\omega^{n/2}}{\sin\frac{\pi n}{N}} \left(\frac{a_q d_q}{b_q c_q}\right)^{n} \tag{7.164}$$

With an appropriate choice of an overall factor, one can finally write the coupling constants in the hamiltonian as

$$\alpha_n = \frac{\exp\left(i\frac{2n-N}{N}\phi\right)}{\sin\frac{\pi n}{N}}, \qquad \overline{\alpha}_n = \kappa' \frac{\exp\left(i\frac{2n-N}{N}\overline{\phi}\right)}{\sin\frac{\pi n}{N}} \tag{7.165}$$

where the "angles" ϕ and $\overline{\phi}$ are defined (mod $N\pi$) by

$$e^{2i\phi/N} = \sqrt{\omega}\frac{a_q c_q}{b_q d_q}, \qquad e^{2i\overline{\phi}/N} = \sqrt{\omega}\frac{a_q d_q}{b_q c_q} \tag{7.166}$$

and are related by the spectral condition

$$\cos\phi = \kappa' \cos\overline{\phi} \tag{7.167}$$

Note that the hamiltonian $h(q)$ depends, in general, on the point q where the derivative of the transfer matrix is evaluated. This is further proof that $R(p,q)$ does not enjoy the difference property.

The hamiltonian (7.148) with the parameters (7.165) is hermitian if, and only if, both ϕ and $\overline{\phi}$ are real:

$$H = H^\dagger \quad \Longleftrightarrow \quad \alpha_n^* = \alpha_{N-n}, \quad \overline{\alpha}_n^* = \overline{\alpha}_{N-n} \tag{7.168}$$

As the reader must have surely noticed, any computation with the chiral Potts model is lengthy and cumbersome, due to the rather involved parametrization of the Boltzmann weights. To grasp some of the physics behind this integrable model, it is therefore advisable to impose some further symmetries to simplify the Boltzmann weights as much as possible. Let us do just this in two cases: the self-dual and super-integrable chiral Potts models.

7.6.3 Self-dual chiral Potts models

As close relatives of the Ising model, the $\mathbb{Z}_N$ invariant models enjoy the important Kramers–Wannier symmetry, which simplifies the study of their phase structure. The Kramers–Wannier duality can be implemented between the statistical weights $\widehat{W}$ and $\overline{W}$ as follows:

$$\text{Kramers–Wannier}: \quad \begin{cases} \widehat{W}(n) \longrightarrow \widehat{W}^D(n) = \overline{W}(n) \\ \overline{W}(n) \longrightarrow \overline{W}^D(n) = \widehat{W}(n) \end{cases} \tag{7.169}$$

The new partition function Z^D constructed from the dual weights is exactly the same as the original one Z, up to perhaps some overall factor. In the alternative spin chain picture, the Kramers–Wannier dual hamiltonian H^D has the same form as H in (7.148), with

$$\alpha_n^D = \overline{\alpha}_n \,, \qquad \overline{\alpha}_n^D = \alpha_n \tag{7.170}$$

At the hamiltonian level, this symmetry can be implemented through the following operator transformation:

$$\begin{aligned} X_i &\to X_i^D = Z_i Z_{i+1}^\dagger \\ Z_i &\to Z_i^D = X_1^\dagger X_2^\dagger \cdots X_{i-1}^\dagger X_i^\dagger \end{aligned} \tag{7.171}$$

which satisfies the following properties:

$$\begin{aligned} &\left(X_i^D\right)^N = \left(Z_i^D\right)^N = \mathbf{1} \\ &X_i^D Z_j^D = \omega^{\delta_{ij}} Z_j^D X_i^D \\ &\left(A_i^D\right)^D = A_{i+1} \,, \quad A = X \text{ or } Z \end{aligned} \tag{7.172}$$

Let us study the special case where the chiral Potts model is self-dual, i.e.

$$\overline{W}(n) = \widehat{W}(n) \qquad \Longleftrightarrow \qquad \alpha_n = \overline{\alpha}_n \qquad \forall n \in \mathbb{Z}_n \tag{7.173}$$

From the parametrization (7.165), we thus find

$$\kappa' = 1 \Longleftrightarrow \kappa = 0 \qquad \text{and} \qquad \phi = \overline{\phi} \Longleftrightarrow c_q = d_q \quad (\forall q) \tag{7.174}$$

Among all the self-dual chiral Potts models, an especially interesting one is the Fateev–Zamolodchikov model corresponding to the parameter choice

$$c_q = d_q = 1 \,, \qquad b_q = \sqrt{\omega} a_q \,, \qquad \phi = \overline{\phi} = 0 \tag{7.175}$$

In this particularly degenerate case, we recover the difference property if we uniformize the spectral curve by an exponential, $a_q = e^{-iu_q}$. The Boltzmann weights yield, in fact, a trigonometric solution to the Yang–Baxter equation:

$$\begin{aligned} W_{pq}^{FZ}(n) &= \prod_{j=1}^{n} \frac{\sin\left[\frac{\pi}{N}\left(j-\frac{1}{2}\right) - \frac{u_p-u_q}{2}\right]}{\sin\left[\frac{\pi}{N}\left(j-\frac{1}{2}\right) + \frac{u_p-u_q}{2}\right]} \\ \overline{W}_{pq}^{FZ}(n) &= \prod_{j=1}^{n} \frac{\sin\left[\frac{\pi}{N}(j-1) + \frac{u_p-u_q}{2}\right]}{\sin\left[\frac{\pi}{N}j - \frac{u_p-u_q}{2}\right]} \end{aligned} \tag{7.176}$$

The Fateev–Zamolodchikov model is critical, and the corresponding conformal field theory (see chapter 8) is a $\mathbb{Z}_N$ parafermionic model, with central extension $c = 2(N-1)/(N+2)$. The much investigated $N = 3$ case yields the critical three-state Potts model, with Virasoro central charge $\frac{4}{5}$. The case $N = 4$ gives the critical self-dual Ashkin–Teller model, which is a rational conformal field theory with $c = 1$. An integrable deformation of the critical Fateev–Zamolodchikov model yields the general self-dual chiral Potts model.

7.6.4 Super-integrable chiral Potts models

These special chiral Potts models obey the constraint $\phi = \overline{\phi} = \pi/2$. From equation (7.167), we see that κ' is free and that $\overline{\alpha}_n = \kappa' \alpha_n$, so the model is self-dual under the transformation $\kappa' \to 1/\kappa'$. We may thus interpret the parameter κ' as a temperature, or, rather, as the ratio of the temperature to the critical temperature. This interpretation is reinforced by a careful look at the rather peculiar properties of super-integrable chiral Potts models. For example, the mass gap is proportional to $(1 - \kappa')$, as in the Ising model. A quick check will convince the reader that the super-integrable chiral Potts Boltzmann weights cannot all be positive, so the model must be interpreted with caution. The ground state energy for temperatures $0 \leq \kappa' < 1$ can be written in terms of the hypergeometric function:

$$\lim_{L\to\infty} \frac{E_L}{L} = -(1+\kappa') \sum_{n=1}^{N-1} F\left(-\frac{1}{2}, \frac{n}{N}; 1; \frac{4\kappa'}{(1+\kappa')^2}\right) \tag{7.177}$$

We see that the ground state is ferromagnetic, and thus that the super-integrable chiral Potts model is a direct generalization of the Ising model (which we recover by setting $N = 2$). To obtain expression (7.177), we assumed that there exists a translationally invariant ground state. This, in turn, leads to the constraint on κ'. When $\kappa' = 1$ and $N \geq 3$, a transition to an incommensurate phase occurs; for $\kappa' > 1$, the original ferromagnetic ground state picks up momentum, and in fact no translationally invariant ground state exists.

In fact, a remarkable symmetry lies behind both the Ising and the super-integrable chiral Potts models; this was discovered by Onsager himself in his revolutionary analytic solution to the two-dimensional Ising model. The Onsager algebra is generated by the operators A_k and G_k ($k = 1, \ldots, L$, where L is the size of the lattice) subject to

$$\begin{aligned} [A_\ell, A_m] &= 4G_{\ell-m} \\ [G_\ell, A_m] &= 2A_{m+\ell} - 2A_{m-\ell} \\ [G_\ell, G_m] &= 0 \end{aligned} \tag{7.178}$$

To check that these generators do commute with the hamiltonian, it suffices to write the chiral Potts hamiltonian (7.148) as

$$H_{\text{super}} = A_0 + \kappa' A_1 \tag{7.179}$$

and then show that, in the super-integrable case, the conditions

$$\begin{aligned} [A_0, [A_0, [A_0, A_1]]] &= 16\,[A_0, A_1] \\ [A_1, [A_1, [A_1, A_0]]] &= 16\,[A_1, A_0] \end{aligned} \tag{7.180}$$

are satisfied. One can then construct out of A_0 and A_1 an infinite set of conserved and commuting currents which enhance the integrability of the model. In fact, using only the Onsager algebra, it is possible to show that the eigenvalues of the hamiltonian assemble into families of the form

$$E = A + B\kappa' + 4\sum_{n=1}^{n_E} m_n\sqrt{1+\kappa'^2+2\kappa'\cos\theta_n} \tag{7.181}$$

where A, B and θ_n are some constants, the family contains $(2s+1)^{n_E}$ energy eigenvalues, and m_n can take the values $-s, -s+1, \ldots, s-1, s$. For the super-integrable chiral Potts model, just as for the Ising model, s takes the value $\frac{1}{2}$, and no examples are known of realizations of the Onsager algebra with $s > \frac{1}{2}$.

The hamiltonian (7.148) with coefficients given by (7.165) is invariant under $\mathbb{Z}_N$, and we might wonder what happens if we relax the integrability constraint (7.167). The basic goal is to elucidate the various phases of the model (parametrized by N, κ', ϕ and $\overline{\phi}$). For our purposes, the most intriguing issue is integrability. In general, the integrability submanifold is quite small, and it does not coincide with phase boundaries. The phase diagram shown in figure 7.7 for the self-dual ($\phi = \overline{\phi}$) $N = 3$ case is of particular interest. An infinity of conserved charges exists along the two lines labeled as integrable and super-integrable. Onsager's algebra accounts for the super-integrability, and we shall see in the following that the quantum group $U_q(\widehat{s\ell}(2))$ is responsible for the integrable line.

7.6.5 The quantum symmetry

In chapter 6 we reviewed in detail a variety of solutions to the Yang–Baxter equation from the quantum group viewpoint. The obvious question to ask now is, does the chiral Potts model admit a quantum group picture, in contrast to the eight-vertex model? The answer can be summarized, affirmatively, in the following short statement: the chiral Potts model is a descendant of the six-vertex model. When the number of states per site of the chiral Potts model (N) and

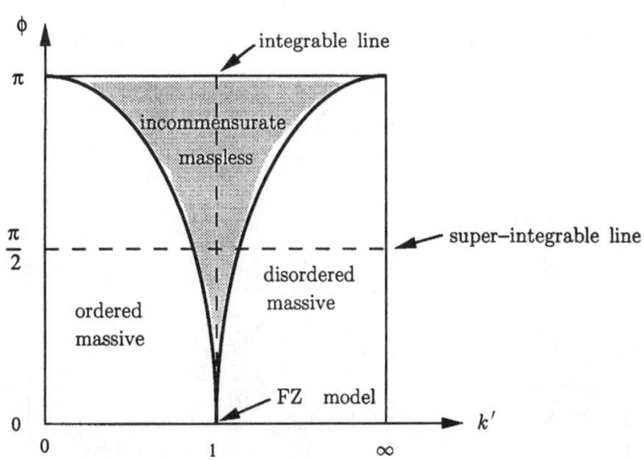

Fig. 7.7. Phase structure for the $\mathbb{Z}_3$ invariant hamiltonian (7.148) with parametrization (7.165) and the self-dual condition ($\phi = \overline{\phi}$).

the anisotropy of the six-vertex model are commensurable, the descent relation follows. This corresponds to the condition $q^N = 1$, with q related in the usual way to the anisotropy of the six-vertex model.

We have seen earlier in this chapter that, when the deformation parameter q of quantum groups is a root of unity, the representation theory of quantum groups, i.e. the spectrum, grows into a surprising structure. We shall now see that it encompasses, in particular, the chiral Potts model. Whereas the six-vertex model and its higher-spin descendants involve only regular representations of $U_q(\widehat{s\ell}(2))$ with integer or half-odd spin, the chiral Potts model calls for fully fledged cyclic representations. The periodicity or cyclicity of these irreducible representations is the quantum group reflection of the $\mathbb{Z}_N$ symmetry of the model. We should expect that the Boltzmann weights of the chiral Potts model play the role of an intertwiner between two cyclic representations.

The identification of the Boltzmann weights as entries of an R-matrix is best obtained in the vertex version of the model. In fact, it will prove convenient to consider a slight generalization of the chiral Potts model, called the checkerboard model: we allow the horizontal rapidities in the dual lattice to alternate between two different values, and similarly for the vertical rapidities. Quite simply, the R-matrix of the model is given by (see figure 7.8)

$$R^{\sigma_4\sigma_3}_{\sigma_1\sigma_2} = W_{p'q}(\sigma_1-\sigma_2)\,\overline{W}_{p'q'}(\sigma_2-\sigma_3)\,W_{pq'}(\sigma_4-\sigma_3)\,\overline{W}_{pq}(\sigma_1-\sigma_4) \tag{7.182}$$

Letting $\xi = (p, p')$ and $\eta = (q, q')$, we may write this R-matrix more succinctly as

$$R(\xi,\eta) = S_{pq'}\left(T_{pq}\otimes T_{p'q'}\right)S_{p'q} \tag{7.183}$$

It is not hard to show that the checkerboard chiral Potts R-matrix preserves the $\mathbb{Z}_N$ charge (7.149), namely

$$R(\xi,\eta)(X\otimes X) = (X\otimes X)\,R(\xi,\eta) \tag{7.184}$$

It would thus seem advisable to choose a basis in which X is diagonal, whereas Z is the shift operator. This can be achieved through a Fourier transform:

$$|e_r\rangle = \frac{1}{\sqrt{N}}\sum_{\sigma=0}^{N-1}\omega^{-r\sigma}\,|\sigma\rangle \tag{7.185}$$

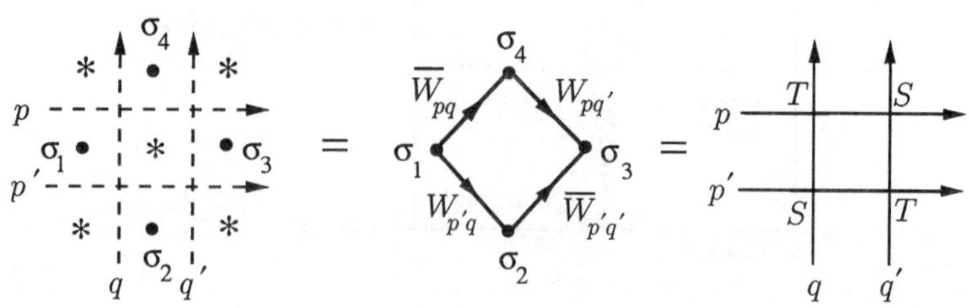

Fig. 7.8 Boltzmann weights (7.182) of the checkerboard chiral Potts model.

whereby

$$X\,|e_r\rangle = \omega^{-r}\,|e_r\rangle\;, \qquad Z\,|e_r\rangle = |e_{r-1}\rangle \tag{7.186}$$

In this basis, the charge conservation (7.184) amounts to the rule

$$R_{r_1 r_2}^{r_1' r_2'}(\xi,\eta) \neq 0 \quad \Rightarrow \quad r_1 + r_2 = r_1' + r_2' \,(\text{mod}\, N) \tag{7.187}$$

which is very reminiscent of the selection rules for the six-vertex model and its higher-spin generalizations, with the crucial difference that now the conservation of the "third component of isospin" holds only modulo N.

In summary, the R-matrix we wish to interpret as an intertwiner of a quantum group has the following properties, which should be accounted for properly in the algebraic framework:

(i) $\mathbb{Z}_N$ conservation law;
(ii) spectral curve isomorphic to $\Gamma_\kappa \otimes \Gamma_\kappa$;
(iii) factorization into four pieces (W's and $\overline{W}$'s).

The first condition (*i*) is automatic once we deal with cyclic irreps. The second condition (*ii*) is much more subtle, and constitutes, in fact, the key to the solution. In section 7.1, we learned that a cyclic irrep of $U_q(s\ell(2))$ with $q^N = 1$ is labeled by three complex parameters, which are the eigenvalues of $E^{N'}$, $F^{N'}$ and $K^{N'}$ ($N' = N$ if N is odd, $N' = N/2$ if N is even). Representations of the infinite-dimensional $U_q(\widehat{s\ell(2)}) = \widehat{U}$ carry, in addition, the usual affine parameter e^u. At first sight, it would seem that we have too many parameters at our disposal; but the intertwiner condition

$$R(\xi,\eta)\left(\pi_\xi \otimes \pi_\eta\right)\,\Delta(g) = \left(\pi_\eta \otimes \pi_\xi\right)\Delta(g)\,R(\xi,\eta) \quad \forall g \in \widehat{U} \tag{7.188}$$

imposes additional constraints on the Casimirs of the affine cyclic irreps π_η and π_ξ. Let us choose for g any of the central elements $E_i^{N'}$, $F_i^{N'}$ and $K_i^{N'}$ ($i = 0,1$), whose eigenvalues in a cyclic irrep ξ are $x_i(\xi)$, $y_i(\xi)$ and $z_i(\xi)$, respectively. A short computation shows that

$$\begin{aligned}
\left(\pi_\xi \otimes \pi_\eta\right)\Delta'\left(E_i^{N'}\right) &= x_i(\xi) + z_i(\xi)x_i(\eta)\\
\left(\pi_\xi \otimes \pi_\eta\right)\Delta'\left(F_i^{N'}\right) &= y_i(\xi)/z_i(\xi) + y_i(\eta)\\
\left(\pi_\xi \otimes \pi_\eta\right)\Delta'\left(K_i^{N'}\right) &= z_i(\xi)z_i(\eta)
\end{aligned} \tag{7.189}$$

Hence, for an intertwiner $R(\xi,\eta)$ to exist between the two cyclic irreps π_η and π_ξ, the Casimirs x_i, y_i and z_i of both representations must satisfy the conditions (recall equations (7.41))

$$\begin{aligned}
\frac{1 - z_i(\xi)}{x_i(\xi)} &= \frac{1 - z_i(\eta)}{x_i(\eta)} = \gamma_{E_i}\\
\frac{z_i(\xi)^{-1} - 1}{y_i(\xi)} &= \frac{z_i(\eta)^{-1} - 1}{y_i(\eta)} = \gamma_{F_i}
\end{aligned} \tag{7.190}$$

with the four γ's constants, independent of the irreps. Recall furthermore that the six Casimirs $x_i(\xi)$, $y_i(\xi)$ and $z_i(\xi)$ are not all independent. Rather, they are subject to

$$z_0 z_1 = 1\,, \qquad x_0 y_0 = x_1 y_1 \tag{7.191}$$

which in turn implies that

$$\gamma_{E_0}\gamma_{F_0} = \gamma_{E_1}\gamma_{F_1} \tag{7.192}$$

Altogether, we have $4 - 3 = 1$ spectral parameter instead of the two required.

To introduce the still missing second free parameter, we need to extend the affine quantum group by two central elements C_i appearing only in the co-product:

$$\begin{aligned} \Delta' E_i &= E_i \otimes \mathbf{1} + C_i K_i \otimes E_i \\ \Delta' F_i &= F_i \otimes K_i^{-1} + C_i^{-1} \otimes F_i \\ \Delta' K_i &= K_i \otimes K_i \\ \Delta' C_i &= C_i \otimes C_i \end{aligned} \tag{7.193}$$

We shall call this central extension of $U_q(\widehat{s\ell}(2))$ $\tilde{U}_q(\widehat{s\ell}(2))$. A cyclic irrep ξ of $\tilde{U}$ is characterized by two more parameters, $c_i = \pi_\xi(C_i)$, in addition to x_i, y_i and z_i $(i = 0, 1)$. However, the existence of an intertwiner between two cyclic irreps of $\tilde{U}_q(\widehat{s\ell}(2))$ requires two more invariants associated with the elements E_0E_1 and F_0F_1, in addition to the suitably modified invariants (7.190). The six invariants are

$$\begin{aligned} \gamma_{E_i} &= \frac{1 - c_i^{N'} z_i}{x_i}\,, & \gamma_{F_i} &= \frac{z_i^{-1} - c_i^{-N'}}{y_i} \\ \gamma_{E_0E_1} &= \frac{1 - c_0 c_1}{\mathrm{tr}(E_0E_1)}\,, & \gamma_{F_0F_1} &= \frac{1 - c_0^{-1} c_1^{-1}}{\mathrm{tr}(F_0F_1)} \end{aligned} \tag{7.194}$$

Choosing $c_0 c_1 = 1$, we are finally left with two independent spectral parameters in the intertwiner $R(\xi, \eta)$ of two cyclic irreducible representations of the centrally extended quantum affine $s\ell(2)$. These two free parameters exist on the spectral curve Γ_γ determined by the values of γ_{E_i} and γ_{F_i}, and we still have to check that this manifold is the same as $\Gamma_\kappa \otimes \Gamma_\kappa$. In order to do this computation, we need an explicit parametrization of the cyclic irreps of $\tilde{U}_q(\widehat{s\ell}(2))$. If we restrict ourselves to the case of odd N, we label a cyclic irrep ξ with the six parameters $(a_0, a_1, b_0, b_1, c_0, c_1)$ as follows:

$$\begin{aligned} \pi_\xi(E_0) &= b_0 Z^{-1} \frac{a_0^2 X^{-1} - 1}{q - q^{-1}} & \pi_\xi(E_1) &= b_1 \frac{a_1^2 X - 1}{q - q^{-1}} Z \\ \pi_\xi(F_0) &= (a_0 a_1 b_0)^{-1} \frac{a_1^2 X - 1}{q - q^{-1}} Z & \pi_\xi(F_1) &= (a_0 a_1 b_0)^{-1} Z^{-1} \frac{a_0^2 X^{-1} - 1}{q - q^{-1}} \\ \pi_\xi(K_0) &= \frac{a_0}{q a_1} X^{-1} & \pi_\xi(K_1) &= \frac{q a_1}{a_0} X \\ \pi_\xi(C_0) &= c_0 & \pi_\xi(C_1) &= c_1 \end{aligned} \tag{7.195}$$

The relationship between these parameters and the ones used earlier on is given by

$$\begin{aligned} z_0 &= \left(\frac{a_0}{a_1}\right)^N & z_1 &= z_0^{-1} \\ x_0 &= b_0^N\left(a_0^{2N}-1\right) & x_1 &= b_1^N\left(a_1^{2N}-1\right) \\ y_0 &= (a_0a_1b_0)^{-N}\left(a_1^{2N}-1\right) & y_1 &= (a_0a_1b_0)^{-N}\left(a_0^{2N}-1\right) \end{aligned} \tag{7.196}$$

and $\omega^2 = q = e^{2\pi i/N}$. To reproduce the curve Γ_κ parametrized by (a,b,c,d) in (7.156), we first introduce the ratios

$$x = \frac{a}{b} \qquad y = \frac{b}{c} \qquad \mu = \frac{d}{c} \tag{7.197}$$

in terms of which Γ_κ is defined by

$$\begin{aligned} x^N + y^N &= \kappa\left(1 + x^N y^N\right) \\ \mu^N &= \frac{\kappa'}{1-\kappa x^N} = \frac{1-\kappa y^N}{\kappa'} \end{aligned} \tag{7.198}$$

with $\kappa^2 + \kappa'^2 = 1$. The map from the intertwiner manifold Γ_γ to the spectral manifold $\Gamma_\kappa \times \Gamma_\kappa$

$$\begin{aligned} \Gamma_\gamma \quad &\to \Gamma_\kappa \times \Gamma_\kappa \\ \xi = (a_0, a_1, b_0, b_1, c_0, c_1) &\mapsto (p, p') = (x, y, \mu;, x', y', \mu') \end{aligned} \tag{7.199}$$

is given by

$$a_0^2 = \frac{y'}{x\mu\mu'} \qquad a_1^2 = \frac{x'\mu\mu'}{y} \qquad c_0a_0a_1 = \frac{qx'}{y} \tag{7.200a}$$

$$b_0 = \kappa_0 x \qquad b_1 = \kappa_1 y \qquad c_1 = c_0^{-1} \tag{7.200b}$$

$$\kappa' = -\gamma_{E_0}\gamma_{F_0} \qquad \kappa_0^N = -\frac{\kappa}{\gamma_{E_0}} \qquad \kappa_1^N = -\frac{\kappa}{\gamma_{E_1}} \tag{7.200c}$$

The net result is that the matrix $R(\xi,\eta)$ in (7.183) satisfies the intertwiner equation (7.188) for the representation (7.195) under the parametrization (7.200).

The Boltzmann weights $W_{pq}(n)$ and $\overline{W}_{pq}(n)$ entering into the definition of the operators S_{pq} and T_{pq} take the following form in terms of the new variables x, y and μ:

$$\begin{aligned} \frac{W_{pq}(n)}{W_{pq}(0)} &= \prod_{j=1}^{n} \frac{\mu_p}{\mu_q}\,\frac{y_q - x_p\omega^j}{y_p - x_q\omega^j} \\ \frac{\widehat{W}_{pq}(n)}{\widehat{W}_{pq}(0)} &= \prod_{j=1}^{n} \frac{\mu_q y_p - \mu_p y_q \omega^{j-1}}{\mu_q x_q - \mu_p x_p \omega^j} \\ \frac{\overline{W}_{pq}(n)}{\overline{W}_{pq}(0)} &= \prod_{j=1}^{n} \mu_p\mu_q\,\frac{x_p\omega - x_q\omega^j}{y_q - y_p\omega^j} \end{aligned} \tag{7.201}$$

As above, the relationship between the deformation parameter q and the root of unity ω is $\omega = q^2$. These expressions are used more often than the original ones (7.153).

The conclusion of this brief overview of the chiral Potts model is that it admits a perfectly adequate description in the framework of (affine) quantum groups. The algebraic structure which is the subject of this book is thus capable of providing us with solutions to the Yang–Baxter equations which are not purely trigonometric, i.e. of genus zero. Let us stress that the higher genus surface exists on the space of irrep labels, and not on the space of the affine parameter of $U_q(\widehat{s\ell}_2)$, as is the case for the eight-vertex model. The quantum group structure of the latter is still somewhat of a mystery, but it is already clear that the difficulty does not arise exclusively from its genus one nature.

A more immediate, and perhaps more difficult, issue is how to exploit the quantum symmetry when describing the spectrum of the chiral Potts model. An important open question how to construct the corner transfer matrix for this model.

7.7 Solving the Yang–Baxter equation

In this and the previous chapter, we have constructed solutions to the Yang–Baxter equation with and without spectral parameters. The tool at work is the quantum group. We have seen that, if there exists a universal $\mathcal{R}$-matrix for the Hopf algebra, then the different $\mathcal{R}$-matrices come from evaluating the universal one in the tensor product of two representations. Similarly, the Yang–Baxter equation for these explicit solutions follows from the universal form (6.23) of the equation in the algebra. There are cases, however, for which the universal $\mathcal{R}$-matrix is not known. Even worse, it might not even exist if Drinfeld's double construction does not apply; one example of such a situation is provided by the chiral Potts model.

So, sometimes we may find ourselves with an $\mathcal{R}$-matrix which intertwines the tensor product of irreps of a Hopf algebra, but is not guaranteed to satisfy the Yang–Baxter equation because the universal $\mathcal{R}$-matrix from which it ought to derive is unknown or non-existent. In these situations, the following procedure, due to the Kyoto group, is quite helpful.

Suppose we consider a Hopf algebra $\mathcal{A}$ and a family of representations π_ξ of $\mathcal{A}$ in a finite-dimensional vector space V, parametrized by some $\xi \in S$, such that the following conditions apply.

(i) The tensor product of representations $\pi_{\xi_1} \otimes \pi_{\xi_2} \otimes \pi_{\xi_2}$ is indecomposable for generic ξ_i. This just means that if an element $f \in \mathrm{End}(V \otimes V \otimes V)$ satisfies

$$\left[f, \left(\pi_{\xi_1} \otimes \pi_{\xi_2} \otimes \pi_{\xi_2}\right) (\Delta^2 a)\right] = 0 \qquad (\forall a \in \mathcal{A}) \tag{7.202}$$

then F is a scalar (Schur's lemma).

(ii) There exists an intertwiner $R(\xi_1, \xi_2)$ among any two representations:

$$R(\xi_1, \xi_2)\ \left(\pi_{\xi_1} \otimes \pi_{\xi_2}\right)(\Delta a) = \left(\pi_{\xi_2} \otimes \pi_{\xi_1}\right)(\Delta a)\ R(\xi_1, \xi_2) \tag{7.203}$$

$(\forall a \in \mathscr{A},\ \forall \xi_1, \xi_2 \in S)$.

(iii) The intertwiner is normalized, $R(\xi, \xi) = \mathbf{1}$.

Under these conditions, the intertwiner $R(\xi_1, \xi_2)$ is bound to satisfy the Yang–Baxter equation

$$\begin{aligned}(R(\xi_2, \xi_3) \otimes \mathbf{1})(\mathbf{1} \otimes R(\xi_1, \xi_3))(R(\xi_1, \xi_2) \otimes \mathbf{1}) \\ = (\mathbf{1} \otimes R(\xi_1, \xi_2))(R(\xi_1, \xi_3) \otimes \mathbf{1})(\mathbf{1} \otimes R(\xi_2, \xi_3))\end{aligned} \tag{7.204}$$

The proof of this theorem is left as an exercise for the reader. To clarify its content, though, several remarks are necessary.

Let us first illustrate the statement that not every intertwining operator (intertwiner for short) satisfies a Yang–Baxter equation. For example, if we took the usual $\mathscr{A} = U_q(s\ell(2))$, all the matrices $\left(\mathscr{R}^{(\frac{1}{2},\frac{1}{2})}\right)^n$ with n a positive integer are intertwiners of $\frac{1}{2} \otimes \frac{1}{2}$, but only $\mathscr{R}^{(\frac{1}{2},\frac{1}{2})}$ and its inverse satisfy the Yang–Baxter equation without spectral parameter. The point here is that the centralizer of the spin-$\frac{1}{2}$ irrep is not trivial or, in other words, that the tensor product $\frac{1}{2} \otimes \frac{1}{2}$ is decomposable. A related problem is the construction of an R-matrix satisfying the Yang–Baxter equation *with* spectral parameter starting from an R-matrix satisfying the Yang–Baxter equation *without* spectral parameter. This is Jones's "baxterization" of R-matrices, discussed briefly in appendix D.

From the above comment, it is clear that the Kyoto theorem requires affine Hopf algebras to be the essential ingredient to produce solutions of the Yang–Baxter equation (with spectral parameters). Indeed, any tensor product of two irreducible representations of an affine Hopf algebra is indecomposable, and thus the solution to (*ii*) is unique.

Exercises

Ex. 7.1 Check that the condition (7.38) is not preserved by the co-adjoint action of section 7.1.3.

Ex. 7.2 Check that the nilpotent R-matrix (7.63) yields

$$R^{0r}_{r0}(\lambda_1, \lambda_2 | u) = \prod_{j=0}^{r-1} \frac{\mathrm{e}^u \lambda_1 \epsilon^{-j} - \mathrm{e}^{-u} \lambda_2 \epsilon^j}{\mathrm{e}^u \lambda_1 \lambda_2 \epsilon^{-j} - \mathrm{e}^{-u} \epsilon^j} \tag{7.205}$$

When $\epsilon^3 = 1$, verify that the hamiltonian derived from (7.63) is indeed (7.78).

Ex. 7.3 Show that the hamiltonians (7.83) are equivalent to a Hubbard model with infinite Coulomb repulsion (i.e. with dynamically incorporated Pauli exclusion principle).

Ex. 7.4 The models at the orbifold points, with hamiltonians (7.83), belong to a family of generalized six-vertex models with lattice variables σ in either a "singlet" ($\sigma = 0$) or a multiplet ($\sigma = 1, 2, \ldots, \mathcal{N}$) transforming as the fundamental irrep of $U(\mathcal{N})$. The non-vanishing Boltzmann weights are

$$\begin{aligned} &\mathcal{R}_{00}^{00}(u) = \mathcal{R}_{ab}^{ba}(u) = \sinh(u + i\gamma) \qquad a, b \in \{1, 2, \ldots, \mathcal{N}\} \\ &\mathcal{R}_{0a}^{0a}(u) = \mathcal{R}_{a0}^{a0}(u) = \sinh u \\ &\mathcal{R}_{0a}^{a0}(u) = \mathcal{R}_{a0}^{0a}(u) = i \sin \gamma \end{aligned} \tag{7.206}$$

When $\mathcal{N} = 1$, this becomes the XXZ model. For $\mathcal{N} = 2$ and $\gamma = \pi/2$, we recover from this $\mathcal{R}$-matrix the hamiltonian (7.83) without the term $\sum \sigma_j^0$, which can be added by hand anyway since it commutes with all the other terms of the hamiltonian. Show that the hamiltonian H associated with the $\mathcal{R}$-matrix (7.206), which acts on the space $\otimes^L \mathbb{C}^{1+\mathcal{N}}$, is given by

$$H_{\mathcal{N},\Delta} = \frac{1}{\sin \gamma} \sum_{j=1}^{L} \left[\sum_{a=1}^{\mathcal{N}} \left(\tau_{a,j}^{+} \tau_{a,j+1}^{-} + \tau_{a,j}^{-} \tau_{a,j+1}^{+} \right) + \Delta \left(\tau_j^z \tau_{j+1}^z - 1 \right) \right] \tag{7.207}$$

where $\tau_a^{\pm}$ are the Pauli matrices $\sigma^{\pm}$ acting on the labels 0 and a:

$$\left(\tau_a^{+} \right)_{mn} = \delta_m^0 \, \delta_n^a \,, \qquad \tau_a^{-} = \left(\tau_a^{+} \right)^{\dagger} \tag{7.208}$$

and τ^z is a "global" σ^z Pauli matrix

$$(\tau^z)_{mn} = 2\delta_m^0 \, \delta_n^0 - \delta_m^n \tag{7.209}$$

not to be confused with $\tau_a^z = \left[\tau_a^{+}, \tau_a^{-} \right]$. Show that $\sum \tau_j^z$ commutes with $H_{\mathcal{N},\Delta}$, and that it can be added to the hamiltonian as the contribution from an external homogeneous magnetic field.

Ex. 7.5 In terms of the representations, it is easy to perform the chain of reductions (specializations) cyclic → semi-cyclic → nilpotent. Does this reduction carry over to the intertwiner R-matrices? To the hamiltonians? A particularly interesting subfamily of nilpotent hamiltonians arises for the regular spin $\lambda = \epsilon^{-1}$. Show that in this case expression (7.84) simplifies somewhat, and compare it with the chiral Potts hamiltonian (7.148) with $\phi = 0$ and $\overline{\phi} = \pi/2$. What is the physical significance of this map?

Ex. 7.6 Verify the conjecture (7.135) for a few simple examples, such as the unlink and the trefoil.

Ex. 7.7 [L. Rozansky and H. Saleur, 'Quantum field theory for multi-variable Alexander–Conway polynomial', *Nucl. Phys.* **B376** (1992) 461.]

The quantum super-group $U_q(g\ell(1,1))$ is generated by two bosonic operators, $\mathcal{N}$ and $\mathcal{E}$, and two fermionic ones, b and $b^\dagger$, subject to the relations

$$[\mathcal{N}, b^\dagger] = b^\dagger\,, \quad [\mathcal{N}, b] = -b\,, \quad \{b, b^\dagger\} = \frac{q^{\mathcal{E}} - q^{-\mathcal{E}}}{q - q^{-1}} \tag{7.210}$$

and $\mathcal{E}$ central. This is the fermionic version of (7.123), whereas the co-product of $U_q(g\ell(1,1))$ is just like (7.124). Construct the universal $\mathcal{R}$-matrix of this algebra, and specialize it to the fermionic "harmonic oscillator" irreps (e,n) of dimension two. From this R-matrix, define a link invariant and show that it is the Alexander–Conway polynomial.

Ex. 7.8 Consider the special case of the $\mathbb{Z}_N$ chiral Potts model, defined by the graph in figure 7.2, with $N = 3$. This graph is exactly the Coxeter graph D_4. What is the relationship between the $N = 3$ chiral Potts model and Pasquier's D_4 graph face model?

Ex. 7.9 Find the one-dimensional hamiltonian associated with the chiral Potts model by expanding the transfer matrix about the point in $\Gamma_\kappa \times \Gamma_\kappa$ where the Boltzmann weights are the identity. Carry out the expansion to second order, and investigate the second hamiltonian structure thus derived. Does the Reshetikhin criterion (exercise 2.8) apply to this case?

Ex. 7.10 Check that the Boltzmann weights (7.153) for the chiral Potts model satisfy the Yang–Baxter equation (7.151) provided all the rapidities p, q and r lie on the same spectral manifold Γ_κ (7.156). In order to do so, perform first the Fourier transform of the weights (7.153) with respect to σ_3, and introduce

$$\begin{aligned} V_{ab} &= \sum_{k=1}^{N} \omega^{bk}\, W_{pr}(a+k)\overline{W}_{qr}(k) \\ \overline{V}_{ab} &= \sum_{k=1}^{N} \omega^{ak}\, \overline{W}_{pr}(k) W_{qr}(b+k) \end{aligned} \tag{7.211}$$

The proof of (7.151) follows from the remark that the product formulae in (7.153) are equivalent to a pair of linear recurrence relations for both V_{ab} and $\overline{V}_{ab}$.

Ex. 7.11 Show that the R-matrix of the six-vertex model satisfies Kyoto's criterion, given in section 7.7. This proves that R^{6v} is a solution to the Yang–Baxter equation.

8

Two-dimensional conformal field theories

8.1 Introduction: critical phenomena

The main method of discovering the various phases in which a quantum system may exist is the study of the system's behavior under scale transformations, which are described by the renormalization group equations. In particular, the fixed points of the scale transformation correspond to the critical points where, since the correlation length becomes infinite, the system enjoys invariance under dilatations.

The renormalization group approach to critical phenomena allows for the explicit computation of critical exponents. These were originally introduced as empirical fitting parameters to describe the scaling behavior of physical magnitudes in the neighborhood of the critical point. Universality means that the critical exponents depend only on the dimensionality and the symmetries of the system. From the point of view of the renormalization group, the critical exponents are just the weights of representations of the dilatation group. Associated with each universality class, characterized by a set of critical exponents, is a set of representations of the dilatation group, whose weights are precisely the above critical exponents. Two systems are in the same universality class if they are related by a renormalization group transformation.

For quantum systems that are invariant under translations and rotations (i.e. homogeneous and isotropic), and with short-range interactions, dilatation invariance implies, in general, invariance under the bigger group of conformal transformations. Therefore, its critical behavior will be described by a quantum field theory invariant under the whole conformal group. This observation, originally due to Polyakov, has dramatic consequences if the dimensionality of the system is two: in this case, the conformal group is infinite dimensional. Before discussing the monumental conformal field theory laid out by Belavin, Polyakov and Zamolodchikov in 1984, we will briefly describe the renormalization group approach to critical phenomena and the meaning of the critical exponents, following the ideas of Kadanoff and Wilson.

8.2 Renormalization group

According to Landau, phase transitions are either of first or second order, depending on how the correlation length ξ behaves when the system approaches the phase transition. If $\xi \to \infty$, the phase transition is of second order; we shall concentrate on this type of transition.

The correlation length ξ is defined in terms of the asymptotic behavior of the two-point correlation function as follows:

$$\Gamma(x_1, x_2)_{|x_1-x_2|\to\infty} \sim \exp\left[-\frac{|x_1 - x_2|}{\xi(g_i, T, \ldots)}\right] \tag{8.1}$$

For a second order phase transition, $\xi \to \infty$, and hence the mass $m = 1/\xi \to 0$, and a massless Goldstone boson appears: the signature of a collective phenomenon with symmetry breaking.

One of the most popular examples of a phase transition is the appearance of spontaneous magnetization in iron (or any other ferromagnet) at temperatures below the Curie point. Denoting by h the external magnetic field, and by $F(h, T)$ the free energy in the thermodynamic limit, the magnetization is defined by

$$M(h, T) = -\frac{\partial F(h, T)}{\partial h} \tag{8.2}$$

For temperatures lower than a critical temperature T_c, the system exhibits spontaneous magnetization:

$$M_0(T) = \lim_{h\to 0^+} M(h, T) \neq 0 \qquad \text{iff} \qquad T < T_c \tag{8.3}$$

A graphical representation of $M(h, T)$ is shown in figure 8.1.

The general behavior of the order parameter $M_0(T)$ around the critical point is well described empirically by a scaling function of the temperature:

$$M_0 \propto (T_c - T)^\beta \tag{8.4}$$

where β is the critical exponent characteristic of the spontaneous magnetization. Other critical exponents exist in the model. An obvious one is associated with the correlation length, whose dependence on the temperature near the phase transition (see figure 8.2) can also be parametrized by a scaling function:

$$\xi(T) \propto (T - T_c)^{-\nu} \tag{8.5}$$

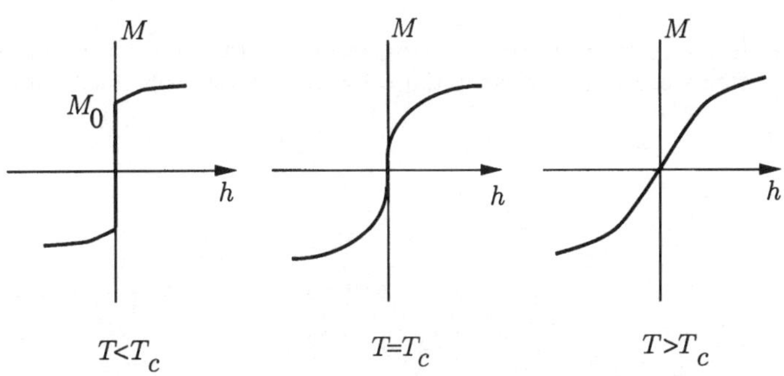

Fig. 8.1 Dependence of magnetization on temperature and external magnetic field.

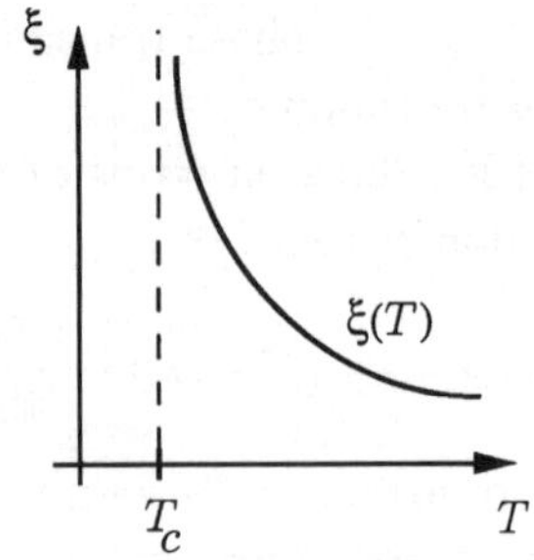

Fig. 8.2. Temperature dependence of the correlation length.

In standard notation, the critical exponents for the specific heat $C = \partial F/\partial T$ and the susceptibility $\chi = \partial M/\partial h$ are

$$C \sim (T - T_c)^{-\alpha}\,, \qquad \chi \sim (T - T_c)^{-\gamma} \tag{8.6}$$

In general, to describe a second order phase transition, we need a set of critical exponents $\{\alpha, \beta, \gamma, \nu, \ldots\}$, the values of which are yielded by experiment and should be predicted by theory. (In passing, note that the critical exponents above are defined by the behavior of the system when $T \to T_c$ $(T > T_c)$; other critical exponents may be defined for $T \to T_c$ $(T < T_c)$. The two values coincide in all known cases, and that is why we shall use only one critical exponent for each physical magnitude.)

A very powerful idea underlies the "universality conjecture": the critical exponents characterizing a phase transition depend only on the spatial dimensionality of the system and its internal symmetry. According to their critical behavior, different systems are thus assembled into universality classes. The critical exponents of the Ising model's universality class in two and three dimensions are shown in table 8.1.

Table 8.1 Critical exponents for the Ising model in two and three dimensions. The $D = 2$ numbers are exact, whereas those for $D = 3$ were obtained with computer simulations.

	$D = 2$	$D = 3$
β	0.125	0.32
ν	1	0.63
γ	1.75	1.24
α	0	0.11

How can we compute these critical exponents? Taking up the ideas of Kadanoff, Wilson developed the renormalization group in the 1970s. The great achievement of the renormalization group approach to critical phenomena was to provide a method for computing critical exponents. The steps in Wilson's renormalization group technique are given in the following.

(i) Start from the hamiltonian of the system $H(g_i, T, \ldots)$ depending on the couplings, temperature, etc.
(ii) Integrate over those fluctuations of the fields in H that are smaller than some spatial cut-off a. In momentum space, integrate out momenta $k \geq a^{-1}$. Thereby obtain an effective hamiltonian H' for the remaining field components.
(iii) Rewrite the effective hamiltonian H' in the same form as the original H, i.e. $H' = H(g_i', \ldots)$. From this, derive the dependence of the new couplings on the old ones $g_i' = g_i'(g_j)$, not necessarily diagonal.
(iv) Repeat step (2) above with a larger cut-off $a' = \lambda a$, and thereby obtain a new effective hamiltonian $H'' = H(g'', \ldots)$. Thus derive the general scaling behavior of the couplings under $a \to \lambda a$: $g_i \to f_\lambda^{(i)}(g_1, g_2, \ldots)$. These scaling operations form the renormalization group.

8.3 Examples

To establish our ideas about the block spin renormalization just presented, let us consider two examples: the one-dimensional Ising model and the gaussian model.

8.3.1 The one-dimensional Ising model

We shall begin by studying the one-dimensional Ising model on a regular lattice by renormalization group methods, and then briefly describe the exact solution to the model.

The spin variables σ_i may take the values ± 1, and the lattice spacing is a:

$$\begin{array}{ccccccccccccccc} \sigma_1 & & \sigma_2 & & \sigma_3 & & & & \sigma_i & & \sigma_{i+1} & & & & \sigma_n \\ \bullet & \underset{a}{-} & \bullet & \underset{a}{-} & \bullet & \underset{a}{-} \cdots & \underset{a}{-} & & \bullet & \underset{a}{-} & \bullet & \underset{a}{-} \cdots & \underset{a}{-} & & \bullet \end{array} \tag{8.7}$$

The partition function is

$$Z = \sum_{\substack{\text{spin} \\ \text{configurations}}} \exp\left[-K \sum_i \sigma_i \sigma_{i+1}\right] \tag{8.8}$$

where the overall parameter is just $K = J/kT$. The correlation function for the spin variables at different sites of the lattice is defined by the expectation value of the two-point function, $\Gamma(x) = \langle \sigma(x)\sigma(0) \rangle - \langle \sigma(x) \rangle \langle \sigma(0) \rangle$.

It is appropriate to view the model as a zero-dimensional system propagating in time. The transfer matrix $T_{\sigma_i\sigma_{i+1}} = \mathrm{e}^{-K\sigma_i\sigma_{i+1}}$ is simply

$$T = \begin{pmatrix} \mathrm{e}^{-K} & \mathrm{e}^{K} \\ \mathrm{e}^{K} & \mathrm{e}^{-K} \end{pmatrix} \tag{8.9}$$

Let us now integrate out the even variables, i.e. scale $a \to 2a$. Following the ideology of the renormalization group, the new effective hamiltonian H' is given by

$$\mathrm{e}^{-H'} = \sum_{\sigma_2\sigma_4\cdots} \mathrm{e}^{-H} = \sum_{\sigma_2\sigma_4\cdots} \exp\left[-K\sum_i \sigma_i\sigma_{i+1}\right] \tag{8.10}$$

which reads, in terms of the transfer matrix, as

$$\begin{aligned} \mathrm{e}^{-H'} &= \sum_{\sigma_2\sigma_4\cdots} \langle\sigma_1|\,T\,|\sigma_2\rangle\,\langle\sigma_2|\,T\,|\sigma_3\rangle\cdots\langle\sigma_i|\,T\,|\sigma_{i+1}\rangle\cdots \\ &= \langle\sigma_1|\,T^2\,|\sigma_3\rangle\,\langle\sigma_3|\,T^2\,|\sigma_5\rangle\cdots \end{aligned} \tag{8.11}$$

Now we must relabel the couplings in such a way that the new transfer matrix T^2 is of the same form as the original one T. Although we started with only one coupling K, there is actually another parameter at our disposal, a constant shift of the hamiltonian representing the vacuum energy. We started with zero shift, but, in general, we may add or subtract a constant $\mu_\circ$ to the hamiltonian without changing the physics involved. Borrowing terminology from general relativity, we shall call $\mu_\circ$ the cosmological constant.

We thus write

$$T^2(K) = \mathrm{e}^{\mu_\circ}T(K') = \mathrm{e}^{\mu_\circ}T(f_2(K)) \tag{8.12}$$

where in the general scaling function $K \to f_\lambda(K)$ we highlight the fact that the scaling here is by a factor of two. The solution to (8.12) is given by

$$f_2(K) = \frac{1}{2}\log(\cosh 2K)\,, \qquad \mu_\circ = f_2(K) + \log 2 \tag{8.13}$$

The renormalization group equations (8.13) for the system yield the dependence of the coupling constant K on the renormalization point, i.e. on the lattice spacing a: the coupling constant changes from K to $f_\lambda(K)$ under the scaling $a \to \lambda a$.

Consider now the system at the critical point. At this point, $\xi \to \infty$, and thus it is immaterial whether the lattice spacing is a or λa. This means that the system at criticality is invariant under dilatations, i.e. it is scale-invariant. The critical point is thus characterized as a fixed point of the renormalization group equation:

$$K_{\text{critical}} = K^* = f_\lambda(K^*) \tag{8.14}$$

We are now in a position to compute the critical exponent ν of the correlation length, parametrized by the coupling K:

$$\xi \sim (K - K^*)^{-\nu} \tag{8.15}$$

Quite generally, the correlation length changes as $\xi \to \xi/\lambda$ when $a \to \lambda a$, i.e. $\xi(f_\lambda(K))$ is just the original correlation length $\xi(K)$ before the scaling, but measured with respect to the new lattice spacing λa. Therefore,

$$\begin{aligned} \xi(f_\lambda(K)) &= \frac{1}{\lambda}\xi(K) = \frac{1}{\lambda}(K-K^*)^{-\nu} \\ &= \left(f_\lambda(K) - f_\lambda(K^*)\right)^{-\nu} = \left(f_\lambda(K) - K^*\right)^{-\nu} \\ &= \left[(K-K^*) f'_\lambda(K)|_{K=K^*}\right]^{-\nu} \end{aligned} \tag{8.16}$$

where we have carried out a Taylor expansion. Hence,

$$\nu = \frac{\log \lambda}{\log f'_\lambda(K^*)} \tag{8.17}$$

This result serves to illustrate the significance of critical exponents as scaling dimensions. In the continuum limit, we shall interpret these parameters as the scaling dimensions of local fields.

For exactly solvable models of the kind studied in previous chapters, the critical exponents can be computed exactly. Even when exact solutions to models are not known, it is always possible to use the renormalization group approach to find the critical exponents, as a perturbative series. We shall next compute explicitly the critical exponent ν for the one-dimensional Ising model, for completeness and comparison. The reader may skip these remarks and proceed to the next example without loss of continuity.

Allowing for an external magnetic field h, the partition function is

$$Z_L(K,H) = \sum_\sigma \exp\left[-K\sum_{i=1}^{L} \sigma_i\sigma_{i+1} - H\sum_{i=1}^{L}\sigma_i\right] \tag{8.18}$$

with $K = J/kT$, $H = h/kT$. We impose periodic boundary conditions $\sigma_{L+1} = \sigma_1$, so actually we consider the Ising model on a circle with L sites. The thermodynamic limit corresponds to $L \to \infty$.

In terms of the transfer matrix

$$T(\sigma_i\sigma_{i+1}) = \exp\left[-K\sigma_i\sigma_{i+1} - H\frac{\sigma_i + \sigma_{i+1}}{2}\right] \tag{8.19}$$

the partition function can be written as

$$Z_L(K,H) = \mathrm{Tr}(T^L) \tag{8.20}$$

The transfer matrix is a simple 2×2 matrix (compare with (8.9) above, where $H = 0$)

$$T = \begin{array}{c} \\ + \\ - \end{array} \begin{array}{c} \begin{array}{cc} \quad + & \qquad - \end{array} \\ \begin{pmatrix} e^{-K-H} & e^{K} \\ e^{K} & e^{-K+H} \end{pmatrix} \end{array} \tag{8.21}$$

which can be diagonalized without much effort:

$$T = M \begin{pmatrix} \lambda_1 & \\ & \lambda_2 \end{pmatrix} M^{-1}, \qquad M = \begin{pmatrix} \cos\phi & -\sin\phi \\ \sin\phi & \cos\phi \end{pmatrix} \tag{8.22}$$

where

$$\cot 2\phi = e^{2K} \sinh H \tag{8.23}$$

The partition function is thus

$$Z_L(K,H) = \lambda_1^L + \lambda_2^L \tag{8.24}$$

The free energy per site in the thermodynamic limit

$$\frac{-1}{kT} F(K,H) = \lim_{L\to\infty} \frac{1}{L} \log Z_L(K,H) = \log \lambda_1 \tag{8.25}$$

depends only on the largest eigenvalue $\lambda_1 > \lambda_2$.

We can now compute the two-point correlation function $g_{ij} =< \sigma_i\sigma_j > - < \sigma_i >< \sigma_j >$. After some simple manipulations, we obtain

$$g_{ij} = \sin^2 2\phi \left(\frac{\lambda_2}{\lambda_1}\right)^{|j-i|} \tag{8.26}$$

where $|j-i|$ is the distance in lattice units between the two lattice sites i and j. Equation (8.26) displays the exponential decay of the correlation function with $|j-i|$:

$$\xi \sim \left(\log \lambda_1/\lambda_2\right)^{-1} \tag{8.27}$$

To obtain the critical exponent, we should consider the behavior of ξ at the critical point, which is defined by the condition $\xi \to \infty$. According to equation (8.27), this corresponds to $\lambda_1 = \lambda_2$. Using (8.21) and (8.22), this amounts in turn to the limit $H = 0$, $T = 0$. In fact, when $H = 0$, we have

$$\lambda_1 = 2\cosh K\,, \qquad \lambda_2 = 2\sinh K \tag{8.28}$$

and therefore $\lambda_1/\lambda_2 = (1+t)/(1-t)$, with $t = \exp(-2K)$. Close to the critical point $t << 1$ ($T \simeq 0$), the correlation function goes like

$$\xi \sim (2t)^{-1} \tag{8.29}$$

and therefore the critical exponent ν is exactly equal to 1.

8.3.2 The gaussian model

Let us now study a simple example from quantum field theory, namely a free scalar field with mass $m_\circ$ and hamiltonian

$$H = \mu_\circ + \frac{1}{2}\int_{|k|<a^{-1}} d^D k (m_\circ^2 + k^2)|\phi(k)|^2 \tag{8.30}$$

where an ultra-violet momentum cut-off a^{-1} has been assumed and $\phi(k)$ are the Fourier components of the field ϕ. The partition function is just $Z = \int d\phi \, \mathrm{e}^{-H}$.

Let us proceed with the renormalization program sketched above and integrate out the fluctuations with $a^{-1} > k > (\lambda a)^{-1}$. This is very easy in the free case we are considering because there is no coupling among Fourier components with different momenta. The effective hamiltonian is given by

$$\mathrm{e}^{-H'} = \int_{\lambda a > |k|^{-1} > a} d\phi(k) \, \mathrm{e}^{-H} \tag{8.31}$$

and thus

$$H' = \mu'_\circ + \frac{1}{2} \int_{|k| < (\lambda a)^{-1}} d^D k \, \left(m_\circ^2 + k^2\right) |\phi(k)|^2 \tag{8.32}$$

In order to recast H' in the same form as H, we first have to scale $k \to k' = \lambda k$ so that the boundary condition recovers its previous form:

$$H' = \mu'_\circ + \frac{1}{2} \int_{|k'| < a^{-1}} d^D k' \, \lambda^{-D} \left(m_\circ^2 + (k')^2 \lambda^{-2}\right) \left|\phi(\lambda^{-1} k')\right|^2 \tag{8.33}$$

Next, we need to renormalize the field variables:

$$\phi \to \phi'(k') = \lambda^{-(2+D)/2} \phi(k) \tag{8.34}$$

Finally, we renormalize the mass term by $m_\circ^2 \to \lambda^2 m_\circ^2$. The effective hamiltonian resulting from these renormalization group operations is exactly of the form (8.30):

$$H' = \mu'_\circ + \frac{1}{2} \int_{|k'| < a^{-1}} d^D k' \, \left((m'_\circ)^2 + (k')^2\right) |\phi'(k')|^2 \tag{8.35}$$

The transformation laws for the mass and the fields are thus

$$\begin{aligned} m_\circ &\longrightarrow m'_\circ = f_\lambda(m_\circ) = \lambda m_\circ \\ \phi(k) &\longrightarrow \phi'(k') = \lambda^{-(2+D)/2} \phi(k) \end{aligned} \tag{8.36}$$

On constructing the field ϕ in x space via a Fourier transformation

$$\phi(x) = \int \frac{d^D k}{(2\pi)^D} \, \mathrm{e}^{ik \cdot x} \phi(k) \tag{8.37}$$

we deduce that the transformation (8.34) implies

$$\phi(x) \longrightarrow \phi'(x') = \lambda^{d_\phi} \phi(x) \tag{8.38}$$

where $x' = x/\lambda$, and $d_\phi = (D-2)/2$ is the "anomalous dimension" of the field ϕ.

At criticality, the system sits at a fixed point of the renormalization group equations (8.36), and hence it is described by field variables satisfying the scaling law

$$\phi(x/\lambda) = \lambda^{d_\phi} \phi(x) \tag{8.39}$$

The anomalous dimensions d_ϕ are essentially the same as the critical exponents in the previous example.

8.4 Operator algebra of a universality class

The previous two examples indicate the general picture. We have a set of parameters g_i characterizing our model: $(m_\circ, \mu_\circ)$ for the gaussian model; $(\mu_\circ, K, H)$ for the one-dimensional Ising model in the presence of an external magnetic field, etc. We associate with each point in the space of parameters a set of local operators $O_i(x)$ playing the role of conjugate extensive variables. A formal hamiltonian density can thus be defined as $\mathscr{H}(x) = \sum g_i O_i(x)$. Using the renormalization group as in the above examples, we can always find the scaling behavior of the couplings g_i under a dilatation $x \to x/\lambda$ (recall that $a \to \lambda a$); this transformation law induces the following change in the hamiltonian:

$$\mathscr{H} = \sum g_i O_i(x) \to \mathscr{H}' = \sum f^i_\lambda(g_j) O'_i(x) \to \cdots \tag{8.40}$$

The transformation law of the local operators $O_i(x)$ is determined by dimensional analysis with the requirement that $\mathscr{H}$ be a well-defined density. We may always diagonalize the renormalization group transformations and find a good (orthogonal under the renormalization group) basis of operators and conjugate couplings. Assume from now on that this diagonalizaton has been carried out.

At criticality, the hamiltonian $\int d^D x \mathscr{H} = H \equiv H^* \to H^*$, and $g^*_i \to g^*_i = f^i_\lambda(g^*_i)$. Near the critical point, we may write $g_i = g^*_i + \delta g_i$, and under a scale transformation we may Taylor-expand to obtain $g_i \to g^*_i + f'^i_\lambda(g^*_i)\delta g_i$. The renormalization group equations allow us to compute the power law behavior of δg_i, i.e. the exponents y_i in $\delta g_i \to \lambda^{y_i}\delta g_i$. Then, since the hamiltonian is invariant, the extensive variables (operators) transform as $O_i \to \lambda^{D-y_i} O_i$. The numbers $d_i = D - y_i$ are known as the critical exponents of the system. We summarize the discussion above in the formulae

$$\begin{aligned} x &\longrightarrow x' = \frac{x}{\lambda} \\ H^* &\longrightarrow H^* \\ \delta g_i &\longrightarrow \lambda^{y_i}\, \delta g_i \\ \mathcal{O}_i(x) &\longrightarrow \mathcal{O}'_i(x') = \lambda^{d_i} \mathcal{O}_i(x)\,, \quad d_i = D - y_i \end{aligned} \tag{8.41}$$

At each second order critical point, the system is characterized by a set of operators O_i scaling according to the "critical exponents" d_i. At the critical point, only those couplings with exponents $y_i > 0$ survive because the couplings scale with exponents $dg_i/d\log\lambda = y_i g_i$, so that $g_i \sim \lambda^{y_i}$. Accordingly, the corresponding operators O_i are classified by their behavior at criticality as either relevant operators ($y_i > 0$), irrelevant operators ($y_i < 0$), or else marginal operators ($y_i = 0$).

The identity operator **1**, conjugate to an overall shift of the hamiltonian density (the cosmological constant), is always included among the operators describing a system at criticality.

8.5 Conformal invariance and statistical mechanics

Let us start by recalling the connection between classical statistical mechanics and quantum field theory. In classical statistical mechanics, the object of interest is the partition function

$$Z = \sum_{\text{configurations}} \exp[-E(c)/kT] \tag{8.42}$$

where $E(c)$ is the classical energy of the configuration c. The analog in quantum field theory is the generating functional for Green functions, Wick-rotated to euclidean spacetime

$$Z = \int d\phi \exp[-S(\phi)/\hbar] \tag{8.43}$$

where $S(\phi)$ is the euclidean action. The main difference between (8.42) and (8.43) is that in quantum field theory the degrees of freedom are continuous fields instead of discrete spin variables, as in classical statistical mechanics. Similarly, sources in quantum field theory are external magnetic fields in classical statistical mechanics, whereas vacuum expectation values in quantum field theory are magnetizations in classical statistical mechanics. The logarithm of Z is the generating functional for connected Green functions in quantum field theory and the free energy in classical statistical mechanics, and its Legendre transform is the generating functional for one-particle irreducible Green functions in quantum field theory, and the thermodynamic potential in classical statistical mechanics.

To connect (8.42) and (8.43), we must first consider the lattice regularization of (8.43). This is achieved by putting the system on a lattice of some definite spacing a which plays the role of an ultra-violet cut-off. The lattice regularization of the euclidean quantum field theory defines a particular statistical system, the transfer matrix of which is derived from the quantum hamiltonian by

$$T = \mathrm{e}^{-aH} \tag{8.44}$$

The continuum limit $a \to 0$ can be obtained, with the help of renormalization group techniques, as the critical point (correlation length $\xi \to \infty$) of this statistical sytem. This limit defines a euclidean quantum field theory of massless fields.

Let us compare the critical point of a classical statistical mechanical model (8.42) with the critical point, i.e. the continuum limit, of the statistical model defined by lattice regularization of a euclidean quantum field theory (8.43). In principle, the two statistical systems are very different from a microscopic point of view. As a consequence of the universality conjecture, however, the critical behavior depends only on the dimensionality and symmetries, and is quite independent of the microscopic details. The critical behavior of each universality class presumably corresponds, therefore, to some euclidean quantum field theory of massless fields. Moreover, the classification of all possible universality classes should be equivalent to the classification of euclidean quantum field theories of massless fields.

A euclidean quantum field theory containing only massless fields is necessarily invariant under scale transformations $x^\mu \to \lambda x^\mu$. It was Polyakov's idea to gauge the scale invariance of systems at a second order phase transition, i.e. to allow for local (as opposed to only global) scale invariance. It should be emphasized that this hypothesis is not a theorem. We shall work out the consequences of the assumed conformal symmetry (gauged dilatation invariance) in the powerful context of local field theory: it does come out rather well, although we do not really understand why dilatation invariance becomes gauged at a second order phase transition in the first place.

To measure the response of the field theoretical system under general co-ordinate transformations $x^\mu \to x^\mu + \alpha^\mu(x)$, we define the energy–momentum tensor $T_{\mu\nu}(x)$ by

$$\delta S = \frac{-1}{2\pi} \int T_{\mu\nu}(x) \partial^\mu \alpha^\nu(x) d^2x \tag{8.45}$$

where S is the action of the field theory appearing in (8.43). Invariance under scale transformations implies that $T_{\mu\nu}(x)$ is traceless:

$$T_\mu{}^\mu = 0 \tag{8.46}$$

In local field theory, this condition is equivalent to invariance under the whole conformal group, i.e. changes of co-ordinates such that the metric tensor scales:

$$g_{\mu\nu} \to \rho(x) g_{\mu\nu} \qquad \text{when} \qquad x^\mu \to \omega^\mu(x) \tag{8.47}$$

For two-dimensional systems, this symmetry is particularly interesting, because the conformal group is infinite-dimensional. Conformal invariance thus produces an infinite number of conservation laws, and at criticality we expect the theory to be integrable.

Since we are only considering two-dimensional systems, the problem of classification of all universality classes reduces to the one of classifying all possible two-dimensional quantum conformal field theories. In order to do this, we must discover the quantum field theoretical meaning of the critical exponents. Before doing so, though, we shall briefly describe some elementary properties of two-dimensional field theories that are invariant under the whole conformal group.

8.6 The two-dimensional conformal group

Recall first some properties of the conformal group in two dimensions. If we introduce complex co-ordinates $z = x^1 + ix^2$, $\overline{z} = x^1 - ix^2$, the conformal group is generated by holomorphic and antiholomorphic transformations

$$z \to \omega(z)\,, \qquad \overline{z} \to \overline{\omega}(\overline{z}) \tag{8.48}$$

all of which leave the line element invariant. The energy–momentum tensor has only two independent components, $T_{zz} = T$ and $T_{\bar{z}\bar{z}} = \overline{T}$ ($T_{z\bar{z}} = T_{\bar{z}z} = \frac{1}{4}T_\mu{}^\mu = 0$), and from the conservation law $\partial^\mu T_{\mu\nu} = 0$ we obtain

$$\partial_{\bar{z}} T = \partial_z \overline{T} = 0 \tag{8.49}$$

Let us stress here that $\overline{T} \neq T^\dagger$, $\overline{\omega} \neq \omega^\dagger$, and quite generally the antiholomorphic quantities are independent of the holomorphic ones.

Take as spacetime a cylinder $S^1 \times \mathbb{R}$, understood to encode periodic spatial boundary conditions, and introduce orthogonal co-ordinates $(\sigma, \tau) \in ([0, 2\pi], \mathbf{R})$. In terms of the light-cone co-ordinates $x^\pm = \tau \pm \sigma$, the Minkowski line element is $ds^2 = dx^+ dx^-$. To form a link with classical statistical mechanics, rotate to euclidean spacetime:

$$x^\pm = \tau \pm \sigma \longrightarrow -i(\tau \pm i\sigma) \tag{8.50}$$

This Wick rotation defines complex co-ordinates $w(\overline{w}) = (\tau \pm i\sigma)$ on the cylinder. With the help of the analytic mapping $z = \mathrm{e}^w$, we can map our system from the cylinder to the complex plane (see figure 8.3). The induced metric on the z plane is the flat one: $ds^2 = dz d\bar{z}$. Surfaces at equal time on the cylinder correspond to circles on the plane around the point $z = 0$, which is the cylinder's infinite past $\tau = -\infty$.

The two-dimensional quantum field theory of interest is characterized by a local energy–momentum tensor, which is traceless with analytic $T(z)$ and antianalytic $\overline{T}(\bar{z})$ components. The infinitesimal transformation of any local quantum field $\phi_i(z, \bar{z})$ under general co-ordinate transformations is given by its commutator with the energy–momentum tensor, defined indeed (see equation (8.45)) as the infinitesimal generator for such transformations. If $\varepsilon(z)$ (respectively, $\bar{\varepsilon}(\bar{z}) \neq \varepsilon^\dagger(\bar{z})$) is the infinitesimal parameter of a holomorphic (respectively, antiholomorphic) transformation (8.48), then

$$\delta_\varepsilon \phi_i(z, \bar{z}) = \left[T_\varepsilon, \phi_i(z, \bar{z})\right], \qquad \delta_{\bar{\varepsilon}} \phi_i(z, \bar{z}) = \left[\overline{T}_{\bar{\varepsilon}}, \phi_i(z, \bar{z})\right] \tag{8.51}$$

where

$$T_\varepsilon = \frac{1}{2\pi i} \oint_{C_\circ} T(z) \varepsilon(z) dz \tag{8.52}$$

where $C_\circ$ is a closed counter-clockwise contour around zero.

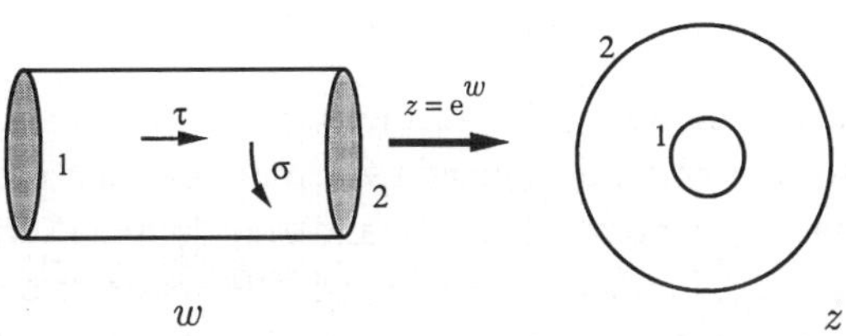

Fig. 8.3 Mapping from the cylinder to the complex plane.

From now on, we only show expressions for the holomorphic sector; the corresponding antiholomorphic formulae are obtained by putting a bar over almost everything.

In the definition (8.52), we used the fact that under $e^w = z$ the fixed-time surfaces are mapped into circles around zero. What used to be time in the cylinder is now the radius; we are carrying out a radial quantization, where normal ordering is effected with respect to the "time-like" distance to the origin. Using (8.52) and Cauchy's theorem, we obtain

$$\delta_\varepsilon \phi_i(z,\bar{z}) = \frac{1}{2\pi i} \oint_{C_z} T(\xi)\varepsilon(\xi)\phi_i(z,\bar{z})d\xi \tag{8.53}$$

where C_z is a closed counter-clockwise contour around the point z.

The group of two-dimensional conformal transformations (8.48) has an infinite number of generators. Its corresponding Lie algebra is isomorphic to two copies of the Lie algebra of vector fields on a circle, Vec(S^1), with generators

$$\ell_n = -z^{n+1}\frac{d}{dz} \qquad n \in \mathbb{Z} \tag{8.54}$$

and commutation relations

$$[\ell_n, \ell_m] = (n-m)\ell_{n+m} \tag{8.55}$$

This is the classical symmetry algebra, the quantum counterpart of which is defined by the Virasoro operators

$$L_n = \oint_{C_\circ} \frac{dz}{2\pi i} z^{n+1} T(z) \tag{8.56}$$

which are the Fourier modes of the energy–momentum tensor,

$$T(z) = \sum_{n\in\mathbb{Z}} L_n z^{-n-2} \tag{8.57}$$

The Virasoro operators define a representation of the Lie algebra Vec(S^1) on the Hilbert space of the two-dimensional field theory. This representation may be projective, as a consequence of possible non-vanishing Schwinger terms (anomalies). A projective representation of an algebra can always be interpreted as a linear representation of some central extension of the algebra. For the algebra Vec(S^1), only a one-parameter family of central extensions exists:

$$[L_n, L_m] = (n-m)L_{n+m} + \frac{c}{12}(n^3-n)\delta_{n+m,0} \tag{8.58}$$

where c is an arbitrary number. The second term in (8.58) is called the central extension, and its form is completely fixed by demanding that the Jacobi identities be satisfied, and that the generators L_1, L_0 and L_{-1} define a subalgebra of (8.58) isomorphic to $s\ell(2,\mathbb{C})$. This last statement will be made clear below when we consider the structure of the vacuum. To form a link with critical phenomena, note that (8.58) is simply the commutator of the Fourier modes of $T(z)$. We may compute the "commutator" $[T, T]$ by using equations (8.51)–(8.53) for $\phi_i = T$. Then the algebra (8.58) and Cauchy's theorem imply the following operator product expansion for TT:

$$T(z)T(w) = \frac{c/2}{(z-w)^4} + \frac{2}{(z-w)^2}\, T(w) + \frac{1}{z-w}\, \partial_w T(w) + (\text{regular terms}) \tag{8.59}$$

and hence the energy–momentum tensor changes under conformal transformations almost as a two-differential, except for a term proportional to the central charge

$$T(w) \underset{w\to z(w)}{\longrightarrow} \left(\frac{dz}{dw}\right)^2 T(z) + \frac{c}{12}\, S(z, w) \tag{8.60}$$

involving the schwartzian derivative

$$S(z, w) = \frac{z'''}{z'} - \frac{3}{2}\left(\frac{z''}{z'}\right)^2 \tag{8.61}$$

with $z' = dz/dw$, etc. These equations imply that the holomorphic energy–momentum tensor $T(z)$ has weight $\Delta_T = 2$, and that it behaves anomalously under generic conformal transformations. Under Möbius transformations, however, $T(z)$ behaves like a proper tensor because $S(z, w) = 0 \iff z = (aw+b)/(cw+d)$.

The transformation law (8.61) has noteworthy consequences, and will play an important role in the analysis of finite-size effects in conformal field theory (section 8.10). Consider again the map in figure 8.3 from a cylinder of circumference L to the complex plane, $w \to z = \exp(2\pi w/L)$. Equation (8.60) yields the following relationship between the energy–momentum on the cylinder, $T_{\text{cyl}}(w)$, and that on the complex plane, $T(z)$:

$$T_{\text{cyl}}(w) = \left(\frac{2\pi}{L}\right)^2 \left[z^2 T(z) - \frac{c}{24}\right] = \left(\frac{2\pi}{L}\right)^2 \left(\sum_{n\in\mathbb{Z}} L_n\, \mathrm{e}^{-2\pi n w/L} - \frac{c}{24}\right) \tag{8.62}$$

Hence, the quantum hamiltonian of the original system on the cylinder is related to the Virasoro operators by

$$H_{\text{cyl}} = \int_0^L \frac{d\sigma}{2\pi} \left[T_{\text{cyl}}(i\sigma) + \overline{T}_{\text{cyl}}(-i\sigma)\right] = \frac{2\pi}{L}\left(L_0 + \overline{L}_0 - \frac{c}{12}\right) \tag{8.63}$$

This equation implies that the central extension c of a conformal field theory manifests itself physically as a Casimir effect. Indeed, if we assume that on the plane $\langle T\rangle = \langle \overline{T}\rangle = 0$, then the vacuum energy of the cylinder depends on its circumference as

$$E_0 = -\frac{\pi c}{6L} \tag{8.64}$$

Similarly, the free energy of a two-dimensional classical statistical system at criticality depends on the size of the system as $1/L$, with the coefficient determined by the value of the central extension c of its universality class. The physical meaning of the central charge is an important factor in establishing the link between two-dimensional conformal field theory and the critical behavior of two-dimensional statistical models.

8.7 Representations of the Virasoro algebra

For non-zero values of c, it is the Virasoro algebra, (8.58), that must be represented on the Hilbert space of states. The Hilbert space of the simplest conformal field theory consists of irreducible (and unitary) representations of the Virasoro algebra, which we now consider.

Irreducible representations of the Virasoro algebra are called highest weight representations because they are characterized by a highest weight state $|\Delta\rangle$ satisfying

$$\begin{aligned} L_n|\Delta\rangle &= 0 \qquad && n > 0 \\ L_0|\Delta\rangle &= \Delta|\Delta\rangle \qquad && \Delta \in \mathbb{C} \end{aligned} \tag{8.65}$$

The representation with highest weight $|\Delta\rangle$ consists of the vector space generated by $|\Delta\rangle$ and all the "descendant" fields obtained by acting with the L_{-n} $(n > 0)$ on the highest weight. The corresponding representation space is called the Verma module $\mathscr{V}^{\Delta}$, and the eigenvalue Δ of L_0 on the highest weight is called the (conformal) weight of the representation. The state $|\Delta\rangle$ is the ground state of L_0 in the Verma module; using (8.58), it is not hard to show that

$$L_0 L_{-n}|\Delta\rangle = (\Delta + n)L_{-n}|\Delta\rangle \qquad (n > 0) \tag{8.66}$$

In conformal field theory, we must consider representations of (8.58) together with those of the Virasoro algebra associated with the antiholomorphic transformations: an irreducible representation is thus fully characterized by a pair of conformal weights $(\Delta, \overline{\Delta})$.

For dilatations $z \to \lambda z$, the transformation law becomes precisely the scaling law found earlier for the local operators $O_i(x)$ describing extensive variables at the critical point. The two-dimensional conformal field theory which we associate with a universality class will contain in its quantum Hilbert space as many irreducible representations of the Virasoro algebra as extensive variable operators; and the weights $(\Delta_i, \overline{\Delta}_i)$ will have to match the critical exponents d_i of these operators by the relation

$$d_i = \Delta_i + \overline{\Delta}_i \tag{8.67}$$

This follows from the fact that $(\ell_0 + \overline{\ell}_0)$ generates dilatations. Similarly, since $i(\ell_0 - \overline{\ell}_0)$ is the generator for rotations, the spin of a state $(\Delta_i, \overline{\Delta}_i)$ is given by

$$s = \Delta_i - \overline{\Delta}_i \tag{8.68}$$

Marginal operators O_i have $d_i = 2$, whereas relevant operators are those with $d_i < 2$, and irrelevant operators are those with $d_i > 2$.

Let us digress briefly to consider the vacuum $|0\rangle = |0\rangle \times |0\rangle$, which is the highest weight vector invariant under dilatations and rotations, i.e. with vanishing conformal weights

$$L_0 |0\rangle = \overline{L}_0 |0\rangle = 0 \tag{8.69}$$

The vacuum representation $|0\rangle$ corresponds naturally to the identity operator. In addition to (8.69), we shall impose that the vacuum be invariant under translations generated by d/dz and $d/d\bar{z}$, i.e. by L_{-1} and $\overline{L}_{-1}$. Therefore, the vacuum state $|0\rangle$ satisfies

$$L_n |0\rangle = \overline{L}_n |0\rangle = 0 \qquad n \geq -1 \tag{8.70}$$

Conditions (8.70) have a nice geometrical meaning. Consider the complex plane compactified to the Riemann sphere by the addition of a point at infinity. Divide the sphere into two hemispheres, as in figure 8.4, and consider the vector fields $\ell_n = -z^{n+1}\frac{d}{dz}$ on the equator. The set of generators L_n annihilating the vacuum corresponds to the set of vector fields which extend holomorphically from hemisphere I to hemisphere II.

In close analogy, we may consider the "out" vacuum as defined by

$$\langle 0| L_n = 0 \qquad n \leq 1 \tag{8.71}$$

which is annihilated by all the Virasoro operators corresponding to vector fields which extend holomorphically to hemisphere I (we use $L_{-n} = L_n^\dagger$). From (8.70) and (8.71), we deduce that the vector fields ℓ_{-1}, ℓ_0 and ℓ_1 are defined globally on the whole sphere, i.e. they are the three conformal Killing vectors of S_2.

From such a geometric interpretation of the symmetries of the vacuum (8.70), we can guess how the vacuum will behave if, instead of working on the complex plane, we consider a field theory on some more complicated two-dimensional manifold, such as, for instance, the Riemann surface depicted in figure 8.5. The natural generalization of (8.70) is obtained by considering only those Virasoro generators which correspond to vector fields defined on the circle in figure 8.5 and which extend holomorphically to region II. This geometric picture can be used

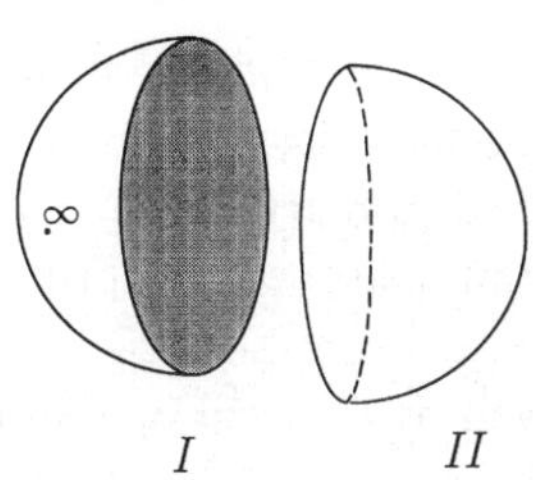

Fig. 8.4 The two halves of the Riemann sphere.

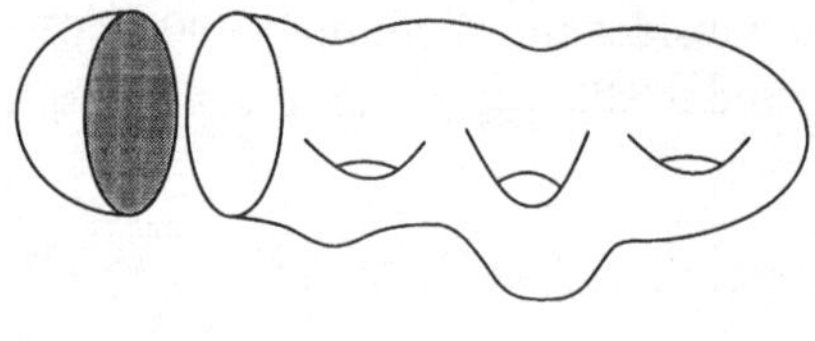

Fig. 8.5 The two halves of a Riemann surface.

to investigate how the ground state is modified by the topology and geometry of "spacetime".

We shall now prove that, associated with each irreducible representation $(\Delta, \overline{\Delta})$ of Virasoro$\otimes\overline{\text{Virasoro}}$, we can find a local quantum field $\Phi_{\Delta,\overline{\Delta}}(z,\overline{z})$ such that

$$\lim_{z,\overline{z}\to 0} \Phi_{\Delta,\overline{\Delta}}(z,\overline{z})\,|0\rangle = |\Delta,\overline{\Delta}\rangle \tag{8.72}$$

This field has the following transformation property:

$$\Phi_{\Delta,\overline{\Delta}}(z,\overline{z}) \longrightarrow \left(\frac{\partial\xi}{\partial z}\right)^{\Delta}\left(\frac{\partial\overline{\xi}}{\partial\overline{z}}\right)^{\overline{\Delta}} \Phi_{\Delta,\overline{\Delta}}(\xi,\overline{\xi}) \tag{8.73}$$

or, infinitesimally,

$$\begin{aligned}\delta_{\varepsilon}\Phi_{\Delta,\overline{\Delta}}(z,\overline{z}) &= \left(\varepsilon(z)\partial + \Delta\varepsilon'(z)\right)\Phi_{\Delta,\overline{\Delta}}(z,\overline{z})\\ \delta_{\overline{\varepsilon}}\Phi_{\Delta,\overline{\Delta}}(z,\overline{z}) &= \left(\overline{\varepsilon}(\overline{z})\overline{\partial} + \overline{\Delta}\overline{\varepsilon}'(\overline{z})\right)\Phi_{\Delta,\overline{\Delta}}(z,\overline{z})\end{aligned} \tag{8.74}$$

Using (8.51), we find

$$T(z)\Phi_{\Delta,\overline{\Delta}}(\xi,\overline{\xi}) = \frac{\Delta}{(z-\xi)^2}\Phi_{\Delta,\overline{\Delta}}(\xi,\overline{\xi}) + \frac{1}{z-\xi}\partial_{\xi}\Phi_{\Delta,\overline{\Delta}}(\xi,\overline{\xi}) \tag{8.75}$$

where we have neglected the terms on the right-hand side which are regular as $z \to \xi$. Therefore,

$$\left[L_n, \Phi_{\Delta,\overline{\Delta}}(z,\overline{z})\right] = \left(z^{n+1}\frac{d}{dz} + (n+1)z^n\Delta\right)\Phi_{\Delta,\overline{\Delta}}(z,\overline{z}) \tag{8.76}$$

which implies, in particular, that

$$\left[L_{-1}, \Phi_{\Delta,\overline{\Delta}}(z,\overline{z})\right] = \partial\Phi_{\Delta,\overline{\Delta}}(z,\overline{z}) \tag{8.77}$$

Hence, to every irreducible representation of Virasoro we associate a local quantum field transforming as a primary field tensor (8.73) under conformal transformations. These fields are called primary fields. The operator product expansion (8.75) of the energy–momentum tensor with a primary field will prove to be very useful later on.

The "in" states $|\Phi\rangle$ are given by the field operators $\Phi(z,\overline{z})$:

$$|\Phi\rangle = \lim_{\tau\to-\infty} \mathrm{e}^{-H\tau}\Phi(\sigma)\,|0\rangle = \lim_{z,\overline{z}\to 0}\Phi(z,\overline{z})\,|0\rangle \tag{8.78}$$

The "out" states $\langle\Phi|$ are obtained from the above by the mapping $z \to 1/z$:

$$\begin{aligned}\langle\Phi| &= \lim_{z,\bar{z}\to 0} \langle 0|\, \Phi(1/z, 1/\bar{z}) z^{-2\Delta}\bar{z}^{-2\bar{\Delta}} \\ &= \lim_{z,\bar{z}\to 0} \langle 0|\, [\Phi(\bar{z}, z)]^\dagger = \left[\lim_{z,\bar{z}\to 0} \Phi(\bar{z}, z)\, |0\rangle\right]^\dagger = |\Phi\rangle^\dagger\end{aligned} \tag{8.79}$$

Hermiticity of the energy–momentum tensor, i.e. $T^\dagger(z) = \bar{z}^{-4}T(1/\bar{z})$, implies $L_n^\dagger = L_{-n}$.

Not any arbitrary set of irreducible representations of Virasoro can be used to define the physical Hilbert space of a two-dimensional conformal field theory. The first and obvious condition they must satisfy has to do with completeness for the set $\{O_i(x)\}$ of local operators, which means that the set of local operators $\{\Phi_{\Delta_i,\bar{\Delta}_i}(z,\bar{z})\}$, associated with the set of Virasoro irreducible representations $\mathscr{V}^{\Delta_i,\bar{\Delta}_i} = \mathscr{V}^{\Delta_i} \times \mathscr{V}^{\bar{\Delta}_i}$ and satisfying (8.72)–(8.76), defines an algebra closed under the operator product expansion. The algebra is generated by the operators $\{\Phi_{\Delta_i,\bar{\Delta}_i}(z,\bar{z})\}$ and their derivatives. The operator product expansion for primary fields can be written as

$$\begin{aligned}\Phi_{\Delta_i,\bar{\Delta}_i}(z,\bar{z})\Phi_{\Delta_j,\bar{\Delta}_j}(w,\bar{w}) = \sum_k C_{ij}^{\ k}(z-w, \bar{z}-\bar{w})\Phi_{\Delta_k,\bar{\Delta}_k}(w,\bar{w}) \\ + \text{derivatives}\end{aligned} \tag{8.80}$$

The dependence of the "structure functions" $C_{ij}^{\ k}(z,\bar{z})$ on the local co-ordinates is fixed by requiring the same behavior under conformal transformations of both sides in equation (8.80). The result is

$$C_{ij}^{\ k}(z,\bar{z}) = z^{\Delta_k-\Delta_i-\Delta_j}\, \bar{z}^{\bar{\Delta}_k-\bar{\Delta}_i-\bar{\Delta}_j}\, C_{ij}^{\ k} \tag{8.81}$$

This general expression can be checked against the particular case found above, from which it is easy to see, in obvious notation, that

$$C_{T\Phi_i}^{\Phi_j} = \Delta_i\delta_{ij}\,, \quad C_{TT}^{T} = 2\,, \quad C_{TT}^{\mathbf{1}} = c/2 \tag{8.82}$$

Equation (8.80) can be interpreted intuitively as providing information on the decomposition rules for the tensor product of Virasoro irreps. In fact, for non-vanishing $C_{ij}^{\ k}$, we expect from (8.75) that the irrep $\mathscr{V}^{\Delta_k,\bar{\Delta}_k}$ will appear in the tensor product $\mathscr{V}^{\Delta_i,\bar{\Delta}_i} \otimes \mathscr{V}^{\Delta_j,\bar{\Delta}_j}$. Thinking along these lines, it is of interest to point out that completeness in the sense of (8.80) implies that the Virasoro algebra is semi-simple, i.e. that the product of irreducible representations can be fully decomposed in irreducible representations. In section 8.8, we shall study these and related issues in detail

So far, we have been concerned with the simplest conformal field theories, namely those where each of the holomorphic and antiholomorphic parts of the conformal fields generate irreps of Virasoro when acting on the vacuum. The more general situation allows for the conformal fields to arrange themselves into irreps of some chiral algebra $\mathscr{A} \supset$Virasoro, an infinite-dimensional algebra generated by local chiral fields. Almost all the results discussed for Virasoro generalize to an arbitrary chiral algebra; in particular, the representation theory for general chiral algebras closely resembles that for pure Virasoro.

In order to define a two-dimensional conformal field theory we must specify two chiral algebras: one containing holomorphic and the other one antiholomorphic local quantum fields. It is the rational conformal field theories, whose Hilbert space of states can be decomposed into a finite sum of irreducible representations of the chiral algebra, that are of most interest to us. As emphasized in section 8.4, the critical point of a universality class of two-dimensional systems is described by a conformal field theory, the Hilbert space of states of which contains as many irreducible representations of a chiral algebra as there are different critical exponents required to describe its scaling behavior.

A chiral algebra is defined as an algebra of local chiral operators $\mathscr{A} \equiv \{\mathscr{O}_i(z)\}$ (similarly for antichiral $\overline{\mathscr{A}} \equiv \{\overline{\mathscr{O}}_{\bar{i}}(\bar{z})\}$) closed under the operator product expansion, containing the identity, and having as descendant the conformal energy–momentum tensor $T(z)$. We shall assume that the algebra $\mathscr{A}$ also contains all derivatives of the operators $\mathscr{O}_i(z)$. The requirement that any chiral algebra contains at least the identity and the energy–momentum tensor implies that Virasoro is always a well-defined subalgebra. The simplest example of such a structure is, of course, the Virasoro algebra itself.

The operator product expansion among the operators generating the chiral algebra can be written as

$$\mathscr{O}_i(z)\mathscr{O}_j(0) = \sum_k C_{ij}{}^k(z)\mathscr{O}_k(0) \tag{8.83}$$

where locality is imposed by requiring the "structure constants" $c_{ij}{}^k(z)$ of the chiral algebra to be single-valued holomorphic functions. The structure constants in the operator product expansion determine a complete set of operators describing a universality class. In chapter 10 we shall focus on the paradigmatic case of affine Lie algebras (Kac–Moody algebras), which are the natural generalization to a holomorphic context of the usual (finite) Lie algebras. We shall study Kac–Moody algebras after deepening our understanding of the Virasoro algebra.

Let us briefly summarize what we have already learned about conformal field theories. A theory is characterized, first, by the central charge c. The Hilbert space of states is defined as a direct sum of irreps of Virasoro$\otimes\overline{\text{Virasoro}}$, each one characterized by the conformal weights $(\Delta, \overline{\Delta})$. With each irrep we associate a primary field $\Phi_{\Delta\overline{\Delta}}(z, \bar{z})$ with scaling dimension $(\Delta + \overline{\Delta})$. In order to construct a well-defined theory, the first additional requirement we impose is completeness, in the sense that the algebra of operators generated by the primary fields and their derivatives closes. These two principles, i.e. conformal invariance and completeness, are not enough to define a theory. In fact, we shall impose two additional postulates: invariance under (a) duality and (b) modular transformations. The first one will introduce some restrictions on the values of the conformal weights $(\Delta, \overline{\Delta})$ for some primary fields, and will provide a way to fix the values of the structure constants in the operator product expansion (8.80). The postulate of modular invariance will improve the one of completeness: not every set of irreps closed under the operator product expansion satisfies the constraint of invariance under modular transformations. Both postulates, duality and modular invariance, have a clear geometric origin: they ensure invariance of the theory under global (and not only local) diffeomorphisms. We shall study these two postulates in chapter 9. We state here that, if the chiral algebra of relevance for a conformal field theory is larger than Virasoro, the physical principles behind the possible values of the structure constants in the operator product expansion are, just as in the pure Virasoro case, irreducibility, duality and modular invariance.

8.8 Decoupling of null vectors

We wish to find irreducible representations $|\Delta\rangle \otimes |\overline{\Delta}\rangle$ of Virasoro$\otimes\overline{\text{Virasoro}}$, which we will associate with the extensive variables $O_i(x)$ relevant for the description of critical phenomena. We consider, for simplicity, only one chiral part, say the holomorphic one.

The Verma module $\mathscr{V}^\Delta$ of highest weight Δ contains the highest weight state $|\Delta\rangle$ ($L_{n>0}\,|\Delta\rangle = 0$, $L_0\,|\Delta\rangle = \Delta\,|\Delta\rangle$) and all its descendants, which are obtained by repeatedly acting with L_{-n} $(n > 0)$ on $|\Delta\rangle$. Quite generally, the monomial $P(L_{-n}) = \prod_{n>0} L_{-n}^{r_n}$ $(r_n \in \mathbb{N})$ creates a state $|\mathscr{X}\rangle = P(L_{-n})\,|\Delta\rangle$, the level $\ell = \sum nr_n$ of which measures the "distance" in the Verma module from $|\mathscr{X}\rangle$ to $|\Delta\rangle$ (figure 8.6). More precisely,

$$L_0\,|\mathscr{X}\rangle = (\Delta + \ell)\,|\mathscr{X}\rangle \tag{8.84}$$

The number of states at a given level ℓ increases with the level, and is, quite simply, the number of possible ways, $\pi(\ell)$, that ℓ can be split into positive integers. This number $\pi(\ell)$ is called the partition of ℓ, and it can be read off from

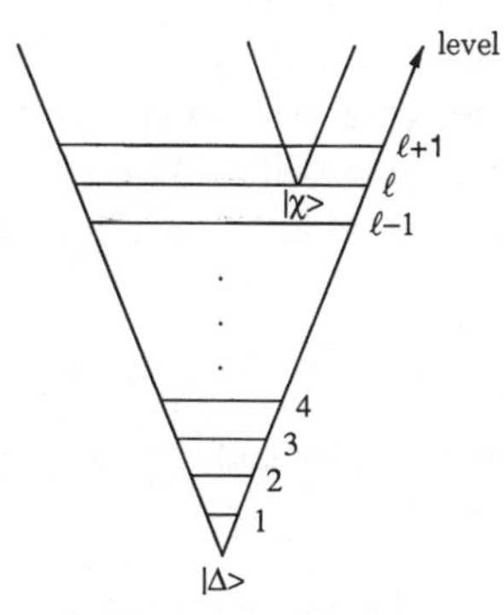

Fig. 8.6 Verma module $\mathscr{V}^\Delta$ and a descendant $|\mathscr{X}\rangle$ at level ℓ.

the generating functional

$$\prod_{n\geq 1}(1-x^n)^{-1} = \sum_{\ell\geq 0}\pi(\ell)x^\ell \tag{8.85}$$

For arbitrary weight Δ, the Verma module $\mathscr{V}^\Delta$ is generically irreducible. It may happen, however, that among the states in $\mathscr{V}^\Delta$ we find a primary field, i.e. a descendant $|\mathscr{X}\rangle$ of $|\Delta\rangle$ which satisfies $L_n|\mathscr{X}\rangle = 0$ $(n > 0)$. Such a state is called a null vector, because its norm is zero. Moreover, since the norm we have defined (based on the $SL(2)$ invariant vacuum $|0\rangle$) is such that $L_n^\dagger = L_{-n}$, the state $|\mathscr{X}\rangle$ is, in fact, orthogonal to all states in $\mathscr{V}^\Delta$, including itself. (It is also orthogonal to all states in the Hilbert space if the theory is unitary.) Indeed, let $|\psi\rangle$ be any state in $\mathscr{V}^\Delta$, necessarily of the form

$$|\psi\rangle = \sum_{n>0}\beta_n L_{-n}^{s_n}|\Delta\rangle \qquad \beta_n \in \mathbb{C},\ s_n \in \mathbb{N} \tag{8.86}$$

Then, $<\psi|\mathscr{X}>= \sum_{n>0}\langle\Delta|\,\beta_n^\dagger L_n^{s_n}|\mathscr{X}\rangle = 0$, and, in particular, $<\mathscr{X}|\mathscr{X}>= 0$.

Clearly, Verma modules containing null states are reducible representations of Virasoro. Although, at first sight, reducible representations seem worse than irreducible ones, they end up producing very interesting theories. The underlying philosophy is very close to usual gauge field theory. There, null states (such as the longitudinal component of the photon) signal the existence of a gauge symmetry, and the quantization of such theories must ensure that null states are not physical, for instance by balancing them out with Faddeev–Popov ghosts. Unitarity is the key word behind the requirement that states with zero norm (or, worse, negative norm) decouple from the physical spectrum, i.e. that all the correlation functions mixing physical states (of positive norm) and unphysical ones vanish.

In the case at hand, enforcing the decoupling of null vectors amounts to producing irreducible representations from reducible ones, without any reference to unitarity. Essentially, we shall subtract from the original Verma module the Verma modules generated by null states. Because the lowering Virasoro generators act as derivatives, $L_{-n} \sim \partial^n/\partial z^n$, the condition that null states decouple from the physical spectrum will yield differential equations, which can be viewed as Ward identities for the underlying conformal symmetry. Hence, irreps whose Verma modules contain null vectors will provide us with very powerful dynamical equations, much to our advantage. Let us develop these ideas in some detail.

8.8.1 The Kac formula

The first question is how to find descendants of Δ which, in turn, are primary fields. Construct a basis

$$\{|\mathscr{X}^{(k_1,k_2,\ldots}>\} = \{\prod_{n\geq 1} L_{-n}^{k_n}|\Delta\rangle\,, k_n \in \mathbb{N}, k_1 \leq k_2 \leq \cdots\} \tag{8.87}$$

for the descendants of Δ at a given level $\ell = \sum nk_n$. Compute then the $\pi(\ell)$-dimensional matrix of scalar products $K^{(\ell)}_{\mathbf{kk}'} = < \mathscr{X}^{\mathbf{k}}|\mathscr{X}^{\mathbf{k}'} >$. Clearly, there are as many null states at level ℓ as zero eigenvalues of $K^{(\ell)}$. A basis-independent quantity which reveals the existence of null descendants is the determinant $\det K^{(\ell)}$. This determinant was calculated by Kac, who found that if Δ is of the form

$$\Delta_{mm'} = \Delta_0 + \frac{1}{4}(\alpha_+ m + \alpha_- m')^2 \qquad m, m' \in \mathbb{N} - \{0\} \tag{8.88}$$

where

$$\begin{aligned} \Delta_0 &= \frac{1}{24}(c-1) \\ \alpha_\pm &= \frac{\sqrt{1-c} \pm \sqrt{25-c}}{\sqrt{24}} \end{aligned} \tag{8.89}$$

then there exists a null vector in $\mathscr{V}^{\Delta_{mm'}}$ at level $\ell = mm'$. Note that the only parameter in (8.88) is the central charge c.

We can check Kac's result for the simple case of level two:

$$\begin{aligned} K^{(2)} &= \begin{pmatrix} \langle\Delta|\, L_2 L_{-2}\,|\Delta\rangle & \langle\Delta|\, L_1^2 L_{-2}\,|\Delta\rangle \\ \langle\Delta|\, L_2 L_{-1}^2\,|\Delta\rangle & \langle\Delta|\, L_1^2 L_{-1}^2\,|\Delta\rangle \end{pmatrix} \\ &= \begin{pmatrix} 4\Delta + c/2 & 6\Delta \\ 6\Delta & 4\Delta(1+2\Delta) \end{pmatrix} \end{aligned} \tag{8.90}$$

If $\det K^{(2)} = 0$, then either $\Delta = [5 - c \pm \sqrt{(1-c)(25-c)}]/16$ or $\Delta = 0$. Only the Verma modules with these highest weights contain null descendants at level two. The first two solutions match the Kac formula as Δ_{12} and Δ_{21}. The last one comes from the level one null state, $L_{-1}\,|0\rangle = 0$. Quite generally, in $\det K^{(\ell)}$ we obtain all the zeroes due to lower-level null states, in addition to the new ones of level ℓ.

To proceed further, we need the descendant field operators $\mathscr{X}(z)$ which, acting on the vacuum $|0\rangle$, produce the descendant $|\mathscr{X}\rangle$ of the highest weight $|\Delta\rangle$. If $|\mathscr{X}\rangle = P(L_{-n})\,|\Delta\rangle$, where $P(L_{-n})$ is any polynomial in the L_{-n} $(n > 0)$, then the operator $\mathscr{X}(z)$ producing the descendant field $|\mathscr{X}\rangle = \lim_{z\to 0} \mathscr{X}(z)\,|0\rangle$ is

$$\mathscr{X}(z) = P\left(\frac{1}{2\pi i}\oint_z \frac{T(\zeta)}{(\zeta - z)^{n-1}} d\zeta\right)\Phi_\Delta(z) \tag{8.91}$$

where $\Phi_\Delta(z)$ is the primary field corresponding to $|\Delta\rangle$, i.e. $|\Delta\rangle = \lim_{z\to 0}\Phi_\Delta(z)\,|0\rangle$. Descendant fields derived in one shot from the highest weight, i.e. descendants of the form $L_{-r}(z)\Phi_\Delta(z)$, appear in the operator product expansion

$$T(z)\Phi_\Delta(\zeta) = \sum_{r>0} (z-\zeta)^{-2+r} L_{-r}(z)\Phi_\Delta(\zeta) \tag{8.92}$$

whereas other descendant fields $\Phi^{\mathbf{r}}_\Delta(z)\Phi^{(r_1, r_2 \ldots)}_\Delta(z)$ appear in the expansion of the operator product $T(z_1)T(z_2)\ldots T(z_n)\Phi_\Delta(z)$, provided $n = \sum r_i$.

The operator product expansion for primary fields (8.80) presented earlier can now be completed to include descendants. Exceptionally, we write this expression with both the holomorphic and antiholomorphic parts:

$$\Phi_{\Delta_i,\overline{\Delta}_i}(z,\overline{z})\Phi_{\Delta_j,\overline{\Delta}_j}(\zeta,\overline{\zeta}) = \sum_{k,\{\mathbf{r},\mathbf{r}'\}} C_{ij}^{k\,\mathbf{r},\mathbf{r}'}(z-\zeta,\overline{z}-\overline{\zeta})\Phi^{\mathbf{r},\mathbf{r}'}_{\Delta_k,\overline{\Delta}_k}(\zeta,\overline{\zeta}) \tag{8.93}$$

where the structure functions $C_{ij}^{k\,\mathbf{r},\mathbf{r}'}(z-\zeta,\overline{z}-\overline{\zeta})$ are completely determined by those for the primary fields (8.81):

$$C_{ij}^{k\,\mathbf{r},\mathbf{r}'}(z,\overline{z}) = z^{\Delta_k-\Delta_i-\Delta_j+\sum r_p}\ \overline{z}^{\overline{\Delta}_k-\overline{\Delta}_i-\overline{\Delta}_j+\sum r'_p}\ C_{ij}^{k}\ \beta_{ij}^{k\,\mathbf{r}}\ \overline{\beta}_{ij}^{k\,\mathbf{r}'} \tag{8.94}$$

The factors β and $\overline{\beta}$ can be computed using conformal invariance, and depend only on the conformal weights Δ_i, Δ_j and Δ_k and on the central charge c.

Let us now try to exploit conformal invariance to obtain information on the correlation functions among primary fields, defined as the vacuum expectation values (again we consider the holomorphic part only)

$$\langle\Phi_1(z_1)\Phi_2(z_2)\cdots\Phi_n(z_n)\rangle = \langle 0|\,\Phi_1(z_1)\Phi_2(z_2)\cdots\Phi_n(z_n)\,|0\rangle \tag{8.95}$$

An intuitive way of interpreting these correlation functions is by thinking of a physical scattering process like the one in figure 8.7, where the complex plane has been compactified to the Riemann sphere. The figure represents a physical scattering process in time, with the "in" vacuum $|0\rangle$ at $t=-\infty$, and four insertions of fields Φ_i at different time and space positions.

8.8.2 Conformal Ward identities

Equation (8.75) is the operator form of the conformal Ward identities. Indeed, we may evaluate the correlation function $\langle T(z)\Phi_1(z_1)\cdots\Phi_n(z_n)\rangle$ on the sphere, using the analyticity properties of $T(z)$ and equation (8.75). As a function of z, $T(z)$ behaves like a meromorphic quadratic differential (i.e. $\Delta=2$) on the sphere. The only singularities appear when z approaches any of the insertion points $z_1,\ldots,z_n$,

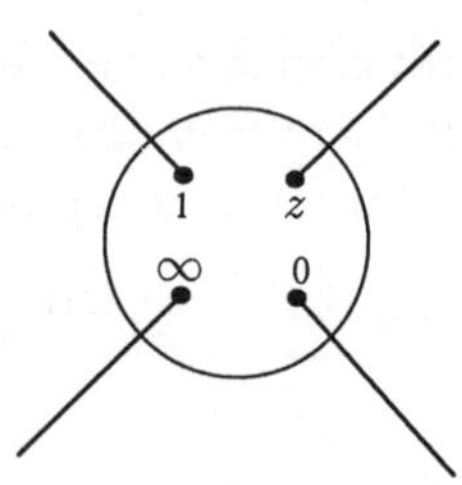

Fig. 8.7 Four-particle scattering process on the sphere.

and they are determined by (8.75). Thus, we find

$$\begin{aligned}&\langle T(z)\Phi_1(z_1)\cdots\Phi_n(z_n)\rangle \\ &\quad = \sum_{j=1}^{n}\left[\frac{\Delta_j}{(z-z_j)^2}+\frac{\partial/\partial z_j}{(z-z_j)}\right]\langle\Phi_1(z_1)\cdots\Phi_n(z_n)\rangle\end{aligned} \tag{8.96}$$

Note that equation (8.96) is exact, whereas (8.75) is only approximate in the sense that regular terms are disregarded – of course, it is exact with regard to singularities. This is why we must pay attention to the derivation of (8.96) from (8.75).

Once we know the correlation functions among primary fields, the correlation functions involving descendants follow. For example,

$$\begin{aligned}&\left\langle\left[L_{-k}\Phi_{\Delta_1}\right](z_1)\Phi_{\Delta_2}(z_2)\cdots\Phi_{\Delta_n}(z_n)\right\rangle \\ &\quad = \frac{1}{2\pi i}\oint_{z_1}\frac{dz}{(z-z_1)^{k-1}}\langle T(z)\Phi_{\Delta_1}(z_1)\Phi_{\Delta_2}(z_2)\cdots\Phi_{\Delta_n}(z_n)\rangle \\ &\quad = \mathscr{L}_{-k}(z_1)\langle\Phi_{\Delta_1}(z_1)\Phi_{\Delta_2}(z_2)\cdots\Phi_{\Delta_n}(z_n)\rangle\end{aligned} \tag{8.97}$$

where the differential operator $\mathscr{L}_{-k}(z_1)$ is defined ($k \geq 1$) as

$$\mathscr{L}_{-k}(z_1) = (-1)^k\sum_{j=2}^{n}\left\{\frac{(k-1)\Delta_j}{(z_1-z_j)^k}+\frac{1}{(z_1-z_j)^{k-1}}\frac{\partial}{\partial z_j}\right\} \tag{8.98}$$

As a general feature, any correlation function involving descendants is obtained by acting with the appropriate differential operator on the correlation function of primary fields.

8.8.3 Minimal models

We are now ready to study the case of a Verma module $\mathscr{V}^{\Delta_{mm'}}$ containing a null state $|\mathscr{X}\rangle$. When we require that the correlation functions involving $\mathscr{X}(z)$ vanish, we will obtain differential equations for the correlations of primary fields.

Consider the simple case $m = 1$, $m' = 2$ in Kac's formula (8.88). The Verma modula $\mathscr{V}^{\Delta_{12}}$ contains a null vector at level two. The state

$$|\mathscr{X}_{12}\rangle = (L_{-2} - [2(2\Delta_{12}+1)/3]^{-1}L_{-1}^2)\,|\Delta_{12}\rangle \tag{8.99}$$

is indeed null, and its descendant field is

$$\mathscr{X}_{12}(z) = \left(L_{-2} - \frac{3}{2(2\Delta_{12}+1)}L_{-1}^2\right)\Phi_{\Delta_{12}}(z) \tag{8.100}$$

The decoupling of this null vector means that

$$\langle\mathscr{X}_{12}(z)\Phi_{\Delta_1}(z_1)\cdots\Phi_{\Delta_n}(z_n)\rangle = 0 \tag{8.101}$$

which can be written explicitly as

$$\left\{\mathscr{L}_{-2}(z) - \frac{3}{2(2\Delta_{12}+1)}\mathscr{L}_{-1}^2(z)\right\}\langle\Phi_{\Delta_{12}}(z)\Phi_{\Delta_1}(z_1)\cdots\Phi_{\Delta_n}(z_n)\rangle = 0 \tag{8.102}$$

Of course, if one of the Δ_i is also of the Kac form, then the decoupling of its null vector will give rise, in turn, to another differential equation.

Let us look in detail at the two-point function, for which (8.102) reads

$$\left[\frac{3\,\partial^2/\partial z^2}{2(2\Delta_{12}+1)} - \frac{\Delta}{(z-z_1)^2} - \frac{\partial/\partial z_1}{(z-z_1)}\right]\langle\Phi_{\Delta_{12}}(z)\Phi_\Delta(z_1)\rangle = 0 \tag{8.103}$$

which should be compared with the operator product expansion of two primary fields already encountered above,

$$\Phi_{\Delta_{12}}(z)\Phi_\Delta(z_1) = \sum_{\Delta'} C_{\Delta_{12}\Delta}^{\Delta'}(z-z_1)^\beta\Big[\Phi_{\Delta'}(z_1) + (z-z_1)L_{-1}\Phi_{\Delta'}(z_1) + \cdots\Big] \tag{8.104}$$

where $\beta = \Delta' - \Delta - \Delta_{12}$. In the limit $(z-z_1) \to 0$, the decoupling condition (8.103) implies

$$\frac{3\beta(\beta-1)}{2(2\Delta_{12}+1)} - \Delta + \beta = 0 \tag{8.105}$$

with only two solutions, Δ'_+ and Δ'_-, for Δ'. Thus, if $|\mathscr{X}_{12}\rangle$ is to decouple, then the coefficient $C_{\Delta_{12}\Delta}^{\Delta'}$ may be non-vanishing only for $\Delta'{=}\Delta'_+$ or Δ'_-.

If Δ is also of the Kac form, say $\Delta_{mm'}$, then things really start to click: the two allowed values for Δ' belong also to the Kac table, namely $\Delta'_{(\pm)} = \Delta_{m,m'\pm1}$. We can encode this information in the "fusion rule"

$$[\Phi_{12}] \times [\Phi_{mm'}] = [\Phi_{m,m'-1}] + [\Phi_{m,m'+1}] \tag{8.106}$$

which is a consequence of the decoupling of the null vector (at level two) of Φ_{12}. We shall treat fusion algebras in some detail at the end of this section, but let us emphasize right away that fusion rules such as the above are a statement about the decomposition of the product of Verma modules in the theory. They do not reveal anything about the values of the structure constants $C_{ij}^{\ k}$, but, rather, about the vector space structure of the tensor product.

To get a feeling for the additional constraint imposed by the decoupling of the null vector of $\Delta_{mm'}$, take $m = 2$, $m' = 1$; the resulting system of two differential equations can be neatly expressed as two fusion rules,

$$\begin{aligned}[\Phi_{12}] \times [\Phi_{21}] &= [\Phi_{20}] + [\Phi_{22}] \\ [\Phi_{12}] \times [\Phi_{21}] &= [\Phi_{02}] + [\Phi_{22}]\end{aligned} \tag{8.107}$$

whose simultaneous compatibility implies uniquely that

$$[\Phi_{12}] \times [\Phi_{21}] = [\Phi_{22}] \tag{8.108}$$

More generally, imposing that the null vectors of both $\Phi_{m_i m'_i}$ $(i = 1,2)$ decouple from the two-point function implies the fusion algebra

$$[\Phi_{m_1 m'_1}] \times [\Phi_{m_2 m'_2}] = \sum_{\ell=|m_1-m_2|+1}^{m_1+m_2-1}\ \sum_{\ell'=|m'_1-m'_2|+1}^{m'_1+m'_2-1} [\Phi_{\ell\ell'}] \tag{8.109}$$

where both ℓ and ℓ' increase in steps of two. In particular, on setting $m_i = (2j_i+1)$ and $\Phi_{m_i 1} = \Phi_{j_i}$, this fusion rule reads exactly like the $SU(2)$ law for composition of angular momenta

$$[\Phi_{j_1}] \times [\Phi_{j_2}] = \sum_{|j_1-j_2|}^{j_1+j_2} [\Phi_{j_3}] \tag{8.110}$$

Some of the powerful consequences of decoupling null vectors are already evident. If the primary fields of our theory are all of the Kac form, i.e. they all have null descendants, then the algebra closes. Unfortunately, as is clear from the fusion rules (8.109), an infinite number of primary fields are necessary to close the algebra, because, by multiplying fields over and over again we can reach arbitrarily high values of the Kac labels m and m'. We have learned that to describe the critical behavior of statistical mechanical systems we need a certain *finite* number of extensive operators, which we want to identify with primary fields. So, for physical reasons, we would like to be able to truncate the expansion on the right-hand side of (8.109). In the more familiar instance of the "thermal subalgebra" (8.110), such a truncation would amount to the existence of a maximal allowed spin.

It turns out that such a truncation takes place when a very simple condition is satisfied by the central charge c, the only parameter in the theory; this condition is

$$c = 1 - \frac{6(p'-p)^2}{pp'} \tag{8.111}$$

where p and p' are mutually prime strictly positive integers, and $p < p'$. This is why we are naturally led to the study of theories with $c \leq 1$. It is clear from Kac's formula (8.88) and (8.89) that for $c \leq 1$ or $c \geq 25$ the conformal weights are real, but are complex for $1 < c < 25$ or $c \notin \mathbb{R}$. The region $c \geq 25$ is also very interesting in connection with the Liouville theory, which we shall not consider here. Condition (8.111) reads, in terms of the quantities defined by (8.89), as $\alpha_-/\alpha_+ = -p/p'$. These theories are called minimal models or, more precisely, (p, p') minimal models; the central charge (8.111) and all the weights in Kac's formula (8.88) are rational numbers:

$$\Delta_{m,m'} = \frac{(mp' - m'p)^2 - (p'-p)^2}{4pp'} \qquad \begin{array}{l} 0 < m < p \\ 0 < m' < p' \end{array} \tag{8.112}$$

Conformal field theories with a finite number of primary fields are termed rational, because the central charge and the conformal weights are rational numbers: the simplest examples thereof are the (p, p'), or minimal, models.

Letting

$$Q = \frac{p}{p'-p} \tag{8.113}$$

the central extension (8.112) and the weights (8.113) are given by the more elegant expressions:

$$c = 1 - \frac{6}{Q(Q+1)}$$
$$\Delta_{mm'} = \frac{1}{4Q(Q+1)}\left[(Q(m-m')+m)^2 - 1\right] \tag{8.114}$$

When $p' = p+1$ ($Q \in \mathbb{N}$, $Q \geq 3$), the minimal model is unitary: the central charge and the conformal weights are real and positive. Unitarity is crucial whenever an S-matrix interpretation for the theory is necessary, for instance in string theory, but it is not a meaningful requirement for classical statistical mechanics. The field content of the first few unitary minimal models is shown in table 8.2. The simplest rational and non-unitary conformal field theory is the $(2,5)$ minimal model, with $c = -\frac{22}{5}$. In addition to the identity, this theory contains only one other primary field, with conformal weight $-\frac{2}{5}$. This conformal field theory describes the Lee–Yang edge singularity in a critical imaginary field, and it is at the basis of the recent attempt by Polyakov to understand the universality of turbulence in two dimensions.

Minimal models enjoy the surprising symmetry $\Delta_{mm'} = \Delta_{p-m,p'-m'}$, called duality for historical "stringy" reasons. Since the physical properties of primary fields depend exclusively on their conformal weight, we shall boldly identify $\Phi_{mm'}$ with $\Phi_{p-m,p'-m'}$. This is a very strong condition, and is justified because the (p,p') models have no symmetry other than Virasoro, i.e. there is no quantum number able to distinguish between two fields other than the conformal weight. Because of the condition that the Kac labels should be strictly positive integers, the primary field content is thus restricted to $\Delta_{mm'}$ with $0 < m < p$ and $0 < m' < p'$ or, to avoid double counting, $0 < m < m' < p'$ and $0 < m < p$. The (strong) requirement

Table 8.2 Primary field content of the first three non-trivial unitary minimal models.
Shown here are the weights of the primary fields, organized according to the labels $(m,m') \leftrightarrow (p-m, p'-m')$. The unitary minimal model with $Q = 2$ is trivial in the sense that its only primary field is the identity, and its central charge is zero.

$m' \setminus m$	1	2	1	2	3	1	2	3	4
4									$\frac{1}{8}$
3					$\frac{1}{10}$			$\frac{1}{15}$	$\frac{2}{3}$
2		$\frac{1}{16}$		$\frac{3}{80}$	$\frac{3}{5}$		$\frac{1}{40}$	$\frac{21}{40}$	$\frac{13}{8}$
1	0	$\frac{1}{2}$	0	$\frac{7}{16}$	$\frac{3}{2}$	0	$\frac{2}{5}$	$\frac{7}{5}$	3
	$Q=3$		$Q=4$			$Q=5$			
	Ising		tricritical Ising			3-state Potts			

that two primary fields related by this symmetry be completely equivalent not only reduces the field content of the theory by one-half, but also ensures that the fusion rules truncate "from above":

$$[\Phi_{m_1 m_1'}] \times [\Phi_{m_2 m_2'}] = \sum_{\ell=|m_1-m_2|+1}^{M} \; \sum_{\ell'=|m_1'-m_2'|+1}^{M'} [\Phi_{\ell\ell'}] \qquad (8.115)$$

$$\text{where} \quad \begin{cases} M = \min(m_1 + m_2 - 1, 2p - m_1 - m_2 - 1) \\ M' = \min(m_1' + m_2' - 1, 2p' - m_1' - m_2' - 1) \end{cases}$$

and, again, both ℓ and ℓ' increase in steps of two. Letting $m_1 = m_2 = 1$, $k = p' - 2$ and defining, as above, $m_i' = 2j_i + 1$, (8.115) reads more simply as

$$[j_1] \otimes [j_2] = \sum_{|j_1-j_2|}^{\min(j_1+j_2, k-j_1-j_2)} [j] \qquad (8.116)$$

To illustrate these ideas, let us work out the minimal model with $p = 3$ and $p' = 4$, which describes the critical behavior of the Ising model's universality class. The Kac labels in the allowed weights (those with Verma modules containing null vectors) $\Delta_{mm'}$ are m=1, 2 and m'=1, 2, 3, so that there are only three primary fields, $\mathbf{1}$, σ and ε, with conformal weights $\Delta_{11} = \Delta_{23} = 0$, $\Delta_{12} = \Delta_{22} = \frac{1}{16}$ and $\Delta_{21} = \Delta_{13} = \frac{1}{2}$, respectively. The fusion rules of the Ising model can be derived from the general expression (8.109) and the reflection symmetry as:

$$\begin{array}{lll} [\mathbf{1}] \times [\mathbf{1}] = [\mathbf{1}] & [\mathbf{1}] \times [\sigma] = [\sigma] & [\mathbf{1}] \times [\varepsilon] = [\varepsilon] \\ & [\sigma] \times [\sigma] = [\mathbf{1}] + [\varepsilon] & [\sigma] \times [\varepsilon] = [\sigma] \\ & & [\varepsilon] \times [\varepsilon] = [\mathbf{1}] \end{array} \qquad (8.117)$$

Let us check, for instance, the last entry. Using (8.109), valid for arbitrary, i.e. irrational, central charge, we obtain:

$$\begin{aligned} [\varepsilon] \times [\varepsilon] &= [\Phi_{21}] \times [\Phi_{21}] = [\Phi_{11}] + [\Phi_{31}] \\ &= [\Phi_{13}] \times [\Phi_{13}] = [\Phi_{11}] + [\Phi_{13}] + [\Phi_{15}] \\ &= [\Phi_{13}] \times [\Phi_{21}] = [\Phi_{23}] \end{aligned} \qquad (8.118)$$

and simultaneous compatibility of all three possibilities leads to the result (8.117).

8.9 Fusion algebra

Let us begin to formalize the important notion of fusion algebra, which was introduced in the previous section as a mere notational convenience but is of crucial relevance further ahead.

Consider the left-moving sector only of a rational conformal field theory, i.e. a conformal field theory with a finite number of primary fields. The operator product expansion of pairs of primary fields yields the fusion algebra

$$[\Phi_i] \times [\Phi_j] = \bigoplus_k N_{ij}{}^k \, [\Phi_k] \tag{8.119}$$

where the vector space multiplicities $N_{ij}{}^k = N_{ji}{}^k \in \mathbb{N}$ are called the fusion rules of the theory; they count the number of times that the irrep $[\Phi_k]$ appears in the decomposition of the tensor product of $[\Phi_i]$ and $[\Phi_j]$. Because the operator product expansion is associative, $N_{im}{}^n N_{jn}{}^p = N_{jm}{}^n N_{in}{}^p$, which can be written as

$$[N_i \, , \, N_j] = 0 \tag{8.120}$$

in terms of the fusion matrices N_i, with matrix elements $(N_i)_j{}^k = N_{ij}{}^k$. Let us stress that the fusion algebra expresses information about the decomposition of the tensor product of vector spaces, not about the coefficients in this vector space decomposition.

Of special interest are the fusion operations which yield the identity **1**. We define $C_{ij} = N_{ij}{}^{\mathbf{1}}$, and interpret it as a charge conjugation matrix, with which we can move indices up and down. It is not difficult to show that $N_{ijk} = N_{ij}{}^{\ell} C_{\ell k}$ is totally symmetric in the three indices. We have assumed implicitly that the charge conjugation matrix C is non-degenerate, i.e. that every primary field has a unique conjugate field with which it fuses into the identity. This is always true if the theory contains a finite number of primary fields, i.e. if it is rational.

From the associativity of the fusion algebra, and the fact that the fusion matrices provide a representation of the fusion algebra itself (i.e. $N_i N_j = N_{ij}{}^k N_k$), the fusion matrices N_i can all be diagonalized simultaneously: there exists a matrix S such that $S^{-1} N_i S$ is diagonal ($\forall i$). Denoting N_i's jth eigenvalue by $v_i^{(j)}$, the fact that $N_{i\mathbf{1}}{}^j = \delta_i^j$ (fusion with the identity is trivial) implies that $v_i^{(j)} = S_i^j / S_{\mathbf{1}}^j$, and, hence, $N_{ij}{}^k = \sum_m S_j^m \frac{S_i^m}{S_{\mathbf{1}}^m} (S^{-1})_m^k$. In the following, we shall have the occasion to understand the matrix S from a geometric point of view, as a generator for modular transformations. Let us advance that $S^2 = C$. From this single piece of information, it follows that S is symmetric ($S^T = S$) and unitary ($S^* = SC$). The fusion rules are thus very constrained, since they can be written in terms of a unitary symmetric matrix S as

$$N_{ijk} = \sum_m S_i^m S_j^m S_k^m / S_{\mathbf{1}}^m \tag{8.121}$$

This concludes our description of the elementary properties of the fusion algebra.

8.10 Finite-size effects

Let us now study in some detail a practical way to connect statistical models at criticality with conformal field theories. The basic tool at our disposal is the theory of finite-size scaling. We shall not review this theory, but merely quote some of its most relevant results when applied to spin chains whose continuum limit yields conformal field theories.

In section 8.7, we saw that the central extension c is related to a Casimir effect, equation (8.64). To analyze this in more detail, let us consider again the spin chains of chapters 2, 3 and 5, with hamiltonian $H = \sum_{i=1}^{L} H_{i,i+1}$, and periodic or open boundary conditions. At criticality, the hamiltonian H has a gapless spectrum with a linear dispersion law in the vicinity of the Fermi level, $\epsilon(p) \simeq v_F|p - p_F|$, and the theory is described by a conformal field theory. The first problem we shall address is how to compute the central extension of the conformal theory. The ground state energy E_0 of H depends on the size L of the chain (when $L \to \infty$) as

$$E_0 = \begin{cases} L\epsilon_0 - \frac{\pi c v_F}{6L} & \text{for periodic b.c.} \\ L\epsilon_0 + E_S - \frac{\pi c v_F}{24L} & \text{for free or fixed b.c.} \end{cases} \tag{8.122}$$

where E_S is a constant surface energy term. The first expression agrees with (8.64) for $v_F = 1$. The normalization of the Fermi velocity ensures the rotational invariance of the two-dimensional statistical system.

The ground state energy E_0 of the quantum spin chain hamiltonian can be viewed as the free energy per unit length F of a two-dimensional statistical system on an infinitely long strip of width L, with scaling behavior given by (8.122). Alternatively, the partition function of a classical system of finite width with periodic boundary conditions may be interpreted as the path integral for an infinitely long quantum chain at finite temperature $T = L^{-1}$. The computation of the free energy per unit length is carried out by minimizing $F = E - T\mathscr{S}$, where E is the energy of the spin chain and $\mathscr{S}$ is the combinatorial entropy (exercise 3.5). This enables us to derive the central extension c from the low temperature behavior of the free energy:

$$\text{tr } \mathrm{e}^{-H/T} = \mathrm{e}^{-F(T)/T}\,, \qquad F(T) = F_0 - \frac{\pi c T^2}{6 v_F} \tag{8.123}$$

Consequently, the specific heat goes as $C = \pi c T/(3 v_F)$ at low temperatures. The actual equation following from the minimization can be solved exactly in the limit $T \to 0$ with the help of the Bethe equations, giving an expression for c in terms of the dilogarithmic function. The ideas and techniques underlying this second approach to the computation of c constitute the heart of the thermodynamic Bethe ansatz, which is very useful for studying perturbed conformal field theories.

Let us now turn to the scaling dimensions and spins of the primary fields of the theory, following the finite-size approach.

On the one hand, we compute the transformation of the two-point function of a primary field $\Phi_{\Delta,\overline{\Delta}}(z,\overline{z})$ under the conformal mapping $z \to w = (L/2\pi)\log z$, which maps the plane to a cylinder of circumference L. With the help of the transformation laws (8.73) and (8.74), we find

$$\left\langle \Phi_{\Delta,\overline{\Delta}}(w,\overline{w})\Phi_{\Delta,\overline{\Delta}}(0,0)\right\rangle_{\text{cyl}} = \left(\frac{2\pi}{L}\right)^{2(\Delta+\overline{\Delta})} \left(e^{2\pi w/L}\right)^{\Delta} \left(e^{2\pi\overline{w}/L}\right)^{\overline{\Delta}} \times \left\langle \Phi_{\Delta,\overline{\Delta}}(e^{2\pi w/L}, e^{2\pi\overline{w}/L})\Phi_{\Delta,\overline{\Delta}}(1,1)\right\rangle_{\text{plane}} \tag{8.124}$$

We write $w = \tau + i\sigma$, and take into account that

$$\left\langle \Phi_{\Delta,\overline{\Delta}}(z_1,\overline{z}_1)\Phi_{\Delta,\overline{\Delta}}(z_2,\overline{z}_2)\right\rangle = \frac{C_{\Delta\Delta 1}}{z_{12}^{2\Delta}\overline{z}_{12}^{2\overline{\Delta}}} \tag{8.125}$$

where $C_{\Delta\Delta 1}$ is a structure constant or, better, an operator product expansion constant. Then the right-hand side of (8.125) becomes

$$\frac{\left(\frac{\pi}{L}\right)^{2(\Delta+\overline{\Delta})} C_{\Delta\Delta 1}}{\left(\sinh\frac{\pi w}{L}\right)^{2\Delta}\left(\sinh\frac{\pi\overline{w}}{L}\right)^{2\overline{\Delta}}} = \frac{\left(\frac{2\pi}{L}\right)^{2d} C_{\Delta\Delta 1}}{e^{2\pi d\tau/L}\, e^{2\pi s\sigma/L}}\left[1 + O(x^{-\tau}, x^{\sigma})\right] \tag{8.126}$$

where the short-hand $x = e^{-2\pi/L}$ has been used. On the other hand, the analog to the computation (8.124) for the hamiltonian system, where τ is the time, gives

$$\langle\Phi(w)\Phi(0)\rangle = \sum_n \langle 0|\,\Phi(\sigma)\,|n\rangle\; e^{-(E_n-E_0)\tau}\,\langle n|\,\Phi(0)\,|0\rangle \tag{8.127}$$

where now Φ should be interpreted as an operator acting on the Hilbert space of the spin chain.

Comparing (8.126) with (8.127), we find that to a primary field $\Phi_{(\Delta,\overline{\Delta})}$ of the conformal field theory there corresponds an energy eigenvalue

$$E_{\Delta,\overline{\Delta}} = E_0 + \frac{2\pi d}{L} = E_0 + \frac{2\pi}{L}(\Delta + \overline{\Delta}) \tag{8.128}$$

Analogously, to match the second exponential in (8.126), we need the momentum operator for the spin chain. The result is that for every primary field there is a momentum eigenvector with eigenvalue $2\pi s/L = (2\pi/L)(\Delta - \overline{\Delta})$. The universality class of the critical limit of the XXZ chain is the gaussian model, a theory with one two-dimensional free scalar field.

The identification (8.128) allows us to find the scaling dimensions of the primary fields of the conformal field theory from the spectrum of the spin chain hamiltonian. We summarize these results in table 8.3.

Next, we summarize, without proof, some results concerning the value of the central extension associated with the continuum limit of some spin chains.

Table 8.3 Some basic definitions of the lattice model and the conformal field theory describing its critical limit.

Lattice model	Conformal field theory
ground state energy	central extension
energy spectrum	scaling dimensions
momentum spectrum	spins

(i) The continuum limit of the closed and periodic XXZ spin chain with anisotropy $-1 \leq \Delta \leq 1$ is described by a conformal field theory with $c = 1$.

(ii) The continuum limit of the quantum group invariant open spin chain of section 3.7,

$$H = \frac{1}{2}\left[\sum_{j=1}^{L-1}\left(\sigma_j^x\sigma_{j+1}^x + \sigma_j^y\sigma_{j+1}^y + \frac{q+q^{-1}}{2}\sigma_j^z\sigma_{j+1}^z\right)\right. \\ \left. + \frac{q-q^{-1}}{2}\left(\sigma_1^z - \sigma_L^z\right)\right] \tag{8.129}$$

at the special point where

$$q = \exp\frac{i\pi}{Q+1} \tag{8.130}$$

has central charge

$$c = 1 - \frac{6}{Q(Q+1)} \tag{8.131}$$

It is therefore one of the unitary minimal models discussed in this chapter.

(iii) The continuum limit of the isotropic spin chain

$$H = \sum_{j=1}^{L}\sum_{n=1}^{2S} a_n\left(\vec{S}_j \cdot \vec{S}_{j+1}\right)^n \tag{8.132}$$

where $\vec{S}_j$ are $S \times S$ matrices representing the generators of $SU(2)$, is a conformal field theory with central extension

$$c = \frac{3S}{S+1} \tag{8.133}$$

We shall see in chapter 10 that this conformal field theory is a Wess–Zumino model, the chiral algebra of which is $A_1^{(1)}$, and all representations are of level $k = 2S$.

Result (*ii*) justifies returning to the discussion, at the end of section 3.7, concerning the space of the low lying excitations for these models. The decoupling of null vectors in minimal models, implying the truncation of the fusion algebra, should have a natural correspondence in the physics of the quantum invariant chain. One way to understand the truncation for minimal models is to replace the Hilbert space $\mathscr{H}_{(n)}$ of section 3.7 by $\oplus\left[W^{n,j}\otimes V^j\right]$, where we decompose $\otimes^n V^{\frac{1}{2}}(q)$ with the help of (8.116), having set $k = Q-1$. The dimension of the space of n elementary excitations is reduced accordingly, and there are no physical states with spin $j > \frac{1}{2}(Q-1)$. What kind of particle-like structure would you expect from these excitations?

Exercises

Ex. 8.1 Prove the following expression for the schwartzian derivative:

$$S(y,x) = \left(\frac{dz}{dx}\right)^2 [S(y,z) - S(x,z)] = \left(\frac{dy}{dx}\right)^2 S(x,y) \tag{8.134}$$

and use it to show that the law (8.60) for changes of co-ordinates is consistent under the composition $x \to y \to z$.

Ex. 8.2 Landau–Ginsburg lagrangians [A.B. Zamolodchikov, *Sov. J. Nucl. Phys.* **44** (1986) 529].

Consider the lagrangian for a two-dimensional scalar field

$$\mathscr{L} = \frac{1}{2}\partial\phi\bar{\partial}\phi + g\phi^{2(m-1)} \tag{8.135}$$

We should expect that a conformal field theory is associated with a non-trivial fixed point of the renormalization group flow of this theory. At the fixed point, the operator algebra of the theory (8.135) contains the operators

$$\mathbf{1}, \phi, \phi^2, \phi^3, \ldots \tag{8.136}$$

satisfying the operator product expansion

$$\phi(z,\bar{z})\phi^n(0) \sim (z\bar{z})^{d_{n+1}-d_n-d_1}\phi^{n+1}(0) \tag{8.137}$$

where d_n is the anomalous dimension of the operator ϕ^n. In this exercise, we shall establish the connection between a minimal model and the field theory (8.135) at its fixed point. After identifying

$$\phi = \Phi_{2,2} \tag{8.138}$$

in Kac's notation, check that (8.137) is consistent with the identifications

$$\phi^p = \begin{cases} \Phi_{p+1,p+1} & \text{if } 1 \le p \le m-2 \\ \Phi_{p-m+3,p-m+2} & \text{if } m-1 \le p \le 2m-4 \end{cases} \tag{8.139}$$

or, equivalently,with the help of the symmetry $\Delta_{p,q} = \Delta_{m-p,m+1-q}$ in a minimal model of central charge $c = 1 - 6/(m + m^2)$,

$$\phi^p = \begin{cases} \Phi_{m-p-1,m-p} & \text{if } 1 \leq p \leq m-2 \\ \Phi_{2m-p-3,2m-p-1} & \text{if } m-1 \leq p \leq 2m-4 \end{cases} \tag{8.140}$$

These identifications indicate that (8.135) is the field theory description of the mth unitary minimal model.

Ex. 8.3 A "b–c system of spin j" consists of two Grassman or anticommuting fields, a field b of conformal weight j and a field c of weight $(1-j)$, with singular operator product expansion

$$b(z)c(\zeta) = -c(\zeta)b(z) = \frac{1}{z-\zeta} \tag{8.141}$$

and energy–momentum tensor

$$T_{bc} = -jb\partial c - (1-j)c\partial b \tag{8.142}$$

Check that the Virasoro modes which follow from this energy–momentum tensor are

$$L_m^{(bc)} = \sum_n [m(j-1) - n] b_{m+n} c_{-n} \tag{8.143}$$

and that the central charge is

$$c_{bc} = -2(6j^2 - 6j + 1) \tag{8.144}$$

It is amusing to note that usual Dirac fermions are just a particular case of b–c systems, with $j = \frac{1}{2}$ and $c = 1$. Since Dirac fermions consist of two chiral fermions, the latter have a central charge $c = \frac{1}{2}$ and enjoy the energy–momentum tensor

$$T(z) = \frac{1}{2} \ :\psi(z)\partial\psi(z): \tag{8.145}$$

with operator product expansion

$$\psi(z)\psi(\zeta) = \frac{1}{z-\zeta} + \text{regular terms} \tag{8.146}$$

Ex. 8.4 Similarly, a "spin-j β–γ system" consists of a bosonic field β of weight j and a bosonic field γ of weight $(1-j)$, with

$$\beta(z)\gamma(\zeta) = \gamma(\zeta)\beta(z) = \frac{1}{z-\zeta} \tag{8.147}$$

and

$$T_{\beta\gamma} = j\beta\partial\gamma + (1-j)\gamma\partial\beta \tag{8.148}$$

Check that the central charge of (8.148) is the opposite of (8.144).

Ex. 8.5 Find the bosonization map from a spin-j b–c system to a set of two-dimensional scalar quantum fields $\phi_i(z)$ such that the fields b and c can be expressed in terms of exponentials of ϕ_i, and such that the energy–momentum tensor $T_{\{\phi_i\}}$ has the same central charge as the original b–c system.

Ex. 8.6 Using a bosonization map analogous to that of the previous exercise, show that a spin-j β–γ system is equivalent to a b–c system of spin 1 and a free scalar field with a charge at infinity.

Ex. 8.7 Verify the Kac formula (8.88) at level three. Try to prove the formula in general.

Ex. 8.8 Zamolodchikov's c theorem [A. B. Zamolodchikov, 'Infinite additional symmetries in two-dimensional conformal quantum field theory', *Teo. Mat. Fiz.* **65** (1985) 347; 'Renormalization group and perturbation theory about fixed points in two-dimensional field theory', *Yad. Fiz.* **46** (1987) 1819; 'Irreversibility of the flux of the renormalization group in a two-dimensional field theory', *JETP Lett.* **43** (1989) 730].

In conformal field theory, we have

$$\bar{\partial} T = \partial \overline{T} = 0 \tag{8.149}$$

as a consequence of the tracelessness condition $T^{\mu}_{\mu} = 0$. Away from the critical point, the conservation of the energy–momentum tensor reads as follows:

$$\begin{aligned} \bar{\partial} T(z) &= -\frac{1}{4}\partial\left(T^{\mu}_{\mu}\right) \equiv -\frac{1}{4}\partial\Theta \\ \partial \overline{T}(\bar{z}) &= -\frac{1}{4}\bar{\partial}\left(T^{\mu}_{\mu}\right) \equiv -\frac{1}{4}\bar{\partial}\Theta \end{aligned} \tag{8.150}$$

Let $\mathscr{L}_{\mathrm{CFT}}$ denote the lagrangian of a conformal field theory, and consider the perturbed lagrangian

$$\mathscr{L} = \mathscr{L}_{\mathrm{CFT}} + \lambda \int d^2 z \Phi(z, \bar{z}) \tag{8.151}$$

where $\Phi(z, \bar{z})$ is some primary field of conformal weight $\Delta \neq 1$. Prove that, for the theory with lagrangian $\mathscr{L}$,

$$\bar{\partial} T(z) = \lambda(1 - \Delta)\partial\Phi \tag{8.152}$$

Define now the following two-point functions:

$$\begin{aligned} F(z, \bar{z}) &= z^4 \langle T(z, \bar{z}) T(0, 0)\rangle \\ G(z, \bar{z}) &= z^3 \bar{z} \langle T(z, \bar{z}) \Theta(0, 0)\rangle \\ H(z, \bar{z}) &= z^2 \bar{z}^2 \langle \Theta(z, \bar{z}) \Theta(0, 0)\rangle \end{aligned} \tag{8.153}$$

Define Zamolodchikov's c function as

$$C = 2F - G - \frac{3}{8}H \tag{8.154}$$

and prove that its "equation of motion" is

$$z\,\bar{z}\,C' \;=\; -\frac{3}{4}\,H \tag{8.155}$$

Interpret the physical meaning of this result.

Ex. 8.9 [G. von Gehlen, V. Rittenberg and H. Ruegg, *J. Phys* **A18** (1985) 107; M. Henkel, *J. Phys* **A20** (1987) 995.]

Consider the XY spin chain hamiltonian in an external field

$$H = -\frac{1}{2}\sum_{j=1}^{L}\left[\frac{1+\eta}{2}\sigma_j^x\sigma_{j+1}^x + \frac{1-\eta}{2}\sigma_j^y\sigma_{j+1}^y + t\,\sigma_j^z\right] \tag{8.156}$$

which is critical when $t = 1$. Imposing periodic boundary conditions, compute the ground state energy

$$E_0 = -\sum_{j=0}^{L-1}\sqrt{\sin^2\frac{\pi(2j+1)}{2L} - (1-\eta^{-2})\sin^4\frac{\pi(2j+1)}{2L}} \tag{8.157}$$

Expanding for large L, show that the critical free energy density $f = E_0/L$ is given by

$$f = -\frac{1}{\pi}\left(1+\frac{\arccos\eta}{\eta\sqrt{1-\eta^2}}\right) - \frac{\pi}{12}\frac{1}{L^2} + \frac{7\pi^3}{960}\left(\eta^{-2}-\frac{4}{3}\right)\frac{1}{L^4} + O(L^{-6}) \tag{8.158}$$

and thus that the central charge of the conformal field theory describing the critical limit is $\frac{1}{2}$. Compute the same quantity f for antiperiodic boundary conditions, and compare.

9

Duality in conformal field theories

In chapter 8, we characterized a conformal field theory by its symmetry algebra, namely the Virasoro algebra or, more generally, the chiral algebra. We have seen that in the minimal models of type (p, p'), with central charge $c < 1$, the existence of degenerate representations of this algebra restricts enormously the operator content of the theory. The construction and classification of conformal field theories can thus be formulated as mathematical problems in the representation theory of infinite-dimensional chiral algebras. This chapter is devoted to a pursuit of this more formal approach to conformal field theories, along the very same lines as the "operator formalism" of string theory.

The operator formalism of conformal field theories produces some simple diagrammatics to describe physical states and their correlation functions. The building blocks for such diagrams are two-dimensional orientable Riemann surfaces, with topology classified by the genus or number of handles and by the number of punctures or local operator insertions. Having the diagrams at hand, one is immediately compelled to analyze the behavior of states and correlators under "dual transformations", symmetry operations which eventually call for the interpretation of the Riemann surface as the world-sheet of an extended object, the quantum string. Under a duality transformation, for instance, a four-particle correlator on the sphere is mapped to another four-particle correlator. If we think of the correlator as an amplitude for the scattering of two particles into two particles, then through a duality transformation we may map the amplitude in the s channel to that in the t or in the u channels (s, t and u are the Mandelstam variables; see figure 9.1). A dual theory, such as string theory, admits different factorizations of one single physical process; duality transformations relate these various factorizations, and the dual theory is invariant under their action.

Although the minimal (p, p') models fulfil all the requirements imposed in chapter 8 to describe critical behavior, two intriguing questions arise immediately.

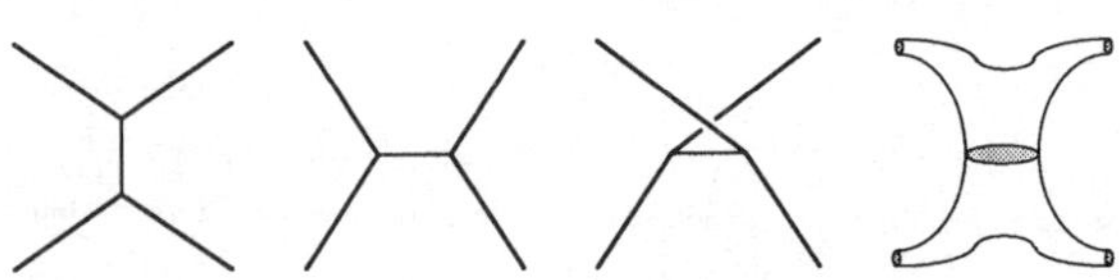

Fig. 9.1. The s, t and u channels of a $2 \to 2$ scattering in field theory, and their common "stringy" equivalent, which, topologically, is a sphere with four punctures.

First, how do we relate the holomorphic and antiholomorphic parts? In order to ensure duality invariance of a conformal field theory, one has to consider the full correlation functions, i.e. both their holomorphic and antiholomorphic parts. Duality imposes restrictions on the way that these two pieces may be assembled together to construct monodromy-invariant physical correlation functions, or modular-invariant partition functions. Secondly, could we not describe criticality with a thermal subalgebra only? In the Ising model with fusion rules (8.117), what goes wrong if we restrict the primary field content to $\{\mathbf{1}, \varepsilon\}$, which closes under fusion? We shall deal with these questions now, and we will see that the answers are related.

The structure of this chapter is as follows. First, we shall introduce the basic notions of monodromy transformations, conformal blocks and chiral vertex operators. Then, we shall explain the technique for sewing Riemann surfaces together. This will lead us to the study of the braiding and fusion of conformal blocks. In particular, we shall investigate the polynomial equations at genus zero, which establish a fruitful bridge between conformal field theories and the face models of chapter 5. This connection is further clarified by the relationship between conformal field theories and towers of algebras. Finally, we shall study the polynomial equations at genus one, and give a proof of Verlinde's theorem, which relates the fusion rules of a rational conformal field theory to the monodromy matrix S of modular transformations.

9.1 Monodromy invariance

Let us consider a generic four-point amplitude

$$A^{ijk\ell|\bar{\imath}\bar{\jmath}\bar{k}\bar{\ell}}(z_1, \bar{z}_1; \ldots; z_4, \bar{z}_4) = \langle 0| \Phi_{i\bar{\imath}}(z_1, \bar{z}_1) \cdots \Phi_{\ell\bar{\ell}}(z_4, \bar{z}_4) |0\rangle \tag{9.1}$$

where $\Phi_{i\bar{\imath}}(z, \bar{z})$ is the primary field associated with the irreducible representation $(i, \bar{\imath})$ of Virasoro $\otimes$ $\overline{\text{Virasoro}}$.

We are assuming that the fields $\Phi_{i,\bar{\imath}}$ are all mutually local, that is to say that the transport of one field around another one (i.e. the monodromy) does not produce any phase factors. For this reason, the correlator (9.1) must be single-valued in its variables: in order to interpret (9.1) as a physical correlation function, we must ensure its invariance under monodromy transformations. Note that there are times when the fields in a correlator are not mutually local; this happens, for instance, when a correlator contains both order and disorder fields, with non-trival monodromy. We shall not consider this possibility here, and we concentrate on correlators of mutually local fields.

To investigate the monodromy transformations, let us first use $SL(2)$ invariance to fix the values of three of the insertion points, say z_1, z_2 and z_3 (and also $\bar{z}_1$, $\bar{z}_2$ and $\bar{z}_3$, of course). Next, consider a non-contractible path γ in $\mathbb{C} - \{z_1, z_2, z_3\}$ with base point z_4. This path is characterized by a continuous function $z_4^\gamma : [0, 1] \to \mathbb{C} - \{z_1, z_2, z_3\}$ such that $z_4^\gamma(0) = z_4^\gamma(1) = z_4$ (see figure 9.2). Associated with this path γ is the monodromy transformation M_γ of the four-point function (9.1); it is defined by

$$A^{ijk\ell|\bar{\imath}\bar{\jmath}\bar{k}\bar{\ell}}\left(z_1,\bar{z}_1;z_2,\bar{z}_2;z_3,\bar{z}_3;z_4^\gamma[1],\bar{z}_4^\gamma[1]\right)$$
$$= M_\gamma\, A^{ijk\ell|\bar{\imath}\bar{\jmath}\bar{k}\bar{\ell}}\left(z_1,\ldots,\bar{z}_3;z_4^\gamma[0],\bar{z}_4^\gamma[0]\right) \tag{9.2}$$

Invariance under monodromy transformations means simply that $M_\gamma = 1$ for any path γ or, in other words, that physical amplitudes are single-valued functions on the complex sphere.

To familiarize ourselves with monodromy transformations, consider the three-point function

$$A^{ijk|\bar{\imath}\bar{\jmath}\bar{k}}(z,\bar{z}) = \langle 0|\,\Phi_{\bar{\imath}\bar{\imath}}(\infty,\infty)\Phi_{j\bar{\jmath}}(z,\bar{z})\Phi_{k\bar{k}}(0,0)\,|0\rangle \tag{9.3}$$

Using the operator product expansion (8.93), we obtain

$$A^{ijk|\bar{\imath}\bar{\jmath}\bar{k}}(z,\bar{z}) = z^{\Delta_i-\Delta_j-\Delta_k}\,\bar{z}^{\overline{\Delta}_{\bar{\imath}}-\overline{\Delta}_{\bar{\jmath}}-\overline{\Delta}_{\bar{k}}}C^{i\bar{\imath}}_{j\bar{\jmath},k\bar{k}} \tag{9.4}$$

Now consider the path γ based at z and going around zero counter-clockwise. The corresponding monodromy transformation M_γ is obtained by letting $z \to e^{2\pi i}z$ in (9.4):

$$M_\gamma = \exp[2\pi i(\Delta_i - \Delta_j - \Delta_k - \overline{\Delta}_{\bar{\imath}} + \overline{\Delta}_{\bar{\jmath}} + \overline{\Delta}_{\bar{k}})] \tag{9.5}$$

Therefore, monodromy invariance requires that the difference between the outgoing (∞) and incoming (0 and z) spins $s = \Delta - \overline{\Delta}$ is an integer. A similar analysis for two- and one-point functions shows that a necessary condition for invariance under monodromy transformations is that the spins of all primary fields are integers.

We have found a general feature of physical amplitudes, namely that the holomorphic and antiholomorphic contributions are not independent: they must combine appropriately to yield well-defined monodromy invariants.

We have seen in section 8.8 that one way to characterize the chiral components of a generic n-point correlation function is to solve the differential equations

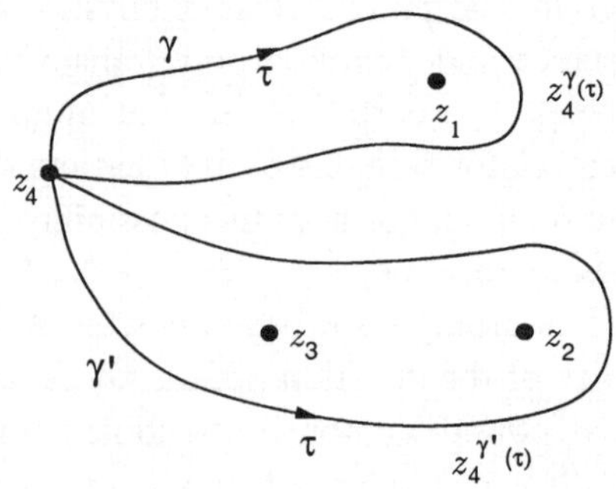

Fig. 9.2 Examples of non-contractible paths in $\mathbb{C} - \{z_1, z_2, z_3\}$ with base point z_4.

following from the decoupling of null vectors. In fact, the decoupling equations are completely determined by the null vectors in the Verma modules of the primary fields in the amplitude, affecting the holomorphic or antiholomorphic components separately. The dependence on z (respectively $\bar{z}$) of the four-point amplitude (9.1), for instance, is constrained by the differential equations of order N enforcing the decoupling of the null vectors of level N in the Verma modules of i, j, k and ℓ (and also $\bar{\imath}$, $\bar{\jmath}$, $\bar{k}$, $\bar{\ell}$). Consider the space of (anti)holomorphic functions which are solutions to the decoupling equations. The dimension of this space is bounded above by the minimal order N of all the decoupling equations, i.e. by the lowest level at which a null vector occurs in either of $\mathcal{V}^i$, $\mathcal{V}^j$, $\mathcal{V}^k$ or $\mathcal{V}^\ell$. Now, if $f^{ijk\ell}(z_1, z_2, z_3, z_4)$ is a solution to the decoupling equations, then so is $M_\gamma f^{ijk\ell}(z_1, z_2, z_3, z_4)$, for any monodromy transformation M_γ. More precisely, the space of solutions to the decoupling equations defines a representation space for the monodromy transformations. We can thus expand the four-point amplitude in a basis $f_\lambda^{ijk\ell}(z_1, z_2, z_3, z_4)$ of the space of solutions to the decoupling equations:

$$\begin{aligned} A^{ijk\ell|\overline{ijk\ell}}&(z_1, \bar{z}_1; z_2, \bar{z}_2; z_3\bar{z}_3; z_4, \bar{z}_4) \\ &= \sum_{\lambda\bar{\lambda}} c_{\lambda\bar{\lambda}} f_\lambda^{ijk\ell}(z_1, z_2, z_3, z_4) f_{\bar{\lambda}}^{\overline{ijk\ell}}(\bar{z}_1, \bar{z}_2, \bar{z}_3, \bar{z}_4) \end{aligned} \tag{9.6}$$

where the constants $c_{\lambda\bar{\lambda}}$ will be determined in terms of the monodromy transformations of the chiral components f, by requiring monodromy invariance of the physical amplitude.

In summary, we have arrived at the factorized representation (9.6) of the four-point correlation function from requiring both the decoupling of null vectors and the invariance under monodromy transformations. Even if there were no null vectors to decouple, however, equation (9.6) would still be true.

9.2 Conformal blocks and chiral vertex operators

In the above discussion, the chiral components $f_\lambda^{ijk\ell}$ were defined as a basis of the space of solutions to the decoupling equations. The dimensionality of this space is given by the number of possible intermediate channels, namely $\sum_p N_{ij}^p N_{pk}^\ell$. (Recall our derivation of the fusion algebra from the discussion on null vectors.) Let us now address the construction of these chiral components.

It is clear from the operator product expansion (8.93) that the amplitude (9.1) can be written in a form similar to (9.6), with the constants $c_{\lambda\bar{\lambda}}$ replaced by the product of two structure constants, one from each holomorphic and antiholomorphic vertex of the operator algebra. The dependence of the n-point chiral amplitude on the co-ordinates is partially determined by conformal invariance. In particular, by a conformal transformation, we can always fix three of the insertion points to be at 0, 1 and ∞:

$$z_1 = \bar{z}_1 = 0\,, \qquad z_2 = \bar{z}_2 = 1\,, \qquad z_n = \bar{z}_n = \infty \tag{9.7}$$

The non-trivial dependence of the n-point amplitude that we now want to constrain with the decoupling equations is therefore reduced to $(n-3)$ complex co-ordinates. This is termed the dependence of the amplitude on its "moduli". When $n = 4$, for instance, we may choose as modular co-ordinate the harmonic ratio

$$\eta = \frac{z_{12}\, z_{34}}{z_{13}\, z_{24}} \tag{9.8}$$

where $z_{ij} = z_i - z_j$. With the help of the operator product expansion (8.94), and assuming for simplicity that $N_{ij}^{\ k} \in \{0, 1\}$, we find

$$A^{ijk\ell|\overline{ijk\ell}}(\eta, \overline{\eta}) = \sum_{p\overline{p}} C_{i,j}^{p} C_{k,\ell,p}\, \mathscr{F}_{p}^{ijk\ell}(\eta)\, \overline{\mathscr{F}}_{\overline{p}}^{\overline{ijk\ell}}(\overline{\eta}) \tag{9.9}$$

where the conformal blocks $\mathscr{F}_{p}^{ijk\ell}(\eta)$ are defined as

$$\begin{aligned} \mathscr{F}_{p}^{ijk\ell}(\eta) =& \eta^{\Delta_p - \Delta_k - \Delta_\ell} \sum_{P_{\{r\}}} \eta^r\, \beta_{k\ell}^{p\ \{\mathbf{r}\}} \\ &\times \frac{\langle 0|\, \Phi_i(\infty)\Phi_j(1) P_{\{r\}}(L_{-n})\Phi_p(0)\, |0\rangle}{\langle 0|\, \Phi_i(\infty)\Phi_j(1)\Phi_p(0)\, |0\rangle} \end{aligned} \tag{9.10}$$

Comparing (9.6) with (9.9), we establish the important result that the structure constants of the operator algebra $C_{ij}^{\ k}$ are determined from the condition of monodromy invariance for the physical amplitudes. Therefore, the operator product expansion encodes both the information on decoupling of null vectors and on monodromy invariance.

The representation (9.10) of the chiral components of the amplitude sheds light on the physical meaning of the dimension of the space of solutions to the decoupling equations. In fact, the basis (9.10) of this space indicates, quite physically, that each conformal block is associated with a possible intermediate channel p. The picture for the multi-valued chiral components of the amplitude $\mathscr{F}_{p}^{ijk\ell}(\eta)$ derived from the representation (9.10) is already familiar to us:

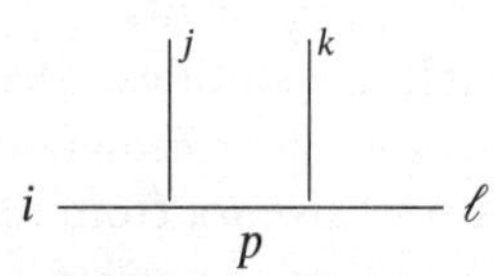

(9.11)

From the point of view of standard quantum field theory, the chiral process represented by the conformal block (9.11) could be interpreted as a Feynman diagram for some cubic theory. In fact, we can introduce Feynman rules for the propagator and the cubic vertex of this make-believe chiral theory quite nicely. Let us start with the vertex, which characterizes completely the chiral component of a three-point amplitude. Let us allow for more general fusion rules than just 0 and 1, i.e. let us assume that the three-point vertices are labeled by some $\alpha = 1, \ldots, N_{ij}^{k}$. Also, we fix the co-ordinates of the three insertion points to be 0, z and ∞. The left-over freedom from conformal invariance completely determines the dependence on z of the three-point amplitude, equation (9.4). The three-point conformal block is thus

$$f^{ijk}_{(\alpha)} =$$ 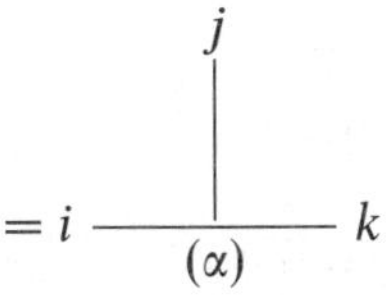

(9.12)

The chiral three-point functions $f^{ijk}_{(\alpha)}(z)$ define a basis for the space of solutions to the decoupling of null vectors in the Verma modules $\mathscr{V}^i$, $\mathscr{V}^j$ and $\mathscr{V}^k$. The dimension of this space is precisely given by the fusion rule N^k_{ij}. To represent chiral n-point functions, we thus need to indicate in general the label for each intervening three-point vertex, as in the following four-point correlator:

$$\mathscr{F}^{ijk\ell}_{p;(\alpha,\beta)} =$$ 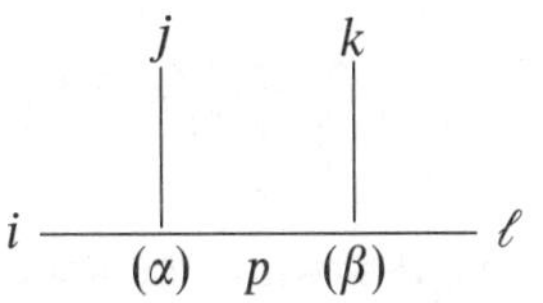

(9.13)

Even when the vertex label (α) is not shown explicitly, it should be understood.

We associate with each solution $f^{ijk}_{(\alpha)}(z)$ of the decoupling equations an operator

$$\Phi^{k\ (\alpha)}_{ij}(z) : \mathscr{V}^i \otimes \mathscr{V}^j \longrightarrow \mathscr{V}^k \tag{9.14}$$

satisfying

$$\langle k|\, \Phi^{k\ (\alpha)}_{ij}(z)\, |i\rangle \otimes |j\rangle = f^{ijk}_{(\alpha)}(z) \qquad \lambda = 1, \ldots, N^k_{ij} \tag{9.15}$$

where $|i\rangle$, $|j\rangle$ and $|k\rangle$ are the highest weight vectors of $\mathscr{V}^i$, $\mathscr{V}^j$ and $\mathscr{V}^k$, respectively.

The operators $\Phi^{k\ (\alpha)}_{ij}(z)$ are called the chiral vertex operators. From their definition (9.15) we see that they satisfy the following equation of motion:

$$\left[\frac{d}{dz}\Phi^{k\ (\alpha)}_{ij}(z)\right] \big(|u\rangle \otimes |v\rangle\big) = \Phi^{k\ (\alpha)}_{ij}(z) \left[|u\rangle \otimes \rho^j(L_{-1})\,|v\rangle\right] \tag{9.16}$$

where we denote by ρ^j the representation j of the Virasoro generators, and $|u\rangle$, $|v\rangle$ belong to the Verma modules of $|i\rangle$ and $|j\rangle$, respectively.

The chiral vertex operators are just the chiral version of the operator product expansion of primary field in a conformal field theory. We could be tempted to identify the chiral vertex operators with the intertwiners of the chiral algebra, but this identification is wrong because the co-multiplication rule for the Virasoro algebra $\Delta L_n = L_n \otimes \mathbf{1} + \mathbf{1} \otimes L_n$ implies that the central extensions add up, which is not the case for chiral vertex operators, since $\mathscr{H}_i$, $\mathscr{H}_j$ and $\mathscr{H}_k$ are Verma modules associated to the same central extension. Nevertheless, following Moore and Seiberg we may introduce a "geometrical co-multiplication" given by the contour deformation of Virasoro. Indeed, denoting by $L^{(\infty)}_n$ the Virasoro operators acting on the Hilbert space $\mathscr{H}_k$ at $z = \infty$, we have

$$L^{(\infty)}_n = \oint_{C_\infty} \frac{d\zeta}{2\pi i} \zeta^{n+1} T(\zeta) \tag{9.17}$$

Since $T(\zeta)$ is analytic, we may deform the contour at infinity around zero and z, the insertion points of $\mathscr{H}_i$ and $\mathscr{H}_j$:

$$\begin{aligned} L_n^{(\infty)} &= \oint_{C_0} \frac{d\zeta}{2\pi i} \zeta^{n+1} T(\zeta) + \oint_{C_z} \frac{d\zeta}{2\pi i} \zeta^{n+1} T(\zeta) \\ &= L_n^{(0)} + \sum_{k=0}^{\infty} \binom{n+1}{k} z^{n+1-k} L_{k-1}^{(z)} \end{aligned} \tag{9.18}$$

The chiral vertex operators commute with contour deformations, so we reach the intertwining condition

$$\begin{aligned} L_n^{(\infty)} \Phi_{ij}^k(z) \, |i\rangle \otimes |j\rangle &= \Phi_{ij}^k(z) \Big[L_n^{(0)} \, |i\rangle \otimes |j\rangle \\ &\quad + \sum_{k=0}^{\infty} \binom{n+1}{k} z^{n+1-k} \, |i\rangle \otimes L_{k-1}^{(z)} \, |j\rangle \Big] \end{aligned} \tag{9.19}$$

This equation motivates the following "geometric" co-multiplication of Virasoro:

$$\Delta_z^G L_n = L_n \otimes \mathbf{1} + \sum_{k=0}^{\infty} \binom{n+1}{k} z^{n+1-k} \, \mathbf{1} \otimes L_{k-1} \tag{9.20}$$

which has the property that the central extension of $\Delta_z^G L_n$ is the same as that of $L_n \otimes \mathbf{1}$. In section 10.4.1, we shall discuss the intertwiners of a Kac–Moody algebra and relate them to the chiral vertex operators of this section.

The process described by the three-point amplitude can be viewed as taking place in the complex plane. However, it is more useful to work on the Riemann sphere obtained by compactifying the plane at infinity. In this way, once we have chosen three representations i, j and k of Virasoro, we can interpret the decoupling equations as a procedure for associating the Riemann sphere with three punctures P_1, P_2 and P_3 (and co-ordinates $z(P_1) = 0$, $z(P_2) = z$, $z(P_3) = \infty$) with the holomorphic solutions $f_{(\alpha)}^{ijk}(z)$, and therefore the chiral vertex operators $\Phi_{ij}^{k(\alpha)}(z)$. The decoupling equations allow us to associate a chiral vertex operator to the geometric data characterizing a three-punctured Riemann sphere. As we shall see next, these chiral vertex operators are the basic ingredients with which solutions to the decoupling equations for many-point correlation functions can be found.

9.3 Sewing

Once we have defined the chiral vertex operators, a diagrammatic construction of arbitrary chiral n-point functions requires us to compose chiral vertex operators in such a way that the decoupling of null vectors is preserved. Moreover, the composition of chiral vertex operators must yield the co-ordinate dependence required by conformal invariance. To define the composition of chiral vertex operators, we shall use a sewing construction, based on the geometrical fact that any Riemann surface can be obtained by sewing three-punctured Riemann spheres. We shall promote the geometrical sewing to a composition of chiral vertex operators such that the decoupling of null vectors is automatically preserved.

Consider a Riemann sphere with three punctures P_i (i=1, 2, 3), and define local co-ordinates z_i around each of the punctures such that $z_i(P_i) = 0$. We need transition functions f_{ij} among different co-ordinate patches in order to have a manifold structure on the z sphere, $z_i = f_{ij}(z_j)$ $(i \neq j)$. It is convenient to choose cyclic transition functions given by

$$z_i = \frac{1}{1 - z_{i+1}}\,, \qquad z_{i+1} = 1 - \frac{1}{z_i} \tag{9.21}$$

A useful property of this choice is that, for any local co-ordinate z_i, the three punctures are located at the canonical 0, 1 and ∞:

$$z_i(P_i) = 0\,, \qquad z_i(P_{i+1}) = 1\,, \qquad z_i(P_{i+2}) = \infty \tag{9.22}$$

We have used the convention that $P_{i+3} = P_i$.

Let us consider now, quite generally, the sewing of two Riemann surfaces Σ_1 and Σ_2 with punctures P_i and $\tilde{P}_j$. The result of sewing together these two punctured surfaces (figure 9.3) is a new Riemann surface $\Sigma_1 *_{(i,j;q)} \Sigma_2$, obtained by gluing (identifying) a local disk around the puncture P_i of Σ_1 to a local disk around $\tilde{P}_j$ of Σ_2. This gluing operation $*_{(i,j;q)}$ is performed using the transition function $q = z_i z_j$, where z_i and z_j are the local co-ordinates around P_i and $\tilde{P}_j$, respectively.

By direct counting, we see that, if p_i is the genus (number of handles) and n_i the number of punctures of Σ_i, then the sewed surface $\Sigma_1 *_{(i,j;q)} \Sigma_2$ has genus $(p_1 + p_2)$ and has $(n_1 + n_2 - 2)$ punctures. The importance of sewing is that any orientable Riemann surface can be obtained by sewing Riemann spheres with three punctures (figure 9.4). Any punctured Riemann surface is thus completely characterized by the sewing parameters q, the number of which depends only on the genus p and on the number of punctures n:

$$\text{number of sewing parameters} = 3p - 3 + n \tag{9.23}$$

Formula (9.23) shows that the sewing parameters are good local co-ordinates of the moduli space of Riemann surfaces. In particular, different local co-ordinates on the moduli space will correspond to different sewing representations of the

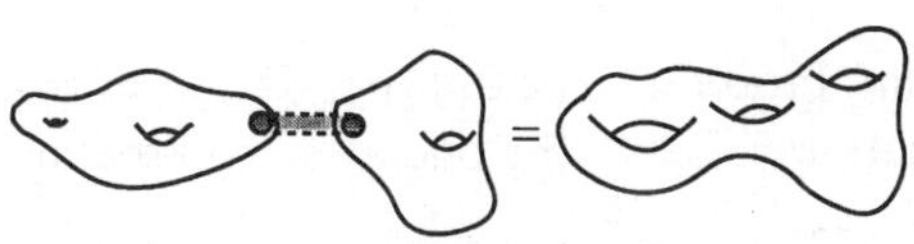

Fig. 9.3 Sewing of two Riemann surfaces.

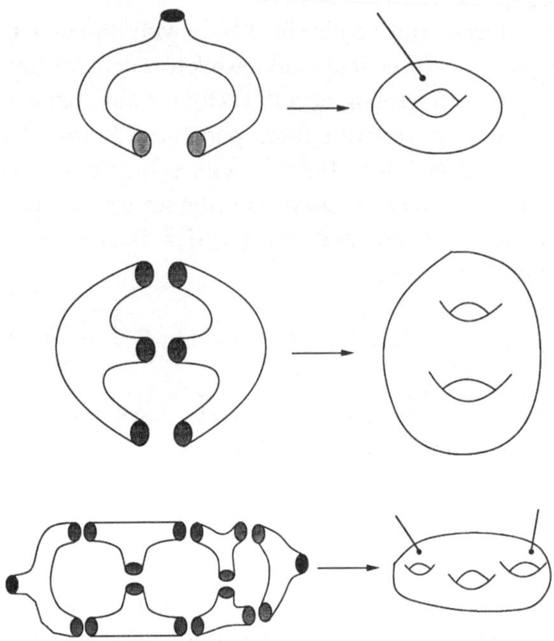

Fig. 9.4 Some examples of sewing three-punctured spheres.

same Riemann surface, i.e. different ways of sewing three-punctured spheres to construct the same surface. As a simple example, we may consider the Riemann sphere with four punctures. According to (9.23), the dimension of the moduli space is, in this case, one. In fact, the Riemann sphere with four punctures can be obtained by sewing two Riemann spheres with three punctures, so the number of required sewing parameters is, indeed, one.

Let us now use this geometrical sewing as a composition law for chiral vertex operators. The z dependence of the four-point chiral amplitude $f_\lambda^{ijk\ell}(z)$ reflects its dependence on the moduli. By sewing chiral vertex operators, we may obtain solutions to the decoupling equations in which the dependence on the moduli co-ordinate is replaced by a dependence on the sewing co-ordinate. Concentrate on the sewing shown in figure 9.5. To start with, the states $|\alpha\rangle \in \mathscr{H}^{(i)}$ and $|\beta\rangle \in \mathscr{H}^{(j)}$ are inserted at the points P_1 and P_2 with z_1 co-ordinates 0 and 1, respectively. Acting with the chiral vertex operator $\Phi_{ij}^m(1)$ produces a state in $\mathscr{H}^{(m)}$ at the point P with $z_1(P) = \infty$. This state can be written as

$$\Phi_{ij}^m(1)\Big[|\alpha\rangle \otimes |\beta\rangle\Big] \in \mathscr{H}^{(m)} \tag{9.24}$$

Following the sewing procedure, we apply to the local co-ordinates around P the sewing transformation $zz' = q_s$, the operator representation of which is

$$\sum_{|\tilde{p}\rangle} q_s^{L_0 - c/24}\, |\tilde{p}\rangle\, \langle \tilde{p}|\, \left[\Phi_{ij}^m(1)\left(|\alpha\rangle \otimes |\beta\rangle\right)\right] \tag{9.25}$$

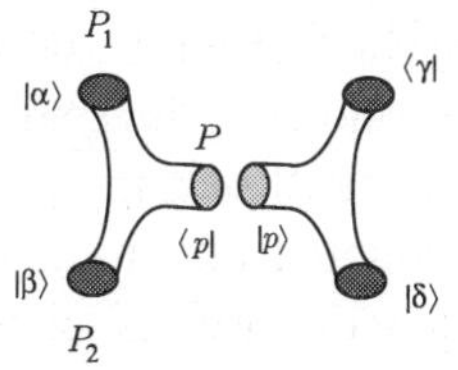

Fig. 9.5

where the sum extends over a basis of the irrep $\mathscr{H}^{(m)}$. The factor $q_s^{L_0-c/24}$ implements the dilatation $z \to q_s z$, and can be understood as the propagator. The generator of dilatations on the plane is L_0, but in the sewing construction we are effectively working on the cylinder, not the plane, and therefore we must also transform the dilatation generator L_0 from the plane to the cylinder. Recall that, under the change of coordinates $z = \mathrm{e}^{\tau+i\sigma} = \mathrm{e}^{w}$, the energy–momentum tensor changes as

$$T_{\text{cyl}}(w) = z^2 T_{\text{plane}}(z) - \frac{c}{24} \tag{9.26}$$

so that

$$(L_0)_{\text{cyl}} = (L_0)_{\text{plane}} - \frac{c}{24} \tag{9.27}$$

which explains the factor in (9.25) above.

To obtain the final expression for the conformal block, we just apply the second chiral vertex operator on the states (9.25):

$$\langle\gamma|\,\Phi^k_{\ell m}(1)\left\{|\delta\rangle\otimes\left(\sum_{|\tilde{p}\rangle} q_s^{L_0-c/24}\,|\tilde{p}\rangle\,\langle\tilde{p}|\,\left[\Phi^m_{ij}(1)\left(|\alpha\rangle\otimes|\beta\rangle\right)\right]\right)\right\} \tag{9.28}$$

where we have already projected onto a final state $|\gamma\rangle \in \mathscr{H}^{(k)}$. Recall that $|\alpha\rangle \in \mathscr{H}^{(j)}$, $|\beta\rangle \in \mathscr{H}^{(j)}$ and $|\delta\rangle \in \mathscr{H}^{(\ell)}$.

Expression (9.28) is automatically a solution to the decoupling equations, and it thus yields the desired diagrammatic representation of the four-point chiral amplitudes $f_\lambda^{ijk\ell}(z)$. It is clear that a basis of solutions to the decoupling equations is one to one related with the possible intermediate channels. In order to match the z dependence of the chiral amplitude with the sewing dependence of (9.28), we must find the relationship between z and q_s. This can be done with the help of the transition functions (9.21). In the coordinates of the "1" patch, the result is

$$z_1(P_1) = 0\,, \quad z_1(P_2) = 1\,, \quad z_1(P_3) = 1 - 1/q_s\,, \quad z_1(P_4) = \infty \tag{9.29}$$

and therefore $z = 1 - 1/q_s$.

Let us stress that the sewing construction replaces the direct solution of the differential equations derived from requiring the decoupling of null vectors. Instead of obtaining the chiral amplitudes $f^{\{i_j\}}_{\{\alpha_k\}}(\{z_\ell\})$ by solving the decoupling equation, the sewing construction starts from the same geometric data and reaches the same chiral amplitude in the following three steps, which constitute the logic of the sewing construction. First, we dissect the punctured Riemann surface on which the correlator is to be evaluated into three-punctured Riemann spheres; this is geometrical unsewing. Secondly, we associate a chiral vertex operator to each Riemann sphere and perform their algebraic sewing, with the help of the sewing transformations to glue them where they meet. Thirdly, to find the chiral amplitude in its standard form, we identify the sewing parameters as in (9.29) above.

To close this discussion of the sewing representation of chiral amplitudes, we summarize the basic "Feynman rules" involved:

(i) chiral vertex operators are defined for the three-holed sphere with punctures at 0, 1 and ∞;
(ii) the dependence on the sewing parameters appears through insertions of $q^{L_0-c/24}$, which implement the dilatation accompanying the sewing change of co-ordinates $zz' = q$ and play the role of propagators;
(iii) we must sum over all possible intermediate states belonging to the same irreducible representation.

At this point, the reader might wonder what would happen if we used a different sewing representation for the same chiral process. As already mentioned, the same Riemann sphere can be represented by sewing three-holed spheres in different ways. In particular, for the Riemann sphere with three punctures, we may use three different sewings corresponding to the various Mandelstam channels (figure 9.6).

Before we can uncover the effect on the chiral amplitude of changing the sewing, we must obtain the same four-punctured Riemann sphere in different

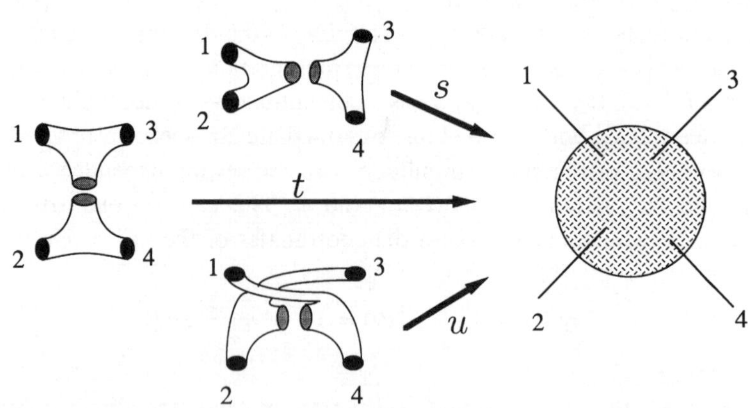

Fig. 9.6. The three different sewing representations of the Riemann sphere with four punctures.

ways. Using the transition functions (9.21) and (9.22), we find for the t sewing

$$z_1(P_1) = 0\,, \quad z_1(P_2) = 1\,, \quad z_1(P_3) = \frac{1}{1-q_t}\,, \quad z_1(P_4) = \infty \tag{9.30}$$

whereas for the u sewing we obtain

$$z_1(P_1) = 0\,, \quad z_1(P_2) = \frac{1}{1-q_u}\,, \quad z_1(P_3) = \infty\,, \quad z_1(P_4) = 1 \tag{9.31}$$

The different sewing procedures yield the same Riemann surface if and only if the values of the moduli are equal. In this case, the only modulus is the harmonic ratio (9.8), which should be equal for the three sets of co-ordinates (9.29), (9.30) and (9.31). Hence,

$$\eta \;=\; \frac{z_{12}\, z_{34}}{z_{13}\, z_{24}} \;=\; \frac{q_s}{q_s - 1} \;=\; 1 - q_t \;=\; \frac{1}{q_u} \tag{9.32}$$

From a physical viewpoint, it is important to stress that the various co-ordinates we have introduced are one to one related with the various crossing channels, and therefore the transition functions reflect the usual duality transformations on the moduli space.

9.4 Braiding and fusion

It is clear from the sewing construction that different sewings define different solutions to the decoupling equations and, therefore, induce a change of basis in the space of chiral amplitudes. Denoting by $\mathscr{F}^{ijk\ell}_{m}(q_s)$ the sewing amplitude (9.28), these changes of basis can be written in a matrix form:

$$\begin{aligned} \mathscr{F}^{ijk\ell}_{m}(q_s) &\longrightarrow \sum_n B^{\pm}_{mn} \begin{bmatrix} j & k \\ i & \ell \end{bmatrix} \mathscr{F}^{ijk\ell}_{n}(q_u) \\ \mathscr{F}^{ijk\ell}_{m}(q_s) &\longrightarrow \sum_n F_{mn} \begin{bmatrix} j & k \\ i & \ell \end{bmatrix} \mathscr{F}^{ijk\ell}_{n}(q_t) \end{aligned} \tag{9.33}$$

These changes of sewing co-ordinates are usually called s–u duality or braiding (B) and s–t duality or fusing (F). Physical amplitudes (9.6) are invariant under them, because of conformal invariance and the single-valuedness of physical amplitudes. This means that the $B^{\pm}$- and F-matrices defining the sewing transformations are unitary with respect to the "metric" defined by the constants $C_{\lambda\bar{\lambda}}$ in (9.6). In (9.33), we have distinguished between two kinds of braiding matrices, B^+ and B^-, in order to differentiate two different transformations: the duality transformation $s \to u$ involves an analytic continuation from the region with $|\eta| < 1$ to that where $|\eta| > 1$, and this continuation can be carried out in two different ways depending on the chosen braiding.

The two fundamental duality transformations can be depicted in terms of conformal blocks as follows:

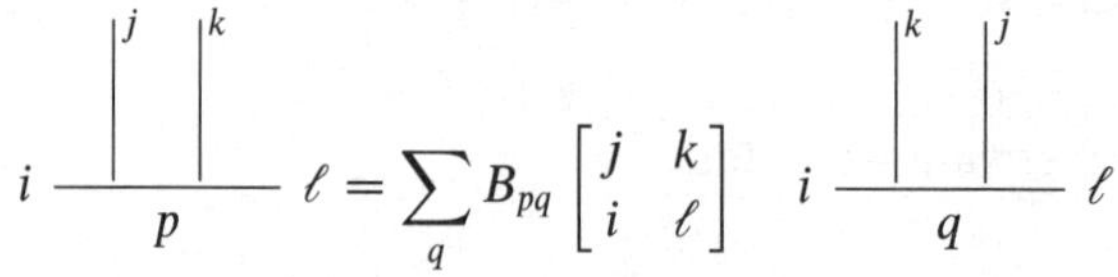

(9.34)

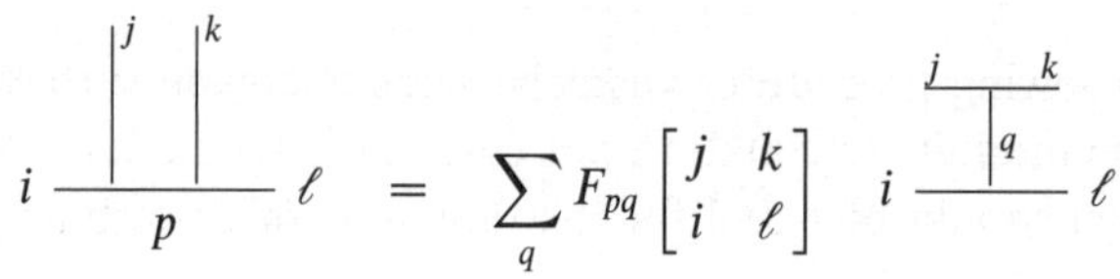

(9.35)

In equation (9.34), it is perhaps clearer to think of the external lines i, k, j, ℓ as strings attached to fixed points away from the diagram. Then, the conformal block on the left-hand side calls for a crossing of the lines labeled j and k: this is the braiding. It is understood that the line of the conformal block labeled k goes under the line j. If line j went over line k, then we would be representing σ^{-1} instead of σ, which is not the case. The operator B^* is not the same thing as B, although the two are certainly related: any B^* operation appears like a B operation if you turn the page and look through the paper.

The following commutative diagram, with σ_{ij} meaning the interchange of the vector spaces attached to lines i and j, is also of interest:

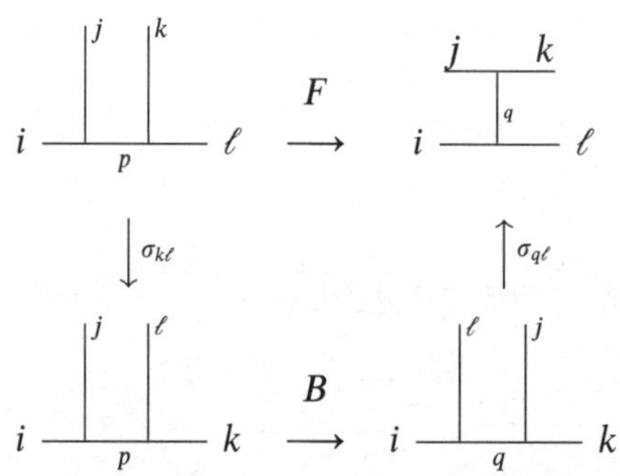

(9.36)

Until now, our construction of physical correlation functions in conformal field theory has been based on two principles:

- decoupling of null vectors, and
- invariance under monodromy transformations.

Inspired by the Riemann–Hilbert problem, we may expect to replace the information about decoupling of null vectors with the actual monodromy matrices. The basic data would be thus the B- and F-matrices, precisely. In conformal field theory, these matrices satisfy the following equations (the sum over repeated

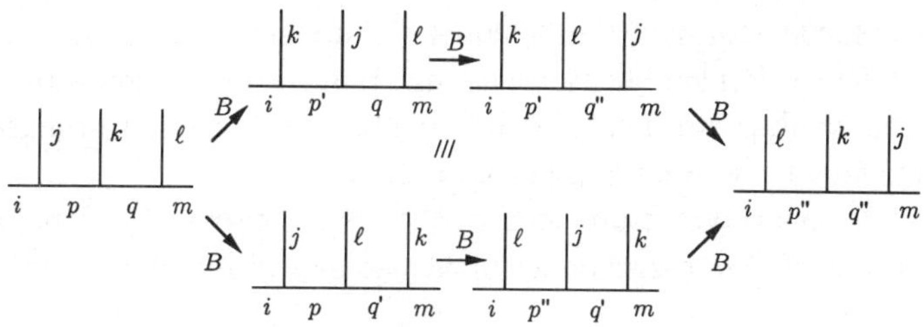

Fig. 9.7 The hexagonal equation (9.37).

indices is understood):

$$B_{p'p''}\begin{bmatrix} k & \ell \\ i & q'' \end{bmatrix} B_{qq''}\begin{bmatrix} j & \ell \\ p' & m \end{bmatrix} B_{pp'}\begin{bmatrix} j & k \\ i & q \end{bmatrix} = B_{q'q''}\begin{bmatrix} j & k \\ p'' & m \end{bmatrix} B_{pp''}\begin{bmatrix} j & \ell \\ i & q' \end{bmatrix} B_{qq'}\begin{bmatrix} k & \ell \\ p & m \end{bmatrix} \tag{9.37}$$

$$F_{qq''}\begin{bmatrix} p' & \ell \\ i & m \end{bmatrix} F_{pp'}\begin{bmatrix} j & k \\ i & q \end{bmatrix} = F_{q'p'}\begin{bmatrix} j & k \\ q'' & \ell \end{bmatrix} F_{pq''}\begin{bmatrix} j & q' \\ i & m \end{bmatrix} F_{qq'}\begin{bmatrix} k & \ell \\ p & m \end{bmatrix} \tag{9.38}$$

These equations correspond to the closed loops of sewing transformations represented in figure 9.7 and figure 9.8, respectively. Note that the summation over indices is restricted by the fusion rules of the theory. Equations (9.37) and (9.38) are part of the polynomial equations characterizing any conformal field theory, and they are called the hexagonal and pentagonal equations, respectively. Note that the pentagonal equation is very close to an identity we found for quantum $6j$ symbols, equation (6.88). In chapter 11, we shall see that this connection is not accidental.

Let us briefly reconsider equations (9.37) and (9.38) from the perspective of integrable models. More precisely, let us analyze the similarities between equation (9.37) and the Yang–Baxter equation for graph face models. To this effect,

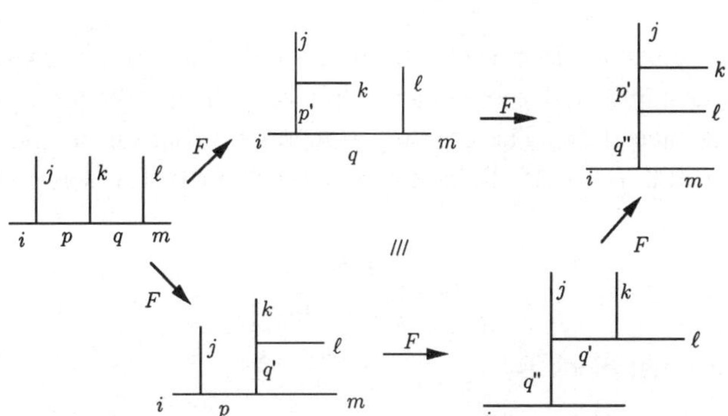

Fig. 9.8 The pentagonal equation (9.38).

we must first associate formally an "auxiliary" face model to a given rational conformal field theory. For simplicity, we assume that we are dealing with theories where the charge conjugation matrix is the identity, i.e. where all primary fields Φ are self-conjugate, i.e. where $(\forall\Phi)$, $[\Phi] \times [\Phi] \ni \mathbf{1}$.

The clue to this connection is provided by the fusion matrix N_{ij}^{k} of the rational conformal field theory. We associate an incidence matrix $\Lambda^{(k)}$ with each irrep k, such that

$$\left(\Lambda^{(k)}\right)_i^j = N_{ki}^{j} \tag{9.39}$$

The auxiliary graph face model is defined after using as lattice variables the different Virasoro irreps in the Hilbert space of the rational conformal field theory, with incidence matrix (9.39).

In order to relate the Yang–Baxter operators of this auxiliary face model to the braiding matrices (9.33), we shall consider conformal blocks with all but one of the external lines saturated by the same irrep k which we chose to define the incidence matrix of the auxiliary face model. These conformal blocks are thus of the following very particular form:

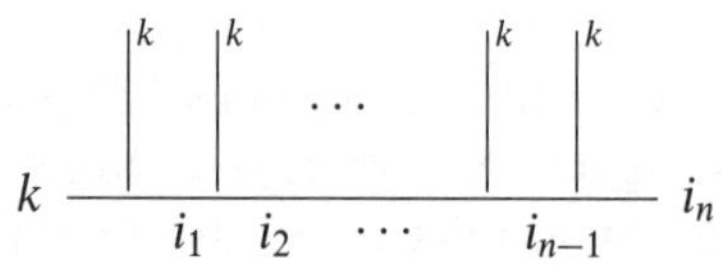

(9.40)

For this special family of conformal blocks, the set of internal channels define a quantum state of the auxiliary face model. More formally, the internal channels of these conformal blocks of type k correspond univocally with the paths of the Bratelli diagram defined by the incidence matrix $\Lambda^{(k)}$. Thus, every conformal block is associated uniquely with one face state:

$\mathcal{F}^{k\ldots k\, i_n}_{i_1\cdots i_{n-1}} \equiv$ 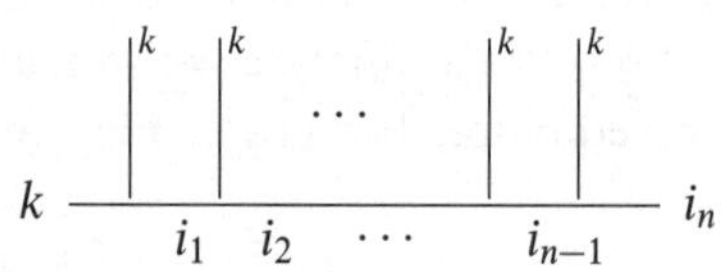 $\longrightarrow |i_1 \cdots i_n\rangle$ (9.41)

Let us now consider in parallel both the effect of an s–u crossing transformation on the conformal blocks, and that of a Yang–Baxter operator on the corresponding face state. The crossing s–u transformation on the external lines in positions j and $j+1$ of the conformal block in (9.41) is represented by the B-matrix

$$B_{i_j i'_j}\begin{bmatrix} k & k \\ i_{j-1} & i_{j+1} \end{bmatrix} \tag{9.42}$$

The new conformal block is

$$\sum_{i'_j} B_{i_j i'_j}\begin{bmatrix} k & k \\ i_{j-1} & i_{j+1} \end{bmatrix} \mathcal{F}^{k\ldots k\, i_n}_{i_1\cdots i'_j \cdots i_{n-1}} \tag{9.43}$$

and the induced effect on the associated face states

$$|i_1 \cdots i_j \cdots i_n\rangle \longrightarrow B_{i_j i'_j}\begin{bmatrix} k & k \\ i_{j-1} & i_{j+1} \end{bmatrix} |i_1 \cdots i'_j \cdots i_n\rangle \tag{9.44}$$

can be interpreted as the result of acting with a Yang–Baxter operator. Denoting by X_j the operator which effects the transformation in (9.44), we obtain from (9.37) the braid group relation

$$X_i X_{i+1} X_i = X_{i+1} X_i X_{i+1} \tag{9.45}$$

which is the Yang–Baxter equation for Yang–Baxter operators. The effect of *s*–*u* crossing on face states is thus

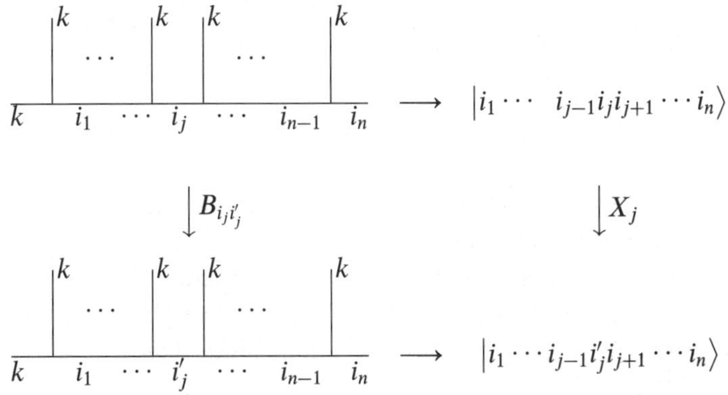

(9.46)

9.5 Conformal field theories and towers of algebras

A distinctive feature of the trigonometric solution to the Yang–Baxter equation is the use of the Temperley–Lieb–Jones algebra (5.13). Recall from chapter 5 the algebraic meaning of the Temperley–Lieb–Jones algebras as conditional expectations of a tower of algebras. Now we shall perform a similar construction in the context of rational conformal field theories. To define the rational conformal field theory tower of algebras, we will proceed as follows. We shall define the space $V^k_{i_n}$ generated by the conformal blocks (9.40). This space is completely characterized by the field k and the "spectator" i_n. The dimension of this space is given by the fusion matrix

$$d^k_{i_n} = \dim V^k_{i_n} = \left(\left[\Lambda^{(k)}\right]^n\right)^{i_n}_k \tag{9.47}$$

and is the number of possible intermediate channels.

The $V^k_{i_n}$ are representation spaces of the duality transformations defined above. Any "duality operation" x is represented on $V^k_{i_n}$ by a matrix $x^{i_n} \in \mathrm{End}\,(V^k_{i_n}) = \mathrm{Mat}_{d^k_{i_n}}(\mathbb{C})$:

$$(x\mathscr{F})^{i_n}_{i_1 \cdots i_{n-1}} = \sum_{i'_1, \ldots, i'_{n-1}} x^{i_n}_{i_1 \cdots i_{n-1}; i'_1 \cdots i'_{n-1}} \mathscr{F}^{i_n}_{i'_1 \cdots i'_{n-1}} \tag{9.48}$$

For different spectators, we obtain different representations of the same duality transformation. We may collect all these transformations together to form a multi-matrix algebra

$$A^n_k = \bigoplus_{i_n} \mathrm{Mat}_{d^k_{i_n}}(\mathbb{C}) \tag{9.49}$$

The multiplication in A^n_k is just the ordinary matrix multiplication. The multi-matrix algebras A^n_k assemble naturally into an infinite tower of algebras

$$A^1_k \subset A^2_k \subset \cdots \tag{9.50}$$

The inclusion map $\pi^n : A^{n-1}_k \to A^n_k$ is defined by promoting any operator $x \in A^{n-1}_k$ to an operator $\pi^n(x) \in A^n_k$ which leaves untouched the nth extra field k. The explicit expression is

$$\pi^n(x)^{i_n}_{i_1\cdots i_{n-1};i'_1\cdots i'_{n-1}} = \delta(i_{n-1}, i'_{n-1}) N^{i_n}_{k i_{n-1}} x^{i_{n-1}}_{i_1\cdots i_{n-2};i'_1\cdots i'_{n-2}} \tag{9.51}$$

The inclusion $A^{n-1}_k \subset A^n_k$ is thus characterized by the matrix Λ in (9.39) which was derived from the fusion matrices.

It is easy to check that, as a consequence of the rationality of conformal field theories, the tower of algebras (9.50) becomes isomorphic after some level $n_\circ$ to the one obtained in chapter 5 by the fundamental construction, because $A^{n+2}_k = \mathrm{End}_{A^n_k} A^{n+1}_k$ for $n \geq n_\circ$.

Summarizing, so far we have proven that any primary field k of a rational conformal field theory generates a tower of multi-matrix algebras, which is the model for the Bratelli diagram of the matrix $\Lambda^j_i = (N_k)^j_i$. In simpler terms, we are interpreting the duality transformations as a "model" for the fusion algebra.

In chapter 5, we proved some general results for towers of algebras obtained by the fundamental construction. In particular, we learned that these towers always enjoy a Markov trace, and that their conditional expectations generate the Temperley–Lieb–Jones algebra. Next, we give a purely conformal field theoretical interpretation of the Temperley–Lieb–Jones algebra derived from the tower (9.50). After that, the B-matrices of the rational conformal field theory should be easily reproduced with the help of the relation between the Yang–Baxter operators and Temperley–Lieb–Jones generators studied in chapter 5.

A direct consequence of (5.44) is

$$(e_n\mathcal{F})^{i_m}_{i_1\cdots i_{m-1}} = \delta(i_{n+1}, i_{n-1}) \sum_{i'_n} \frac{\sqrt{t^n_{i_n} t^n_{i'_n}}}{t^{n-1}_{i_{n-1}}}$$

$\times$ 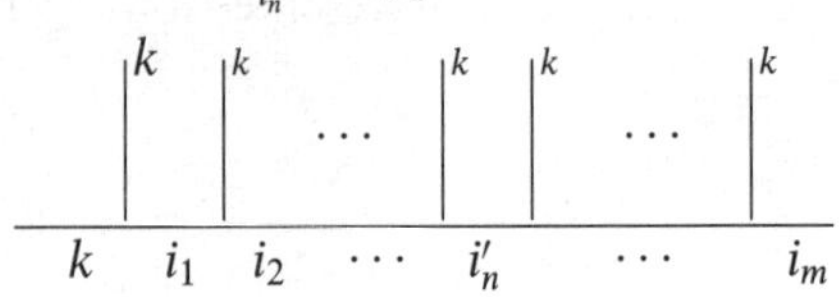

(9.52)

where $t^n_{i_m} = t^n(i_m)$, and the functional t^n defines the Markov trace on the tower (9.50) (see equations (5.30) and (5.45)):

$$\beta t^{2n} = \Lambda^{(k)}\Lambda^{(k)T} t^{2n}\,, \quad \beta t^{2n+1} = \Lambda^{(k)T}\Lambda^{(k)} t^{2n+1} \tag{9.53}$$

The parameter β is the customary highest eigenvalue of the incidence matrix $\Lambda^{(k)} = N_k$. Using now the duality relation (9.38), we define the following operation e_j on the conformal blocks. It satisfies all the requirements on the conditional expectations of the tower of algebras (9.50), and thus this operation represents a generator of the Temperley–Lieb–Jones algebra:

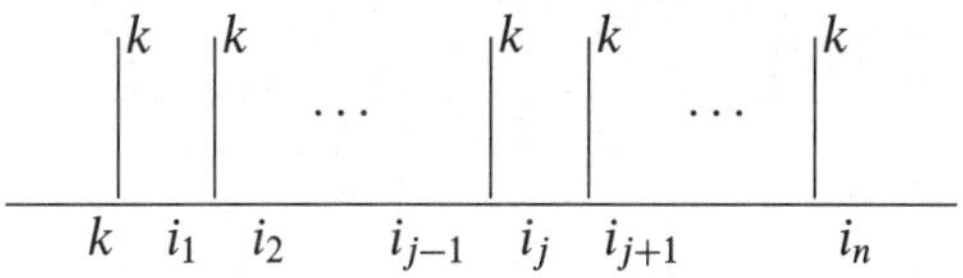

$$= \sum_{i'_j} F_{i_j i'_j}\begin{bmatrix} k & k \\ i_{j-1} & i_{j+1}\end{bmatrix}$$

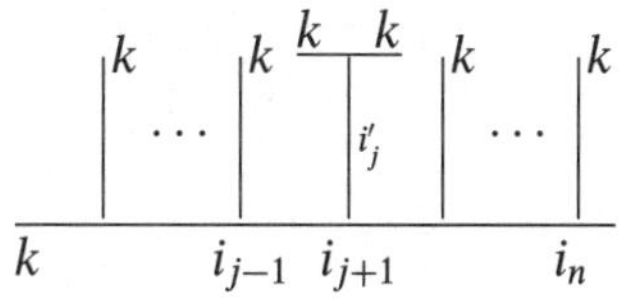

$$\xrightarrow{e_j} F_{i_j\mathbf{1}}\begin{bmatrix} k & k \\ i_{j-1} & i_{j+1}\end{bmatrix} \delta(i_{j-1}, i_{j+1})$$

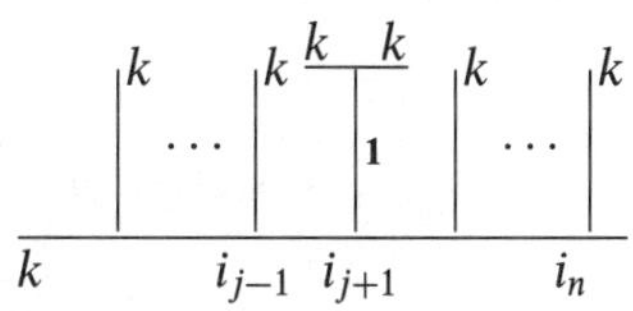

$$= F_{i_j\mathbf{1}}\begin{bmatrix} k & k \\ i_{j-1} & i_{j+1}\end{bmatrix} F_{\mathbf{1}i'_j}\begin{bmatrix} k & i_{j+1} \\ k & i_{j-1}\end{bmatrix} \delta(i_{j-1}, i_{j+1})$$

$$\times \tag{9.54}$$

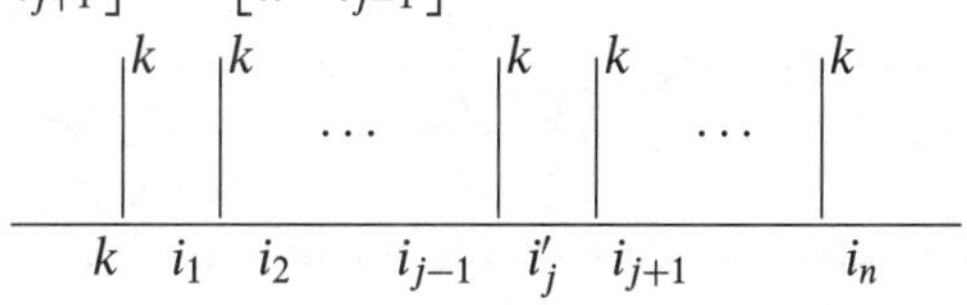

In words, we say that e_n is the projection operator on the identity $\mathbf{1}$ in the nth t channel. The operators e_j defined by (9.54) automatically satisfy

$$\begin{aligned} e_i^2 &= e_i \\ e_i e_j &= e_j e_i \qquad |i-j| \geq 2 \end{aligned} \tag{9.55}$$

Moreover, using (9.38) we obtain

$$e_n e_{n\pm1} e_n = F_k^2 e_n \tag{9.56}$$

where F_k is an important quantity, defined as

$$F_k = \epsilon^{\mathbf{1}}_{kk} F_{\mathbf{11}}\begin{bmatrix} k & k \\ k & k\end{bmatrix} \tag{9.57}$$

Here, $\epsilon^{\mathbf{1}}_{kk}$ is $+1$ or -1 depending on whether the identity irrep $\mathbf{1}$ appears in the symmetric or the antisymmetric part of the operator product expansion

$k \otimes k = \mathbf{1} + \cdots$. We will see that $1/F_k$ turns out to be the "quantum dimension" of the representation k. The graphical proof of (9.56) is sketched below:

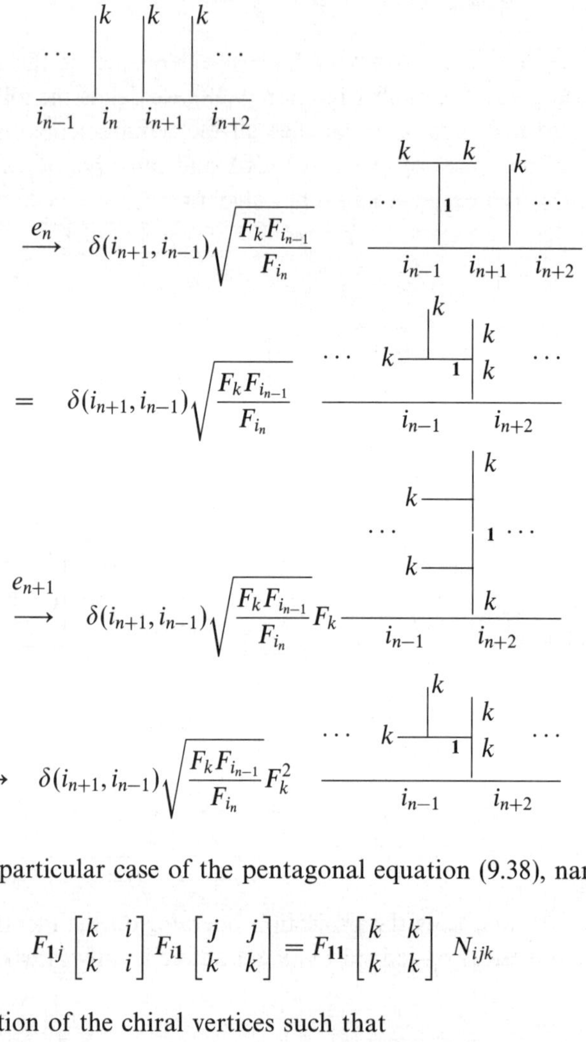

(9.58)

We have used a particular case of the pentagonal equation (9.38), namely

$$F_{\mathbf{1}j}\begin{bmatrix} k & i \\ k & i \end{bmatrix} F_{i\mathbf{1}}\begin{bmatrix} j & j \\ k & k \end{bmatrix} = F_{\mathbf{1}\mathbf{1}}\begin{bmatrix} k & k \\ k & k \end{bmatrix} N_{ijk} \tag{9.59}$$

and a normalization of the chiral vertices such that

$$F_{\mathbf{1}i}\begin{bmatrix} j & k \\ j & k \end{bmatrix} = F_{i\mathbf{1}}\begin{bmatrix} j & j \\ k & k \end{bmatrix} = \sqrt{\frac{F_j F_k}{F_i}}\, N_{ijk} \tag{9.60}$$

Comparing (9.56) with the defining relations of the Temperley–Lieb–Jones algebra (5.13), we obtain the interesting result

$$\beta^{-1} = F_k^2 \tag{9.61}$$

Note that F_k appears twice in the computation above because there are two s–t duality transformations involved in the proof.

Furthermore, if we compare (9.54) with (9.52), we obtain

$$F_{i_j 1}\begin{bmatrix} k & k \\ i_{j-1} & i_{j+1} \end{bmatrix} F_{1 i'_j}\begin{bmatrix} k & i_{j+1} \\ k & i_{j-1} \end{bmatrix} \delta(i_{j-1}, i_{j+1}) = \frac{\sqrt{t^n_{i_j} t^n_{i'_j}}}{t^{n-1}_{i_{j-1}}} \tag{9.62}$$

which implies that

$$t^n_i = \frac{F^n_k}{F_i} \tag{9.63}$$

The two equations (9.61) and (9.62) condense the close connection between the fusion matrices of a rational conformal field theory and the Temperley–Lieb–Jones algebra of the associated tower of multi-matrix algebras. Before delving into a more careful study of (9.61) and (9.62), let us summarize this section.

(i) We have associated to a rational conformal field theory a tower of algebras satisfying the requirements of the fundamental construction. It is essential that the conformal field theory is rational.
(ii) The Temperley–Lieb–Jones algebra of conditional expectations is defined in terms of duality operations. To prove that they do satisfy the Temperley–Lieb–Jones relations, we need to use the pentagonal equations for the fusion matrices.
(iii) The uniqueness of the Temperley–Lieb–Jones algebra allows us to find unexpected relationships between the fusion matrices and the Markov trace of the tower; the latter is completely determined by the fusion algebra.

9.6 Genus one polynomial equations

Let us now determine the properties of a conformal field theory on a torus, instead of on a sphere or a plane. A toroidal two-dimensional spacetime is equivalent to a cylinder with periodic boundary conditions. At any fixed time, space is like a circle. Time is also like a circle at any fixed space location. Hence, the partition functions we shall find by working on a torus are essentially those of field theory at finite temperature. We shall define our conformal theories on a cylinder, and then construct the partition functions with periodic boundary conditions in time.

The basic object we wish to consider is the partition function at a finite temperature $T = \beta^{-1}$, defined as

$$\begin{aligned} Z(\beta) &= \text{tr } e^{-\beta H_{\text{cyl}}} \\ &= \text{tr } e^{-\beta\left(L_0 + \overline{L}_0 - c/12\right)} \end{aligned} \tag{9.64}$$

The trace in this expression is taken over the whole Hilbert space of the conformal field theory: it involves all the primary states and descendants of all the irreps of the chiral algebra. Recalling that the momentum on the cylinder $P_{\text{cyl}} = L_0 - \overline{L}_0$ is the spin operator on the plane (or the momentum operator for the underlying spin chain, as worked out in section 8.10), we may generalize (9.64) as follows:

$$Z(q,\overline{q}) = \mathrm{tr}\ \mathrm{e}^{-2\pi\,\mathrm{Im}(\tau)\,H_{\mathrm{cyl}}+2\pi i\,\mathrm{Re}(\tau)\,P_{\mathrm{cyl}}} = \mathrm{tr}\,q^{L_0-c/24}\overline{q}^{\overline{L}_0-c/24} \tag{9.65}$$

Having introduced the nome $q = \mathrm{e}^{2\pi i\tau}$, the limit (9.64) is recovered from (9.65) when $\tau = i\beta$. Note that the operator $q^{L_0-c/24}$ has already appeared as a "propagator" when we constructed states in the sewing formalism, equation (9.25). The trace in (9.65) means effectively that we are working on a torus. A torus can be built by modding out a two-dimensional lattice $\mathbb{Z}a \oplus \mathbb{Z}b$ from the complex plane ($a, b \in \mathbb{C} - \{0\}$) or, more intuitively, by identifying opposite sides of a parallelogram, one of whose side lengths we normalize to unity. The various possible tori are thus parametrized by a complex number τ (Im $\tau > 0$), the modulus of the torus or modular parameter, which measures the other side's length and the relative twist we introduce when we curl up the parallelogram into a torus. Often, $q = \exp(2\pi i\tau)$ is used instead of τ. This q should not be confused with the deformation parameter of quantum groups.

The chiral contribution of each primary field to the partition function (9.65) will be

$$\mathscr{X}_i(q) = \mathrm{tr}_{[\Phi_i]}q^{L_0-c/24} \tag{9.66}$$

which are the standard Virasoro characters. Since we work on the cylinder, the hamiltonian is $(L_0 - c/24)$. Note also that the trace is taken over the whole Virasoro irrep associated with the primary field Φ_i. The characters defined by (9.66) are to the partition function what the chiral amplitudes or conformal blocks are to the physical amplitude. Geometrically, expression (9.66) can be understood as a pure sewing propagator, i.e. as the "chiral Feynman diagram" corresponding to the bubble vacuum to vacuum amplitude. In analogy with equation (9.6), we may propose the following representation of the partition function:

$$Z(q,\overline{q}) = \sum_{i,\overline{j}} h_{i\overline{j}}\mathscr{X}_i(q)\mathscr{X}_{\overline{j}}(\overline{q}) \tag{9.67}$$

The sum extends over all the irreps contributing to the Hilbert space of physical states.

When we studied the genus zero situation, we found that the constants $c_{\lambda\overline{\lambda}}$ in (9.6) were determined by imposing invariance under monodromy transformations, which meant invariance under global diffeomorphisms of the punctured Riemann sphere. It is thus reasonable to expect that the coefficients $h_{i\overline{j}}$ in (9.67) are fixed by requiring invariance of the partition function under global diffeomorphisms of the torus without punctures. They are called modular transformations, and turn out to be generated by

$$\begin{aligned} S &: \tau \to -1/\tau \\ T &: \tau \to \tau + 1 \end{aligned} \tag{9.68}$$

These two operations generate the group $PSL_2(\mathbb{Z})$, which is

$$SL(2,\mathbb{Z}) = \{f : z \to (az+b)/(cz+d)\ , \quad ad - bc \neq 0\ ; \quad a,b,c,d \in \mathbb{Z}\}$$

modulo the $\mathbb{Z}_2$ symmetry $(a,b,c,d) \to (-a,-b,-c,-d)$. Geometrically, it is useful to view the elementary lattice vectors a and b on the plane as closed non-contractible circles on the torus. The transformation S changes the cycles (a,b) into $(-b,a)$, whereas the transformation T sends (a,b) to $(a,a+b)$.

The first obvious condition for the partition function to be modular-invariant is that the characters $\mathcal{X}_i(q)$ define a representation space of the modular transformations:

$$\begin{aligned} S &: \mathcal{X}_i \to S_i^j \mathcal{X}_j \\ T &: \mathcal{X}_i \to e^{2\pi i(\Delta_i - c/24)} \mathcal{X}_i \end{aligned} \tag{9.69}$$

For these matrices to generate a representation of the modular group $PSL(2,\mathbb{Z})$ on the space of characters, they must satisfy

$$S^2 = (ST)^3 = C \tag{9.70}$$

where C is the charge conjugation matrix (see section 8.9). The diagonal transformation law for T follows automatically from the definition (9.66) of the characters, but we still need to do some more work to find the matrix S in (9.69). The metric $h_{i\bar{j}}$ in the partition function (9.67) is determined by demanding modular invariance of the partition function of the model.

For a given irrep of the Virasoro algebra with central extension c and conformal weight Δ, the character (9.66) becomes

$$\mathcal{X}_{c,\Delta}(q) = q^{-c/24+\Delta} \sum_{N \geq 0} d_N\, q^N \tag{9.71}$$

where d_N is the number of states at level N in the Verma module $\mathcal{V}^\Delta$, i.e. the dimension of the subspace of $\mathcal{V}^\Delta$ with an eigenvalue of L_0 equal to $\Delta + N$. If $\mathcal{V}^\Delta$ has no null vectors, a basis of $\mathcal{V}^\Delta$ is given by

$$\{L_{-k_1} \cdots L_{-k_m} |\Delta\rangle \qquad 1 \leq k_1 \leq k_2 \cdots\} \tag{9.72}$$

which is obtained by applying any polynomial in L_{-n} $(n > 0)$ to the highest weight state. The character of $\mathcal{V}^\Delta$ is then easily seen to be

$$\mathcal{X}_{c,\Delta}(q) = \frac{q^{-c/24+\Delta}}{\prod_{n=1}^{\infty}(1-q^n)} \tag{9.73}$$

If, more interestingly, there are null vectors, then we must subtract the submodules generated by them to obtain the character of the irreducible representation associated to a given primary field. Let us illustrate this subtraction procedure by the computation of the character for a Virasoro primary field $\Phi_{mm'}$ of a (p,p') minimal model, with $0 < m < p$ and $0 < m' < p'$. Their Verma modules $\mathcal{V}^{mm'}$

contain an infinite number of null vectors. Letting

$$\Delta_{mm'} = \frac{1}{4pp'}\left[\left(mp' - m'p\right)^2 - \left(p - p'\right)^2\right] \tag{9.74}$$

and using Kac's theorem, which states the existence of a null vector in $\mathscr{V}^{mm'}$ at level mm' (when both $m, m' > 0$), together with the useful formula

$$\Delta(m, m') = \Delta(p - m, p' - m') = \Delta(-m, -m') = \Delta(p + m, p' + m') \tag{9.75}$$

it is not difficult to obtain the Rocha-Caridi formula for the Virasoro characters

$$\mathscr{X}_{m,m'}(q) = \frac{q^{-c/24}}{\prod_{n\geq 1}(1 - q^n)} \sum_{k\in\mathbb{Z}} \left(q^{a(k)} - q^{b(k)}\right) \tag{9.76}$$

with

$$\begin{aligned} a(k) = \Delta(m + 2pk, m') &= \frac{\left(2pp'k + mp' - m'p\right)^2 - \left(p - p'\right)^2}{4pp'} \\ b(k) = \Delta(m + 2pk, -m') &= \frac{\left(2pp'k + mp' + m'p\right)^2 - \left(p - p'\right)^2}{4pp'} \end{aligned} \tag{9.77}$$

Using (8.111), (8.112) and Dedekind's function

$$\eta(\tau) = \mathrm{e}^{\pi i\tau/12} \prod_{\ell\geq 1} \left(1 - \mathrm{e}^{2\pi i\ell\tau}\right) \tag{9.78}$$

we find, quite explicitly,

$$\begin{aligned} \mathscr{X}_{c(p,p'),\Delta(m,m')}(q) = \frac{1}{\eta(\tau)} \sum_{\ell\in\mathbb{Z}} \Bigg(& q^{(2\ell pp' + mp' - m'p)^2/(4pp')} \\ & - q^{(2\ell pp' + mp' + m'p)^2/(4pp')} \Bigg) \end{aligned} \tag{9.79}$$

where

$$q = \mathrm{e}^{2\pi i\tau} \tag{9.80}$$

The modular properties of Dedekind's η function

$$\begin{aligned} T &: \eta(\tau) \longrightarrow \eta(\tau + 1) = \mathrm{e}^{i\pi/12}\eta(\tau) \\ S &: \eta(\tau) \longrightarrow \eta(-1/\tau) = \sqrt{-i\tau}\,\eta(\tau) \end{aligned} \tag{9.81}$$

allow us to find the behavior of the characters $\mathscr{X}_{c,\Delta}(q)$ under modular transformations. Writing $N = 2pp'$ and $\lambda = mp' - m'p \pmod N$, we obtain

$$\begin{aligned} T &: \mathscr{X}_{c,\lambda}(\tau) \to \mathscr{X}_{c,\lambda}(\tau+1) = \exp\left[2\pi i \left(\frac{\lambda^2}{2N} - \frac{1}{24}\right)\right] \mathscr{X}_{c,\lambda}(\tau) \\ S &: \mathscr{X}_{c,\lambda}(\tau) \to \mathscr{X}_{c,\lambda}(-1/\tau) = \frac{1}{\sqrt{N}} \sum_{\lambda' \in \mathbb{Z}_N} \exp\left[\frac{2\pi i \lambda\lambda'}{N}\right] \mathscr{X}_{c,\lambda'}(\tau) \end{aligned} \tag{9.82}$$

Before proceeding, let us highlight a crucial difference between the genus one and genus zero cases. The chiral components of physical amplitudes define representation spaces for the monodromy transformations. The dimension of these spaces is fixed by the order of the differential equations enforcing the decoupling of null vectors, i.e. by the fusion algebra. This means that any set of primary fields closed under fusion yields well-defined genus zero physical amplitudes. In contrast, at genus one this is not the case. The thermal subalgebras of minimal models, i.e. the subalgebras generated by fields of the form $\{\Phi_{1m}\}$ or $\{\Phi_{n1}\}$, constitute a set of primary fields which closes under fusion but not under modular transformations. For the Ising model, the identity and ε close under the fusion algebra (8.117), but the order parameter σ is needed in order to have a set of fields closed under modular transformations. From (9.82), we obtain the modular-invariant partition function for the Ising model:

$$Z_{\text{Ising}} = |\mathscr{X}_{\mathbf{1}}|^2 + |\mathscr{X}_\sigma|^2 + |\mathscr{X}_\varepsilon|^2 \tag{9.83}$$

which contains contributions from all three operators. We might thus reasonably presume that modular invariance will provide us with a criterion for characterizing the completeness of the operator algebra.

Having emphasized the distinction between the fusion algebra and modular invariance, we may appreciate the beauty of the discovery by Verlinde of the relationship between the fusion matrix N_{ij}^k and the modular matrix S_i^j, already displayed without proof in the previous chapter, equation (8.121). The main result we wish to prove is that the S-matrix diagonalizes the fusion algebra. In order to do so, we shall introduce two operators, $\Phi_i(a)$ and $\Phi_i(b)$, on the genus one chiral amplitudes, which will effect what we shall call operation (a) and operation (b). Operation (a) amounts to first inserting the identity $\mathbf{1} = i \times \bar{\imath}$ for some primary field i, then moving the i field all the way around the cycle a, and finally annihilating again i with $\bar{\imath}$. The conjugate fields are annihilated back only after one of them has scanned along the cycle a the global nature of the two-dimensional space. Operation (b) is exactly the same, but the tour is taken along the cycle b of the torus instead. We shall assume that $i = \bar{\imath}$.

Let us see the explicit action of these operators on characters. We start with operation (a):

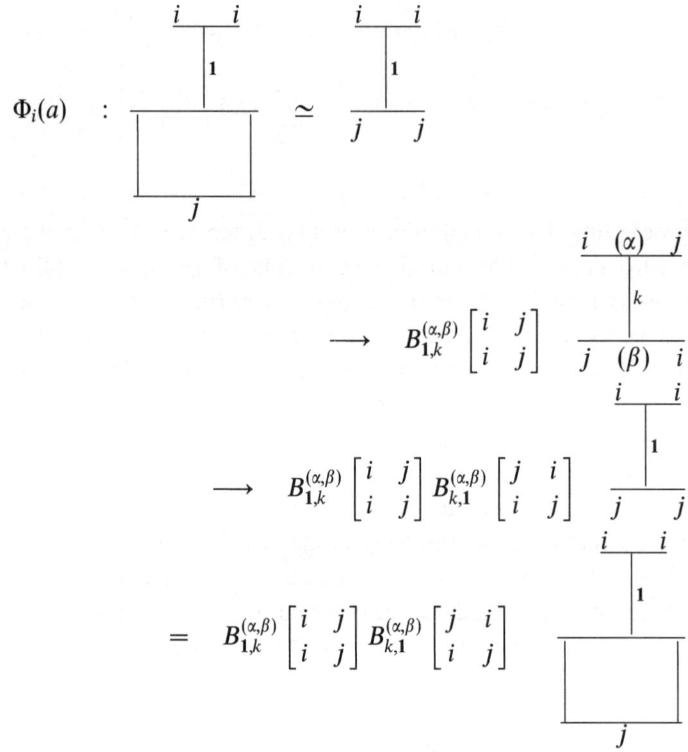

(9.84)

The Greek indices label the various possible couplings, $\alpha, \beta{=}1,\ldots, N_{ij}^{k}$. The closed line carrying the irrep j indicates the genus one nature of the chiral amplitude: we open up the torus (i.e. the box) to manipulate it, then we close it at the end.

Similarly, we can define operation (b):

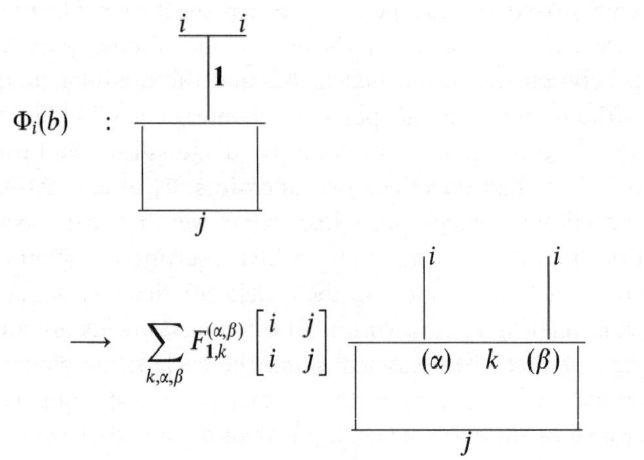

(9.85)

$$\longrightarrow \sum_{k,\alpha,\beta} F^{(\alpha,\beta)}_{\mathbf{1},k}\begin{bmatrix} i & j \\ i & j \end{bmatrix} \; F^{(\beta,\alpha)}_{j,\mathbf{1}}\begin{bmatrix} i & i \\ k & k \end{bmatrix}$$ 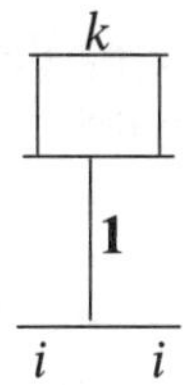

For clarity, let us note that on starting from

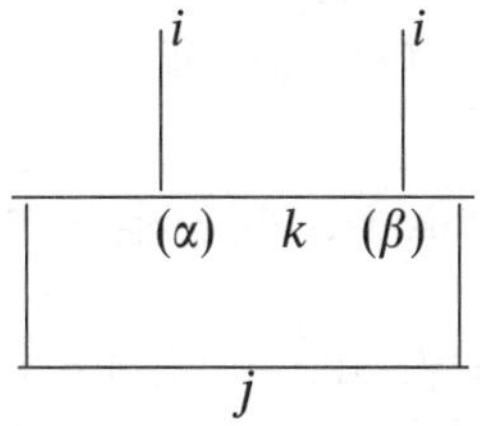

(9.86)

we may go either to

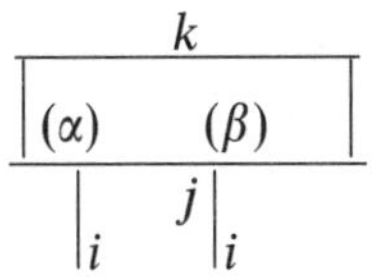 $\longrightarrow \; F^{(\alpha,\beta)}_{j\mathbf{1}}\begin{bmatrix} k & k \\ i & i \end{bmatrix}$ 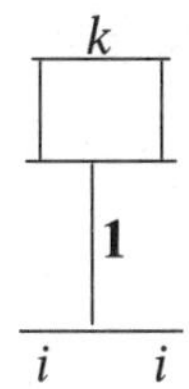 (9.87)

or to

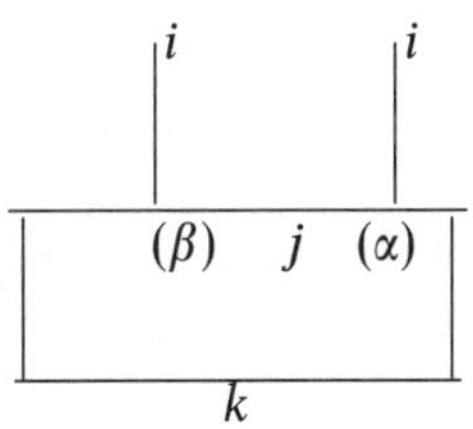 $\longrightarrow \; F^{(\beta,\alpha)}_{j\mathbf{1}}\begin{bmatrix} i & i \\ k & k \end{bmatrix}$ 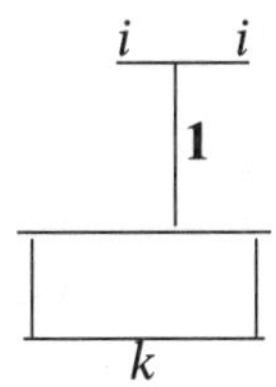 (9.88)

The expressions resulting from (9.87) or (9.88) are actually equal.

It is convenient to normalize operation (b) such that

$$\Phi^{(\text{norm})}_i(b)\,\mathscr{X}_{\mathbf{1}} = \mathscr{X}_i \tag{9.89}$$

As it stands, operation (b) yields

$$\Phi_i(b)\,\mathscr{X}_{\mathbf{1}} = F_{\mathbf{11}}\begin{bmatrix} i & i \\ i & i \end{bmatrix}\mathscr{X}_i \tag{9.90}$$

and thus the normalized operation (b) on characters is

$$\Phi^{(\text{norm})}_i(b)\,\mathscr{X}_j = \left(\sum_{k,\alpha,\beta} \frac{F^{(\alpha,\beta)}_{\mathbf{1},k}\begin{bmatrix} i & j \\ i & j \end{bmatrix} F^{(\beta,\alpha)}_{j,\mathbf{1}}\begin{bmatrix} i & i \\ k & k \end{bmatrix}}{F_{\mathbf{1},\mathbf{1}}\begin{bmatrix} i & i \\ i & i \end{bmatrix}}\right)\mathscr{X}_k \tag{9.91}$$

After normalizing the operator $\Phi_i(a)$ with the same normalization as $\Phi_i(b)$, we can use (9.84) to define the operation (a) on the characters:

$$\Phi_i^{(\text{norm})}(a)\,\mathscr{X}_j = \left(\sum_{k,\alpha,\beta} \frac{B^{(\alpha,\beta)}_{\mathbf{1},k}\begin{bmatrix} i & j \\ i & j \end{bmatrix} B^{(\alpha,\beta)}_{k,\mathbf{1}}\begin{bmatrix} j & i \\ i & j \end{bmatrix}}{F_{\mathbf{1},\mathbf{1}}\begin{bmatrix} i & i \\ i & i \end{bmatrix}}\right)\mathscr{X}_j$$
$$\equiv \lambda_i^{(j)}\mathscr{X}_j \qquad (9.92)$$

Notice that $\Phi_i(a)$ does not modify the characters $\mathscr{X}_j$, but simply multiplies them by the factor $\lambda_i^{(j)}$.

Some similarities between (9.91) and (9.59) are already apparent. In fact, they are the same except for the additional three-point vertex labels (α) and (β) in (9.91). To evaluate the coefficient on the right-hand side of equation (9.91), we shall use the one-loop version of (9.38), shown below:

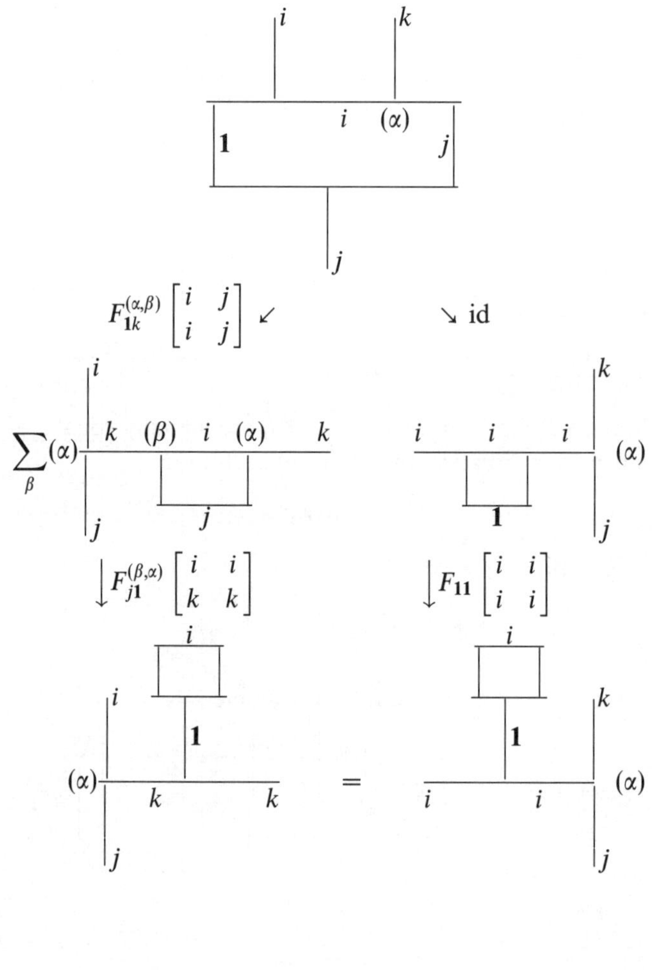

(9.93)

In formulae, the duality transformations at one loop (9.93) reduce to

$$\sum_{\beta=1}^{N_{ij}^k} F_{\mathbf{1}k}^{(\alpha,\beta)} \begin{bmatrix} i & j \\ i & j \end{bmatrix} F_{j\mathbf{1}}^{(\beta,\alpha)} \begin{bmatrix} i & i \\ k & k \end{bmatrix} = F_{\mathbf{11}} \begin{bmatrix} i & i \\ i & i \end{bmatrix} \tag{9.94}$$

and therefore

$$\sum_{\alpha,\beta=1}^{N_{ij}^k} \frac{F_{\mathbf{1}k}^{(\alpha,\beta)} \begin{bmatrix} i & j \\ i & j \end{bmatrix} F_{j\mathbf{1}}^{(\beta,\alpha)} \begin{bmatrix} i & i \\ k & k \end{bmatrix}}{F_{\mathbf{11}} \begin{bmatrix} i & i \\ i & i \end{bmatrix}} = N_{ij}^k \tag{9.95}$$

which implies

$$\Phi_i^{(\text{norm})}(b)\, \mathscr{X}_j = N_{ij}^k\, \mathscr{X}_k \tag{9.96}$$

Note that (9.59) is just the simplified version of (9.95), for the case when no labels are needed for the three-point vertices, i.e. when N_{ij}^k are all 0 or 1.

Since the modular transformation S interchanges cycles a and b, we have

$$S\, \Phi_i^{(N)}(b)\, S^{-1} = \Phi_i^{(N)}(a) \tag{9.97}$$

Acting on characters, with the help of (9.69) (and also (9.84) to (9.95)), we find that

$$\sum_k N_{ijk}\, S_{k\ell} = S_{j\ell}\, \lambda_i^{(\ell)} \tag{9.98}$$

Choosing now $j = \mathbf{1}$ and $k = i$ yields

$$S_{i\ell} = S_{\mathbf{1}\ell}\, \lambda_i^{(\ell)} \tag{9.99}$$

and thus, using the fact that $S^2 = \mathbf{1}$ for $C = 1$, we obtain finally

$$N_{ijk} = \sum_n \frac{S_{in} S_{jn} S_{nk}}{S_{\mathbf{1}n}} \tag{9.100}$$

which is the quantitative content of Verlinde's theorem on the link between the fusion algebra and the one-loop modular transformations.

Equation (9.100) is the expression needed to clarify the results of the previous section, namely equations (9.61) and (9.62). The role of the Temperley–Lieb–Jones parameter β should also be clearer now. Comparing equations (9.53) and (9.100),and recalling that the identity **1** corresponds to the spin-0 label, it follows that

$$\beta = \left(\frac{S_{k\mathbf{1}}}{S_{\mathbf{11}}} \right)^2 \tag{9.101}$$

and

$$t_m^n = \frac{S_{\mathbf{1}m}}{S_{\mathbf{11}}} \left(\frac{S_{\mathbf{11}}}{S_{k\mathbf{1}}} \right)^n \tag{9.102}$$

Hence, from (9.61) and (9.63) we conclude that

$$t(j) = \left(F_{11} \begin{bmatrix} j & j \\ j & j \end{bmatrix} \right)^{-1} F_{11} \begin{bmatrix} k & k \\ k & k \end{bmatrix} \tag{9.103}$$

and thus

$$F_i \equiv \epsilon_{ii}^1 F_{11} \begin{bmatrix} i & i \\ i & i \end{bmatrix} = \frac{S_{11}}{S_{1i}} \tag{9.104}$$

This, together with equation (9.100), contains most of the relationship between duality at genera one and zero of rational conformal field theories.

From the derivation above, it is also clear that a substantial part of the content of Verlinde's theorem is already present in the results derived from the tower of algebras structure of rational conformal field theories.

Exercises

Ex. 9.1 To be quite rigorous, the definition of chiral vertex operators involves not only the equation of motion (9.16) but also the intertwiner condition

$$\rho^k(A)\Phi_{ij}^{k\,(\alpha)}(z)\left(|u\rangle \otimes |v\rangle\right) = \Phi_{ij}^{k(\alpha)}\left\{\rho^i \otimes \rho^j \Delta_{0,z}(A)\left(|u\rangle \otimes |v\rangle\right)\right\} \tag{9.105}$$

where we denote the representation j of the chiral algebra operator A by $\rho^j(A)$, and $\Delta_{0,z}$ is the co-multiplication of A. Using complex analysis, try to define the co-product of the Virasoro generators L_n, and check that (9.105) holds.

Ex. 9.2 Prove equation (9.23) for a variety of examples. Note that it predicts no moduli for a torus with no punctures, which is wrong. What are the (few) exceptions to the formula (9.23)?

Ex. 9.3 Write explicitly the relationship between the F-matrix and the B-matrices depicted in (9.36). Do not forget the phases coming from the elementary braidings $\sigma_{k\ell}$ and $\sigma_{q\ell}$.

Ex. 9.4 Compare the hexagonal equation (9.37) for braiding matrices of the type in equation (9.42) with the Yang–Baxter equation (5.6) for face models, in the braid limit $u, v \to \infty$. What is the precise relationship between conformal field theory braiding matrices and face Boltzmann weights such that both equations are equivalent?

Ex. 9.5 Check the validity of the pentagonal equation (9.38) with the help of the following expressions for the fusion matrices of the $\widehat{SU(2)}_k$ Wess–Zumino model:

$$F_{11}\begin{bmatrix} j & j \\ j & j \end{bmatrix} = \frac{(-1)^{2j}}{[2j+1]}$$

$$F_{1j}\begin{bmatrix} j_1 & j_2 \\ j_1 & j_2 \end{bmatrix} = F_{j1}\begin{bmatrix} j_1 & j_1 \\ j_2 & j_2 \end{bmatrix} \tag{9.106}$$

$$= (-1)^{j_1+j_2-j}\, N_{j_1 j_2 j} \sqrt{\frac{[2j+1]}{[2j_1+1]\,[2j_2+1]}}$$

where, as always, $[x] = \left(q^x - q^{-x}\right) / \left(q - q^{-1}\right)$, with $q = \exp i\pi/(k+2)$.

Ex. 9.6 Letting $\mathscr{H}_i$ denote the Verma module associated with an irrep i, we may define formally the "dimension" d_i of $\mathscr{H}_i$ as

$$d_i = \frac{\dim \mathscr{H}_i}{\dim \mathscr{H}_1} = \lim_{\tau\to 0} \frac{\mathrm{tr}_{\mathscr{H}_i}\, e^{2\pi i\tau(L_0 - c/24)}}{\mathrm{tr}_{\mathscr{H}_1}\, e^{2\pi i\tau(L_0 - c/24)}} = \lim_{\tau\to 0} \frac{\mathscr{X}_i(\tau)}{\mathscr{X}_1(\tau)} \tag{9.107}$$

Using the modular transformation S, show that

$$d_i = \frac{S_{i1}}{S_{11}} \tag{9.108}$$

Ex. 9.7 Note that the "dimension" d_i of the previous exercise is just the square root of the Temperley–Lieb–Jones parameter β for the tower of algebras constructed from the primary field Φ_i. Compute these numbers for an $\widehat{SU(2)}_k$ Wess–Zumino model, and show that they coincide with the "quantum dimensions", i.e. that $d_j = [2j+1]$. Show furthermore that these numbers satisfy the following "fusion rules":

$$d_i\, d_j = \sum_k N_{ij}^k\, d_k \tag{9.109}$$

Ex. 9.8 Repeat the previous exercise for the $c < 1$ minimal models.

Ex. 9.9 Show that in a *rational* conformal field theory, i.e. a conformal field theory with a finite number of primary fields, the central extension and the conformal weights of all fields are rational numbers.

Ex. 9.10 Using the expression (9.79) for the Virasoro characters in the simplest (2, 3) minimal model with $c = 0$ and only the identity ($\Delta = 0$) as primary field, prove Euler's pentagonal identity

$$\prod_{n\geq 1}(1 - q^n) = \sum_{\ell\in\mathbb{Z}} (-1)^\ell q^{\ell(3\ell+1)/2} \tag{9.110}$$

Ex. 9.11 Consider the Virasoro characters of the Ising model, i.e. put $p = 3$ and $p' = 4$ in (9.76). With the help of Jacobi's triple product identity, obtain the following expressions:

$$\begin{aligned}\mathscr{X}_{\substack{0\\1/2}} &= q^{-\frac{1}{48}}\left[\prod_{n\geq 0}\left(1+q^{n+\frac{1}{2}}\right) \pm \prod_{n\geq 0}\left(1-q^{n+\frac{1}{2}}\right)\right] \\ \mathscr{X}_{1/16} &= q^{-\frac{1}{48}+\frac{1}{16}}\prod_{n\geq 1}(1+q^n)\end{aligned} \tag{9.111}$$

This last expression serves to establish the equivalence between the critical Ising model and a free Majorana–Weyl fermion. As an application, prove the modular invariance of (9.83).

Ex. 9.12 Consider a rational conformal field theory with only two primary fields, the identity **1** and another field Φ, with fusion rule

$$\Phi \times \Phi = \mathbf{1} + n\Phi \qquad n \in \mathbb{N} \tag{9.112}$$

The eigenvalues of the fusion matrix N_Φ (recall that $(N_\Phi)_i^{\ j} = N_{\Phi i}^{\ j}$) satisfy

$$\lambda^2 = 1 + n\lambda \tag{9.113}$$

Parametrizing the modular matrices as

$$S = \begin{pmatrix} \cos\theta & \sin\theta \\ \sin\theta & -\cos\theta \end{pmatrix}, \qquad T = \mathrm{e}^{-i\pi c/12}\begin{pmatrix} 1 & 0 \\ 0 & \mathrm{e}^{2\pi i\Delta_\Phi} \end{pmatrix} \tag{9.114}$$

deduce from Verlinde's theorem that $\tan\theta = \lambda$. Next, from $(ST)^3 = \mathbf{1}$, derive that

$$\cos(2\pi\Delta_\Phi) = -\frac{n\lambda}{2}, \qquad 12\Delta_\Phi - c = 2 (\mathrm{mod}\ 8) \tag{9.115}$$

Use the solutions to these equations to classify all conformal field theories with only two primary fields. Find examples corresponding to $n = 0, 1$. Is $n = 2$ allowed? For $n = 1$, compute the dimensions (9.47) of the associated tower of algebras and show that they are given by the Fibonacci numbers.

Ex. 9.13 The modular matrix S for an $\widehat{SU(2)}_k$ Wess–Zumino model is given by

$$S_{jj'} = \sqrt{\frac{2}{k+2}}\,\sin\left[\pi\,\frac{(2j+1)(2j'+1)}{k+2}\right] \qquad 0 \leq j, j' \leq \frac{k}{2} \tag{9.116}$$

whereas the conformal weights are $\Delta_j = j(j+1)/(k+2)$. Check the equations $S^2 = (ST)^3 = \mathbf{1}$ of the modular group in this case, as well as Verlinde's theorem for the fusion rules.

Ex. 9.14 Find the modular matrix S for the (p, p') minimal models with $c < 1$, and establish a relationship with that of the previous exercise under the identifications $k = p - 2$ and $k = p' - 2$.

Ex. 9.15 [A. Capelli, C. Itzykson and J.B. Zuber, 'Modular invariant partition functions in two dimensions', *Nucl. Phys.* **B280 [FS18]** (1987) 445; 'The ADE classification of minimal and $A_1^{(1)}$ conformal invariant theories', *Commun. Math. Phys.* **113** (1987) 1.]

Writing the partition function (9.67) as $Z = \mathscr{X}^\dagger h \mathscr{X}$, show that the requirements on the matrix h for Z to enjoy modular invariance can be written as

$$T^\dagger h T = S^\dagger h S = h \tag{9.117}$$

where S and T are the modular matrices in (9.69). Consider now the Wess–Zumino model based on $\widehat{SU(2)}_k$. Using the results from exercise 9.13, show that $h = \mathbf{1}$ is always a solution to (9.117). This yields the "diagonal" partition function $Z = \sum |\mathscr{X}_j|^2$. The authors above succeeded in classifying all the non-diagonal modular invariant partition functions for $\widehat{SU(2)}_k$ and the unitary minimal models. Their result is very interesting because the allowed partition functions are in one to one relation to the simply-laced Dynkin diagrams of semi-simple Lie algebras, i.e. to the A, D and E Coxeter graphs. For the Wess–Zumino case, the A series corresponds to the diagonal partition functions (level $k \leftrightarrow A_{k+1}$); the D series arises only if the level is even (level $k \leftrightarrow D_{(k+4)/2}$); and the three exceptional graphs arise only when $k + 2$ is 12, 18 or 30 (these are the Coxeter numbers of the E series).

10

Coulomb gas representation

10.1 Free and Feigin–Fuks scalar fields

Let us start with a short detour to introduce the simplest example of a conformal field theory, provided by a free massless scalar field X, with energy–momentum tensor

$$T(z) = -\frac{1}{4} : [\partial X(z)]^2 : \tag{10.1}$$

Since we consider only the z-dependent part, we are actually dealing with the chiral half of a free field. Note that the classical equation of motion for a free field in two dimensions, which is just the wave equation, is satisfied by $X(z, \overline{z}) = X(z) + \overline{X}(\overline{z})$ where X and $\overline{X}$ are arbitrary functions. The two-point function is the two-dimensional logarithmic propagator

$$< X(z)X(\zeta) >= -2\log(z-\zeta) \tag{10.2}$$

from which it follows that

$$< \partial X(z)\partial X(\zeta) >= \frac{-2}{(z-\zeta)^2} \tag{10.3}$$

Another common normalization includes $\frac{1}{2}$ instead of $\frac{1}{4}$ in (10.1). This amounts to the field redefinition $X \to \sqrt{2}X$.

The two-point function (10.3) allows us to give meaning to the normal ordered expression in (10.1):

$$: [\partial X(z)]^2 : = \lim_{z\to\zeta} \left[\partial X(z)\partial X(\zeta) + \frac{2}{(z-\zeta)^2} \right] \tag{10.4}$$

In general, a normal ordered product is the singular limit of an operator product expanded as a Laurent series, from which we subtract the divergent terms. Using the operator product expansion $X(z)X(\zeta) = -2\log(z-\zeta) + \cdots$ and the Wick theorem, it is easy to check that the holomorphic energy–momentum tensor (10.1) defines a conformal theory with central charge $c = 1$.

What is the primary field content of this theory? It is clear from the two-point function (10.2) that $X(z)$ is not a primary field. The reader may check that $\partial X(z)$ is, nevertheless, a primary field with conformal weight $\Delta_{\partial X} = 1$. To build some more primary fields, it is useful to dissect the free field X: its expansion in modes is given by (see appendix F)

$$X(z) = \sqrt{2}\left(q - ip\log z + i\sum_{n\neq 0}\frac{a_n}{n}z^n\right) \tag{10.5}$$

where q, p and the "oscillators" a_n satisfy the "stringy" relations

$$[q,p] = i\,, \qquad [a_n, a_m] = n\,\delta_{n+m,0} \tag{10.6}$$

Using these last two equations, it is not hard to verify (10.3) above.

Remarkably, the normal ordered exponentials of the free field

$$\begin{aligned} V_\alpha(z) &\equiv :\mathrm{e}^{i\alpha X(z)}: \\ &= \mathrm{e}^{i\sqrt{2}\alpha q}z^{\sqrt{2}\alpha p}\exp\left(\sqrt{2}\alpha\sum_{n>0}\frac{a_{-n}}{n}z^n\right)\exp\left(-\sqrt{2}\alpha\sum_{n>0}\frac{a_n}{n}z^{-n}\right) \end{aligned} \tag{10.7}$$

are also primary fields, of conformal weight $\Delta_\alpha = \alpha^2$. Notice the reflection symmetry in the conformal weights of these exponentials, $\Delta_\alpha = \Delta_{-\alpha}$. The operator product expansion of these operators can be computed with the help of the Campbell–Hausdorf formula:

$$:\mathrm{e}^{i\alpha X(z)}::\mathrm{e}^{i\beta X(\zeta)}:=(z-\zeta)^{2\alpha\beta}\ :\mathrm{e}^{i\alpha X(z)+i\beta X(\zeta)}: \tag{10.8}$$

It is then a straightforward exercise to evaluate the two-point function of the vertex operators (10.7):

$$\begin{aligned} \langle V_\alpha(z)V_\beta(\zeta)\rangle &= (z-\zeta)^{2\alpha\beta}\left\langle :\mathrm{e}^{i\alpha X(z)+i\beta X(\zeta)}:\right\rangle \\ &= (z-\zeta)^{2\alpha\beta}\left\langle \mathrm{e}^{i(\alpha+\beta)q\sqrt{2}}\right\rangle = \frac{\delta_{\alpha+\beta,0}}{(z-\zeta)^{2\alpha^2}} \end{aligned} \tag{10.9}$$

We may think of α and β as charges: it is a general feature that the only non-vanishing correlation functions of exponentials are those for which the total charge vanishes. Repeated use of (10.8) yields

$$\langle V_{\alpha_1}(z_1)\cdots V_{\alpha_n}(z_n)\rangle = \begin{cases} \prod_{i<j}(z_i-z_j)^{2\alpha_i\alpha_j} & \text{if } \sum\alpha_i = 0 \\ 0 & \text{otherwise} \end{cases} \tag{10.10}$$

The operator that measures the $U(1)$ charge α of the vertex operator $V_\alpha(z)$ is just $Q = (4\pi)^{-1}\oint dz\,\partial X(z)$:

$$Q\,|\alpha\rangle = \oint\frac{dz}{4\pi}\partial X(z)\ \mathrm{e}^{i\alpha X(0)}\,|0\rangle = \alpha\,|\alpha\rangle \tag{10.11}$$

A theory described by the energy–momentum tensor (10.1) has, thus, a continuous infinity of primary fields. This enormous degeneracy can be reduced by modifying the boundary conditions.

For instance, we may compactify the field X on a circle, i.e. require $X(z) = X(z) + 2\sqrt{2}\pi R$. In addition to the field ∂X, only the single-valued exponentials with $\alpha = \frac{n}{\sqrt{2}R}$ ($n \in \mathbf{Z}$) remain primary. We shall not deal with these theories here; suffice it to say that they often lead to conformal field theories whose chiral algebra is larger than pure Virasoro.

Another possible way of changing the boundary conditions is to add a boundary term, or total derivative, to the energy–momentum tensor:

$$T(z) = -\frac{1}{4} : [\partial X(z)]^2 : +i\alpha_\circ \partial^2 X(z) \tag{10.12}$$

The central charge of this theory is $c = 1 - 24\alpha_\circ^2$. For $\alpha_\circ$ real, the central extension is thus reduced with respect to the original value. The total derivative in (10.12) can be understood as the addition of the "background charge" $-2\alpha_\circ$ at infinity. The conformal weight of the exponential $V_\alpha(z)$, which is still a primary field for any value of α:

$$\Delta_\alpha = \alpha(\alpha - 2\alpha_\circ) \tag{10.13}$$

A free field with a background charge as in (10.12) is called a Feigin–Fuks field, and it provides us with the simplest example of a Coulomb gas representation for a conformal field theory, because the insertions of vertex operators mimic $U(1)$ point charges in a background field. The reflection symmetry found above survives, but takes into account the presence of the background charge:

$$\Delta_\alpha = \Delta_{2\alpha_\circ - \alpha} \tag{10.14}$$

The identity, in particular, can be written either as $\mathbf{1}$ or as $V_{2\alpha_\circ}$. This duality shows that the space of states is larger than strictly necessary.

Note that, in the modified theory (10.12), the operator ∂X is no longer a primary field (see exercise 10.10). Also, the theory seems not to be unitary, because the energy–momentum (10.12) is not hermitian. We shall see shortly that, for special values of $\alpha_\circ$, the theory is indeed unitary, and the infinite set of primary fields truncates to a finite subset.

Due to the background charge at infinity, the correlation functions of vertex operators will vanish unless their total charge compensates for the background charge:

$$\langle V_{\alpha_1}(z_1) \cdots V_{\alpha_n}(z_n) \rangle_{2\alpha_\circ} \neq 0 \quad \Rightarrow \quad \sum_{i=1}^{n} \alpha_i = 2\alpha_\circ \tag{10.15}$$

This can be checked easily by expanding the free field X in modes and looking at the zero mode pieces (momentum conservation is always assumed). The above correlator, if non-zero, is still equal to $\prod_{i<j}(z_i - z_j)^{2\alpha_i\alpha_j}$. The non-vanishing two-point functions are, explicitly,

$$\langle V_\alpha(z)\, V_{2\alpha_\circ - \alpha}(\zeta) \rangle_{2\alpha_\circ} = \frac{1}{(z-\zeta)^{2\alpha(\alpha - 2\alpha_\circ)}} \tag{10.16}$$

This can be formalized by noting that the presence of the boundary term in (10.12) modifies the "out"-states. In particular, the "out"-vacuum still enjoying

$SL(2)$ invariance for the theory at hand is

$$ {}_{2\alpha_\circ}\langle 0| = \frac{\langle 0|\, V_{-2\alpha_\circ}(\infty)}{\langle 0|\, V_{-2\alpha_\circ}(\infty)\, V_{2\alpha_\circ}(0)\,|0\rangle} \tag{10.17} $$

When $\alpha_\circ$ vanishes, an out-going state Φ_α is realized by the vertex operator $V_{-\alpha}(\infty)$, whereas, more generally with $\alpha_\circ \neq 0$, the out-states are of the form $V_{-\alpha+2\alpha_\circ}(\infty)$ (see the exercises).

We are ready to start unveiling the magic of the Coulomb gas. Let us pause for a moment and return to the decoupling equation (8.102) for a level-two null vector, which reads

$$ \left(\rho\frac{\partial^2}{\partial z^2} - \sum_{i=1}^{n}\frac{\Delta_i}{(z-z_i)^2} - \sum_{i=1}^{n}\frac{1}{z-z_i}\frac{\partial}{\partial z_i}\right)\langle \Phi_\delta(z)\Phi_{\Delta_1}(z_1)\cdots\Phi_{\Delta_n}(z_n)\rangle = 0 \tag{10.18} $$

where $\delta = \Delta_{12}$ or Δ_{21} and $\rho^{-1} = (4\delta+2)/3$. In the special case of a four-point function ($n = 3$), the differential operator in (10.18) simplifies to

$$ \begin{aligned} \rho\frac{d^2}{dz^2} + \sum_{i=1}^{3}\left(\frac{1}{z-z_i}\frac{d}{dz} - \frac{\Delta_i}{(z-z_i)^2}\right) + \frac{\delta+\Delta_1+\Delta_2-\Delta_3}{(z-z_1)(z-z_2)} \\ + \frac{\delta+\Delta_1+\Delta_3-\Delta_2}{(z-z_1)(z-z_3)} + \frac{\delta+\Delta_2+\Delta_3-\Delta_1}{(z-z_2)(z-z_3)} \end{aligned} \tag{10.19} $$

where we have traded the derivatives $\partial/\partial z_i$ in (10.18) for $\partial/\partial z$ with the help of the conformal Ward identities (8.96) derived from the $SL(2,\mathbb{C})$ invariance of the vacuum (see exercise 10.10),

$$ \mathscr{L}_k\,\langle \Phi_\delta(z)\Phi_{\Delta_1}(z_1)\cdots\Phi_{\Delta_n}(z_n)\rangle = 0 \qquad \text{for} \quad k = \pm 1, 0 \tag{10.20} $$

We may now use $SL(2)$ invariance of the vacuum to fix three co-ordinates $z_1 \to 0$, $z_2 \to 1$, $z_3 \to \infty$ through the conformal mapping $z \to x = (z-z_1)(z_2-z_3)/(z-z_3)(z_2-z_1)$, so the decoupling equation becomes

$$ \left[\rho\frac{d^2}{dx^2} + \left(\frac{1}{x} + \frac{1}{x-1}\right)\frac{d}{dx} + \frac{\delta+\Delta_1+\Delta_2-\Delta_3}{x(x-1)} - \frac{\Delta_2}{(x-1)^2} - \frac{\Delta_1}{x^2}\right]G(x) = 0 \tag{10.21} $$

with

$$ G(x) = \langle \Phi_{\Delta_3}(\infty)\Phi_{\Delta_2}(1)\Phi_\delta(x)\Phi_{\Delta_1}(0)\rangle \tag{10.22} $$

Redefining the dependent variable (the correlation function) of this equation by the rescaling $G(x) = x^{\lambda_1}(x-1)^{\lambda_2}y(x)$, where λ_i are the roots of $\rho\lambda_i^2+(1-\rho)\lambda_i-\Delta_i = 0$, the decoupling equation finally becomes

$$ x(1-x)y'' + [c-(a+b+1)x]y' - aby = 0 \tag{10.23} $$

In the above, a, b and c (with $b > a$) are the roots of

$$ \begin{cases} a+b+1 = 2(\rho^{-1}+\lambda_1+\lambda_2) \\ ab = 2\lambda_1\lambda_2 + (\lambda_1+\lambda_2+\delta+\Delta_1+\Delta_2-\Delta_3)/\rho \\ c = \rho^{-1}+2\lambda_2 \end{cases} \tag{10.24} $$

Note that the parameters a, b and c in (10.23) depend only on the conformal weights of the primary fields in the original correlation function.

Equation (10.23) is Riemann's hypergeometric equation. It has two independent solutions, which, for generic values of the parameters, can be chosen as

$$\begin{aligned} y_1 &= F(a,b;c;x) \\ y_2 &= (-x)^{1-c}F(a-c+1,b-c+1;2-c;x) \end{aligned} \tag{10.25}$$

where F is the hypergeometric function

$$F(a,b;c;x) = \frac{\Gamma(c)}{\Gamma(a)\Gamma(b)} \sum_{n\geq 0} \frac{\Gamma(a+n)\Gamma(b+n)}{\Gamma(c+n)} \frac{z^n}{n!} \tag{10.26}$$

whose integral representation is

$$F(a,b,c;x) = \frac{\Gamma(c)}{\Gamma(b)\Gamma(c-b)} \int_0^1 dt\, t^{b-1}(1-t)^{c-b-1}(1-xt)^{-a} \tag{10.27}$$

The simple decoupling just considered of a level-two null descendant in a four-point correlation function has led us to the well-known hypergeometric functions. In general, the decoupling equations for arbitrary correlation functions may be very complicated, but their solution always admits an integral representation, due to Dotsenko and Fateev, akin to the one just shown. In order to obtain actual numerical results, the contour integrals still have to be evaluated, but many properties of the correlation functions can be found without actually carrying out the full integrals. At any rate, more information can be extracted about the physical system from the integral representations of the solutions to the decoupling equations than from the decoupling equations themselves.

10.2 Screening charges in correlation functions

The main virtue of the Coulomb gas representation of rational conformal field theories is that it yields expressions for the correlation functions of primary fields which are, by construction, integral solutions to the decoupling equations. For example, the "Coulomb gas correlation functions" of vertex operators made from a single Feigin–Fuks field are solutions to the decoupling equations of minimal models. Let us stress, then, that the point of the Coulomb gas representation is to provide correlation functions which automatically satisfy the conformal Ward identities.

The important technical ingredients in the construction of correlation functions in the Coulomb gas representation of rational conformal field theories are the "screening currents". The conformal weight of a holomorphic screening current J is always one, whereby it can be integrated without ambiguity over a contour in the holomorphic variable z to yield a "screening charge" Q. The computation of (holomorphic) correlation functions then calls for evaluating these contour integrals.

Let us concentrate on an example and try to compute the four-point function $\langle(\Phi_\Delta)^4\rangle$ of the same primary field $\Phi_\Delta(z)$. (It is understood that in order to obtain the physical four-point function, we have to multiply the holomorphic part by its associated antiholomorphic part.) In the Coulomb gas, we may represent Φ_Δ by the vertex operators V_α or $V_{2\alpha_\circ-\alpha}$, where $\Delta = \alpha(\alpha - 2\alpha_\circ)$. Because of the background charge at ∞, the $SL(2)$ invariant out-vacuum has some structure, as epitomized in the above discussion regarding equation (10.16). We may also exploit this $SL(2)$ invariance to fix three of the four insertion co-ordinates. Thus, the holomorphic correlation function of interest would seem to be

$$\langle V_{-\alpha+2\alpha_\circ}(\infty)V_\alpha(1)V_\alpha(z)V_\alpha(0)\rangle_{2\alpha_\circ} \tag{10.28}$$

By the charge conservation derived from looking at the zero modes, it would seem that the correlator (10.28) should vanish unless $\alpha = 0$. This is true, but the actual (holomorphic) correlation function might not be given precisely by (10.28). Notably, we are free to introduce other operators into the correlator (10.28) to balance the charge, while leaving the all-important conformal properties unchanged. If an operator insertion $\mathcal{O}$ of zero conformal weight exists which does balance the charge, the correlation function $\langle(\Phi_\alpha)^4\rangle$ should then be given by the vacuum expectation of $\mathcal{O}$ and the vertex operators in (10.28). In other words, the physical correlation function would be a more complicated correlator than (10.28).

Operators with non-zero charge but zero conformal weight, the screening charges, arise very naturally in the Coulomb gas. Indeed, if we look for a primary field of the form $: e^{i\alpha X(z)} :$ with conformal weight one, i.e. $\Delta_\alpha = \alpha(\alpha - 2\alpha_\circ) = 1$, we find two solutions

$$S_\pm(z) = V_{\alpha_\pm}(z) =: \exp i\alpha_\pm X(z) : \tag{10.29}$$

whose charges satisfy

$$\begin{aligned} 2\alpha_\circ &= \alpha_+ + \alpha_- \\ \alpha_+\alpha_- &= -1 \end{aligned} \tag{10.30}$$

This means that the operator differential forms $S_\pm(z)dz$, of charge $\alpha_\pm$, are invariant under conformal transformations. To make sense out of the differential, we integrate the screening currents $S_\pm$, thus obtaining the screening charges

$$Q_\pm = \oint S_\pm(t)dt \tag{10.31}$$

The most general form of the operator insertion $\mathcal{O}$ is thus a (finite) product of screening charges Q_+ and Q_-. In order to compute the correlation function of four identical fields V_α in the Coulomb gas, we should therefore evaluate

$$\left\langle V_{-\alpha+2\alpha_\circ}(\infty)V_\alpha(z)V_\alpha(1)Q_+^{m-1}Q_-^{m'-1}V_\alpha(0)\right\rangle_{2\alpha_\circ} \qquad m, m' \in \mathbb{N} - \{0\} \tag{10.32}$$

which coincides, as shown, with the integral representation for the solution to the decoupling equations, provided the integration contours in $Q_\pm$ are the same ones as those required above in the discussion of the hypergeometric function (and its generalizations, of course).

In the Coulomb gas formalism, the important Kac formula follows as a deceivingly simple exercise. Indeed, (10.32) may be non-vanishing only if the total charge is

$$(2\alpha_\circ - \alpha) + 3\alpha + (m-1)\alpha_+ + (m'-1)\alpha_- = 2\alpha_\circ \tag{10.33}$$

This calls for the insertion of $(m-1)$ Q_+'s and $(m'-1)$ Q_-'s. The solution to (10.33) is

$$\alpha_{mm'} = \frac{1-m}{2}\alpha_+ + \frac{1-m'}{2}\alpha_- \tag{10.34}$$

We thus see that the conformal weights

$$\Delta_{mm'} = -\alpha_\circ^2 + \frac{1}{4}(m\alpha_+ + m'\alpha_-)^2 \tag{10.35}$$

of the vertex operators $V_{mm'} = \exp[i\alpha_{mm'}X(z)]$ reproduce the Kac formula (8.88).

Note that the values of m and m' in (10.32) which make the whole expression neutral are unique. On the other hand, there might be several inequivalent contours yielding a well defined (single-valued) correlation function. Since these contour integrals just ensure the decoupling conditions *ab initio,* the different independent contours correspond one to one to the different linearly independent solutions of the decoupling equations. In turn, each of these corresponds to a different intermediate primary in the scattering process $|\alpha\rangle\,|\alpha\rangle \to |\alpha\rangle\,|\alpha\rangle$ or, equivalently, to a different conformal block.

In the example of the hypergeometric function, for instance, we saw that there were two "independent" contours, reflecting the fact that the differential equation it solved was of second order. This means also that there were two independent channels for the scattering.

The simple four-point function above contains almost all the general features of any correlation function in the Coulomb gas representation of conformal field theories. If the fields in the correlation function are different, the number of screening charges we must insert (i.e. the order of the decoupling differential equation) varies depending on what field we choose to declare out-going. This translates into the compact Coulomb gas formalism the existence of different simultaneous decoupling equations when several fields in the correlator have null descendants.

As an illustration, consider the three-point function of Kac fields

$$\left\langle V_{m_1m_1'} V_{m_2m_2'} V_{m_3m_3'} \right\rangle_{2\alpha_\circ} \tag{10.36}$$

If we view $V_{m_1m_1'}$ as out-going , then we compute

$$_{2\alpha_\circ}\left\langle V_{m_1m_1'}\middle| Q_+^{r_+}Q_-^{r_-}V_{m_2m_2'}(1)\middle| V_{m_3m_3'}\right\rangle \tag{10.37}$$

If this does not vanish, then the fusion rule $N_{23}^1 = N_{m_1m_1',m_2m_2',m_3m_3'}$ is not zero (it is one). The condition that the total charge be balanced is

$$2\alpha_\circ - \alpha_{m_1m_1'} + \alpha_{m_2m_2'} + \alpha_{m_3m_3'} + r_+\alpha_+ + r_-\alpha_- = 2\alpha_\circ \tag{10.38}$$

where $r_\pm \in \mathbb{N}$. We could equally well view $V_{m_2m_2'}$ or $V_{m_3m_3'}$ as out-going; hence, we find two more equations such as (10.38):

$$\begin{aligned} 2\alpha_\circ - \alpha_{m_2m_2'} + \alpha_{m_3m_3'} + \alpha_{m_1m_1'} + r_+'\alpha_+ + r_-'\alpha_- &= 2\alpha_\circ \\ 2\alpha_\circ - \alpha_{m_3m_3'} + \alpha_{m_1m_1'} + \alpha_{m_2m_2'} + r_+''\alpha_+ + r_-''\alpha_- &= 2\alpha_\circ \end{aligned} \tag{10.39}$$

For these conformal field theories, the charge conjugation matrix is the identity, and thus

$$N_{12}^3 = N_{23}^1 = N_{13}^2 = N_{123} \tag{10.40}$$

Requiring the three charge balance equations to be compatible (with all the r's natural), and supposing that the correlators (10.37) satisfying these conditions are all non-vanishing, one obtains

$$1 + m_1 \le m_2 + m_3\ , \qquad 1 + m_1' \le m_2' + m_3' \tag{10.41}$$

and four more equations by cyclicity in the indices, which are simply the fusion rules (8.109). Furthermore, if $\alpha_-/\alpha_+ = -p/p'$, $(p,p') = 1$, then $\Delta_{mm'} = \Delta_{p-m,p'-m'}$. If we enforce absolutely the identification $V_{mm'} = V_{p-m,p'-m'}$, the fusion rules truncate to those for minimal (p,p') models, equation (8.115), and the theory becomes rational: the number of Q_+ and Q_- in a correlation function become bounded above.

The existence of a null descendant of $V_{mm'}$ at level mm' can also be shown explicitly. It is necessary to identify the Verma modules $\mathscr{V}^\alpha$ and $\mathscr{V}^{2\alpha_\circ-\alpha}$, once more to absolute limits. Noting that

$$\begin{aligned} 2\alpha_\circ - \alpha_{mm'} &= \alpha_{-m,-m'} \\ 2\alpha_\circ - \alpha_{mm'} - m\alpha_+ &= \alpha_{m,-m'} \\ 2\alpha_\circ - \alpha_{mm'} - m'\alpha_- &= \alpha_{-m,m'} \end{aligned} \tag{10.42}$$

it is not hard to show that

$$\begin{aligned} \mathscr{X}_{mm'}^+ &= Q_+^m V_{m,-m'} \\ \mathscr{X}_{mm'}^- &= Q_-^n V_{-m,m'} \end{aligned} \tag{10.43}$$

are indeed null vectors of $\mathscr{V}^{2\alpha_\circ-\alpha}$, of level mm'.

Before concluding our brief introduction to the Coulomb gas, let us verify how easily we can reproduce the hypergeometric function (10.27) so laboriously derived from the massaged decoupling equation (10.18). The simple case of a primary field with a null vector at level two calls for only one screening charge insertion to balance the charge. The Coulomb gas expression of the correlator

$$G(x) = \langle \Phi_{\Delta_3}(\infty)\Phi_{\Delta_2}(1)\Phi_{\Delta_{2,1}}(x)\Phi_{\Delta_1}(0)\rangle \tag{10.44}$$

is given by

$$\begin{aligned} G(x) &= \oint dt \, \langle V_{\alpha_3}(\infty)V_{\alpha_2}(1)V_{\alpha_{2,1}}(x)V_{\alpha_1}(0)S_+(t)\rangle_{2\alpha_\circ} \\ &= x^{2\alpha_{2,1}\alpha_1}(x-1)^{2\alpha_{2,1}\alpha_2} \oint_C dt \, t^{2\alpha_+\alpha_1}(1-t)^{2\alpha_+\alpha_2}(t-x)^{2\alpha_+\alpha_{2,1}} \end{aligned} \tag{10.45}$$

where we have assumed the charge neutrality condition

$$\alpha_1 + \alpha_2 + \alpha_3 + \alpha_{2,1} + \alpha_+ = 2\alpha_\circ \tag{10.46}$$

We immediately see that the rescaling $G(x) = x^{\lambda_1}(x-1)^{\lambda_2}y(x)$ is contained in (10.45) with the identification $\lambda_i = -\alpha_+\alpha_i$ $(i = 1, 2)$, which indeed satisfy the equation $\rho\lambda_i^2 + (1-\rho)\lambda_i - \Delta_i = 0$ with $\rho = \alpha_-^2$. Finally, the two solutions y_1 and y_2 can be derived from (10.45) by choosing the contour C to run from 1 to ∞ or from 0 to x, respectively. Different independent choices of the contour of integration are in one to one correspondence with different intermediate channels for the scattering or, if you prefer, different conformal blocks. To connect with our earlier discussion, we must effect the identifications

$$a = \alpha_+^2\,, \quad b = \alpha_+^2 - 1 - 2\alpha_+(\alpha_1 + \alpha_2)\,, \quad c = \alpha_+^2 - 2\alpha_+\alpha_1 \tag{10.47}$$

which indeed satisfy (10.24) provided the charge neutrality condition holds. We have thus proved in this example that the Coulomb gas produces an accurate integral representation of the null vector decoupling equations.

10.2.1 Braiding matrices: an explicit example

Once we have the integral representation of the decoupling equations, we can easily compute the braiding matrices which were introduced formally in the previous chapter. The braiding matrices will act on the space of solutions to the decoupling equations. To obtain these matrices, we must extend analytically solutions with the arguments ordered as $|z_1| > |z_2| > |z_3| > \ldots$ into solutions where two consecutive points are in the reverse order. We present the explicit calculation for the Ising model below.

The Coulomb gas representation of the three primary fields of the Ising model is

$$\mathbf{1} = 1\,, \quad \epsilon = e^{i\alpha_{21}X}\,, \quad \sigma = e^{i\alpha_{12}X} \tag{10.48}$$

or their charge conjugates. Using (10.45), we find

$$\langle \epsilon\epsilon\epsilon\epsilon\rangle = \frac{1}{z_{13}z_{24}}\,[\eta(1-\eta)]^{\frac{2}{3}} \times \begin{cases} [\eta(1-\eta)]^{-\frac{5}{3}}\,F(-2, -\frac{1}{3}, -\frac{2}{3}; \eta) \\ F(\frac{4}{3}, 3, \frac{8}{3}; \eta) \end{cases} \tag{10.49}$$

where the two different results correspond to an integration contour from 0 to η, and from 1 to ∞, respectively. Quite generally,

$$\eta = \frac{z_{12}\, z_{34}}{z_{13}\, z_{24}} \tag{10.50}$$

A priori, the existence of two independent contours indicates that there are two channels for the $\langle\epsilon\epsilon\epsilon\epsilon\rangle$ correlator. Nevertheless, in the second expression, $t = 0$ is not a branch point, and, therefore, the second contour does not really contribute, in agreement with the fusion rule $[\epsilon] \times [\epsilon] = [\mathbf{1}]$. The conformal block $\langle\epsilon\epsilon\epsilon\epsilon\rangle$ is therefore given by

$\langle\epsilon\epsilon\epsilon\epsilon\rangle =$ 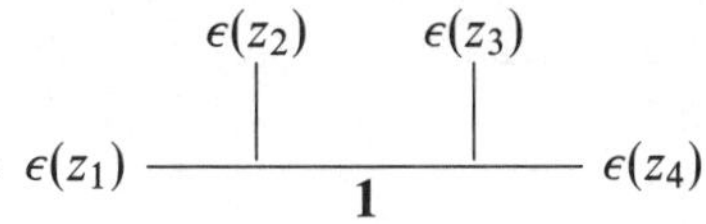

$$= \frac{1}{z_{13}\, z_{24}} \frac{1 - \eta + \eta^2}{\eta(1-\eta)} \tag{10.51}$$

Similarly, we find also

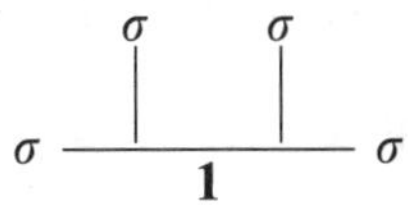

$$= \frac{(1-\eta)^{\frac{3}{8}}}{(z_{13}\, z_{24}\, \eta)^{\frac{1}{8}}} \sqrt{\frac{1+\sqrt{1-\eta}}{2(1-\eta)}}$$

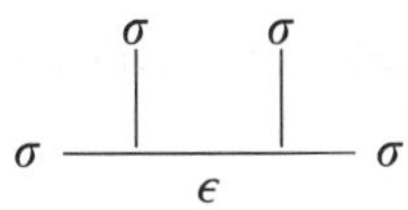

$$= \sqrt{2}\frac{(1-\eta)^{\frac{3}{8}}}{(z_{13}\, z_{24}\, \eta)^{\frac{1}{8}}} \sqrt{\frac{1-\sqrt{1-\eta}}{1-\eta}}$$

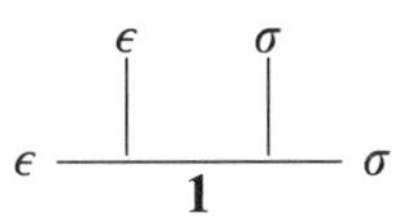

$$= z_{12}^{-1}\, z_{34}^{-\frac{1}{8}} \frac{1 - \frac{1}{2}\eta}{\sqrt{1-\eta}}$$

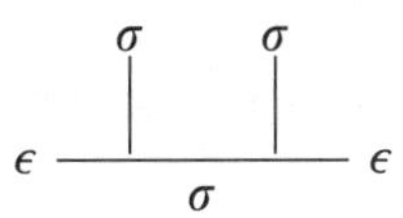

$$= \left(\frac{z_{14}}{z_{13}\, z_{24}}\right)^{\frac{1}{8}} \frac{1}{z_{14}} \frac{1+\eta}{\sqrt{\eta}(1-\eta)^{\frac{1}{8}}}$$

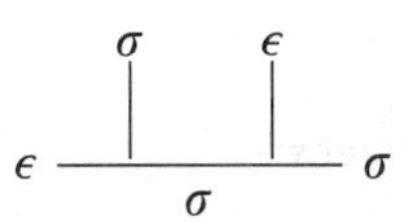

$$= z_{13}^{-1} z_{24}^{-\frac{1}{8}} \frac{1-2\eta}{\sqrt{\eta(1-\eta)}} \tag{10.52}$$

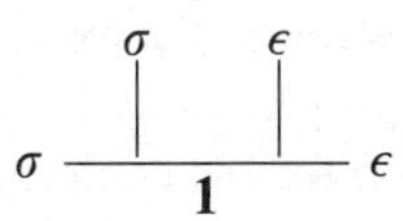

$$= \frac{1}{z_{12}^{\frac{1}{8}} z_{34}} \frac{1-\eta/2}{\sqrt{1-\eta}}$$

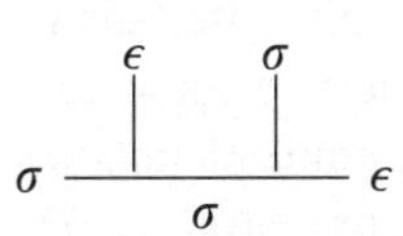

$$= \frac{1}{z_{13}^{\frac{1}{8}} z_{24}} (1-2\eta)\sqrt{\eta(1-\eta)}$$

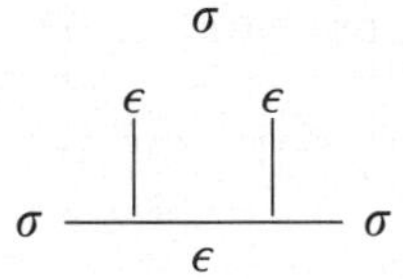

$$= \frac{z_{14}}{z_{14}^{\frac{1}{8}} z_{13}\, z_{24}} \frac{1+\eta}{\sqrt{\eta}(1-\eta)}$$

The braiding of the legs 2 and 3 is obtained by the analytic continuation $z_{23} \to e^{i\pi} z_{32}$. Using the same notation as in the previous chapter, we find

$$
\begin{aligned}
&B_{1\sigma}\begin{bmatrix} \epsilon & \sigma \\ \epsilon & \sigma \end{bmatrix} = -\frac{1}{2}\,e^{-i\pi/2} &\qquad& B_{\sigma\sigma}\begin{bmatrix} \sigma & \sigma \\ \epsilon & \epsilon \end{bmatrix} = e^{-i\pi/8} \\
&B_{\sigma 1}\begin{bmatrix} \sigma & \epsilon \\ \epsilon & \sigma \end{bmatrix} = -2\,e^{-i\pi/2} && B_{1\sigma}\begin{bmatrix} \sigma & \epsilon \\ \sigma & \epsilon \end{bmatrix} = -\frac{1}{2}\,e^{-i\pi/2} \\
&B_{\sigma 1}\begin{bmatrix} \epsilon & \sigma \\ \sigma & \epsilon \end{bmatrix} = -2\,e^{-i\pi/2} && B_{\sigma\sigma}\begin{bmatrix} \epsilon & \epsilon \\ \sigma & \sigma \end{bmatrix} = e^{-i\pi} \\
&B_{11}\begin{bmatrix} \sigma & \sigma \\ \sigma & \sigma \end{bmatrix} = \frac{1}{\sqrt{2}}\,e^{i\pi/8} && B_{1\epsilon}\begin{bmatrix} \sigma & \sigma \\ \sigma & \sigma \end{bmatrix} = \frac{1}{2\sqrt{2}}\,e^{-3i\pi/8} \\
&B_{\epsilon 1}\begin{bmatrix} \sigma & \sigma \\ \sigma & \sigma \end{bmatrix} = \sqrt{2}\,e^{-3i\pi/8} && B_{\epsilon\epsilon}\begin{bmatrix} \sigma & \sigma \\ \sigma & \sigma \end{bmatrix} = \frac{1}{\sqrt{2}}\,e^{i\pi/8}
\end{aligned}
\tag{10.53}
$$

An immediate application of the above formulae is the determination of the operator product expansion coefficients of the Ising model. The only non-vanishing ones are $C^{\mathbf{1}}_{\mathbf{11}} = 1$, $C^{\mathbf{1}}_{\epsilon\epsilon}$, $C^{\mathbf{1}}_{\sigma\sigma}$ and $C^{\epsilon}_{\sigma\sigma}$. We may choose, however, $C^{\mathbf{1}}_{\epsilon\epsilon} = C^{\mathbf{1}}_{\sigma\sigma} = 1$, which amounts to the normalization of the two-point functions

$$
\langle \epsilon(z,\bar{z})\epsilon(w,\bar{w})\rangle = \frac{1}{|z-w|^2} \qquad \langle \sigma(z,\bar{z})\sigma(w,\bar{w})\rangle = \frac{1}{|z-w|^{\frac{1}{4}}} \tag{10.54}
$$

To determine the only remaining non-trivial operator product expansion, $C^{\epsilon}_{\sigma\sigma}$, we write the four-point function $\langle \sigma^4 \rangle$ as in equation (9.9):

$$\langle\sigma\sigma\sigma\sigma\rangle = \left(C^{\mathbf{1}}_{\sigma\sigma}\right)^2 \left| \quad \right|^2$$

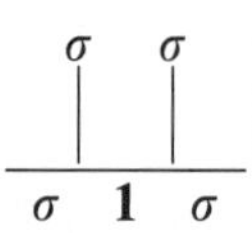

$$+ \left(C^{\epsilon}_{\sigma\sigma}\right)^2 \left| \quad \right|^2 \tag{10.55}$$

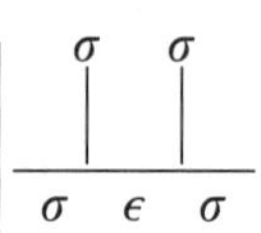

Requiring invariance under braiding of the external legs, one easily finds

$$
C^{\epsilon}_{\sigma\sigma} = \frac{1}{2} \tag{10.56}
$$

10.2.2 Contour techniques

It could seem from the previous paragraph that we need the explicit form of the conformal blocks to obtain their transformation properties under monodromy or braiding. Nevertheless, if an integral representation for the conformal blocks is at our disposal, then it is not necessary to carry out the integrals for such matters, and simple contour manipulations suffice. To illustrate these contour techniques for computing braiding and fusing matrices in some detail, let us consider a generic conformal block with only one screening operator, say Q_+:

$$
\begin{aligned}
\mathscr{F}_C^{\alpha_1\alpha_2\alpha_3\alpha_4}(z) &= \int_C dt\, \langle V_{\alpha_1}(\infty)V_{\alpha_2}(1)V_{\alpha_3}(z)V_{\alpha_4}(0)S_+(t)\rangle \\
&= (1-z)^{2\alpha_2\alpha_3} z^{2\alpha_3\alpha_4} I_C(a,b,c;z)
\end{aligned}
\tag{10.57}
$$

with

$$I_C(a,b,c;z) = \int_C dt\, t^a (t-1)^b (t-z)^c \tag{10.58}$$

and

$$a = 2\alpha_+\alpha_4\,, \quad b = 2\alpha_+\alpha_2\,, \quad c = 2\alpha_+\alpha_3\,, \quad d = 2\alpha_+\alpha_1 \tag{10.59}$$

so that $a+b+c+d = -2$ by charge conservation.

In the s channel $(1+2 \to 3+4)$, we choose the following contours:

$$\begin{aligned} I_1^{(s)}(a,b,c;z) &= \int_1^\infty dt\, t^a (t-1)^b (t-z)^c \\ I_2^{(s)}(a,b,c;z) &= \int_0^z dt\, t^a (1-t)^b (z-t)^c \end{aligned} \tag{10.60}$$

with $|z| < 1$. The logarithm is determined by choosing the leaf with $\arg(t-z) \in (-\pi, \pi)$. We assume for simplicity in what follows that z and the integration variable t are on the real axis.

An alternative basis for the conformal blocks, in the t channel this time, is given by

$$\begin{aligned} I_1^{(t)}(a,b,c;z) &= \int_{-\infty}^0 dt\, (-t)^a (1-t)^b (z-t)^c \\ I_2^{(t)}(a,b,c;z) &= \int_z^1 dt\, t^a (1-t)^b (t-z)^c \end{aligned} \tag{10.61}$$

Using the change of variables $t \to 1-t$, we see that

$$I_i^{(t)}(a,b,c;z) = I_i^{(s)}(b,a,c;1-z) \tag{10.62}$$

Finally, a basis for the u channel conformal blocks is given by

$$\mathscr{F}_C^{\alpha_1\alpha_2\alpha_3\alpha_4}(z) = (z-1)^{2\alpha_2\alpha_3} z^{2\alpha_3\alpha_4} I_i^{(u)}(a,b,c;z) \tag{10.63}$$

with

$$\begin{aligned} I_1^{(u)}(a,b,c;z) &= \int_z^\infty dt\, t^a (t-1)^b (t-z)^c \\ I_2^{(u)}(a,b,c;z) &= \int_0^1 dt\, t^a (1-t)^b (z-t)^c \end{aligned} \tag{10.64}$$

where now $|z| > 1$.

Having defined the basis in the three possible channels of a four-point scattering, we wish to relate them through s–t and s–u duality.

We achieve s–t duality by means of a deformation of the contours as shown schematically below. The contour of integration from 1 to ∞

$I_1^{(s)} =$ 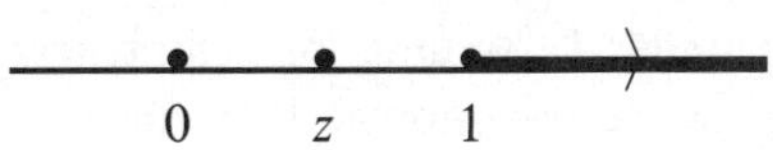 (10.65) is equivalent to a contour from 1 to $-\infty$ either by turning the contour counter-clockwise or clockwise. In the first case, the singularities at 1, z and 0 end up under the contour:

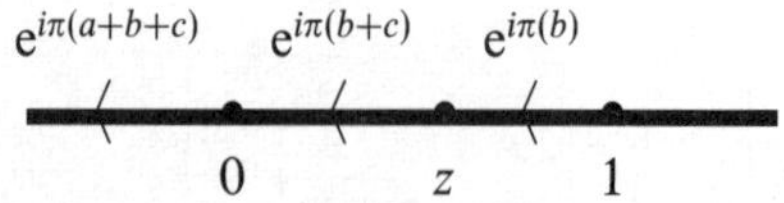

(10.66a)

whereas in the second case they end up on top of the contour:

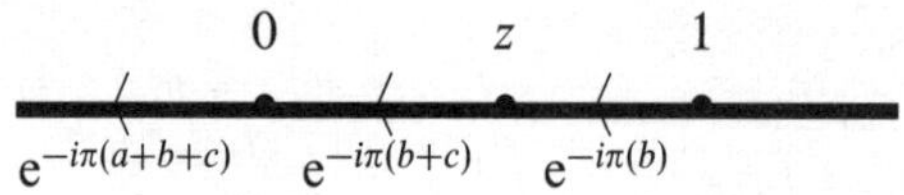

(10.66b)

Except for the singularities, where the contour undergoes an infinitesimal deformation to be slightly above or slightly under, the contour is always on the real axis. The various phases indicate the factor by which the integral in that interval must be multiplied. In the first case, the screening operator on the contour is braided over the operator insertions at 1, z and 0, whereas in the second case it braids below them. The phases arise, explicitly, from the reversal of ordering in the monomials

$$\begin{aligned}(t-\zeta)^\alpha &\to \begin{cases} e^{i\pi\alpha}(\zeta-t)^\alpha & t \text{ turns counter-clockwise over } \zeta \\ e^{-i\pi\alpha}(\zeta-t)^\alpha & t \text{ turns clockwise under } \zeta \end{cases} \\ t>\zeta &\to \zeta>t \end{aligned} \tag{10.67}$$

Multiplying now $I_1^{(s)}$ by $e^{\mp i\pi(b+c)}$ and subtracting the two results, it follows from (10.66) and (10.61) that

$$I_1^{(s)}(a,b,c;z) = \frac{\sin\pi a}{\sin\pi(b+c)}\, I_1^{(t)}(a,b,c;z) - \frac{\sin\pi c}{\sin\pi(b+c)}\, I_2^{(t)}(a,b,c;z) \tag{10.68}$$

A similar calculation may be carried out for $I_2^{(s)}$. The fusing matrix $F\begin{bmatrix}\alpha_2 & \alpha_3\\ \alpha_1 & \alpha_4\end{bmatrix}$ defined by

$$\mathscr{F}^{\alpha_1\alpha_2\alpha_3\alpha_4}_{s,i}(z) = \sum_{j=1}^{2} F_{ij}\begin{bmatrix}\alpha_2 & \alpha_3\\ \alpha_1 & \alpha_4\end{bmatrix} \mathscr{F}^{\alpha_1\alpha_2\alpha_3\alpha_4}_{t,j}(z) \tag{10.69}$$

turns out to be

$$F\begin{bmatrix}\alpha_2 & \alpha_3\\ \alpha_1 & \alpha_4\end{bmatrix} = \frac{1}{\sin\pi(b+c)}\begin{pmatrix}\sin\pi a & -\sin\pi c\\ \sin\pi d & -\sin\pi b\end{pmatrix} \tag{10.70}$$

To compute braiding matrices, we have to perform the analytic extension of conformal blocks with $|z|<1$ to the region $|z|>1$. Again, taking the point z (with $0<z<1$) to another z with $1<z$ can be done in two ways: either the point goes over or under 1. Assume for definiteness that we perform the analytic continuation keeping the operator at 1 on top of the operator at z, i.e. we take

the point z under 1. Then $I_1^{(s)}$ becomes the integral

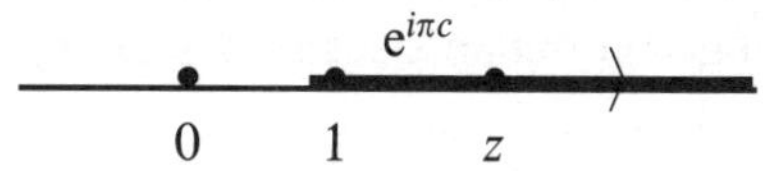

(10.71)

where the contour goes over the point z. Under the same continuation, the other s channel conformal block, with a contour from 0 to z,

$I_2^{(s)} =$ 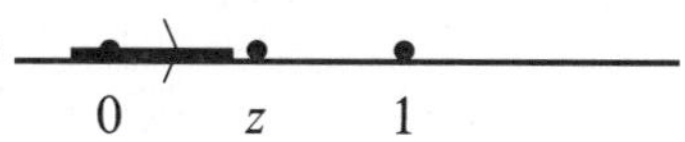

(10.72)

becomes

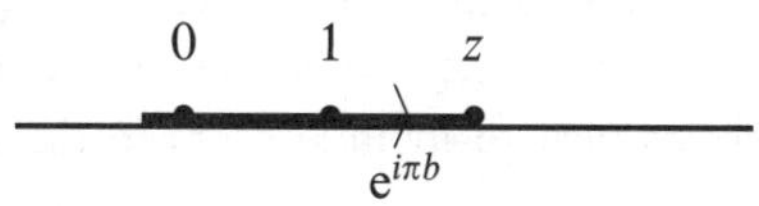

(10.73)

where the contour goes from 0 to z and under 1.

Now the integrals (10.71) and (10.73) between two singular points can be treated as in (10.66) in order to obtain the integrals (10.64) corresponding to the u channel.

The final result for the braiding matrix $B\begin{bmatrix}\alpha_2 & \alpha_3\\ \alpha_1 & \alpha_4\end{bmatrix}$ defined by

$$\sigma\, \mathscr{F}_{s,i}^{\alpha_1\alpha_2\alpha_3\alpha_4}(z) = \sum_{j=1}^{2} B_{ij}\begin{bmatrix}\alpha_2 & \alpha_3\\ \alpha_1 & \alpha_4\end{bmatrix} \mathscr{F}_{u,j}^{\alpha_1\alpha_2\alpha_3\alpha_4}(z) \tag{10.74}$$

is the following:

$$B\begin{bmatrix}\alpha_2 & \alpha_3\\ \alpha_1 & \alpha_4\end{bmatrix} = \frac{e^{2\pi i\alpha_2\alpha_3}}{\sin\pi(c+d)}\begin{pmatrix} e^{-i\pi d}\sin\pi c & e^{i\pi c}\sin\pi a\\ -e^{i\pi b}\sin\pi d & -e^{-i\pi a}\sin\pi b\end{pmatrix} \tag{10.75}$$

Acting with σ^{-1} instead on the s channel conformal blocks, we obtain the complex conjugate of B:

$$\sigma^{-1}\, \mathscr{F}_{s,i}^{\alpha_1\alpha_2\alpha_3\alpha_4}(z) = \sum_{j=1}^{2} B_{ij}^{*}\begin{bmatrix}\alpha_2 & \alpha_3\\ \alpha_1 & \alpha_4\end{bmatrix} \mathscr{F}_{u,j}^{\alpha_1\alpha_2\alpha_3\alpha_4}(z) \tag{10.76}$$

This concludes our explicit example. If the conformal blocks involve more than one integral, the computation of the braiding and fusing matrices becomes somewhat more complicated, but it still involves only elementary complex analysis and combinatorics.

10.3 Lagrangian approach

Let us sharpen our understanding of the Coulomb gas representation of minimal models by looking at it from a different angle. Since we have a holomorphic energy–momentum tensor $T(z)$ for the Feigin–Fuks field X, we may consider the lagrangian from which it derives. It is necessary to use both the holomorphic and antiholomorphic stress tensors, which we assume are identical for simplicity.

The two-dimensional lagrangian density describing one full Feigin–Fuks field is

$$\mathscr{L}_\circ = \partial X(z,\bar{z})\bar{\partial} X(z,\bar{z}) + i\alpha_\circ R(z,\bar{z}) X(z,\bar{z}) \tag{10.77}$$

The curvature scalar R can be concentrated at infinity, taking advantage of the conformal invariance of (10.77):

$$R(z,\bar{z}) = 2\partial\bar{\partial}\,\delta(z-\infty)\,\delta(\bar{z}-\infty) \tag{10.78}$$

(On a generic orientable Riemann surface other than the sphere, the factor on the right-hand side is $2(1-h)$, with h the genus.) The screening currents are still very mysterious, so it is perhaps best to include them in the lagrangian:

$$\mathscr{L} = \mathscr{L}_\circ + \mathrm{e}^{i\alpha_+ X(z,\bar{z})} + \mathrm{e}^{i\alpha_- X(z,\bar{z})} \tag{10.79}$$

and treat them as perturbative interactions in the evaluation of correlations of local vertex operators V_i. Note that these two insertions are the most general conformally invariant ones, with $\alpha_\pm$ fixed in terms of $\alpha_\circ$ by (10.30). Setting

$$Q_\pm = \int dz d\bar{z}\ \mathrm{e}^{i\alpha_\pm X(z,\bar{z})} = \int d^2z\ S_\pm(z,\bar{z}) \tag{10.80}$$

we write

$$\begin{aligned}\left\langle \prod V_i(z_i,\bar{z}_i)\right\rangle_{\alpha_\circ} &= \int \mathscr{D}X \exp\left(-\int d^2z \mathscr{L}\right) \prod V_i(z_i,\bar{z}_i) \\ &= \int \mathscr{D}X \exp\left[-\int d^2z\,(\mathscr{L}_\circ + S_+ + S_-)\right] \prod V_i(z_i,\bar{z}_i) \\ &= \sum_{\substack{m\geq 0\\ n\geq 0}} \frac{1}{m!\,n!} \int \mathscr{D}X \exp\left(-\int d^2z \mathscr{L}_\circ\right) Q_+^m Q_-^n \prod V_i(z_i,\bar{z}_i)\end{aligned} \tag{10.81}$$

Now, from the classical equation of motion, $X(z,\bar{z}) = x_\circ + X(z) + \overline{X}(\bar{z})$ and $\mathscr{D}X = \mathscr{D}\varphi\,\mathscr{D}\overline{X} dx_\circ$. The integration over the zero mode $x_\circ$ can be done immediately. After that, each term in the series can be factorized into a holomorphic part times an antiholomorphic part. Accordingly, the two-dimensional surface integrals become contour integrals (in z or $\bar{z}$). However, from the integral over $x_\circ$, it is easy to see that only those terms in the series (10.81) with balanced charge will be non-zero:

$$2\alpha_\circ = m\alpha_+ + n\alpha_- + \sum_i \alpha_i \tag{10.82}$$

When $\alpha_\circ$ is real (i.e. for the usual (p,p') minimal models), only one term of the series is non-zero.

10.4 Wess–Zumino models

In this section, we shall use affine Lie algebras to define conformal field theories. This just means that we shall assume the chiral algebra to be a Kac–Moody algebra (a very brief review of Kac–Moody algebras has been given in section 6.12). Conformal field theories with a Kac–Moody algebra as its chiral algebra are called Wess–Zumino, Wess–Zumino–Witten, or Wess–Zumino–Novikov–Witten models.

10.4.1 The Knizhnik–Zamolodchikov equation

Starting with the generators J^x_n of $\hat{\mathscr{G}}$ ($x \in \mathscr{G}$), we define the currents

$$J^x(z) = \sum_{n\in\mathbb{Z}} J^x_n z^{-n-1} \tag{10.83}$$

The Sugawara construction yields, for a level k representation of the affine algebra, the energy–momentum tensor of the corresponding conformal field theory in terms of these currents:

$$\begin{aligned} T(z) &= \frac{1}{2(k+h^*)} \sum_x : J^x(z) J^x(z) : \\ &= \frac{1}{2(k+h^*)} \sum_x \left(\lim_{\zeta\to z} J^x(\zeta) J^x(z) - \frac{k}{(\zeta - z)^2} \right) \end{aligned} \tag{10.84}$$

where h^* is the dual Coxeter number of $\mathscr{G}$ (up to a normalization, this is just the quadratic Casimir evaluated in the adjoint representation; see table 10.1). The Virasoro generators are thus

$$L_m = \frac{1}{2(k+h^*)} \sum_x \sum_{n\in\mathbb{Z}} : J^x_{m+n} J^x_{-n} : \tag{10.85}$$

and they generate the Virasoro algebra with central extension

$$c = \frac{k \dim \mathscr{G}}{k + h^*} \tag{10.86}$$

For simplicity, we shall limit ourselves throughout to the simplest Kac–Moody algebra, $\hat{\mathscr{G}} = \widehat{SU(2)}$. Once we choose a level k, we will refer to the corresponding Wess–Zumino model as $\widehat{SU(2)}_k$.

Irreps of $\widehat{SU(2)}$ are characterized by a spin j and a level k. Irreps of level zero are finite dimensional. Given three irreps (j_1, k), (j_2, k) and $(j_3, 0)$, we consider the intertwining operator

$$\Phi^{(j_2,j_3)}_{(j_1;k)}(z) : V^{(j_1,k)} \longrightarrow V^{(j_2,k)} \otimes V^{(j_3)}(z)\, z^{-\Delta(j_1)-\Delta(j_3)+\Delta(j_2)} \tag{10.87}$$

Table 10.1 Dimensions and dual Coxeter numbers of the semi-simple Lie algebras $\mathscr{G}_\ell$ of rank ℓ.
The central extension of the Wess–Zumino model based on the affine $\hat{\mathscr{G}} = \mathscr{G}^{(1)}$ with representations of level k is given by (10.86).

$\mathscr{G}$		dim $\mathscr{G}$	h^*
A_ℓ	$\ell \geq 1$	$\ell(\ell+2)$	$\ell+1$
B_ℓ	$\ell \geq 2$	$\ell(2\ell+1)$	$2\ell-1$
C_ℓ	$\ell \geq 3$	$\ell(2\ell+1)$	$\ell+1$
D_ℓ	$\ell \geq 4$	$\ell(2\ell-1)$	$2\ell-2$
G_2		14	4
F_4		52	9
E_6		78	12
E_7		133	18
E_8		248	30

where $\Delta(j_i)$ are the conformal weights relative to the Virasoro generator L_0 defined by the Sugawara construction,

$$\Delta(j) = \frac{j(j+1)}{k+2} \tag{10.88}$$

Given a vector in the dual to $V^{(j_3)}$, $v \in V^{(j_3)*}(z)$, we define the operator

$$\begin{aligned} \Phi_v(z) : V^{(j_1,k)} &\longrightarrow V^{(j_2,k)} \otimes \mathbb{C}(z)\, z^{-\Delta(j_1)-\Delta(j_3)+\Delta(j_2)} \\ |\cdot\rangle_{j_1} &\mapsto \Phi_v(z)\,|\cdot\rangle_{j_1} = {}_{j_3}\langle v|\, \Phi^{(j_2,j_3)}_{(j_1;k)}(z)\,|\cdot\rangle_{j_1} \end{aligned} \tag{10.89}$$

In pictures, we draw a double line to denote the infinite-dimensional modules $V^{(j)}$ of level k:

$\Phi_v(z) \quad = \quad \text{out} \leftarrow j_2$ 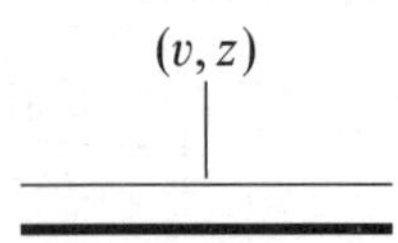 $j_1 \leftarrow \text{in}$ (10.90)

When $j_1 = 0$, the operators $\Phi_v(z)$ define the primary fields of the Wess–Zumino model $\widehat{SU(2)}_k$.

The operators $\Phi_v(z)$ satisfy the differential equation

$$(k + h^*)\frac{d}{dz}\Phi_v(z) = \sum_x :J^x(z)\Phi_{xv}(z): \tag{10.91}$$

where xv represents the action of the element $x \in \mathscr{G}$ on the vector vS.

With the help of the operators $\Phi_v(z)$, we define the correlators of the theory:

$$\Psi_{v_1,\dots,v_n}(z_1,\dots,z_n) = \langle\cdot|\,\Phi_{v_1}(z_1)\cdots\Phi_{v_n}(z_n)\,|\cdot\rangle \tag{10.92}$$

$\Psi(z_1,\dots,z_n) = \langle\cdot|$ 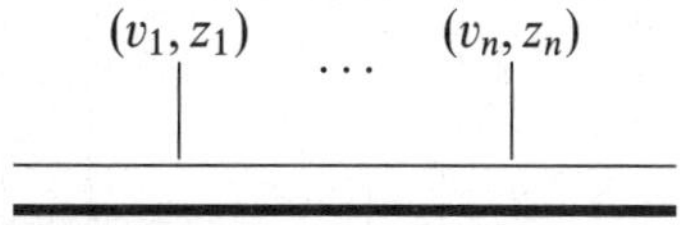 $|\cdot\rangle$ (10.93)

Applying now the equation of motion (10.91) to the correlators (10.92), we obtain the celebrated Knizhnik–Zamolodchikov equation

$$(k+h^*)\frac{d}{dz_i}\Psi(z_1,\dots,z_n) = \sum_{j\neq i}\frac{\Omega_{ij}}{z_i - z_j}\Psi(z_i,\dots,z_n) \tag{10.94}$$

where we have used the Sugawara representation (10.85) of $T(z)$ and the operator product

$$J^x(z)\Phi_v(\zeta) = \frac{1}{z-\zeta}\Phi_{xv}(\zeta) \tag{10.95}$$

The matrix Ω is given by

$$\Omega_{ij} = \sum_x \mathbf{1}\otimes\cdots\otimes\mathbf{1}\otimes T^x_{(i)}\otimes\mathbf{1}\otimes\cdots\otimes\mathbf{1}\otimes T^x_{(j)}\otimes\mathbf{1}\otimes\cdots\otimes\mathbf{1} \tag{10.96}$$

with $T^x_{(i)}$ the generator T^x of $SU(2)$ acting on $V^{(j_i)}(z_i)$.

It is interesting to note that consistency of the Knizhnik–Zamolodchikov equation follows from the fact that

$$\frac{\Omega_{ij}}{z_i - z_j} = r_{ij}(z_i - z_j) \tag{10.97}$$

satisfies the classical Yang–Baxter equation (6.130). Interpreting $r_{ij}(z_i - z_j)$ in (10.94) as a connection for the Friedan–Shenker bundle, i.e. the vector bundle of conformal blocks, the classical Yang–Baxter equation is equivalent to the flatness condition. Let us stress also that the whole meaning of equation (10.94) is already contained in the equation of motion (10.91) for the intertwining operators.

The duality structure of the Wess–Zumino model is determined by the monodromy properties of the Knizhnik–Zamolodchikov equation. More precisely, the space of solutions to (10.94) defines irreducible representations of the braid group. In fact, given a solution $\Psi(z_1,\dots,z_n)$ of (10.94) with $|z_1| > |z_2| > \cdots > |z_n|$, we can obtain, by analytic continuation, another solution with $|z_i| < |z_{i+1}|$, all other z_j being kept in the original order. The braiding matrix connecting these

two solutions is defined, as usual, by

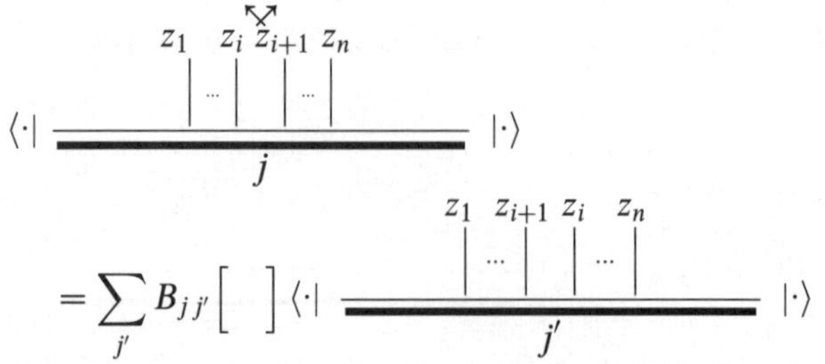

(10.98)

where the B-matrices satisfy the hexagonal equation. This equation is an example of the s–u duality displayed in equation (9.33). In the next chapter, we shall work out more explicitly these braiding matrices, showing that they can be understood as the $6j$ symbols of $U_q(s\ell(2))$ with $q = \exp \pi i/(k+2)$, where k is the level of the baseline in (10.98).

Note that in the previous section we derived the braiding matrices from the decoupling equations, whereas now we have started from the Knizhnik–Zamolodchikov equation (10.94). In fact, this equation can be interpreted also as a decoupling condition for the mixed Virasoro and Kac–Moody null vector:

$$\left(L_{-1} - \frac{1}{2(k+h^*)} \sum_x J^x_{-1} J^x_0 \right) |v_\circ\rangle = 0 \tag{10.99}$$

where $|v_\circ\rangle$ is the highest weight vector.

It is important to stress the difference between the intertwiner operators (10.87) and the chiral vertex operators we have used in chapter 9. The latter originate in the framework of the geometric data consisting of a Riemann surface equipped with local co-ordinates and three punctures P_i such that $z(P_1) = 0$, $z(P_2) = z$ and $z(P_3) = \infty$. Furthermore, given three irreps (i, j, k) of the chiral algebra (the Virasoro algebra in chapter 9) with a fixed value of the central extension, the operators $\Phi^k_{ij}(z)$ are defined such that $\langle\gamma|\, \Phi^k_{ij}(z)\, |\alpha\rangle \otimes |\beta\rangle$ is precisely the three-point function if $|\alpha\rangle$, $|\beta\rangle$ and $|\gamma\rangle$ are the highest weight vectors. The operator $\Phi^k_{ij}(z)$ maps the Verma modules $\mathscr{V}^i \otimes \mathscr{V}^j$ into the Verma module $\mathscr{V}^k$, and satisfies the equation of motion (9.16). And yet, strictly speaking, it is not an intertwining operator for the Virasoro algebra, as explained at the end of section 9.2. In fact, the intertwiner condition

$$\Phi^k_{ij}(z)\rho^i \otimes \rho^j(\Delta L_n) = \rho^k(L_n)\Phi^k_{ij}(z) \tag{10.100}$$

for the co-multiplication $\Delta L_n = L_n \otimes \mathbf{1} + \mathbf{1} \otimes L_n$ would require that the sum of the central extensions of the three Verma modules vanish. In contrast, the operator $\Phi^{(j_2,j_3)}_{(j_1;k)}(z)$ certainly intertwines between the irrep (j_1, k) and $(j_2, k) \otimes (j_3, 0)$, i.e.

$$\Phi^{(j_2,j_3)}_{(j_1;k)}(z)\rho^{(j_1,k)}(x) = \rho^{(j_2,k)} \otimes \rho^{(j_3,0)}(\Delta x)\Phi^{(j_2,j_3)}_{(j_1;k)}(z) \tag{10.101}$$

for any generator x of $\widehat{SU(2)}$ such that $\Delta x = x \otimes \mathbf{1} + \mathbf{1} \otimes x$. Precisely because $\Phi^{(j_2,j_3)}_{(j_1;k)}(z)$ is an intertwiner operator, one of the levels in the tensor product $j_2 \otimes j_3$ must be zero. We are thus forced to use two infinite-dimensional irreps of level k and one finite-dimensional irrep of level zero. Furthermore, the geometric interpretation in the construction of the chiral vertex operators is replaced here by the use of the irrep $V^j(z)$ introduced in section 6.12. A relationship between chiral vertex operators and intertwiners does exist, however, and it is given by the formula

$$\Phi^{j_2}_{j_1 j_3}(z)\left(|\alpha\rangle_{j_1} \otimes |v\rangle_{j_3}\right) = {}_{j_3}\langle v|\, \Phi^{(j_2,j_3)}_{(j_1;k)}(z)\, |\alpha\rangle_{j_1} \tag{10.102}$$

10.4.2 Free field representation of Wess–Zumino models

This section relies on the lagrangian approach to field theories, for which the main postulate is locality. Some familiarity with the path integral formulation of gauge theories is presumed.

The idea is to define a theory whose fields take values on a group manifold. In four spacetime dimensions, the homotopy group π_5 is a non-trivial topological invariant, which may contribute to the action as the integral (over a fictitious extra variable) of a total derivative. The addition to the action is thus a boundary term, which modifies the energy density but not the equations of motion. If the non-linear σ-model used to describe pions (Wess–Zumino model) is supplemented by this topological (Novikov–Witten) term, then solitons appear in the spectrum. Interestingly, the solitons are fermionic, whereas the original $\vec{\pi}$ and σ fields were bosonic. It is natural to interpret these fermionic solitons or skyrmions as nucleons. More generally, a non-linear σ-model describing mesons will include also baryons as solitons of the theory, provided the action includes the topological term.

In two dimensions, the relevant topological invariant is π_3. The action for a two-dimensional Wess–Zumino model is the generalization to group-valued fields of the usual non-linear σ-model, including a kinetic term and the topological $\pi_3(\mathcal{G})$. The lagrangian density is thus

$$\mathcal{L} = \frac{-k}{4\pi}\left\{\frac{1}{2}\mathrm{Tr}(\Omega^{-1}\partial\Omega)^2 + \frac{i}{3}\mathrm{Tr}\int \epsilon^{\mu\nu\rho}(\Omega^{-1}\partial_\mu\Omega)(\Omega^{-1}\partial_\nu\Omega)(\Omega^{-1}\partial_\rho\Omega)\right\} \tag{10.103}$$

where $\Omega(x_0, x_1) = \Omega(z, \bar{z})$ is an element of the group manifold. This model is conformally invariant; in fact, the variables can always be factorized into holomorphic and antiholomorphic parts:

$$\Omega(z,\bar{z}) = \Omega_L(\bar{z})\Omega_R(z) \tag{10.104}$$

The Wess–Zumino lagrangian (10.103), with the Witten–Novikov topological term included, is invariant under independent left and right gauge transformations, $\Omega(z,\bar{z}) \rightarrow U_L(\bar{z})\Omega(z,\bar{z})U_R(z)$, where $U_{L,R}$ are each generated by a Kac–Moody algebra.

Following Gauss, it is convenient to decompose the group element $\Omega(z)$ (we drop the subscript R where no confusion may arise) as a product of three matrices: one diagonal, one upper triangular, with ones on the diagonal, and one lower triangular, also with ones on the diagonal. These three factors are in one to one correspondence with the Cartan, raising and lowering operators of the algebra.

In order to lighten the notation, we shall most often deal with the simplest Lie algebra, namely $\mathscr{G} = SU(2)$. The Gauss decomposition for this case is explicitly

$$\Omega = \begin{pmatrix} 1 & \psi \\ 0 & 1 \end{pmatrix} \begin{pmatrix} e^{\varphi} & 0 \\ 0 & e^{-\varphi} \end{pmatrix} \begin{pmatrix} 1 & 0 \\ \gamma & 1 \end{pmatrix} \tag{10.105}$$

Here and throughout, all exponentials of fields, as well as any other function of fields defined through its power series (log, sin, etc.), are understood to be normal-ordered.

The lagrangian (10.103) can be rewritten in terms of the fields ψ, φ and γ. It is best, however, to effect the change of variables

$$\beta = k\, e^{-2\varphi} \partial\psi \tag{10.106}$$

In form language, this holomorphic field $\beta = \tilde{\beta}_1 - i\tilde{\beta}_2$ comes from $\tilde{\beta} = k\, e^{-2\varphi} d\psi$.

The quadratic part of the $SU(2)$ Wess–Zumino lagrangian becomes

$$\frac{-1}{4\pi} \{k(\partial_\mu\varphi)^2 + \tilde{\beta}_\mu \partial_\mu \gamma\} \tag{10.107}$$

whereas the topological term yields

$$\frac{-i}{4\pi} \int d(\tilde{\beta} d\gamma) = \frac{-i}{4\pi} \epsilon_{\mu\nu} \tilde{\beta}_\mu \partial_\nu \gamma \tag{10.108}$$

After the Gauss decomposition and the change of variables (10.106), one half of the lagrangian (10.103) reads

$$\mathscr{L} = \frac{-1}{4\pi} (\beta\bar{\partial}\gamma + k\partial\varphi\bar{\partial}\varphi) \tag{10.109}$$

The other half is identical, but with the antiholomorphic fields coming from the Gauss decomposition of Ω_L. The subtleties associated with the possible mismatch of the zero modes of the scalars ϕ and its "antiholomorphic" partner would lead us too far into the delicacies of modular invariance. For the time being, we concentrate on the chiral lagrangian (10.109) and analyze only the holomorphic half of the theory.

The change of variables (10.106) induces also a non-trivial jacobian, det $e^{-2\varphi}\partial$, which adds to the lagrangian (through the usual Faddeev–Popov procedure) the contribution

$$-\frac{1}{4\pi} \{2\partial\varphi\bar{\partial}\varphi + \varphi R\} \tag{10.110}$$

where the curvature scalar is $R = \partial\bar{\partial} \log g$, where g is the induced two-dimensional metric in the conformal gauge. The net result for the action density is thus

$$\mathscr{L} = \frac{-1}{4\pi}\left[\beta\bar{\partial}\gamma + (k+2)\partial\varphi\bar{\partial}\varphi + R\varphi\right] \tag{10.111}$$

It is convenient to rescale:

$$\varphi = \frac{-i}{\sqrt{2(k+2)}}\phi \tag{10.112}$$

Also, we may choose a flat metric g everywhere except for a single point z_R, where all the curvature R is concentrated as a singularity. We shall take $z_R = \infty$. Accordingly, the linear coupling of the field to the curvature in the lagrangian reflects itself in the familiar boundary term or charge at infinity in the enery-momentum tensor:

$$T(z) = \beta\partial\gamma - \frac{1}{2}(\partial\phi)^2 - \frac{i}{\sqrt{2(k+2)}}\partial^2\phi \tag{10.113}$$

The singular operator product expansions among the various fields are therefore

$$\begin{aligned} \phi(z)\phi(\zeta) &= -\log(z-\zeta) \\ \beta(z)\gamma(\zeta) &= \frac{1}{z-\zeta} \end{aligned} \tag{10.114}$$

The holomorphic part of a Wess–Zumino model can thus be described by a free $j=1$ β–γ pair (β and γ are commuting Virasoro primary fields of conformal weights one and zero, respectively (see exercise 8.2), and a free scalar with a charge at infinity related to the level k, which is in fact the level of the Kac–Moody algebra. Unitarity implies that the exponential of the action in the path integral is single-valued; hence, the level k is an integer.

The β–γ system behaves very much like a ghost sector. Indeed, the theory whose energy–momentum tensor contains only the Cartan scalars (Toda field theory) is still conformal but not gauge-invariant anymore: no ghosts, no gauged symmetry. It is as if the Cartan scalars embodied the full structure, which must then be adorned with raising and lowering operators (the modes of $\beta(z)$ and $\gamma(z)$) in group space to enhance the symmetry.

Obviously, the conformal energy–momentum tensor (10.113) does not exhibit the Kac–Moody symmetry in any explicit way. However, from the original group elements Ω, the matrix of currents $K = \sum_{a=1}^{3} K^a\sigma^a$ is given by the Knizhnik–Zamolodchikov formula

$$J = (k+2)g^{-1}dg \tag{10.115}$$

which implies, in terms of the free fields,

$$
\begin{aligned}
J^+(z) &= \frac{i}{\sqrt{2}}\beta(z) \\
J^3(z) &= i\sqrt{\frac{k+2}{2}}\,\partial\phi(z) - :\beta(z)\gamma(z): \\
J^-(z) &= \frac{i}{\sqrt{2}}\left[:\beta(z)\gamma^2(z): - i\sqrt{2(k+2)}\,\gamma(z)\partial\phi(z) - k\partial\gamma(z)\right]
\end{aligned}
\tag{10.116}
$$

It is a good exercise to check, with the help of the operator product expansions (10.114), that the currents in (10.116) do indeed satisfy an $SU(2)_k$ Kac–Moody algebra. Also, the energy–momentum tensor (10.113) is again reproduced by the Sugawara construction,

$$
T(z) = \frac{1}{2(k+2)}\left(2 :J^+(z)J^-(z): + :J^3(z)J^3(z):\right) \tag{10.117}
$$

In this Coulomb gas construction, invented by Wakimoto, the vertex operator, which generates the primary field of the Verma module of spin j, is given by

$$
V_j(z) =: \exp i\sqrt{\frac{2}{k+2}}\, j\,\phi(z) : \tag{10.118}
$$

It has conformal weight $\Delta_j = j(j+1)/(k+2)$, as can be easily proven from (10.113) and (10.114). Acting with $\oint J^-(z)dz$ on (10.118), we generate a spin-j multiplet of fields, labeled by $m = -j, -j+1, \ldots, j-1, j$:

$$
V_{j,m}(z) = \gamma(z)^{j-m} \; : \exp i\sqrt{\frac{2}{k+2}}\, j\,\phi(z) : \tag{10.119}
$$

The theory defined by (10.113) does not display the affine symmetry explicitly, but has the great virtue of being a sum of free field lagrangians. This is the Coulomb gas representation for a conformal field theory whose chiral algebra is the affine $\widehat{SU(2)}$ with representations of level k. It will allow us to find the irreducible highest weight modules very simply, automatically ensuring the decoupling of null vectors.

Recall that we needed screening charges for the computation of correlation functions in the Coulomb gas picture of a conformal field theory. In this lagrangian approach to Wess–Zumino models (i.e. to rational conformal field theories whose chiral algebra is a Kac–Moody algebra), the screening charges appear naturally as a consequence of locality. Indeed, let us consider once more the change of variables (10.106), whose jacobian already gave rise to a shift of the central extension $k \to k+2$, and to the charge at infinity for the scalar field. Since in the original formulation of the Wess–Zumino model the group elements $\Omega(z)$ were single-valued, the fields ψ, φ and γ are also single-valued, but not necessarily so the field β. The single-valuedness of ψ, which can be expressed as the mathematical condition that $\oint d\psi = 0$ along any closed contour, translates into a global condition involving the field β:

$$\oint_C \beta(z) \exp -i\sqrt{\frac{2}{k+2}}\phi(z)\, dz = 0 \tag{10.120}$$

In terms of the screening current $S = \beta \exp[-i\phi\sqrt{2/(k+2)}]$, these constraints can be taken into account by adding to the measure of the generating functional the following δ function insertions:

$$\begin{aligned}\prod_C \delta\left(\oint_C S(z)\, dz\right) &= \prod_C \int d\lambda\, e^{i\lambda \oint_C S(z)\, dz} \\ &= \sum_{n\geq 0} \frac{1}{n!} \prod_C \int d\lambda \left(i\lambda \oint_C S(z)\, dz\right)^n \end{aligned} \tag{10.121}$$

where the product is taken over all topologically different closed contours on the (punctured) Riemann surface. To any given correlation function, only a finite number of terms in the sum (10.121) contribute. Observe that the conformal weight of the exponential in S is zero, and all the conformal weight of S comes from the field β, trivial under monodromy.

10.4.3 The Goddard–Kent–Olive construction

Interestingly enough, from the Sugawara construction (10.85) of the Virasoro generators in terms of the Kac–Moody currents we may obtain the unitary series of minimal models described in chapter 8, in terms of two Kac–Moody algebras.

The idea is to choose a pair of affine Lie algebras $\hat{g}$, $\hat{h}$ such that $\hat{h}$ is a subalgebra of $\hat{g}$. Through the Sugawara construction, we obtain the Virasoro generators $L_n^{(g)}$ and $L_n^{(h)}$. The operators

$$K_n = L_n^{(g)} - L_n^{(h)} \tag{10.122}$$

commute with the subalgebra $\hat{h}$ and define a new realization of the Virasoro algebra with central extension

$$c_K = c_{\hat{g}} - c_{\hat{h}} \tag{10.123}$$

To reproduce the central extension for minimal models, we may take, for instance,

$$\hat{g} = \widehat{SU(2)}_k \otimes \widehat{SU(2)}_1\,, \qquad \hat{h} = \widehat{SU(2)}_{k+1} \tag{10.124}$$

so that

$$c_K = 1 - \frac{6}{(k+2)(k+3)} = 1 - \frac{6}{Q(Q+1)} \tag{10.125}$$

with $Q = k+2$, in the notation of (8.114).

The above construction, due to Goddard, Kent and Olive (GKO), allows us to view the Virasoro characters of minimal models as branching coefficients, and to find the modular matrix of the Virasoro characters in terms of the modular transformations of Kac–Moody characters. The main idea is the following. Let us consider an irrep of the Lie algebra $\hat{g}$ in (10.124), namely

$$V^{(j_1,k)} \otimes V^{(j_2,1)} \tag{10.126}$$

If $\hat{h}$ is embedded diagonally in $\hat{g}$, then we may decompose this irrep into irreps of $\hat{h}$ as follows:

$$V^{(j_1,k)} \otimes V^{(j_2,1)} = \bigoplus_{j_3} W^{(j_1,j_2;j_3)} \otimes V^{(j_3,k+1)} \tag{10.127}$$

Since the GKO Virasoro generators $K_n = L_n^{(g)} - L_n^{(h)}$ commute with $\hat{h}$, they act on the space $W^{(j_1,j_2;j_3)}$. Their characters are thus

$$b_{j_1,j_2;j_3}(q) = \mathrm{tr}_{W^{(j_1,j_2;j_3)}}\left(q^{-c_K/24+K_0}\right) \tag{10.128}$$

With the help of (10.127), we find the following relationship between Kac–Moody characters:

$$\mathscr{X}_{j_1,k}(q)\mathscr{X}_{j_2,1}(q) = \sum_{j_3} b_{j_1,j_2;j_3}(q)\, \mathscr{X}_{j_3,k+1}(q) \tag{10.129}$$

where

$$\mathscr{X}_{j,k}(q) = \mathrm{tr}_{V^{(j,k)}}\left(q^{-c_k/24+L_0^{(k)}}\right) \tag{10.130}$$

where $c_k = 3k/(k+2)$, and $L_0^{(k)}$ is the Virasoro operator of $\widehat{SU(2)}_k$. The explicit expression for these characters is similar to the one obtained in chapter 8 for the Virasoro modules:

$$\mathscr{X}_{j,k}(q) = \frac{1}{\eta^3}\sum_{\ell\in\mathbb{Z}}(N\ell+2j+1)q^{(N\ell+2j+1)^2/(2N)} \tag{10.131}$$

with $N = 2(k+2)$.

Under the modular transformation $S : \tau \to -1/\tau$, the characters (10.131) transform as follows:

$$\begin{aligned}\mathscr{X}_{j,k}(-1/\tau) &= \sum_{\ell=0}^{k/2}\sqrt{\frac{2}{k+2}}\sin\frac{\pi(2j+1)(2\ell+1)}{k+2}\,\mathscr{X}_{\ell,k}(\tau)\\ &= \sum_{\ell=0}^{k/2} S_{j\ell}^{(k)}\,\mathscr{X}_{\ell,k}(\tau)\end{aligned} \tag{10.132}$$

From (10.129) we thus obtain for the GKO Virasoro characters the remarkable identity

$$b_{j_1,j_2;j_3}(-1/\tau) = \sum_{\substack{0\le\ell_1\le k/2\\ 0\le\ell_2\le\frac{1}{2}\\ 0\le\ell_3\le\frac{k+1}{2}}} S_{j_1\ell_1}^{(k)} S_{j_2\ell_2}^{(1)} S_{j_3\ell_3}^{(k+1)}\, b_{\ell_1,\ell_2;\ell_3}(\tau) \tag{10.133}$$

Note that the characters for irreps with $j \leq k/2$ close under modular transformations.

10.5 Magic corner transfer matrix

In chapters 1, 2 and 7, we discussed the concept of the transfer matrix and its associated one-dimensional hamiltonian. Two different cases were considered, the row to row transfer matrix (chapters 1 and 2) and the diagonal transfer matrix (chapter 7). They proved most useful in computations involving vertex and face models, respectively. They are, of course, quite identical conceptually: given a classical statistical mechanical model on a square lattice, we may reinterpret it as the result of the time evolution of a one-dimensional system, and time may flow either from one row to the next or diagonally in the square lattice. In either case, a one-dimensional state in the thermodynamic limit is represented by a string of lattice variables labeled by the integers $\mathbb{Z}$. In contrast, the hamiltonian associated with the corner transfer matrix introduced in chapter 5 acts on a semi-infinite lattice, labeled by $\mathbb{N} - \{0\}$. Baxter's corner transfer matrix is a rich and subtle concept. We now return to its semi-infinite hamiltonians and study their intriguing relationship with the Goddard–Kent–Olive (GKO) Virasoro characters.

To start with, let us consider a face model over the A_Q graph

$$\underset{1}{\bullet}\!\!-\!\!-\!\!-\underset{2}{\bullet}\!-\!\!-\!\!-\cdots-\!\!-\!\!-\underset{Q}{\bullet} \tag{10.134}$$

Lattice variables may thus take Q different values, and the variables on two neighboring sites must differ by one. Denoting the lattice variable at the ith site of the semi-infinite one-dimensional lattice by σ_i ($1 \leq i < \infty$), this means that

$$\sigma_i \in \{1, 2, \ldots, Q\} \qquad\qquad \sigma_i - \sigma_{i+1} = \pm 1 \tag{10.135}$$

Let us associate to a configuration $\{\sigma_i\}$ the energy

$$E(\{\sigma_i\}) = \sum_{j\geq 1} j\varepsilon\left(\sigma_j, \sigma_{j+1}, \sigma_{j+2}\right) \tag{10.136}$$

with

$$\varepsilon\left(\sigma_j, \sigma_{j+1}, \sigma_{j+2}\right) = \frac{|\sigma_j - \sigma_{j+2}|}{4} \tag{10.137}$$

The energy (10.136) is positive definite, so, in the configurations minimizing it, all the even sites take a given value and all odd sites take the same value plus one (or minus one): the energy for all such configurations is zero. Therefore, the system has $2(Q-1)$ degenerate ground states, labeled by an ordered pair of lattice variables (b, c) linked by the graph (10.134).

Let us now try to compute expectation values in the various ground states. By this we mean that we consider local operators on the semi-infinite lattice, subject to the condition that, for sufficiently large i, the state coincides with a (b, c) ground state. We may think of each ground state as providing asymptotic boundary conditions on the open end of the semi-infinite lattice. An interesting quantity is the probability $P(a;b,c)$ of having lattice variable a on the first site, with (b,c) boundary conditions. Clearly,

$$P(a;b,c) = \sum_{\{\sigma_i\}(a;b,c)} \exp\left(-E(\{\sigma_i\})/k_B T\right) \tag{10.138}$$

where the sum is over all states such that

$$\sigma_1 = a \qquad \text{and} \quad \exists m > 1\,,\ \forall j > m \quad \begin{cases} \sigma_{2j-1} = b \\ \sigma_{2j} = c \end{cases} \tag{10.139}$$

The magic result about these "local state probabilities" that we can quote without proof is the following. For the GKO construction, with $\hat{h} \subset \hat{g}$ given by (10.124),

$$P(a;b,c) = b_{j_1,j_2;j_3}(q) \tag{10.140}$$

provided we identify

$$\begin{aligned} Q &= k+2 \\ q &= \mathrm{e}^{-1/k_B T} \end{aligned} \qquad \begin{aligned} 2j_1 &= \frac{b+c-3}{2} \\ 2j_2 &= \frac{b-c+1}{2} \\ 2j_3 &= a-1 \end{aligned} \tag{10.141}$$

This construction provides an interesting, fruitful and not completely understood link between face models on finite graphs and minimal models in conformal field theory.

Exercises

Ex. 10.1 With the expressions (10.53) at hand, check the hexagonal equation (9.37) for this particular case of the (3,4) minimal model. The monodromy matrices for the analytic continuation of z_2 around z_3 are the square of (10.53).

Ex. 10.2 The sine-Gordon model of exercise 3.3, with lagrangian

$$\mathscr{L} = (\partial\phi)^2 + \lambda :\cos\beta\phi: \tag{10.142}$$

can be interpreted as a deformation of the $c = 1$ conformal field theory of a free scalar field. Compute the conformal weight of the operator $:\cos\beta\phi:$. For what values of β is this operator relevant, i.e. of conformal dimension < 2? Using the results of exercise 3.3, discover for what values of XXZ anisotropy Δ is the operator $:\cos\beta\phi:$ relevant.

Ex. 10.3 Show in detail how the surface integral in (10.81) can be swapped for contour integrals.

Ex. 10.4 Kac's formula indicates that there exist rational theories with $\alpha_\circ$ real, i.e. $0 \leq c \leq 1$. In fact, there exist rational theories with $\alpha_\circ$ purely imaginary, i.e. in the interval $24 \leq c < 25$. Using the formula (10.81) in this latter case to evaluate correlation functions calls for a negative number of screenings. Assuming that the identity

$$\prod_{i=m}^{n} f(i) = \frac{\prod_{i=1}^{n} f(i)}{\prod_{i=1}^{m-1} f(i)} \tag{10.143}$$

is used to define formally the products $\prod_{i=m}^{n}$ with $n < m$, find all four-point functions for the theory with central charge $c = 49/2$.

Ex. 10.5 Using the definition (10.89) of primary fields, prove the following operator product expansion:

$$J^x(z)\Phi_v(\zeta) = \frac{\Phi_{xv}(\zeta)}{z-\zeta} \tag{10.144}$$

with $x \in \mathscr{G}$. Derive from this result the Knizhnik–Zamolodchikov equation (10.94).

Ex. 10.6 The correlators (10.92) can be interpreted, in string theory, as tree-level amplitudes: the affine parameters z_i of the finite-dimensional representations of $A_1^{(1)}$ take values on the Riemann sphere. Let us compute the one-loop amplitudes, i.e. let us assume that z_i exist on a torus of periods $(1, \tau)$. Set $q = \exp 2\pi i\tau$ and define

$$\Psi^q(z_1, \ldots, z_n) = \operatorname{tr}_V \left(q^{L_0}\Phi_1(z_1)\cdots\Phi_n(z_n)\right) \tag{10.145}$$

where we assume that the level k source space of Φ_n coincides with the target space of Φ_1. Obtain the Knizhnik–Zamolodchikov equation obeyed by Ψ^q, and check whether its kernel is an elliptic solution to the classical Yang–Baxter equation for $\widehat{SU(2)}$.

Ex. 10.7 Generalization of the Wakimoto construction [A. Gerasimov, A. Marshakov, A. Morozov, M. Olshanetskiĭ and S. Shatashvili, 'Wess–Zumino–Witten model as a theory of free fields', *Int. J. Mod. Phys.* **A5** (1990) 2495].

Let us discuss the Coulomb gas description of a Wess–Zumino model based on an untwisted affine Kac–Moody algebra $\mathscr{G}^{(1)}$, along the same lines as in section 10.4.2 for $A_1^{(1)}$. If $d_\mathscr{G}$ denotes the dimension of $\mathscr{G}$, and if $r_\mathscr{G}$ denotes its rank, then the free field content of the conformal field theory consists of $(d_\mathscr{G} - r_\mathscr{G})/2$ free β–γ pairs of spin 1, and $r_\mathscr{G}$ free real scalar fields $\vec{\phi}$ with a boundary term (charge at infinity). The energy–momentum tensor is thus the sum of the free terms plus the background contribution:

$$T(z) = \sum_{\alpha\in\Delta_+} :\beta_\alpha\partial\gamma_\alpha: -\frac{1}{2} :\left(\partial\vec{\phi}(z)\right)^2: -\frac{i}{v}\vec{\rho}\cdot\partial^2\vec{\phi}(z) \tag{10.146}$$

Here, $v = \sqrt{k+h^*}$, k is the level of the Kac–Moody algebra, $h^* = \vec{\theta}\cdot(\vec{\theta}+2\vec{\rho})/\vec{\theta}^2$ is the dual Coxeter number, $\vec{\theta}$ is the highest root, $2\vec{\rho}$ is the sum of all positive roots of $\mathcal{G}$ and we adopt the convention that $\vec{\theta}^2 = 2$. The singular operator product expansions are

$$\gamma_\alpha(z)\beta_{\alpha'}(\zeta) = \frac{\delta_{\alpha\alpha'}}{z-\zeta}\,, \quad (\vec{a}\cdot\vec{\phi}(z))(\vec{b}\cdot\vec{\phi}(\zeta)) = -(\vec{a}\cdot\vec{b})\log(z-\zeta) \tag{10.147}$$

As in the case of $\widehat{SU(2)}_k$, the primary fields associated to an irrep of $\mathcal{G}$ labeled by its highest weight vector $\vec{\lambda}$ is given by the vertex operator $V_{\vec{\lambda}}(z) =: \exp \frac{i}{v}\vec{\lambda}\cdot\vec{\phi}(z)$.

- Show that the total central charge is

$$c = \frac{k\, d_{\mathcal{G}}}{k+h^*} \tag{10.148}$$

[Hint: prove first the Freudenthal–de Vries strange formula $\vec{\rho}^{\,2} = d_{\mathcal{G}}h^*/12$.]

- Show that the conformal weight of $V_{\vec{\lambda}}(z)$ is

$$\Delta_{\vec{\lambda}} = \Delta_{-\vec{\lambda}-2\vec{\rho}} = \frac{\vec{\lambda}\cdot(\vec{\lambda}+2\vec{\rho})}{2(k+h^*)} \tag{10.149}$$

- Prove the depth rule

$$\vec{\lambda}\cdot\vec{\theta} \le k \tag{10.150}$$

The Cartan currents are quite general. If $\vec{\mu}$ is a vector in the Cartan space,

$$J^{\vec{\mu}}(z) = -\sum_{\alpha\in\Delta_+}(\vec{\alpha},\vec{\mu})\,\beta_\alpha\gamma_\alpha + iv\,\vec{\mu}\cdot\partial\vec{\phi} \tag{10.151}$$

We shall denote

$$H^a(z) = J^{\vec{\alpha}_a}(z) \qquad a - 1,\ldots,r_{\mathcal{G}} \tag{10.152}$$

The non-diagonal currents are best described in the Chevalley basis. There are $r_{\mathcal{G}}$ Cartan degrees of freedom, so there are $r_{\mathcal{G}}$ raising currents J^{+a} and $r_{\mathcal{G}}$ lowering currents J^{-a} ($a = 1,\ldots,r_{\mathcal{G}}$).

Let us now specialize to $A^{(1)}_{N-1}$. The free field content of $\widehat{SU(N)}$ is ϕ_c ($1 \le c \le N-1$), $\beta_{b,a}$ and $\gamma_{a,b}$ ($1 \le a < b \le N$). Here, $\phi_c = \vec{\alpha}_c\cdot\vec{\phi}$, with $\vec{\alpha}_c$ a simple root, and

$$\phi_a(z)\phi_b(\zeta) = -\vec{\alpha}_a\cdot\vec{\alpha}_b\log(z-\zeta)\,, \quad \beta_{ba}(z)\gamma(\zeta)_{cd} = \frac{\delta_{bd}\delta_{ac}}{z-\zeta} \tag{10.153}$$

• Prove that the following currents, in addition to (10.152), do generate affine $SU(N)$ (normal ordering is implicit throughout):

$$J^{+a} = \beta_{a+1,a} + \sum_{b=a+2}^{N} \gamma_{a+1,b}\beta_{ba} \tag{10.154}$$

$$J^{a}_{-}(z) = \gamma_{a,a+1}\left\{\sum_{b=1}^{a-1}\gamma_{b,a}\beta_{a,b} - \sum_{b=1}^{a}\gamma_{b,a+1}\beta_{a+1,b}\right\} - \sum_{b=1}^{a+1}\gamma_{b,a+1}\beta_{ab}$$
$$+ \sum_{b=a+2}^{N}\gamma_{a,b}\beta_{b,a+1} - iv\gamma_{a,a+1}\partial\phi_a - (v^2 - a - 1)\partial\gamma_{a,a+1}$$

Remember that you must also verify the Serre relations.
The $r_{\mathcal{g}}$ screening currents are of the form

$$S^a(z) = \left(\beta_{a+1,a} + \sum_{b=1}^{a-1}\beta_{a+1,b}\gamma_{b,a}\right) e^{-i\phi_a/v} \tag{10.155}$$

• Prove the following:

$$\begin{aligned} &[J^{+a}_n, S^b(z)] = [H^a_n, S_b(z)] = 0 \\ &[J^{-a}_n, S^b(z)] = \delta_{ab}\, v^2 \frac{d}{dz}\left(z^n\, e^{-i\phi_a/v}\right) \end{aligned} \qquad (n > 0) \tag{10.156}$$

Ex. 10.8 Find the map between the Goddard–Kent–Olive Virasoro modules $W^{(j_1,j_2;j_3)}$ in (10.127) and the Verma modules of the minimal model with central extension (10.123). In actual terms, find the relationship between $(j_1, j_2; j_3)$ and (m, m').

Ex. 10.9 In exercise 9.10, we used the Virasoro characters for $c = 0$ to prove Euler's pentagonal identity. Using now (10.131) for $k = j = 0$, prove Jacobi's triple product identity

$$\eta(\tau)^3 = \sum_{\ell\in\mathbb{Z}}(4\ell + 1)\exp\left(\pi i\tau(4\ell + 1)^2/4\right) \tag{10.157}$$

Ex. 10.10 Show that, in the Coulomb gas, $\partial X(z)$ is not a primary field if $\alpha_\circ \neq 0$, and that its transformation law under the conformal mapping $w \to z(w)$ picks up an anomalous term:

$$\partial X(w) \to z'\partial X(z) + 2i\alpha_\circ\frac{z''}{z'} \; ; \quad z' = \frac{dz}{dw} \tag{10.158}$$

Use this law to show that the Coulomb charge

$$Q = \frac{1}{4\pi}\oint \partial X(z)dz \tag{10.159}$$

transforms as $Q \to 2\alpha_\circ - Q$ when $w = 1/z$.

Ex. 10.11 Derive from the $s\ell(2,\mathbb{C})$ invariance of the vacuum the following Ward identities:

$$\begin{aligned} L_{-1} &\to \sum_{i=1}^{n} \frac{\partial}{\partial z_i} \langle \Phi(z_1)\cdots\Phi(z_n)\rangle = 0 \\ L_0 &\to \sum_{i=1}^{n} \left(z_i \frac{\partial}{\partial z_i} + h_i \right) \langle \Phi(z_1)\cdots\Phi(z_n)\rangle = 0 \\ L_1 &\to \sum_{i=1}^{n} \left(z_i^2 \frac{\partial}{\partial z_i} + 2z_i h_i \right) \langle \Phi(z_1)\cdots\Phi(z_n)\rangle = 0 \end{aligned} \tag{10.160}$$

Ex. 10.12 Using the $s\ell(2,\mathbb{C})$ invariance of the vacuum, show that

$$\langle \Phi_\delta(z)\Phi_{\Delta_1}(z_1)\Phi_{\Delta_2}(z_2)\Phi_{\Delta_3}(z_3)\rangle = \frac{1}{F}\langle \Phi_{\Delta_3}(\infty)\Phi_{\Delta_2}(1)\Phi_\delta(x)\Phi_{\Delta_1}(0)\rangle \tag{10.161}$$

where

$$\begin{aligned} F &= (z-z_3)^{2\delta}\,(z_2-z_1)^{\delta+\Delta_1+\Delta_2-\Delta_3}\,(z_1-z_3)^{\Delta_1+\Delta_3-\Delta_2-\delta}\,(z_2-z_3)^{\Delta_2+\Delta_3-\Delta_1-\delta} \\ x &= \frac{(z-z_1)(z_2-z_3)}{(z-z_3)(z_2-z_1)} \end{aligned} \tag{10.162}$$

This equation establishes the relationship between the correlator (10.18) and $G(x)$ in (10.21). To obtain the full physical answer, incorporate the antiholomorphic part of the correlator.

Ex. 10.13 Using both conformal invariance and the Coulomb gas, we obtain

$$\left\langle \prod_{i=1}^{3} V_{\alpha_i}(z_i) \right\rangle = \prod_{i<j} z_{ij}^{-\Delta_{ij}} = \prod_{i<j} z_{ij}^{2\alpha_i\alpha_j} \tag{10.163}$$

with $z_{ij} = z_i - z_j$, $\Delta_{ij} = \Delta_i + \Delta_j - \Delta_k$, and $\Delta_i = \alpha_i(\alpha_i - 2\alpha_\circ)$. Show that this identity is correct.

Ex. 10.14 Consider in some more detail the integral representation of the hypergeometric function. Instead of using contours from one vertex operator insertion point (singularity) to another, we could take a contour on the complex plane which avoids the singularities. The integrand of (10.27), or, more precisely, the function $t^b(1-t)^{c-b}(1-xt)^{-a}$, should return to its original value after going around the integration contour. The integral (10.27) has singularities due to the field insertions at $t = 0$, 1 and x^{-1}, so, for (10.27) to be a solution to (10.23), the integration contours should be chosen to enclose singularities in such a way that the integral is single-valued. Contours which ensure this enclose two singularities and go around each of them twice, once clockwise and once counter-clockwise.

Show that such a contour, called a Pochhammer contour, can be deformed (figure 10.1) and then opened from branch point to branch point, provided we take into account the phase differences between the various Riemann sheets:

$$\oint_P dz z^{\alpha-1}(1-z)^{\beta-1} = \left(1-\mathrm{e}^{2\pi i\alpha}\right)\left(1-\mathrm{e}^{2\pi i\beta}\right)\int_0^1 dz z^{\alpha-1}(1-z)^{\beta-1} \tag{10.164}$$

Find a basis for the four-point conformal blocks involving only one closed analytic contour in the s, t and u channels.

Ex. 10.15 Consider the affine quantum group $U_q(\widehat{s\ell}(2))$, two infinite-dimensional irreps $V^{j_1,k}$ and $V^{j_2,k}$ of level k, and a finite-dimensional irrep $V^{j_3}(z)$ of level 0 defined as in (6.154). Let us define q-intertwiner operators

$$\Phi^{j_2,j_3}_{j_1;k}(z) : V^{j_1,k} \to V^{j_2,k} \otimes V^{j_3}(z) \tag{10.165}$$

with the following pictorial representation:

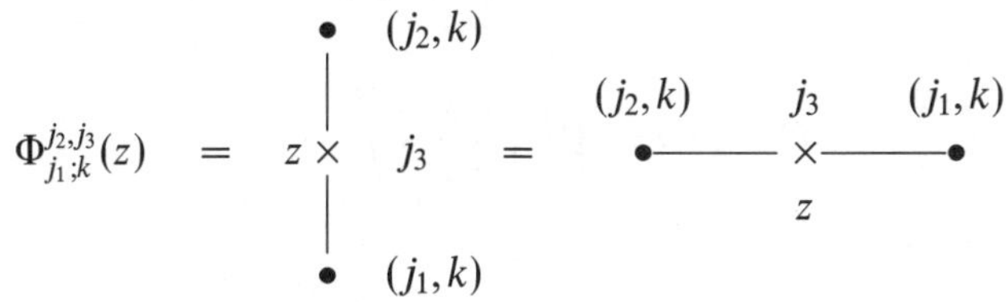

(10.166)

Assume that $j_3 = \frac{1}{2}$. The equivalent to Baxter's equation of chapter 4 becomes now

$$R^{(\frac{1}{2},\frac{1}{2})}(z_1/z_2)\Phi^{j_1,\frac{1}{2}}_{j_2;k}(z_1)\Phi^{j_2,\frac{1}{2}}_{j_3;k}(z_2) = \sum_{j_4} B\left(\begin{array}{cc|c} j_1 & j_4 & \\ j_2 & j_3 & z_1/z_2 \end{array}\right)\ \Phi^{j_1,\frac{1}{2}}_{j_4;k}(z_12)\Phi^{j_4,\frac{1}{2}}_{j_3;k}(z_1) \tag{10.167}$$

where $R^{(\frac{1}{2},\frac{1}{2})}(z)$ is the six-vertex intertwiner of chapter 2 and the B-matrices are the elliptic solution to the face model of type A_{k+1}, whose elliptic parameter is q^{k+2}.

Using the q-intertwiners to compute correlators, we can construct a q-Wess–Zumino model, whose duality data is an elliptic solution to the Yang–Baxter equation.

Define the q-correlators $\Psi(z_1, z_2, \ldots, z_n)$ as in (10.92) and (10.93) with the identity in the incoming and out-going base-line states, and $j_1 = j_2 = \cdots = j_n = 1/2$. Verify that (10.167) yields the monodromy structure of the q-Knizhnik–

Fig. 10.1 Deforming Pochhammer's contour into an open contour between singularities.

Zamolodchikov equation

$$\begin{aligned}\Psi(z_1,\ldots,pz_i,\ldots,z_n) &= R^{(\frac{1}{2},\frac{1}{2})}_{i,i-1}(pz_i/z_{i-1})\cdots R^{(\frac{1}{2},\frac{1}{2})}_{i,1}(pz_i/z_1)\\ &\times R^{(\frac{1}{2},\frac{1}{2})}_{i,n}(z_i/z_n)\cdots R^{(\frac{1}{2},\frac{1}{2})}_{i,i+1}(z_i/z_{i+1})\,\Psi(z_1,\ldots,z_i,\ldots,z_n)\end{aligned} \qquad (10.168)$$

where

$$p = q^{-2(k+2)} \qquad (10.169)$$

and, as usual, $R^{(\frac{1}{2},\frac{1}{2})}_{ij}$ denotes the action of the q-intertwiner $R^{(\frac{1}{2},\frac{1}{2})}$ on the ith and jth spaces. Compare this equation with the Bethe equations for a family of spin waves whose phase shifts are determined by $R^{(\frac{1}{2},\frac{1}{2})}$.

Compare (10.168) with the Knizhnik–Zamolodchikov equation (10.94), and try to find the quantization rules to go from (10.94) to (10.168). [Hint: use exercise 6.1.]

Ex. 10.16 Using Baxter's corner transfer matrix for the XXZ model, verify that the one-point function $\langle 0|\,\sigma^z_0\,|0\rangle$ in the antiferromagnetic regime ($0 < q < 1$) is given by

$$\langle 0|\,\sigma^z_0\,|0\rangle = \prod_{k\geq 1}\left(\frac{1-q^{2k}}{1+q^{2k}}\right)^2 \qquad (10.170)$$

Appendix F
Vertex operators

In this brief appendix, we clarify somewhat the mathematics of the vertex operators (10.7).

Start with the infinite-dimensional Heisenberg algebra

$$H = \mathbb{C}c \bigoplus_{j\in\mathbb{Z}-\{0\}} \mathbb{C}a_j \tag{F10.1}$$

where the generators a_i are paired into creation and annihilation operators, and c is central:

$$\left[a_i, a_j\right] = i\,\delta_{i+j,0}c\,, \qquad [c, a_i] = [c, c] = 0 \tag{F10.2}$$

We may represent this algebra on the space of functions $\mathbb{C}[x_1, x_2, \ldots]$ of countably infinitely many variables:

$$\begin{aligned} \rho^B : H &\longrightarrow \mathbb{C}[x_1, x_2, \ldots] \\ a_j &\mapsto \frac{\partial}{\partial x_j} \qquad j \geq 1 \\ a_{-j} &\mapsto jx_j \qquad j \geq 1 \end{aligned} \tag{F10.3}$$

With respect to the scalar product in $\mathbb{C}[x_1, x_2, \ldots]$ defined by

$$\langle P|Q\rangle = P\left(\frac{\partial}{\partial x_1}, \frac{\partial}{\partial x_2}, \ldots\right) Q(x_1, x_2, \ldots)\Big|_{x_1=x_2=\cdots=0} \tag{F10.4}$$

it is true that

$$\left[\rho^B(a_i)\right]^* = \rho^B(a_{-i}) \tag{F10.5}$$

Compare the scalar product (F10.4) with the one used in exercise 6.2.

The standard Fock space representation of H relies on a vacuum vector $|0\rangle$ such that

$$\begin{aligned} a_j\,|0\rangle &= 0 \qquad j \geq 1 \\ c\,|0\rangle &= |0\rangle \end{aligned} \tag{F10.6}$$

We may extend this Fock space representation to one of the extended algebra

$$\widehat{H} = \mathbb{C}c \bigoplus_{j\in\mathbb{Z}} \mathbb{C}a_j \tag{F10.7}$$

where $[a_0, a_i] = [a_0, c] = 0$, setting

$$a_0 \, |0\rangle = 0 \tag{F10.8}$$

For any $w \in \mathbb{C}$, we define a representation V^w of $\widehat{H}$ as follows:

$$\begin{aligned} a_0 \, |w\rangle &= w \, |w\rangle \\ a_j \, |w\rangle &= 0 \qquad j \geq 1 \\ c \, |w\rangle &= |w\rangle \end{aligned} \tag{F10.9}$$

The parameter w can be interpreted as the momentum of the center of mass of the string.

We may realize the Virasoro algebra in terms of the generators of $\widehat{H}$. For arbitrary $\alpha_\circ \in \mathbb{C}$, we define

$$L_n(\alpha_\circ) = -\sqrt{2}\alpha_\circ(n+1)a_n + \frac{1}{2}\sum_{k \in \mathbb{Z}} : a_k a_{n-k} : \tag{F10.10}$$

where the normal ordering means that we always write the operator with a positive index to the right of one with a negative index. It is not hard to check that $L_n(0)$ satisfy the Virasoro algebra with central charge equal to one:

$$[L_m(0), L_n(0)] = (m-n)L_{m+n}(0) + \delta_{m+n,0}\frac{m^3-m}{12}\mathbf{1} \tag{F10.11}$$

Similarly, $L_m(\alpha_\circ)$ define a representation of the Virasoro algebra with central charge equal to $1 - 24\alpha_\circ^2$:

$$[L_m(\alpha_\circ), L_n(\alpha_\circ)] = (m-n)L_{m+n}(\alpha_\circ) + \delta_{m+n,0}\frac{m^3-m}{12}\left(1 - 24\alpha_\circ^2\right)\mathbf{1} \tag{F10.12}$$

We will denote by $V^{w,\alpha_\circ}$ this representation of the Virasoro algebra obtained from (F10.9) and (F10.10).

We are ready to define the vertex operators. For arbitrary $\alpha \in \mathbb{C}$, define

$$Q^{\pm}(\alpha, z) = \exp\left(\mp\alpha\sqrt{2}\sum_{n\geq 1}\frac{a_{\pm n}}{n}z^{\mp n}\right) \tag{F10.13}$$

Define furthermore an operator $x_\circ$ such that

$$e^{\alpha\sqrt{2}x_\circ} \, |0\rangle = |\alpha\rangle \tag{F10.14}$$

The vertex operator is

$$V_\alpha(z) = e^{i\alpha\sqrt{2}x_\circ} \, z^{\alpha p\sqrt{2}} \, Q^-(\alpha, z)Q^+(\alpha, z) \tag{F10.15}$$

In section 10.1, we denoted a_0 and $x_\circ$ by p and q, respectively.

From the definition, it is clear that the vertex operator $V_\alpha(z)$ acts between two

Fock space representations of $\widehat{H}$ with the same central charge:

$$V_\alpha(z) : V^{w,\alpha_\circ} \longrightarrow W^{w+\alpha,\alpha_\circ} \tag{F10.16}$$

The reader familiar with one of the most beautiful branches of theoretical physics, namely string theory, will recognize in this appendix all the elements used in the construction of the Fubini–Veneziano vertex operators.

11

Quantum groups in conformal field theory

11.1 The hidden quantum symmetry

Let us start our discussion by considering the Wess–Zumino models whose chiral algebra is $\widehat{SU(2)}_k$, i.e. we consider the Kac–Moody algebra $\widehat{SU(2)}$ and intertwiner operators of level k (see section 10.4.1 for the definitions). There exists a deep and fascinating connection between two apparently different mathematical structures, the Kac–Moody algebra $\widehat{SU(2)}$ and the quantum group $U_q(s\ell(2))$. Two salient motivations for this connection are worth mentioning:

(i) The braiding matrices for the conformal field theory with chiral algebra $\widehat{SU(2)}_k$ can be identified with the quantum $6j$ symbols of $U_q(s\ell(2))$ provided the deformation parameter q and the level k are related by

$$q = \exp \frac{i\pi}{k+2} \tag{11.1}$$

(ii) The truncation of the fusion algebra of Wess–Zumino models to spins $j \leq k/2$.

In this introduction, we will first elaborate on these two approaches to linking Kac–Moody algebras and quantum groups. Then we shall sketch a third, more rigorous, approach, which we shall develop later on in the chapter. Although we talk, for convenience, of Kac–Moody algebras or Virasoro, the program applies to conformal field theories with any chiral algebra.

As discussed in section 10.4.1, the correlators of the $\widehat{SU(2)}_k$ Wess–Zumino model satisfy the Knizhnik–Zamolodchikov equations

$$(k+2)\frac{d}{dz_i}\Psi(z_1,\ldots,z_N) = \sum_{j\neq i} \frac{\Omega_{ij}}{z_i - z_j}\Psi(z_1,\ldots,z_N) \tag{11.2}$$

where $\Omega_{ij}/(z_i - z_j)$ is the classical Yang–Baxter r-matrix for $SU(2)$. Through analytic continuation, we find a representation of the braid group on the space of solutions to (11.2). It turns out that this representation is isomorphic to the one introduced in chapter 6 for the centralizer of $U_q(s\ell(2))$ on the space of Clebsch–Gordan coefficients, provided q and k are related by (11.1). In this first approach, the connection between the Kac–Moody algebra and the quantum group can be schematized as in table 11.1. In order to invert the sequence from Kac–Moody algebras to quantum groups, we must invoke some sort of Riemann–Hilbert problem, namely how to deduce the differential equation from the monodromy of the solution. Some appropriate generalizations of the Schlesinger equations should make the ascending arrow explicit.

The second approach, option (*ii*) above, relies on the truncated fusion algebra of the Wess–Zumino model

$$[j_1] \times [j_2] = \sum_{|j_1-j_2|}^{\min(j_1+j_2,k-j_1-j_2)} N_{j_1 j_2}^{\,j} \, [j] \tag{11.3}$$

For $j_2 = 1/2$ the fundamental irrep, we associate with this fusion algebra the Coxeter diagram whose incidence matrix is precisely $N_{j_1 j_2}^{\,j} = N_{j_1 \frac{1}{2}}^{\,j}$:

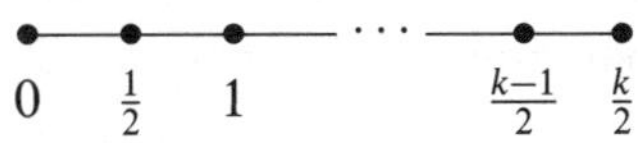

(11.4)

In chapter 5, we learned about the Temperley–Lieb–Jones (TLJ) algebra A_β associated with such a graph, with β fixed by the Coxeter number $k+2$ of the

Table 11.1 First logical path for linking Kac–Moody algebras and quantum groups.

Kac–Moody $\widehat{SU(2)}_k$	$\longleftarrow$	
$\downarrow$		$\uparrow$
KZ equation for level k correlators		$\uparrow$
$\downarrow$		Riemann–Hilbert problem
Braid group representation on the space of solutions		$\uparrow$
$\downarrow$		$\uparrow$
Centralizer		$\uparrow$
$\downarrow$		
$U_q(s\ell(2))$ with $q = e^{i\pi/k+2}$	$\longrightarrow$	

graph:

$$\beta = \left(q + q^{-1}\right)^2 = \left(2\cos\frac{\pi}{k+2}\right)^2 , \qquad q = \exp\frac{i\pi}{k+2} \tag{11.5}$$

Moreover, we know that the centralizer of $U_q(s\ell(2))$ in the tensor product of the spin-$\frac{1}{2}$ irreps is isomorphic to the above Temperley–Lieb–Jones algebra. Thus we manage to connect the Kac–Moody algebra with the quantum group. This second approach is encapsulated in table 11.2.

The two approaches sketched above can be interpreted as two alternative heuristic ways to establish the relation (11.1) between the Kac–Moody level k and the quantum group deformation parameter q. These two paths are certainly not independent. It is easy to unify them once we realize that both the fusion rules and the Knizhnik–Zamolodchikov equation reflect the decoupling of null vectors. Still, it is worth stressing that either of the two ways leads to the quantum group rather indirectly.

The two equivalent approaches just discussed work in the continuum. There are discrete models, however, which exhibit a quantum symmetry from the start. This is the case, notably, for the quantum group invariant hamiltonians of the open spin chains studied in section 3.6, which display the $U_q(s\ell(2))$ symmetry explicitly, and which renormalize to the continuum conformal field theories with $c = 1 - 6/(p^2 + p)$, where $q = \exp i\pi/(p+1)$. It would thus seem that $U_q(s\ell(2))$ is

Table 11.2 Second logical path for linking Kac–Moody algebras and quantum groups.

Kac–Moody $\widehat{SU(2)}_k$
↓
Level k fusion algebra
↓
Coxeter graph A_{k+1}
↓
TLJ algebra A_β with $\beta = [2]_q^2$
↓
Hecke algebra $H(q)$ with $q = e^{i\pi/k+2}$
↓
$U_q(s\ell(2))$

a symmetry of the unitary minimal models. Still, we must be very careful because the quantum group $U_q(s\ell(2))$ does not really act on the Hilbert space of the $c < 1$ conformal field theories. The precise relationship between the conformal states and the low lying excitations of the open spin chain in the thermodynamic limit is $\mathscr{H}_{\mathrm{RCFT}} = \mathscr{H}^{\text{low lying}}_{\text{spin chain}}/U_q(s\ell(2))$. Hence, the quantum group does not act directly on the physical states of the conformal field theory, but rather plays the role of a hidden symmetry. On the other hand, the centralizer of $U_q(s\ell(2))$, i.e. the Hecke algebra, certainly acts on $\mathscr{H}_{\text{spin chain}}$; in particular, the spin chain hamiltonian itself belongs to the centralizer of the quantum group. It is also clear that this centralizer does act on $\mathscr{H}_{\mathrm{RCFT}}$.

Following this line of thought, it could appear suggestive to interpret the Temperley–Lieb–Jones algebra as the discrete analog of the Virasoro algebra. It would thus seem that, in the continuum limit, only the centralizer of the manifest quantum group symmetry on the lattice shows up. We shall see in this chapter that this intuition is not quite right, and that one can exhibit in the continuum the hidden quantum group symmetry underlying a conformal field theory. For the time being, we summarize the discrete version of the rigorous approach to quantum groups in table 11.3.

Recall that the introduction of boundary terms was crucial in order to have a quantum group symmetry in finite chains (see section 3.6). A different proposal has been put forth by the Kyoto group which concerns the study of quantum group invariant chains, directly in the thermodynamic limit. They considered, in particular, an infinitely long XXZ spin chain whose quantum symmetry is the affine $U_q(\widehat{s\ell(2)})$, much bigger than that in table 11.3. Although their approach is very powerful, it does not allow them to modify the boundary conditions in order to obtain models with central charge different from unity. In fact, their method applies only to the massive antiferromagnetic regime $\Delta < 0$, where they have developed the representation theory of $U_q(\widehat{s\ell(2)})$ based on the crystal basis.

In this chapter, we shall concentrate on conformal field theories formulated directly in the continuum, and we shall exhibit explicitly their quantum group

Table 11.3 Discrete approach to quantum groups in conformal field theories.

Open spin chain hamiltonian + boundary terms
↓
Quantum group invariance + centralizer (TLJ)
↓
Virasoro algebra

symmetry. Our (rigorous, continuous) approach is based on the Coulomb gas version of conformal field theories. As can be inferred from the examples of minimal models and Wess–Zumino models, almost all known conformal field theories admit a free field representation. Relying on this, we will show in this chapter how to associate to any chiral algebra a unique and well-defined quantum group. The centralizer of this Hopf algebra will turn out to be represented by the braiding matrices of the decoupling equations. The path we will follow is sketched in table 11.4.

On comparing tables 11.3 and 11.4, we notice striking similarities between the (rigorous) discrete and continuous approaches to the quantum symmetry of conformal field theories.

First, we use a bigger Hilbert space for the realization of the quantum group symmetry. In one case, it is the full Hilbert space of the spin chain. In the other, it is the Fock space of the free field theory, which is certainly larger than the Hilbert space of the conformal field theory itself. In fact, in order to recuperate (in both cases) the physical Hilbert space, one has to advocate a BRS-like construction. After this "gauge fixing" the quantum group becomes invisible to the conformal field theory states. This is why we say that the quantum symmetry is hidden in conformal field theories.

Secondly, and more technically, the explicit appearance of the quantum group symmetry calls for tampering with the boundary of our system, be it by changing the boundary conditions for the physical states of the open spin chain, or by adding a charge at infinity to the conformal plane.

Of course, the rigorous approach we will follow is not independent of the two heuristic ones explained at the beginning of this section. Recall that the free field representation of a chiral algebra is just a clever trick to find the integral representation of the solutions to the decoupling equations. In the first two approaches we focus on the braiding or monodromy properties of these solutions, and thus we obtain only the centralizer of the quantum group. In the third approach, we concentrate on the integral representation of the solution itself instead, and thus unravel fully the hidden Hopf algebra structure.

Table 11.4 Continuous approach to quantum groups in conformal field theories.

Free field lagrangian		Quantum group
+	⟶ Coulomb gas ⟶	+
Background charge		Chiral algebra

11.2 Braiding matrices and quantum $6j$ symbols

The simplest possible way to establish (11.1) is by comparing relation (6.84) for the quantum Clebsch–Gordan coefficients

$$R^{j_1 j_2} K_j^{j_1 j_2} = (-1)^{j_1+j_2-j}\, q^{c_j - c_{j_1} - c_{j_2}}\, K_j^{j_2 j_1} \tag{11.6}$$

with the result of braiding ($z \to e^{i\pi} z$) the operator product of the corresponding primary fields of the $\widehat{SU(2)}_k$ Wess–Zumino model:

$$\Phi_{j_1}(z)\Phi_{j_2}(0) = \sum_j z^{h(j)-h(j_1)-h(j_2)}\, \Phi_j(0) \tag{11.7}$$

Indeed, replacing z by $e^{i\pi}z$ in (11.7) yields the extra factor

$$(-1)^{j_1+j_2-j}\, e^{i\pi[h(j)-h(j_1)-h(j_2)]} \tag{11.8}$$

on the right-hand side of (11.7). The first factor, namely $(-1)^{j_1+j_2-j}$, just indicates whether Φ_j appears symmetrically or antisymmetrically in $\Phi_{j_1} \times \Phi_{j_2}$. Requiring the phase factors in (11.6) and (11.8) to be the same provides us with the desired relation

$$q = \exp \frac{i\pi}{k+2} \tag{11.9}$$

Implicit in this simple argument are the translations shown in table 11.5.

Before justifying this, let us check it by exploring its consequences. Having dealt with the quantum Clebsch–Gordan or $3j$ symbols, it should seem natural to investigate the quantum $6j$ symbols defined by (6.89)

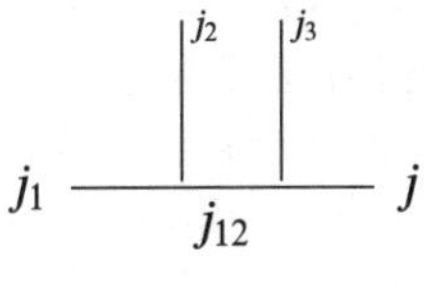

$$= \sum_{j_{23}} \begin{Bmatrix} j_1 & j_2 & j_{12} \\ j_3 & j & j_{23} \end{Bmatrix}_q \tag{11.10}$$

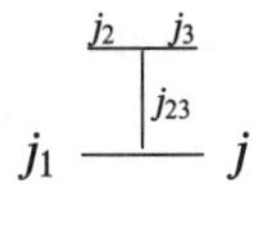

with the fusion matrices (9.35)

Table 11.5 Translation of terms in conformal field theories and quantum groups.

CFT		QG
Braiding	⟷	Acting with R
Operator product expansion	⟷	q-Clebsch–Gordan decomposition
Irreps of chiral algebra	⟷	Irreps of quantum group
Fusion algebra	⟷	Decomposition rules

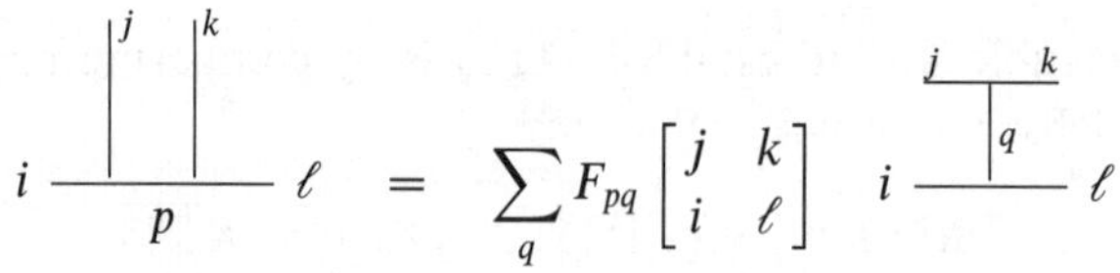

(11.11)

According to table 11.5, we find

$$\begin{Bmatrix} j_1 & j_2 & j_{12} \\ j_3 & j & j_{23} \end{Bmatrix}_q = F_{j_{12}\, j_{23}}\begin{bmatrix} j_2 & j_3 \\ j_1 & j \end{bmatrix} \tag{11.12}$$

We should of course set $q = \exp \pi i/(k+2)$, but it is interesting that (11.12) is true provided we restrict all the irreps in the quantum $6j$ symbol to spins $j \leq k/2$.

Similarly, and less surprisingly, the braiding matrices can be translated from (6.102). As explained in section 6.6, they define the representation of the centralizer of the quantum group on the space of quantum Clebsch–Gordan coefficients. In fact, the explicit relationship between the braiding matrices and the quantum $6j$ symbols is given by (6.102).

Although it is a fact that the braiding matrices connecting different solutions to the Knizhnik–Zamolodchikov equation are just the quantum $6j$ symbols, up to some phases, these quantities are blind to the quantum internal indices (the m's in $|j,m\rangle$). Based on the translations contained in table 11.5, we have proposed a new route to establish this result. And yet, as we shall see below, the entry in the table connecting the fusion algebra with the decomposition rules is the source of many headaches. To implement the translation fully, we should devise a way to reproduce the truncation of the fusion algebra in purely quantum group theoretic terms. The clue comes from the parameter identification (11.1): when the Wess–Zumino model is unitary, the level k is an integer and thus q is a root of unity. In this special situation, discussed in chapter 7, the only finite-dimensional irreps transforming trivially under the whole center are precisely those on which the truncated fusion algebra closes, namely those with spin $j \leq k/2$. Therefore, the translations should be supplemented with some other condition. Most obviously, we could require that the quantum group irreps must be finite dimensional and regular, i.e. with zero eigenvalues for the central elements E^{k+2} and F^{k+2}. This condition eliminates almost all undesirable representations, except for the one with $j = (k+1)/2$, of q-dimension zero. Physically, also, it is not clear why a specific eigenvalue of E^{k+2} or F^{k+2} should be preferred over any other. Fortunately, the correct condition can be implemented much more simply by merely requiring that the whole space of generic irreps (all of dimension $k+2$) is thrown out. The conclusion is thus that the Wess–Zumino model is related not to the quantum group $U_q(s\ell(2))$ itself, but to the quantum group modulo its central Hopf algebra $U_q(s\ell(2))/Z_q$. This quotient is not a Hopf algebra in general, but is what is known as a quasi-Hopf algebra. This minor technicality is not very fruitful from either the conceptual or the computational points of view, so it is the usual quantum group with only the non-generic irreducible representations realized.

A different way to attack the problem of the truncation of the fusion algebra is to reduce our discourse to the peculiarities of the representation theory of the centralizer of $U_q(s\ell(2))$, i.e. the Hecke algebra $H(q)$, for q a root of unity. It follows directly from the Knizhnik–Zamolodchikov equation that the space of conformal blocks with $(n+2)$ external legs, all of them except the last one carrying the fundamental,

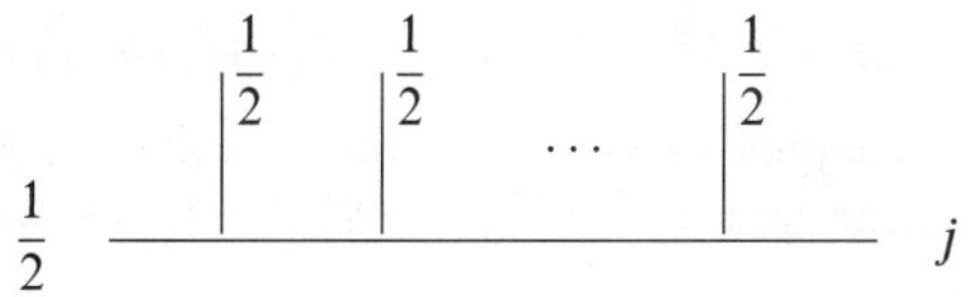

(11.13)

define irreps of $H_n(q)$ with $q = \exp i\pi/(k+2)$. These representations (see section 6.5) are irreducible and unitary iff $j \le k/2$.

The truncation phenomenon is not specific to conformal field theories; we have already encountered it in chapter 5 when dealing with face models. As noted there, when $\omega_\circ = 0$ and the anisotropy γ is $\gamma = \pi/r$, the face Yang–Baxter equation naturally truncates the lattice variables from the graph A_∞ to the Coxeter graph A_{r-1}. We summarize in table 11.6 the various instances where we have encountered the truncated decomposition rule.

Table 11.6 Various places where we have encountered the truncated decomposition rules $j_1 \times j_2 = \sum_{|j_1-j_2|}^{\min(j_1+j_2,k-j_1-j_2)} j$.

Chapter	System	Implementation	Value of k
3	spin-S chain	low lying excitations	$2S$
5	restricted face models $A_r =$ •—•···—• (1 2 r)	lattice variables	$r-1$
6	Hecke algebra, $q^\ell = \pm 1$	unitary irrep dimension	$\ell - 2$
8	(p, p') minimal models	thermal subalgebras	$p-2$, $p'-2$
10	Wess–Zumino models	fusion rules	$k =$ level
3, 7	quantum group invariant hamiltonians, $q^\ell = \pm 1$	low lying excitations	$\ell - 2$

11.3 Ribbon Hopf algebras

The close and interesting relationship between conformal field theories and quantum groups ramifies to Riemann surfaces other than the sphere. At genus one, light can be shed on the matter with the help of the ribbon structure of the underlying quantum group.

Let us start by recalling the main structural properties of conformal field theories on a torus with period ratio τ. The characters $\mathscr{X}_i(\tau) = \mathrm{tr}_i\, e^{2\pi i\tau(L_0-c/24)}$ of primary fields Φ_i with conformal weights Δ_i transform under modular transformations as follows:

$$\mathscr{X}(\tau+1) = e^{2\pi i(\Delta_i-c/24)}\mathscr{X}_i(\tau)\,, \quad \mathscr{X}(-1/\tau) = S_{ij}\mathscr{X}_j(\tau) \tag{11.14}$$

Verlinde's theorem (chapter 9) says that the modular matrix S diagonalizes simultaneously all fusion matrices N_i, $\left(S^{-1}N_iS\right)_{jk} = \delta_{jk}\,\lambda_i^{(j)}$, and, in fact,

$$\lambda_i^{(k)}\lambda_j^{(k)} = N_{ij}^{\ell}\lambda_\ell^{(k)}\,, \quad \lambda_i^{(j)} = \frac{S_{ij}}{S_{1j}}$$
$$(N_i)_j^k = N_{ij}^k = \sum_m \frac{S_{im}S_{jm}S_{mk}}{S_{1m}} \tag{11.15}$$

Using the pentagonal identity (9.38), it is easy to prove that

$$\frac{S_{jk}}{S_{11}} = \sum_\ell e^{2\pi i(\Delta_\ell-\Delta_j-\Delta_k)}\frac{S_{1\ell}}{S_{11}}N_{jk}^\ell \tag{11.16}$$

This is the basic equation that we want to study; it relates duality data at genus zero (the fusion rules and the conformal weights) to the genus one modular matrix.

For Wess–Zumino models based on $\widehat{SU(2)}$ at level k, the modular matrix (10.132) is

$$S_{j\ell} = \sqrt{\frac{2}{k+2}}\,\sin\frac{\pi(2j+1)(2\ell+1)}{k+2} \tag{11.17}$$

Setting $q = \exp i\pi/(k+2)$, (11.16) becomes the following identity between q-numbers:

$$[(2j+1)(2k+1)] = \sum_\ell q^{2(c_\ell-c_j-c_k)}\,[2\ell+1]\,N_{jk}^\ell \tag{11.18}$$

where $c_j = (k+2)\Delta_j$ is the classical $SU(2)$ Casimir in the irrep of spin j.

Just as equations (11.6) and (11.8) were the starting point for the translations at genus zero, we shall now use (11.16) and (11.18) to obtain quantum group translations for conformal field theories at genus one. Interestingly enough, the modular properties of conformal field theory are encoded in the ribbon structure of the associated quantum group.

In order to establish equation (11.16), it is best to start from

$$\mathrm{tr}_{j\otimes k}\,\Delta(a) = \sum_\ell N_{jk}^\ell\,\mathrm{tr}_\ell(a) \qquad \forall a \in U_q(s\ell(2)) \tag{11.19}$$

and then use the fact that $U_q(s\ell(2))$ is a "ribbon Hopf algebra". This just means that there exists a central element $u \in U_q(s\ell(2))$ such that

$$\begin{aligned} \Delta(u) &= \mathscr{R}^{-1}\, \sigma\left(\mathscr{R}^{-1}\right)\, u \otimes u \\ \Delta\left(u^{-1}\right) &= \sigma\left(\mathscr{R}\right)\, \mathscr{R}\left(u^{-1} \otimes u^{-1}\right) \end{aligned} \tag{11.20}$$

where $\mathscr{R}$ is the universal $\mathscr{R}$-matrix and σ is the permutation map. In fact, this distinguished element is given by

$$u = m\,(\gamma \otimes \mathbf{1})\ \sigma\left(\mathscr{R}\right)\,, \qquad u^{-1} = m\left(\mathbf{1} \otimes \gamma^2\right)\ \sigma\left(\mathscr{R}\right) \tag{11.21}$$

For $U_q(s\ell(2))$, the element u in the irrep j is given by

$$\rho^j(u) = q^{-2c_j}\, q^H\,, \qquad \rho^j(u^{-1}) = q^{2c_j}\, q^{-H} \tag{11.22}$$

and thus

$$\mathrm{tr}_j(u) = q^{-2c_j}\ [2j+1]\,, \qquad \mathrm{tr}_j(u^{-1}) = q^{2c_j}\ [2j+1] \tag{11.23}$$

From this we obtain

$$\frac{S_{jk}}{S_{11}} = q^{2\left(c_j+c_k\right)}\ \mathrm{tr}_{j\otimes k}\,\Delta(u) = q^{-2\left(c_j+c_k\right)}\ \mathrm{tr}_{j\otimes k}\,\Delta\left(u^{-1}\right) \tag{11.24}$$

and using (11.20) we finally obtain a quantum group formula for the modular matrix S:

$$\begin{aligned} \frac{S_{jk}}{S_{11}} &= \mathrm{tr}_{j\otimes k}\left[\sigma(\mathscr{R})\mathscr{R}\ q^{-H} \otimes q^{-H}\right] \\ &= \mathrm{tr}_{j\otimes k}\left[\mathscr{R}^{(k,j)}\mathscr{R}^{(j,k)}\ q^{-H} \otimes q^{-H}\right] \end{aligned} \tag{11.25}$$

This expression is simply the Markov trace evaluated on the two-component link $\widehat{\sigma_1^2}$ (two linked circles) where each component is associated to the irreps j and k (see figure 11.1).

We have thus produced a quantum group proof that the modular matrix is a link invariant, a result which can also be obtained by surgery in Witten's Chern–Simons theory. In the powerful quantum group context, this result follows from the co-multiplication rule of the distinguished element u. We leave the

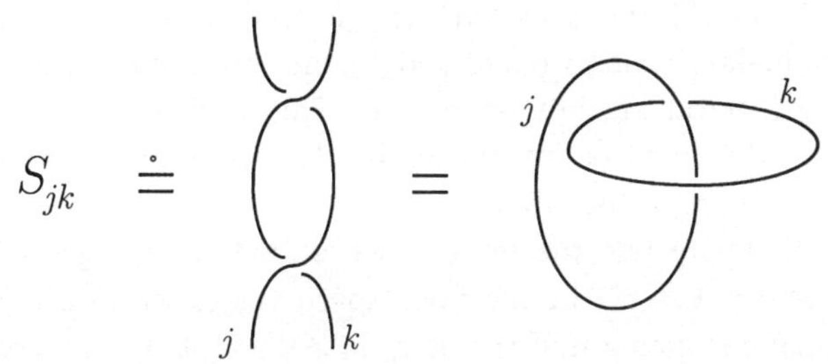

Fig. 11.1. The modular matrix S_{jk} is the value of the link invariant of two linked circles carrying irreps j and k, equation (11.25).

derivation of Verlinde's theorem from the quantum group's quasi-triangularity until the exercises.

11.4 The contour representation of quantum groups

In this section, we shall present in detail the rigorous continuous approach discussed in the introduction 11.1, namely the procedure followed to obtain, via the free field (or Coulomb gas, or Wakimoto, or Feigin–Fuks) representation of the chiral algebra, its associated Hopf algebra. The main ingredient we shall use is a non-local generalization of vertex operators. We will keep the discussion general at first, and then work out the quantum group for some simple chiral algebras.

11.4.1 Screened vertex operators

The basic feature of the free field representation of a chiral algebra is that the Hilbert space of the conformal field theory is enlarged into a bigger space consisting of a tensor product of Fock spaces. These contain null vectors, which should be weeded out. For this purpose, as well as for computing correlation functions, we introduce the screening currents $S_i(z)$ i=1,..,N_S, the basic property of which is that the most singular part of their operator product expansion with the chiral algebra operators $O_n(z)$ is a total derivative, namely:

$$O_n(z)\, S_i(w) = \frac{\partial}{\partial w}\left(\frac{c_{ni}^m}{(z-w)^{\Delta(n)-\Delta(m)}}\, O_m(w)\right) + \text{r.t.} \tag{11.26}$$

where $\Delta(n)$ is the conformal weight of O_n. In particular, the screening operators S_i have conformal weight one, which implies the following equation:

$$T(z)\, S_i(\zeta) = \frac{\partial}{\partial \zeta}\left(\frac{1}{(z-\zeta)} S_i(\zeta)\right) + \text{regular terms} \tag{11.27}$$

These two equations yield the following fundamental result:

$$\left[O_n(z), \int_C dw\, S(w)\right] = \text{boundary terms} \tag{11.28}$$

If the contour C in (11.28) is closed, i.e. if it does not cross any branch cuts, then the boundary terms vanish, and so the screening charges $\int_C dw\, S(w)$ commute with the chiral algebra operators. This is the key property of the screening charges: they may be introduced inside correlators without modifying the conformal properties of the latter.

Another way to appreciate the importance of the screening operators is to construct null vectors. Let $\mathcal{F}_a$ be the Fock space associated to a representation a of the chiral algebra; then a null vector ξ_a in $\mathcal{F}_a$ adopts the general form:

$$\xi_a(z) = \int_{C_1} dt_1 S_{i_1}(t_1) \dots \int_{C_{n_a}} dt_{n_a} S_{i_{n_a}}(t_{n_a}) V_b(z) \tag{11.29}$$

where V_b is a primary field, and the contours $C_1, \ldots, C_{n_a}$ are chosen in such a way as to cancel the boundary terms in (11.28). Whenever the latter choice is possible, the state ξ_a becomes a null vector. This result plays an important role in the construction of quantum groups.

In the Coulomb gas approach to conformal field theories, the contours of integration of the screening operators are an essential part of the theory. This was first shown by Dotsenko and Fateev in their computation of the correlators of the $c < 1$ theories: the choice of the contours fixes the channel of the corresponding conformal blocks. Moreover, the duality properties of these conformal blocks (i.e. s–t duality or s–u duality) can be easily obtained through contour manipulations, without invoking their explicit expressions. This is the easiest way to compute F- and B-matrices, as shown in section 10.2.2. The numerical values of the duality matrices F and B depend, in fact, on very elementary Coulomb gas data, namely on the braiding factors among screening operators S_i and vertex operators V_a:

$$S_i(z_1)\, S_j(z_2) = e^{i\pi\, \Omega_{ij}}\, S_j(z_2) S_i(z_1) \tag{11.30a}$$
$$S_i(z_1)\, V_a(z_2) = e^{i\pi\, \Omega_{ia}}\, V_a(z_2) S_i(z_1) \tag{11.30b}$$
$$V_a(z_1)\, V_b(z_2) = e^{i\pi\, \Omega_{ab}}\, V_b(z_2) V_a(z_1) \tag{11.30c}$$

In these expressions, the braiding is done in the counter-clockwise sense, $z_{12} \to e^{i\pi} z_{21}$. This means that we consider the operator product expansion of the left-hand side for $|z_1| > |z_2|$, then perform an analytic continuation to $|z_1| < |z_2|$ moving z_1 under z_2, and record the phase we pick up in the process on the right-hand side of (11.30). Quite reasonably, we assume that the matrices Ω_{ij} and Ω_{ab} are symmetric.

The richness of the duality matrices emerges in the Coulomb gas approach from the contour structure of the conformal blocks. In other words, under a duality transformation the contours start mixing in complicated ways, producing all kinds of braiding factors of the type described in equations (11.30).

The first step in finding the Hopf algebra from the chiral algebra is to define vector spaces $\mathscr{V}^a$ as families of screened vertex operators e_I^a as follows. Here, a denotes a representation of the chiral algebra realized in the Coulomb gas version. Let $V_a(z)$ be the free field vertex operator associated to the Fock space $\mathscr{F}_a$, and let the index I denote an ordered sequence $\{i_1, \ldots, i_r\}$ of labels of screening operators. The operator e_I^a is given by

$$e_I^a(z) = \int_{C_1} dt_1 S_{i_1}(t_1) \ldots \int_{C_r} dt_r S_{i_r}(t_r) V_a(z) \tag{11.31}$$

where the contours $C_1, \ldots, C_r$ start at infinity, encircle counter-clockwise the point z once and return to infinity, as in figure 11.2. Whenever we label a contour by C, it will mean precisely this geometry. In the customary notation of complex analysis, $\int_C = \int_{i\infty}^{(z+)}$.

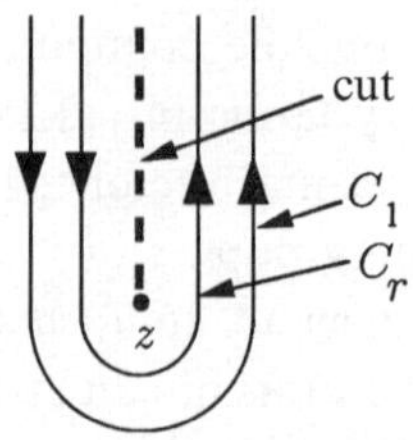

Fig. 11.2 Contours in the screened vertex operator (11.31). In this drawing, $r = 2$.

The integrals in (11.31) are defined on the integrand's region of analyticity; the integrand is obtained by performing the operator product expansions between the screening operators S_i and the vertex V_a. The existence of a branch cut in these operator product expansions leads to the above choice of contours: since we have fixed the conformal gauge, it is at infinity where the curvature is concentrated. We choose the cut to run parallel to the imaginary axis, but any other contour going to infinity, for example a radial one or one parallel to the real axis, would do as well. Our choice makes it easier to draw pictures.

The vertex operator V_a belongs, of course, to the space $\mathscr{V}^a$, namely $V_a = e^a_0$. For definiteness, we restrict ourselves to the case when this vertex operator is a primary field. The remaining vertex operators e^a_I, as can be seen from their definition, have a contour structure which is absent in the vertex e^a_0. To qualify this distinction, we call the vertex e^a_0 a "highest weight vector" of the space $\mathscr{V}^a$; the remaining vertices are called "descendants". This notation will be justified when we discuss the quantum group. Note also that we have defined the space $\mathscr{V}^a$ for every value of $z \in \mathbb{C}$. Sometimes, to make this dependence explicit, we shall write $\mathscr{V}^a(z)$

It is important to realize that we have established a one to one correspondence between highest weight vectors of quantum groups and primary fields of conformal field theories. The main consequence of this identification is the deep relationship between the representation theory of quantum groups and that of conformal field theories. Indeed, let us suppose that a screened vertex operator e^a_I turns out to be a null vector in the conformal field theory sense. Then, since it is a primary field, it becomes a highest weight vector for a new representation of the quantum group. In other words, it is also a null vector in the quantum group sense. This occurs whenever the relative monodromy of the screening operators and the vertex V_a cancel out, indicating that the contour structure effectively collapses, in which case there is no need for the contours to leave the local chart by going off to the point at infinity.

To illustrate these ideas, let us consider the case of a screened vertex operator $e^a_{r_i}$ defined by r_i contours C of the same screening operator S_i. Using elementary

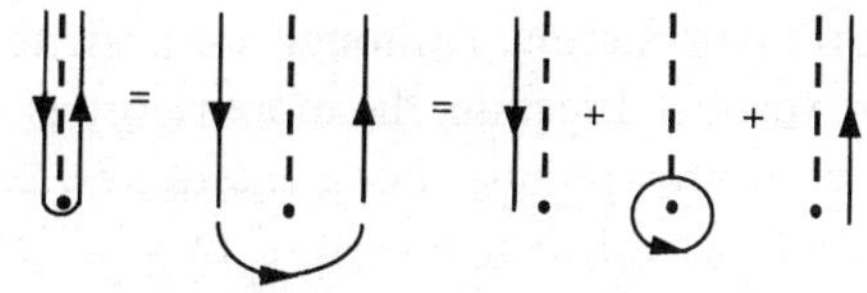

Fig. 11.3. The contour of a screened vertex operator decomposes into three parts. The two vertical components can be brought infinitesimally close to the cut, and the circle around z becomes very small.

complex analysis, we can write $e^a_{r_i}$ as (see figure 11.3):

$$e^a_{r_i}(z) = e^a_{i\ldots i} = \prod_{\ell=0}^{r_i-1}(1 - e^{2\pi i\,\Omega_{ia}}\, q_i^{2\ell})\, \tilde{e}^a_{r_i}(z) \tag{11.32}$$

where

$$q_i = \exp\left(\frac{i\pi\Omega_{ii}}{2}\right) \tag{11.33}$$

and we have introduced the "light-house" screened vertex operators

$$\tilde{e}^a_{r_i}(z) = \int_{i\infty}^{z} dt_1 S_i(t_1)\ldots\int_{i\infty}^{z} dt_{r_i} S_i(t_{r_i}) V_a(z) \tag{11.34}$$

shown in figure 11.4.

The factor in (11.32) comes from the braiding of the screening operators among themselves, as well as with the vertex V_a. Its vanishing indicates the appearance of a null vector. In this case, the subspace of $\mathscr{V}^a$ spanned by $\{e^a_{r_i}\}_0^{n_{i,a}-1}$ has finite dimension equal to $n_{i,a}$, which satisfies the condition

$$e^{2\pi i\Omega_{ia}}\; e^{i\pi(n_{i,a}-1)\,\Omega_{ii}} = 1 \tag{11.35}$$

or equivalently:

$$2\,\Omega_{ia} + \Omega_{ii}\,(n_{i,a}-1) = 0 \mod(2) \tag{11.36}$$

This dimension may become infinite.

The order of the operators in (11.31) is the natural one given by the geometry of the contours: since they do not intersect, no ordering problems show up. On the contrary, the expression (11.34) calls for an ordering prescription, which is

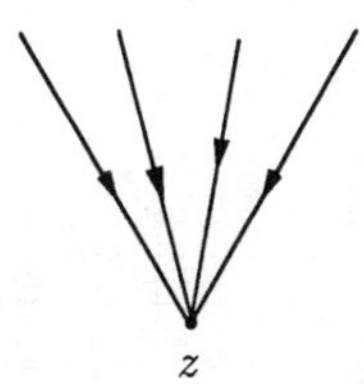

Fig. 11.4. Light-house screened vertex operators (11.34). For clarity, we have fanned out the contours, all of which actually come down straight from $i\infty$ to z.

provided by the natural radial ordering. Although we draw the screened vertices as if time increases in the vertical direction, the more rigorous picture would have the cut departing from z towards infinity along the radius, i.e. along the vector from 0 to z. The contours C would then consist of two straight parts parallel to and separated by the cut, each parametrized by the distance to the origin, and a semi-circle around the puncture. When the radius of this circle goes to zero, the two straight sections of all the contours collapse on each other. That is why the light-house operators must be ordered, whereas the original operators are well defined as such. The path ordering is effected according to $|t_1| > |t_2| > \ldots |z|$ (see figure 11.5):

$$\tilde{e}^a_{r_i}(z) = |[r_i]|_i!\, P(\tilde{e}^a_{r_i}(z)) \tag{11.37}$$

where

$$P(\tilde{e}^a_{r_i}(z)) = \int_{i\infty}^{t_2} dt_1 S_i(t_1) \int_{i\infty}^{t_3} dt_2 S_i(t_2) \cdots \int_{i\infty}^{z} dt_{r_i} S_i(t_{r_i}) V_a(z) \tag{11.38}$$

and $[|x|]_i$ is the q-number

$$[|x|]_i = \frac{1 - q_i^{2x}}{1 - q_i^2} \tag{11.39}$$

In equation (11.32), we have neglected the contributions to the integrals coming from the neighborhood of the point z. This can be justified by analytically continuing the result obtained in (11.34), in close analogy with the Hankel contour integral representation of the Γ function.

If $q_i^p = \pm 1$, it follows from (11.37) that $\tilde{e}^a_{r_i}$, and thus also $e^a_{r_i}$, is a null vector for $r_i \geq p$. This kind of phenomenon takes place in rational conformal field theories: it is the origin of the truncations in the representation theory of the chiral algebra. At the quantum group level, rationality implies the nilpotency of the quantum generators E_i and F_i: $E_i^{p'} = F_i^{p'} = 0$ (where $p' = p$ if p is odd and $p' = p/2$ if p is even; recall chapter 7).

11.4.2 Examples

Before continuing, let us recall briefly the bare bones of the free field representation of minimal models and Wess–Zumino models we previously discussed

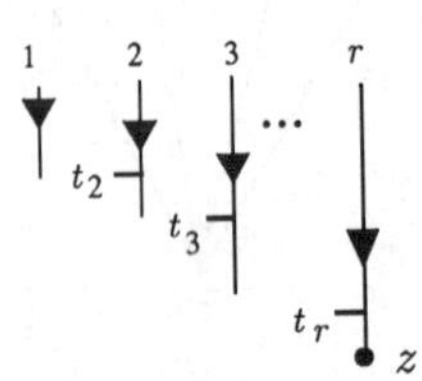

Fig. 11.5 The path ordered light-house screened vertex operators (11.38).

in chapter 10. The following elements should suffice to make the subsequent discussion explicit enough for the industrious reader.

(i) In minimal models, there are two screening operators

$$S_+ =: \exp i\alpha_+\phi(z) :, \quad S_- =: \exp i\alpha_-\phi(z) : \tag{11.40}$$

The central extension is given by $c = 1 - 24\alpha_\circ^2$ and $\alpha_+ + \alpha_- = 2\alpha_\circ$, $\alpha_+\alpha_- = -1$. The vertex operators are of the same form $V_\alpha(z) =: \exp i\alpha\phi(z) :$. The braiding factor of two vertices V_α and V_β is:

$$V_\alpha(z_1)V_\beta(z_2) = e^{2\pi i\alpha\beta}V_\beta(z_2)V_\alpha(z_1) \quad \Rightarrow \quad \Omega_{\alpha\beta} = 2\alpha\beta \tag{11.41}$$

(ii) In the Coulomb gas version of the Wess–Zumino–Witten models based on a simple Lie algebra $\mathcal{G}$ (section 10.4.2 and exercise 10.7), there are $\ell = \text{rank}(\mathcal{G})$ screening currents $S_{\vec{\alpha}_i}$, in one to one correspondence with the simple roots $\vec{\alpha}_i$ of G. The braiding factors among screening operators and highest weight vertex operators $V_{\vec{\lambda}}(z)$ (with $\vec{\lambda}$ in "dynkinese") turn out to be:

$$\begin{aligned} \Omega_{ij} &= \frac{1}{\nu^2}\,\vec{\alpha}_i \cdot \vec{\alpha}_j \\ \Omega_{i\lambda} &= -\frac{1}{\nu^2}\,\vec{\lambda} \cdot \vec{\alpha}_i \\ \Omega_{\lambda\lambda'} &= \frac{1}{\nu^2}\,\vec{\lambda} \cdot \vec{\lambda}' \end{aligned} \tag{11.42}$$

Note how the structure of the classical group of the corresponding Wess–Zumino model enters into the Coulomb gas version: it fixes the braiding factors among the screening operators. More generally, we may interpret the matrix Ω_{ij} as some kind of generalized symmetrized Cartan matrix of the conformal field theory.

Let us apply the general arguments of the previous section to $c < 1$ pure Virasoro theories. The formalism is as in (*i*) above, but we do not presume rationality. We consider three different examples of vertex operators V_α and their associated quantum group spaces $\mathcal{V}^\alpha$.

First, recall the Kac formula

$$\alpha_{m,m'} = \frac{1-m}{2}\alpha_+ + \frac{1-m'}{2}\alpha_- \tag{11.43}$$

with $m, m' \geq 1$. Since by (11.41) S_+ and S_- commute, the space $\mathcal{V}^{\alpha_{m,m'}}$ spanned by $e_{r_+,r_-}^{\alpha_{m,m'}}$ has dimension $m \cdot m'$. This last result follows from the equations

$$\int \prod_{i=1}^{m} dt_i S_+(t_i) V_{\alpha_{m,m'}}(z) = \int \prod_{j=1}^{m'} dt_j S_-(t_j) V_{\alpha_{m,m'}}(z) = 0 \tag{11.44}$$

Secondly, consider the vertex operators $V_{2\alpha_\circ - \alpha_{m,m'}} = V_{\alpha_{-m,-m'}}$ with $m, m' \geq 1$. They give rise to infinite-dimensional spaces $\mathcal{V}^{\alpha_{-m,-m'}}$ for α_+^2 not rational. If

$\alpha_+^2 = p'/p \in \mathbb{Q}$, then $\alpha_{-m,-m'} = \alpha_{p-m,p'-m'}$, and consequently we return to the situation above, where $\mathscr{V}^\alpha$ is finite-dimensional.

Thirdly, the intermediate situation between the two cases above is encountered when $V_{2\alpha_\circ-\alpha_{m,m'}-m\alpha_+} = V_{\alpha_{m,-m'}}$ or $V_{2\alpha_\circ-\alpha_{m,m'}-n\alpha_-} = V_{\alpha_{-m,m'}}$. These vertices give rise to finite-dimensional spaces spanned by $e_{r_+}^{\alpha_{m,-m'}}$ (of dimension m) and $e_{r_-}^{\alpha_{-m,m'}}$ (of dimension m'), respectively. The corresponding null vectors

$$\begin{aligned} \mathscr{X}^+_{m,m'} &= \int \prod_{i=1}^{m} dt_i S_+(t_i) V_{2\alpha_\circ-\alpha_{m,m'}-m\alpha_+}(z) \\ \mathscr{X}^-_{m,m'} &= \int \prod_{j=1}^{m'} dt_j S_-(t_j) V_{2\alpha_\circ-\alpha_{m,m'}-m\alpha_-}(z) \end{aligned} \tag{11.45}$$

belong to the Fock space $\mathscr{F}_{2\alpha_\circ-\alpha_{m,m'}}$.

Let us stress that the finite dimensionality of the quantum group spaces $\mathscr{V}^a$ is intimately connected with the existence of null vectors in the $c < 1$ theory.

11.4.3 The quantum qroup

The vector spaces $\mathscr{V}^a$ spanned by the screened vertex operators $e_I^a(z)$ are the obvious candidates for the representation spaces of the quantum group Q, which we shall define in this section.

First of all, Q is the Hopf algebra generated by a collection of operators $\{E_i, F_i, K_i\}$ associated with the screening currents $\{S_i\}$.

We shall define the operator F_i, acting on the space $\mathscr{V}^a = \mathscr{V}^a(z)$, as creating a contour integral of the screening operator S_i on any screened operator belonging to $\mathscr{V}^a$, namely (see figure 11.6):

$$F_i e_I^a(z) \equiv \int_C dt\ S_i(t)\ e_I^a(z) \tag{11.46}$$

where the contour C has the same shape as those appearing in (11.31).

The co-product ΔF_i is defined by acting on the space $\mathscr{V}^a(z) \otimes \mathscr{V}^b(w)$ exactly as in (11.46):

$$\Delta F_i \left(e^a_{i_1\ldots i_r}(z)\, e^b_{j_1\ldots j_s}(w) \right) = \int_C dt\ S_i(t) e^a_{i_1\ldots i_r}(z)\, e^b_{j_1\ldots j_s}(w) \tag{11.47}$$

where the contour C surrounds both points z and w, as shown in figure 11.7.

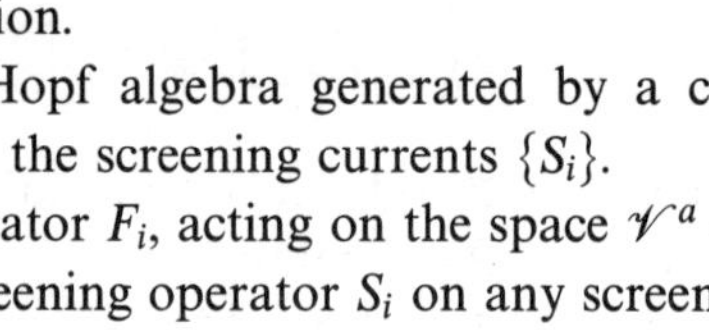

Fig. 11.6. The quantum group lowering generator F_i acts on a screened vertex operator by adding a contour with screening S_i.

On performing a contour deformation as in figure 11.7, and taking proper account of the braidings (11.30) involved in the reshuffling of operators, we deduce that ΔF_i can be written as

$$\Delta F_i \left[e^a_{i_1 \dots i_r}(z)\, e^b_{j_1 \dots j_s}(w) \right] = F_i \left[e^a_{i_1 \dots i_{r_1}}(z) \right] e^b_{j_1 \dots j_{r_2}}(w)$$
$$+ \exp \left[i\pi \left(\Omega_{ia} + \sum_{\ell=1}^{r} \Omega_{ii_\ell} \right) \right] e^a_{i_1 \dots i_r}(z)\, F_i \left[e^b_{j_1 \dots j_s}(w) \right] \qquad (11.48)$$

where the factor in the second summand arises from the braiding of S_i and the vertex $e^a_{i_1 \dots i_r}$.

A convenient way of rewriting (11.48) is:

$$\Delta F_i = F_i \otimes \mathbf{1} + K_i^{-1} \otimes F_i \qquad (11.49)$$

where the operator K_i just measures the braiding of S_i and the screened vertex $e^a_{i_1 \dots i_r}$:

$$K_i e^a_{i_1 \dots i_r} = \exp \left[-i\pi \left(\Omega_{ia} + \sum_{\ell=1}^{r} \Omega_{ii_\ell} \right) \right] e^a_{i_1 \dots i_r} \qquad (11.50)$$

The quantum group generator K_i is typically given by the exponential of the charge associated to the free scalar field ϕ_i,

$$K_i \simeq \exp \oint \partial \phi_i \qquad (11.51)$$

so that the co-multiplication of K_i is, by construction, simply

$$\Delta K_i = K_i \otimes K_i \qquad (11.52)$$

Therefore, (11.49) and (11.52) satisfy the co-associativity condition

$$(\Delta \otimes \mathbf{1})\Delta = (\mathbf{1} \otimes \Delta)\Delta \qquad (11.53)$$

From equations (11.46) and (11.50), we derive the following relations between F_i and K_i:

$$\begin{aligned} K_i F_j &= \mathrm{e}^{-i\pi\Omega_{ij}} F_j K_i \\ K_i K_j &= K_j K_i \end{aligned} \qquad (11.54)$$

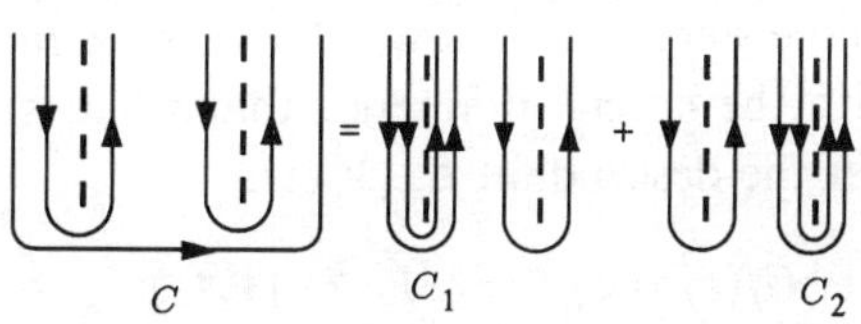

Fig. 11.7. The action of ΔF_i on two screened vertex operators. The contour decomposes as shown.

The operators F_i may satisfy among themselves algebraic relations, a direct analog of the Serre relations for the Kac–Moody generators in the Chevalley basis. These relations follow from the braiding relations

$$S_i S_j = e^{i\pi\Omega_{ij}} S_j S_i \tag{11.55}$$

If the coefficients Ω_{ij} do not satisfy among themselves any rationality conditions, then there are no Serre relations, and therefore equations (11.49) to (11.54) should suffice to define the Hopf algebra Q. On the other hand, if there exist rationality conditions, then there do exist Serre relations. For example, if the rationality condition is $\Omega_{ii} = \Omega_{jj} = -2\Omega_{ij}$, then the implied Serre relation is

$$F_i^2 F_j - [|2|]_q F_i F_j F_i + F_j F_i^2 = 0 \tag{11.56}$$

which is characteristic of $U_q(A_n)$. In the same ideological framework, if the parameter q_i is a root of unity, $q_i^{2p} = 1$, then $F_i^p = 0$ is a Serre-like condition. For rational theories, any of these q-Serre relations severely constrain the dimensionalities of irreducible representations of the quantum group Q.

The operators K_i and F_i defined so far generate an algebra Q_+, which is the analog of the Borel subalgebra of the universal enveloping algebra of a Lie group, typically spanned by lower triangular matrices. To obtain the analog of the full universal enveloping algebra, we need to introduce also the raising operators E_i. In our context, these turn out to be operators which destroy contours, dual to the operators F_i creating contours.

The way to annihilate a contour is to integrate it out. Recall that screening currents have conformal weight one, which implies in particular that

$$\left[L_{-1}, \int_C dt\ S(t)\right] = \int_C dt\ \partial_t S(t) \tag{11.57}$$

Thus, acting with L_{-1} on a screened vertex $e^a_{i_1 \ldots i_r}$ has the effect of integrating $\int dt_i\ S_i(t_i)$ inside $e^a_{i_1 \ldots i_r}$, leaving just boundary terms. These boundary terms allow us to define the operators $\tilde{E}_i$ implicitly:

$$L_{-1} e^a_I(z) = G^a_I(z) - \sum_{i=1}^{N_S} \left(q_i - q_i^{-1}\right) S_i(\infty)\ \tilde{E}_i\ e^a_I(V_a(z)) \tag{11.58}$$

(The term $(1 - q_i^{-2})$ in (11.58) is simply a normalization factor introduced for later convenience; also, we use the notation $\tilde{E}_i$ for the raising operators anticipating a minor redefinition of these operators below.) The first term G is essentially the same thing as $e^a_I(z)$, with the crucial difference that what is screened is not the primary $V_a(z)$ but rather the descendant $L_{-1}V_a(z)$:

$$G^a_I(z) = F_{i_1} F_{i_2} \cdots F_{i_r}\ L_{-1} V_a(z) \tag{11.59}$$

Note that $\tilde{E}_i e^a_I$ is a screened operator with one less contour of the S_i type. Of course, if $i \notin I$ then $\tilde{E}_i e^a_I = 0$ (you cannot destroy what does not exist). In general, from (11.58) we would obtain

$$\tilde{E}_i e_I^a(z) = \rho_{i,I}^a e_{I-\{i\}}^a(z) \tag{11.60}$$

where $\rho_{i,I}^a$ depends on the braiding factors (11.30). For example, acting on the vertices e_r^a (11.32) with r contours of the same type i, we obtain

$$\tilde{E}_i \, e_r^a(z) = \left[|r|\right]_i \frac{1 - q_i^{2(r-1)} \, \mathrm{e}^{2\pi i \Omega_{i,a}}}{q_i - q_i^{-1}} e_{r-1}^a(z) \tag{11.61}$$

Since L_{-1} acts like a derivation, the co-multiplication of L_{-1} is trivial:

$$\Delta L_{-1} = L_{-1} \otimes \mathbf{1} + \mathbf{1} \otimes L_{-1} \tag{11.62}$$

The co-product of $\tilde{E}_i$ follows from that of L_{-1}:

$$\begin{aligned} \Delta L_{-1} \left(e_I^a(z) \, e_J^b(w) \right) &= G_I^a(z) \, e_J^b(w) + e_I^a \, (z) G_J^b(w) \\ &- \sum_{i=1}^{N_S} \left(q_i - q_i^{-1} \right) S_i(\infty) \, \Delta \tilde{E}_i \left(e_I^a(z) \, e_J^b(w) \right) \end{aligned} \tag{11.63}$$

whence

$$\Delta \tilde{E}_i = \tilde{E}_i \otimes \mathbf{1} + K_i^{-1} \otimes \tilde{E}_i \tag{11.64}$$

An important consequence of (11.58) is that the action of the quantum group is not orthogonal to that of the Virasoro algebra. In fact, the contour creation operator F_i does not commute with L_{-1}:

$$[L_{-1}, F_i] e_I^a = -S_i(\infty) \left(1 - K_i^{-2}\right) e_I^a \tag{11.65}$$

This implies, in turn, the following relationship between $\tilde{E}_i$ and F_j:

$$\tilde{E}_i F_j - \mathrm{e}^{i\pi\Omega_{ij}} F_j \tilde{E}_i = \delta_{ij} \frac{1 - K_i^{-2}}{q_i - q_i^{-1}} \tag{11.66}$$

It is convenient to redefine the raising quantum group operators by

$$E_i = K_i \tilde{E}_i \tag{11.67}$$

whereby the commutator (11.66) adopts the more customary form

$$E_i F_j - F_j E_i = \delta_{ij} \, \frac{K_i - K_i^{-1}}{q_i - q_i^{-1}} \tag{11.68}$$

We recover thus the usual quantum group commutation relation between raising and lowering operators. This displays the deep interplay between the quantum group and the chiral algebra of a conformal field theory in its free field representation.

Finally, the relationship between K_i and E_j is given by

$$K_i E_j = \mathrm{e}^{i\pi\Omega_{ij}} E_j K_i \tag{11.69}$$

Let us summarize our discussion. So far, we have defined an algebra Q generated by contour creation operators F_i, contour annihilation operators E_i and monodromy operators K_i satisfying

$$\begin{aligned} K_iK_j &= K_jK_i \\ K_iE_j &= e^{i\pi\Omega_{ij}}E_jK_i \\ K_iF_j &= e^{-i\pi\Omega_{ij}}F_jK_i \\ [E_i, F_j] &= \delta_{ij}\,\frac{K_i - K_i^{-1}}{q_i - q_i^{-1}} \end{aligned} \tag{11.70}$$

in addition to generalized Serre relations among the F_i (and also among the E_i) operators which follow from "rationality" conditions on the Ω_{ij}.

The co-multiplication rules are

$$\begin{aligned} \Delta K_i &= K_i \otimes K_i \\ \Delta E_i &= E_i \otimes K_i + \mathbf{1} \otimes E_i \\ \Delta F_i &= F_i \otimes \mathbf{1} + K_i^{-1} \otimes F_i \end{aligned} \tag{11.71}$$

The antipode of K_i, E_i and F_i can be defined as an operator reversing the contour path; it is given by

$$\begin{aligned} \gamma(K_i) &= K_i^{-1} \\ \gamma(E_i) &= -E_iK_i^{-1} \\ \gamma(F_i) &= -K_iF_i \end{aligned} \tag{11.72}$$

Finally, the co-unit can be viewed as an operator destroying paths:

$$\epsilon(K_i) = 1\,, \qquad \epsilon(E_i) = 0\,, \qquad \epsilon(F_i) = 0 \tag{11.73}$$

With the above definitions, Q becomes a Hopf algebra, i.e. it satisfies the defining relations of section 6.1.

Notice that each screening current S_i gives rise to a quantum group $U_{q_i}(s\ell(2))$ with deformation parameter $q_i = \exp(i\pi\Omega_{ii}/2)$, which is a subalgebra of Q. Hence, the screening currents are a reminder in conformal field theory of the simple roots used by Drinfeld and Jimbo to deform the universal enveloping algebra of semi-simple Lie algebras.

11.4.4 The $\mathcal{R}$-matrix

To a collection of screening currents $\{S_i\}$ with given braiding properties,we have associated a Hopf algebra Q with non-trivial comultiplication Δ. The next step is to implement quasi-triangularity in Q.

As we have learned in chapter 6, a quasi-triangular Hopf algebra $\mathcal{A}$, or quantum group, is a Hopf algebra $\mathcal{A}$ with an element $\mathcal{R}$ in $\mathcal{A} \otimes \mathcal{A}$ which interpolates between the two co-products Δ and $\Delta' = \sigma \circ \Delta$:

$$\Delta'(X)\mathcal{R} = \mathcal{R}\Delta(X) \tag{11.74}$$

and satisfies

$$(\Delta \otimes \mathrm{id})\mathcal{R} = \mathcal{R}_{13}\mathcal{R}_{23} \tag{11.75a}$$

$$(\mathrm{id} \otimes \Delta)\mathcal{R} = \mathcal{R}_{13}\mathcal{R}_{12} \tag{11.75b}$$

$$(\gamma \otimes \mathbf{1})\mathcal{R}_{12} = \mathcal{R}_{21} \tag{11.75c}$$

Equations (11.74) and (11.75) are independent of the representation, i.e. they are defined on the algebra $\mathcal{A}$ itself. That is why $\mathcal{R} \in \mathcal{A} \otimes \mathcal{A}$ is called the universal $\mathcal{R}$-matrix. On the other hand, our construction is based on the representation theory of Q; thus we must evaluate (11.74) and (11.75) on tensor products of representation spaces. For instance, (11.74) on $\mathcal{V}^{a_1} \otimes \mathcal{V}^{a_2}$ becomes

$$\begin{aligned}\left(\Delta(X)^{(a_2,a_1)}\right)^{I_2,I_1}_{I_2',I_1'} \left(\mathcal{R}^{(a_1,a_2)}\right)^{I_1',I_2'}_{I_1'',I_2''} \\ = \left(\mathcal{R}^{(a_1,a_2)}\right)^{I_1,I_2}_{I_1',I_2'} \left(\Delta(X)^{(a_1,a_2)}\right)^{I_1',I_2'}_{I_1'',I_2''}\end{aligned} \tag{11.76}$$

whereas (11.75a) reads as follows:

$$\begin{aligned}\sum_{I'} \left(\mathcal{R}^{(a,a_3)}\right)^{I',I_3'}_{I,I_3} K^{a_1,a_2,I'}_{I_1,I_2,a} \\ = \sum_{I_1',I_2',I_3''} K^{a_1,a_2,I}_{I_1',I_2',a} \left(\mathcal{R}^{(a_1,a_3)}\right)^{I_1,I_3'}_{I_1',I_3''} \left(\mathcal{R}^{(a_2,a_3)}\right)^{I_2,I_3''}_{I_2',I_3}\end{aligned} \tag{11.77}$$

where the K's are quantum Clebsch–Gordan coefficients (see below). This equation expresses the simple fact that braiding and fusing of screened vertex operators commute (figure 11.8).

Our task is thus to find the matrices $\mathcal{R}^{(a_1,a_2)}$ and the quantum Clebsch–Gordan coefficients K such that equations (11.76) and (11.77) are satisfied.

The natural candidate for the intertwiner $\mathcal{R}^{(a_1,a_2)}$ is the braiding matrix of two screened vertex operators $e^{a_1}_{I_1}(z_1)$ and $e^{a_2}_{I_2}(z_2)$ (figure 11.9):

$$e^{a_1}_{I_1}(z_1)\, e^{a_2}_{I_2}(z_2) = \sum_{I_1'I_2'} e^{a_2}_{I_2'}(z_2)\, e^{a_1}_{I_1'}(z_1) \left(\mathcal{R}^{(a_1,a_2)}\right)^{I_1',I_2'}_{I_1,I_2} \tag{11.78}$$

We see immediately that $R^{(a_1,a_2)} = P\mathcal{R}^{(a_1,a_2)}$ provides us with the map

$$R^{(a_1,a_2)}: \quad \mathcal{V}^{a_1}(z_1) \otimes \mathcal{V}^{a_2}(z_2) \longrightarrow \mathcal{V}^{a_2}(z_2) \otimes \mathcal{V}^{a_1}(z_1) \tag{11.79}$$

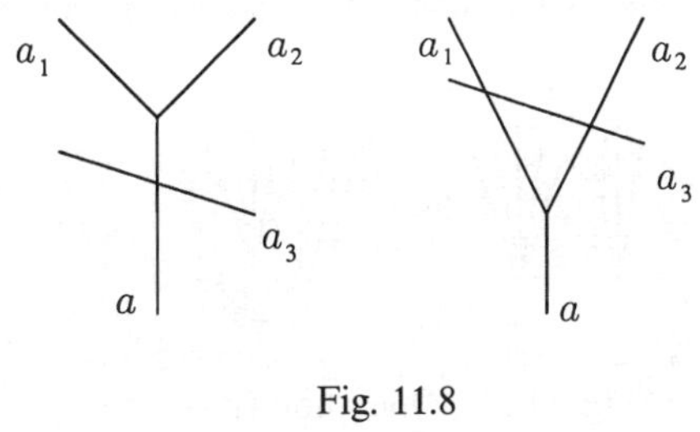

Fig. 11.8

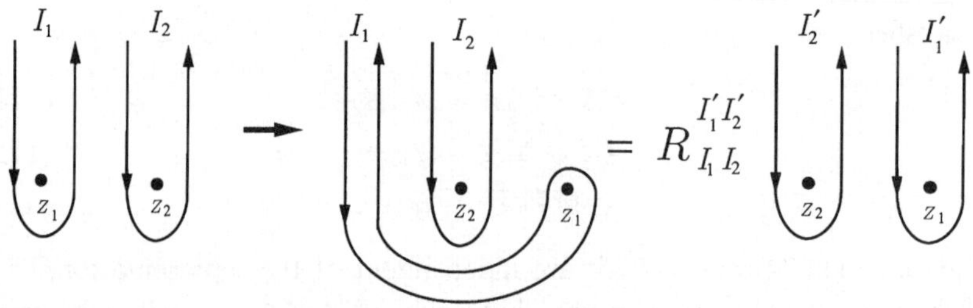

Fig. 11.9 The $\mathscr{R}$-matrix (11.78) in pictures.

which is co-ordinate independent (as usual, $R_{I_1 I_2}^{I'_1 I'_2} = \mathscr{R}_{I_1 I_2}^{I'_2 I'_1}$).

Since the screened vertex operators are not local, when we braid them the contours become deformed, and thus we expect more structure in $\mathscr{R}^{(a_1,a_2)}$ than just pure braiding phases. By purely pictorial representations (figure 11.10), the Yang–Baxter equation is proved:

$$\mathscr{R}^{(a_1,a_2)}\mathscr{R}^{(a_1,a_3)}\mathscr{R}^{(a_2,a_3)} = \mathscr{R}^{(a_2,a_3)}\mathscr{R}^{(a_1,a_3)}\mathscr{R}^{(a_1,a_2)} \tag{11.80}$$

In order to evaluate the result of either $\mathscr{R}_{23}\mathscr{R}_{13}\mathscr{R}_{12}$ or $\mathscr{R}_{12}\mathscr{R}_{13}\mathscr{R}_{23}$ on the original $e_{n_1}^{j_1}(z_1) \otimes e_{n_2}^{j_2}(z_2) \otimes e_{n_3}^{j_3}(z_3)$, we would have to manipulate the contours on the right-hand side of the figure back into the basic contours. This is not necessary, however, to prove the Yang–Baxter equation.

The actual computation of $\mathscr{R}^{(a_1,a_2)}$ is somewhat simplified with the help of the light-house operators $\tilde{e}_I^a(z)$, in terms of which (11.78) becomes

$$\tilde{e}_{I_1}^{a_1}(z_1)\,\tilde{e}_{I_2}^{a_2}(z_2) = \sum_{I'_1 I'_2} \tilde{e}_{I'_2}^{a_2}(z_2)\,\tilde{e}_{I'_1}^{a_1}(z_1)\,\left(\tilde{\mathscr{R}}^{(a_1,a_2)}\right)_{I_1,I_2}^{I'_1,I'_2} \tag{11.81}$$

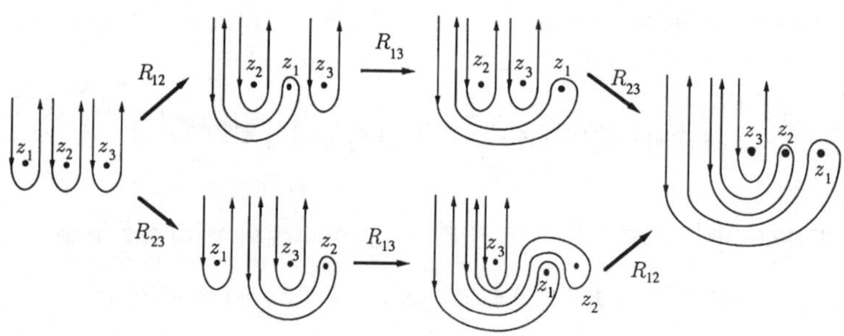

Fig. 11.10. Proof of the Yang–Baxter equation (11.80) in the contour representation of quantum groups.

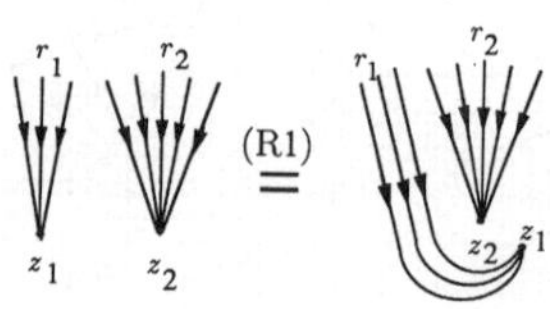

Fig. 11.11 Operation (R1).

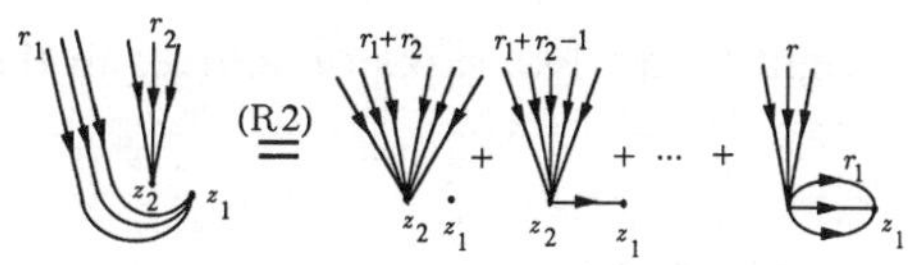

Fig. 11.12 Operation (R2).

The relationship between $\mathscr{R}^{a_1a_2}$ and the light-house intertwiner $\tilde{\mathscr{R}}^{(a_1,a_2)}$ follows immediately from (11.78) and (11.81).

The braiding of the operators $\tilde{e}$ in (11.81) involves the three contour manipulations (R1), (R2) and (R3) shown in figure 11.11, figure 11.12 and figure 11.13, respectively. The operation (R1) yields an overall factor $\exp i\pi\Omega_{a_1a_2}$, while (R2) and (R3) involve the braiding factors $\exp(i\pi\Omega_{ia_2})$ and $\exp(i\pi\Omega_{ij})$.

With this definition of the $\mathscr{R}$-matrix, it is clear that (11.76) is automatically satisfied, since it simply expresses the commutativity between the braiding of screened vertex operators and the action of the quantum group (see figure 11.14). Similarly, as already noted, equation (11.77) is simply the commutativity between braiding and fusing of screened vertex operators, which is also automatic in our construction.

The figures amply illustrate several distinctive features of the $\mathscr{R}$-matrix defined as the braiding matrix between screened contour operators. When we braid the point z_1 down under z_2, we carry its contours I_1 along. Then we decompose these contours, and the only thing which may happen is that z_1 loses some of its contours to z_2. Thus, we capture geometrically the quasi-triangularity or "quantum double" structure of the $\mathscr{R}$-matrix, i.e. the fact that $\mathscr{R} \sim \sum E^n \otimes F^n$. Also, since the total number of contours is conserved throughout the braiding,

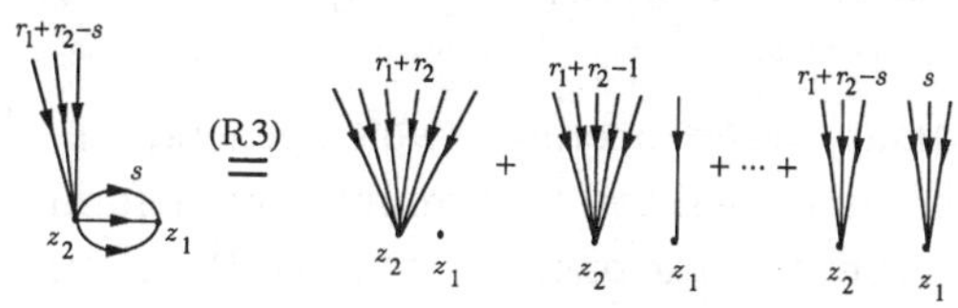

Fig. 11.13 Operation (R3).

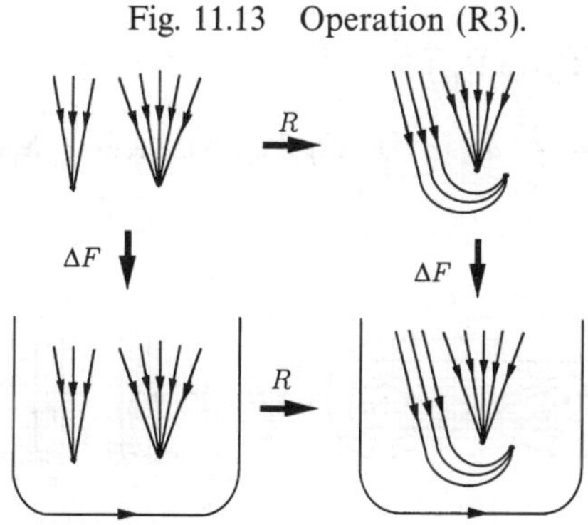

Fig. 11.14

we do not expect any eight-vertex-like sinks or sources in the contour picture of quantum groups.

11.4.5 Chiral vertex operators

We will now show a fundamental relationship between the chiral vertex operators of the conformal field theory and the quantum Clebsch–Gordan coefficients of the underlying quantum group.

Recall from chapter 9 that a chiral vertex operator $\Phi^{a_3}_{a_1,a_2}(z_1, z_2)$ is an intertwiner between three irreducible representations a_1, a_2 and a_3 of the chiral algebra :

$$\Phi^{a_3}_{a_1,a_2}(z_1, z_2) : \mathscr{F}_{a_1} \otimes \mathscr{F}_{a_2} \longrightarrow \mathscr{F}_{a_3} \tag{11.82}$$

where $\mathscr{F}_{a_i}$ $(i = 1, 2, 3)$ are three Fock spaces of the Coulomb gas, located at the points z_1, z_2 and ∞, respectively. Note that the chiral vertex operator has a dependence on both z_1 and z_2, whereas in section 9.2 we put $z_2 = 0$.

Quite analogously, the intertwining operator between three quantum group representations a_1, a_2, and a_3 is given by the quantum Clebsch–Gordan matrix $K^{a_1a_2}_{a_3}$, which exists whenever $\mathscr{V}^{a_3} \subset \mathscr{V}^{a_1} \otimes \mathscr{V}^{a_2}$. If e^a_I is a basis of $\mathscr{V}^a$, then the quantum Clebsch–Gordan coefficients are defined by

$$e^{a_3}_{I_3} = \sum_{I_1,I_2} K^{a_1a_2I_3}_{I_1I_2a_3}\; e^{a_1}_{I_1} \otimes e^{a_2}_{I_2} \tag{11.83}$$

The relationship between $\Phi^{a_3}_{a_1,a_2}(z_1, z_2)$ and $K^{a_1a_2}_{a_3}$ is given in our approach (see figure 11.15) by

$$\Phi^{a_3}_{a_1,a_2}(z_1, z_2)\left(V_{a_1} \otimes V_{a_2}\right) = \sum_{I_1,I_2} K^{a_1a_20}_{I_1I_2a_3}\; e^{a_1}_{I_1}(z_1)\;\; e^{a_2}_{I_2}(z_2) \tag{11.84}$$

Here, V_{a_i} are the primary vertex operators which generate the Fock spaces $\mathscr{F}_{a_i}$.

The coefficients $K^{\cdots}_{\cdots}$ in (11.84) can be computed by deforming the contours in the definition of the chiral vertex operator as in (R3) (see figure 11.15). Indeed, in the Coulomb gas representation, the chiral vertex operators take the form

$$\begin{aligned} &\Phi^{a_3}_{a_1,a_2}(z_1, z_2)\left(V_{a_1} \otimes V_{a_2}\right) \\ &\quad = \int_{C_1} dt_1 \cdots \int_{C_r} dt_r\; V_{a_1}(z_1) S_{i_1}(t_1) \cdots S_{i_r}(t_r) V_{a_2}(z_2) \end{aligned} \tag{11.85}$$

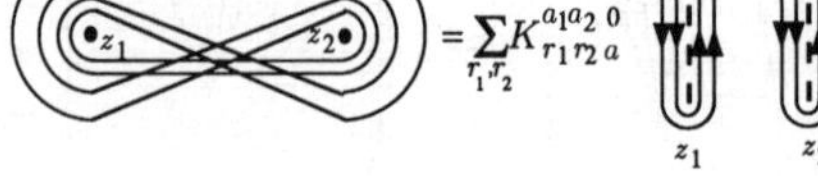

Fig. 11.15 The chiral vertex operator (11.84) in contours.

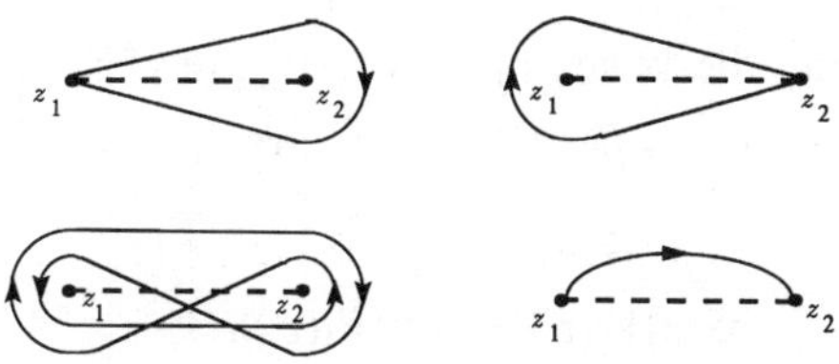

Fig. 11.16. Variety of contours defining the quantum group chiral vertex operator (11.85). Only Pochhammer's contour requires no ordering or regularization.

where the various choices of contours $C_1, \ldots, C_r$ in figure 11.16 yield results differing only by proportionality factors (recall exercise 10.14).

Since the state $\Phi^{a_3}_{a_1,a_2}(z_1, z_2)\left(V_{a_1} \otimes V_{a_2}\right)$ is a primary field of the Fock space $\mathscr{F}_{a_3}$, it automatically becomes a highest weight vector of the quantum group representation $\mathscr{V}^{a_3}$. This explains the superscript 0 in K in (11.84). We can thus generalize equation (11.84), although somewhat abusing the notation, to:

$$e^{a_3}_{I_3}\left[\Phi^{a_3}_{a_1,a_2}(z_1, z_2)\left(V_{a_1} \otimes V_{a_2}\right)\right] = \sum_{I_1,I_2} K^{a_1 a_2 I_3}_{I_1 I_2 a_3}\, e^{a_1}_{I_1}(z_1)\, e^{a_2}_{I_2}(z_2) \tag{11.86}$$

This close relationship between chiral vertex operators and quantum Clebsch–Gordan coefficients brings us back to the fusion rules of rational conformal field theories which were discussed in the introduction to this chapter.

As a final example of the application of quantum groups to conformal field theories, let us consider the effect of braiding the points z_1 and z_2 in (11.84) or (11.86). From chapter 9, we know that this braiding produces the factor

$$\epsilon^{a_3}_{a_1 a_2} \exp i\pi(\Delta_{a_3} - \Delta_{a_1} - \Delta_{a_2}) \tag{11.87}$$

where $\epsilon^{a_3}_{a_1 a_2}$ is the parity of a_3 into $a_1 \otimes a_2$, and Δ_{a_i} are the conformal weights of the primary fields V_{a_i}. Using equations (11.86) and (11.78), we obtain

$$\sum_{I'_1 I'_2} \mathscr{R}^{(a_1,a_2)}_{I_1 I_2, I'_1 I'_2}\; K^{a_1 a_2 I_3}_{I'_1 I'_2 a_3} = \epsilon^{a_3}_{a_1 a_2}\, e^{i\pi(\Delta_{a_3} - \Delta_{a_1} - \Delta_{a_2})}\, K^{a_2 a_1 I_3}_{I_2 I_1 a_3} \tag{11.88}$$

which is the generic form of equation (11.6).

In a similar way, we may compute F- and B-matrices of conformal blocks, which are again given by the representation theory of quantum groups.

11.5 The quantum group of $\widehat{SU(2)}_k$

Let us find rather explicitly the quantum group $U_q(s\ell(2))$ in the free field representation of $\widehat{SU(2)}_k$. This will constitute a simple realization of the ideas presented in the previous section.

The braidings among the vertex operators $V_{j,m} = \gamma^{j-m}\, e^{ij\phi\sqrt{2/(k+2)}}$ of isospin j and the screening current $S = \beta\, e^{-i\phi/\sqrt{2/(k+2)}}$ are

$$\begin{aligned} V_j(z)V_{j'}(z') &= q^{2jj'}V_{j'}(z')V_j(z) \\ S(z)V_j(z') &= q^{-2j}V_j(z')S(z) \\ S(z)S(z') &= q^2 S(z')S(z) \end{aligned} \tag{11.89}$$

where $q = e^{i\pi/(k+2)}$, k is the level of the Kac–Moody representation, and we concentrate on $V_j = V_{j,j}$. Identifying $e_0^j(z) = V_j(z)$, we generate a family of screened vertex operators $e_n^j(z)$ by successive additions of the screening charge on a contour C.

In the expression for the screening current $S = \beta V_{-1}$, only the free field part V_{-1} contributes to the braiding, whereas the main contribution of β is to make the conformal weight of S equal to one. Similarly, the braiding of the vertex operators depends only on the exponentials: any and all factors with β, γ or $\partial\phi$ are inert under braiding and hence do not modify the behavior of the raw exponential under braiding. This implies, notably, that descendant fields have the same braiding properties as their primaries, and hence the quantum group cannot possibly arise from the pure conformal properties of the fields. In other words, the quantum group is a structure different from the conformal one, even if the starting point for the irreducible representations of both the quantum group and the chiral algebra are the same $V_j(z)$. From the chiral algebra viewpoint, $V_j(z)$ is a chiral primary field, whereas, from the quantum group viewpoint, $e_0^j(z) = V_j(z)$ is the state $|j,j\rangle_q$. Let us analyze the structure of $e_n^j(z)$ in some detail.

Consider first $e_1^j(z)$. By Cauchy's theorem, the contribution of the tiny circle around z vanishes if $j \leq (k+1)/2$. For higher values of j, we could, of course, retain only the principal value of the integral. There is no harm, however, in restricting ourselves to $j \leq (k+1)/2$. The integral along C of S can be deformed, therefore, into two straight vertical pieces on top of each other and in opposite directions but on different Riemann sheets, due to the cut. The phase difference due to the cut follows from the braiding between the screening current $S(t)$ and the vertex operator $V_j(z)$. Here, the operator $S(t)$ on the upward straight contour has turned 2π around $V_j(z)$, so the phase difference between the two straight stretches is q^{-4j}. There is also a sign mismatch due to their different orientations. Hence,

$$e_1^j(z) = (1 - q^{-4j})\tilde{e}_1^j \tag{11.90}$$

Without evaluating the integral $\tilde{e}_n^j$, we see that, if $j = 0$, then $e_1^j(z) = 0$. Hence $\dim(\{e_n^0\}) = 1$. The vertex operator $V_0(z) = 1$ is just the identity operator. In the unitary case with $k \in \mathbb{N}$, q is a root of unity, and thus in fact $\dim(\{e_n^{m(k+2)/2}\}) = 1$, for any $m \in \mathbf{Z}$. We shall elaborate on this point below.

Consider next $e_2^j(z)$. In order to find the proportionality factor between $e_2^j(z)$ and $\tilde{e}_2^j(z)$, we must write both $\tilde{e}_2^j(z)$ and the deformed $e_2^j(z)$ properly, as a sum of

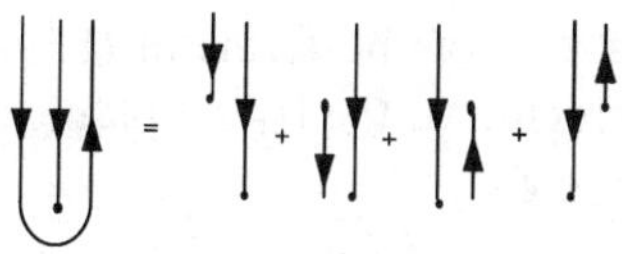

Fig. 11.17 The action of F on $\tilde{e}_1^j$ splits into well-ordered parts.

normal ordered terms. The calculation requires taking into account the various phase factors induced by braiding to achieve a normal order, and also the mismatch $-q^{-4j}$ between the down-coming and up-going straight stretches. It is simplified by thinking of e_2^j as being proportional to $F\tilde{e}_1^j$ (figure 11.17). Explicitly, starting with a well ordered expression for $t_2 > t_1 > z$,

$$\begin{aligned}
e_2^j(z) &= \int_C dt_2 \int_C dt_1 S(t_2)S(t_1)V_j(z) \\
&=: \left[\int_{i\infty}^{z} dt_2 + \int_{z}^{i\infty} dt_2\right]\left[\int_{i\infty}^{z} dt_1 + \int_{z}^{i\infty} dt_1\right] S(t_2)S(t_1)V_j(z) : \\
&= \int_{i\infty}^{z} dt_1 \left[\int_{i\infty}^{t_1} dt_2 S(t_2)S(t_1)V_j(z) + \right. \\
&\qquad + q^2 \int_{t_1}^{z} dt_2 S(t_1)S(t_2)V_j(z) \\
&\qquad + q^{-4j+2} \int_{z}^{t_1} dt_2 S(t_1)S(t_2)V_j(z) \\
&\qquad \left. + q^{-4j+4} \int_{t_1}^{i\infty} dt_2 S(t_2)S(t_1)V_j(z)\right] \\
&+ q^{-4j} \int_{z}^{i\infty} dt_1 \left[\text{same as above}\right]
\end{aligned} \tag{11.91}$$

Now we use also (figure 11.18)

$$\begin{aligned}
\tilde{e}_2^j(z) = \int_{i\infty}^{z} dt_1 \int_{i\infty}^{t_1} dt_2 S(t_2)S(t_1)V_j(z) \\
+ q^2 \int_{i\infty}^{z} dt_1 \int_{t_1}^{z} dt_2 S(t_1)S(t_2)V_j(z)
\end{aligned} \tag{11.92}$$

In order to achieve the result

$$e_2^j(z) = \left(1 - q^{-4j}\right)\left(1 - q^{-4j+2}\right)\tilde{e}_2^j(z) \tag{11.93}$$

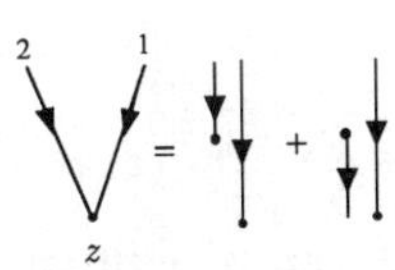

Fig. 11.18 The light-house basis element $\tilde{e}_2^j$ decomposes into well-ordered parts.

we must change variables in one of the terms in (11.92), viewed this time as an integral over a "triangular" region of the (t_1, t_2) plane. Note that (11.93) vanishes for $j = \frac{1}{2} \bmod \left(\frac{k+2}{2}\right)$, with $m \in \mathbf{Z}$.

In general, using the path ordering from radial quantization and similar tricks from complex analysis, the $e_n^j(z)$ operators can be rewritten in terms of light-house vertex operators consisting of n integrals $\int_{i\infty}^{z} dtS(t)$ acting on $V_j(z)$:

$$e_n^j(z) = \left[\prod_{\ell=0}^{\ell=n-1} \left(1 - q^{-4j+2\ell}\right)\right] \tilde{e}_n^j(z) \tag{11.94}$$

Clearly, $n \leq 2j$, and hence dim $\{e_n^j(z)\} = 2j+1$. When k is an integer (or, actually, rational), q is a root of unity, and only in this case can the dimensionality of the vector space $\{e_n^j(z)\}$ be lower than the usual $2j+1$.

The quantum group lowering operator $F = \int_C dtS(t)$ acts among these states simply as

$$Fe_n^j(z) = e_{n+1}^j(z) \tag{11.95}$$

From the definition of the action of F on a single screened vertex operator, we can immediately picture the effect of ΔF on the tensor product of two screened vertex operators, understood simply as two screened vertex insertions at different points of the complex plane: ΔF encloses the two screened vertex operators with one big C contour. The operator ΔF is thus

$$\Delta F \left[e_n^j(z) \otimes e_{n'}^{j'}(z')\right] = \int_C dtS(t) \left[e_n^j(z) e_{n'}^{j'}(z')\right] \tag{11.96}$$

In order to deform the contour C into two contours, one around each screened vertex, the screening current S must braid through the first screened vertex, thus picking up a q-phase:

$$\begin{aligned}(\Delta F) \left[e_n^j(z) \otimes e_{n'}^{j'}(z')\right] &= \int_{i\infty}^{(z+,z'+)} dt\, S(t) e_n^j(z) e_{n'}^{j'}(z') \\ &= \int_{i\infty}^{(z+)} dt\, S(t) e_n^j(z) e_{n'}^{j'}(z') + q^{-2j+2n} \int_{i\infty}^{(z'+)} dt\, e_n^j(z) S(t) e_{n'}^{j'}(z') \\ &= F[e_n^j(z)] \otimes e_{n'}^{j'}(z') + q^{-2j+2n} e_n^j(z) \otimes F[e_{j'}^{n'}(z')]\end{aligned} \tag{11.97}$$

This "braiding" can be checked by introducing both sides of the above equation in any correlation function and seeing that the equality holds.

The $U_q(s\ell(2))$ Cartan generator is

$$H(e_n^j(z)) = 2(j-n) e_n^j(z) = 2m\; e_n^j(z) \tag{11.98}$$

or equivalently, using $K = q^H$,

$$K(e_n^j(z)) = q^{2(j-n)}\; e_n^j(z) = q^{2m}\; e_n^j(z) \tag{11.99}$$

The co-multiplication (11.97) of F can be written conveniently as

$$\Delta F = F \otimes \mathbf{1} + K^{-1} \otimes F \tag{11.100}$$

The asymmetry of Δ thus arises from requiring operator consistency of the contour deformations. Also, the definition of H shows that its co-multiplication is trivial:

$$\Delta H = H \otimes \mathbf{1} + \mathbf{1} \otimes H \quad \Longleftrightarrow \quad \Delta K = K \otimes K \tag{11.101}$$

These generators satisfy $[H,F] = -2F$ or, equivalently, $KF = q^{-2}FK$. The number of contours n is now related to the third component of quantum isospin m by $n = j - m$.

Since we already have the Borel subalgebra $\{F,H\}$, via Drinfeld's quantum double construction, we can find the full Hopf algebra. This means that E can be introduced formally as the adjoint of F, for instance in the (quantum group invariant) two-point function

$$\left\langle \overline{V_j}(z) \middle| F\ V_{j'}(z') \right\rangle = \left\langle E\overline{V_j}(z) \middle| V_{j'}(z') \right\rangle^* \tag{11.102}$$

The presence of the background charge at infinity implies that the out-vacuum $\langle 0|$, and hence the out-state $\langle \overline{V_j}(z)|$, possess a complicated structure. We should thus expect that when E acts on in-states rather than on out-states, some insertions of operators at infinity should show up. As physicists, however, we prefer to find explicitly the raising operator E. It should act on e^j_n by destroying a contour and yielding e^j_{n-1}. Thanks to the Sugawara relationship between the Virasoro and Kac–Moody modes, the operator destroying contours follows from the zero mode f_0 of the Kac–Moody lowering current:

$$[f_0, S(z)] = (k+2)\frac{d}{dz}\left(V_{-1}(z)\right) \tag{11.103}$$

Acting on a screened vertex operator, f_0 yields two distinct terms:

$$f_0 e^j_n(z) = \left\{ \prod_{i=1}^{n} \int_C dt_i \right\} \left\{ \left[f_0, \prod_{i=1}^{n} S(t_i) \right] V_j(z) + \prod_{i=1}^{n} S(t_i) f_0 V_j(z) \right\} \tag{11.104}$$

The second term is in fact a descendant of the highest weight $|j\rangle$, screened n times, and we shall disregard it for the definition of E. It is only the first term which we shall interpret as the quantum group raising operator, because it contains $(n-1)$ contours on the primary V_j. Note that, although the commutator of f_0 with $\int S(t)$ removes the integral, it still leaves the free scalar part of the screening current (an operator) at the boundary of the integration contour, namely at infinity. This is excellent, because the out-vacuum in the Coulomb gas is ${}_{2\alpha_0}\langle 0| = \langle 0|\, V_{-1}(\infty)$. Thus, the construction of the raising operator E from the chiral algebra automatically yields the interpretation of E as the dual to F taking into account the peculiarity of the Coulomb gas out-vacuum. Quite generally, we define

$$f_0 e^j_n(z) = \alpha V_{-1}(\infty)\ \tilde{E}\ e^j_n(z) + \text{descendants} \tag{11.105}$$

with the notation $\tilde{E}$ forecasting a redefinition below, and we find, by straightforward manipulations, that

$$\alpha\tilde{E}e_n^j(z) = -(k+2)\left(1-q^{-4j+2n-2}\right)\frac{1-q^{2n}}{1-q^2}e_{n-1}^j \tag{11.106}$$

whereby $[H,\tilde{E}] = 2\tilde{E}$ or $K\tilde{E} = q^2\tilde{E}K$, regardless of the value of α. We fix this proportionality constant as $\alpha = -(k+2)(1-q^{-2})$ by demanding the correct quantum commutator between $\tilde{E}$ and F:

$$\tilde{E}F - q^2F\tilde{E} = \frac{1-K^{-2}}{1-q^{-2}} \tag{11.107}$$

Then, we have

$$\tilde{E}e_n^j(z) = \frac{1-q^{-2j-2-2m}}{1-q^{-2}}\,\frac{1-q^{2j-2m}}{1-q^2}\,e_{n-1}^j(z) \tag{11.108}$$

Finally, the co-multiplication of $\tilde{E}$ follows from the trivial one for f_0, i.e. from $\Delta f_0 = f_0 \otimes \mathbf{1} + \mathbf{1} \otimes f_0$. It is important to braid the left-over operator $V_{-1}(\infty)$ all the way to the left, whence

$$\Delta\tilde{E} = \tilde{E}\otimes\mathbf{1} + K^{-1}\otimes\tilde{E} \tag{11.109}$$

We may now redefine $E = K\tilde{E}$ to recover $U_q(s\ell(2))$ in its standard form; this manipulation is left as an exercise for the reader.

Summarizing, the full $U_q(s\ell(2))$ has been derived from the Coulomb gas representation of $\widehat{SU(2)}_k$ Kac–Moody. Quite generally, the deformation parameter q of the quantum group hidden in a Wess–Zumino model is related to the level k of the Kac–Moody algebra as in this case, namely

$$q^{2(k+h^*)} = 1 \tag{11.110}$$

For integrable but non-critical models, such as the XXZ model with $\Delta < -1$, this relationship does not hold anymore, i.e. q and k are not related.

There is an important point which we wish to emphasize: the Kac–Moody multiplet $|jm\rangle_{KM} = \gamma^{j-m}V_j$ is different from the quantum group multiplet $|jm\rangle_q = F^{j-m}V_j$. We could have screened, for instance, the state $|j,j-1\rangle_{KM}$ instead of $|jj\rangle_{KM}$, and we would have again obtained a quantum group multiplet. It is convenient to think only of the highest weight $|jj\rangle_{KM}$ as a primary field, and then forget the chiral algebra descendants. Of course, we cannot actually forget them due to the relationship between E and f_0: the quantum group is not orthogonal to the chiral algebra. The dual of $|jm\rangle_q = F^{j-m}V_j$ is ${}_q\langle jm| = F^{-j+m}V_{-1-j}$.

Let us just note, for completeness, that the basis (E,K,F) of $U_q(s\ell(2))$ that we have encountered naturally is not the one used elsewhere in the book. Letting the usual basis be (X^+,H,X^-), the non-diagonal generators are related by a scaling $E = q^{-\frac{H}{2}}X^+$, $F = q^{-\frac{H}{2}}X^-$. In this other basis, the co-multiplication and commutators we have found look more symmetric: $\Delta X^{\pm} = X^{\pm} \otimes q^{\frac{H}{2}} + q^{-\frac{H}{2}} \otimes X^{\pm}$ and

$$[X^+, X^-] = \frac{q^H - q^{-H}}{q - q^{-1}} \tag{11.111}$$

11.5.1 The $\mathscr{R}$-matrix

Let us present an explicit contour computation of the $\mathscr{R}$-matrix $\mathscr{R}^{j_1 j_2}$ for two standard irreps j_1 and j_2 of $U_q(s\ell(2))$,

$$e^{j_1}_{r_1}(z_1) \otimes e^{j_2}_{r_2}(z_2) = \sum_{r'_1, r'_2} e^{j_2}_{r'_2}(z_2) \otimes e^{j_1}_{r'_1}(z_1) \mathscr{R}^{j_1 j_2}_{r'_1 r'_2, r_1 r_2} \tag{11.112}$$

The matrix $\mathscr{R}^{j_1 j_2}$ measures the mismatch induced in the two representations of $U_q(s\ell(2))$ by the co-multiplication Δ, and it satisfies the Yang–Baxter equation. As noted earlier, it is easier to compute the matrix $\tilde{\mathscr{R}}^{j_1 j_2}$ in (11.81) which gives the braiding or commutation relations of the light-house operators $\tilde{e}^{j_i}_{r_i}(z)$. The relationship between the true $\mathscr{R}^{j_1 j_2}$ and the light-house $\tilde{\mathscr{R}}^{j_1 j_2}$ follows from (11.112) and (11.94):

$$\mathscr{R}^{j_1 j_2}_{r'_1 r'_2, r_1 r_2} = \frac{\prod_{x_1=0}^{r_1-1} \left(1 - q^{-4j_1 + 2x_1}\right) \prod_{x_2=0}^{r_2-1} \left(1 - q^{-4j_2 + 2x_2}\right)}{\prod_{y_1=0}^{r'_1-1} \left(1 - q^{-4j_1 + 2y_1}\right) \prod_{y_2=0}^{r'_2-1} \left(1 - q^{-4j_2 + 2y_2}\right)} \tilde{\mathscr{R}}^{j_1 j_2}_{r'_1 r'_2, r_1 r_2} \tag{11.113}$$

Application of the rules (R1), (R2) and (R3) of section 11.4.4 reduces the problem of finding $\tilde{\mathscr{R}}^{j_1 j_2}$ to a combinatorial one. To show more clearly the combinatorial nature of this problem, we shall calculate $\tilde{\mathscr{R}}^{j_1 j_2}$ in the most simple case where all the braiding factors are trivial, i.e.

$$\Omega_{j_1 j_2} = \Omega_{j_1, -1} = \Omega_{j_2, -1} = 0 \qquad \Rightarrow \qquad q = 1 \tag{11.114}$$

(recall that, under braiding, the screening S behaves like V_{-1}). It is obvious that the correct result should be merely

$$\tilde{\mathscr{R}}^{j_1 j_2}_{r'_1 r'_2, r_1 r_2} = \delta_{r'_1 r_1}\, \delta_{r'_2 r_2} \tag{11.115}$$

In the relevant situation, when the braiding factors are not trivial, we shall find a q-analog version of the procedure used to derive (11.115), which goes as follows (we urge the reader to construct the drawings while going through this computation):

Start by applying the rules (R1) and (R2) to $\tilde{e}^{j_1}_{r_1} \otimes \tilde{e}^{j_2}_{r_2}$. The r_1 contours attached to 1 come from infinity, up at the left, go down under 2, and reach 1 to the right. We decompose this into a sum $\sum_{\ell=0}^{r_1} \binom{r_1}{\ell}$ of terms with $r_1 + r_2 - \ell$ contours from infinity to 2, and ℓ contours from 2 to 1. The combinatorial number $\binom{r_1}{\ell}$ gives the number of choices of ℓ contours which, according to the splitting rule (R2), become contours from the point 2 to the point 1. Applying the opening rule (R3) to one such term with ℓ contours from 2 to 1 yields another sum, $\sum_{d=0}^{\ell} (-1)^{\ell-d} \binom{\ell}{d}$ of standard light-house operators, where point 1 (to the right of 2) has d contours and point 2 has $r_1 + r_2 - d$ of them. The sign factor $(-1)^{\ell-d}$ is caused by the change of orientation of the $\ell - d$ lines which return to the point 2 after the opening. Introducing this expression into the previous one, and changing the order of the summands, we obtain

$$\sum_{d=0}^{r_1}\sum_{\ell=d}^{r_1}(-1)^{\ell-d}\binom{r_1}{\ell}\binom{\ell}{d}\,\tilde{e}^{j_2}_{r_1+r_2-d}(z_2)\otimes\tilde{e}^{j_1}_{d}(z_1) \tag{11.116}$$

Finally, on changing variables we arrive at

$$\sum_{n=0}^{r_1}\binom{r_1}{n}\times\sum_{\nu=0}^{n}(-1)^{\nu}\binom{n}{\nu}\,\tilde{e}^{j_2}_{r_2+n}(z_2)\otimes\tilde{e}^{j_1}_{r_1-n}(z_1) \tag{11.117}$$

The result (11.115) follows from the identity

$$(1-1)^n=\sum_{\nu=0}^{n}(-1)^{\nu}\binom{n}{\nu}=\delta_{n,0} \tag{11.118}$$

Let us come back to the case of real interest, with non-trivial braiding factors. The object whose braiding we want to analyze is $\tilde{e}^{j_1}_{r_1}(z_1)\otimes\tilde{e}^{j_2}_{r_2}(z_2)$ or, more explicitly,

$$\int_{i\infty}^{z_1}\prod_{i=1}^{r_1}dt_i\;S(t_i)\;V_{j_1}(z_1)\;\int_{i\infty}^{z_2}\prod_{j=1}^{r_2}ds_j\;S(s_j)\;V_{j_2}(z_2) \tag{11.119}$$

After braiding $V_{j_1}(z_1)$ under $V_{j_2}(z_2)$, and the corresponding deformation of the contours C_1, we obtain (see figure 11.11)

$$q^{2j_1(j_2-r_2)}\int_{i\infty}^{z_1}\prod_{i=1}^{r_1}dt_i\;S(t_i)\;\int_{i\infty}^{z_2}\prod_{j=1}^{r_2}ds_j\;S(s_j)\;V_{j_2}(z_2)V_{j_1}(z_1) \tag{11.120}$$

Splitting the contours, we find

$$\sum_{\ell=0}^{r_1}\left[\left|\begin{matrix}r_1\\ \ell\end{matrix}\right|\right]_q\,q^{2(-j_1+\ell)(-j_2+r_2)}\int_{i\infty}^{z_2}\prod_{i=1}^{r_1+r_2-\ell}dt_i\;S(t_i)\;V_{j_2}(z_2)$$
$$\times\int_{z_2}^{z_1}\prod_{k=1}^{\ell}du_k\;S(u_k)\;V_{j_1}(z_1) \tag{11.121}$$

where

$$\left[\left|\begin{matrix}r_1\\ \ell\end{matrix}\right|\right]_q=\frac{[|r_1|]!}{[|\ell|]!\,[|r_1-\ell|]!}\quad,\qquad [|x|]=\frac{1-q^{2x}}{1-q^2} \tag{11.122}$$

is a q-binomial number. This term arises in (11.121) due to braidings of the screening operators among themselves.

Next, opening the integrals $\int_{z_2}^{z_1}$ yields

$$\sum_{d=0}^{\ell}(-1)^{\ell-d}\left[\left[\begin{matrix}\ell\\ d\end{matrix}\right]\right]_q q^{(\ell-d)(\ell-d-1-2j_2)}\int_{i\infty}^{z_2}\prod_{i=1}^{r_1+r_2-d}dt_i\, S(t_i)V_{j_2}(z_2)$$
$$\times\int_{i\infty}^{z_1}\prod_{j=1}^{d}ds_j\, S(s_j)V_{j_1}(z_1) \tag{11.123}$$

The novel feature of (11.123), apart from the expected modifications of numbers into q-numbers, is the term $q^{(\ell-d)(\ell-d-1)}$, which is due to the reversal of the path ordering of the $\ell-d$ screening operators flying back to the point 2.

Using (11.123) in (11.121), and following the same steps as in the trivial case, we arrive at

$$\sum_{n=0}^{r_1}\left\{\sum_{v=0}^{n}(-1)^v\left[\left[\begin{matrix}n\\ v\end{matrix}\right]\right]_q q^{v(-2j_2+r_2+v-1)}\right\}\left[\left[\begin{matrix}r_1\\ n\end{matrix}\right]\right]_q$$
$$\times\, q^{2(j_1-r_1+n)(j_2-r_2)}\,\tilde{e}^{j_2}_{r_2+n}(z_2)\otimes\tilde{e}^{j_1}_{r_1-n}(z_1) \tag{11.124}$$

The identity we need to simplify (11.124) is the Gauss binomial formula

$$\sum_{v=0}^{n}(-1)^v\left[\left[\begin{matrix}n\\ v\end{matrix}\right]\right]_q z^v q^{v(v-1)}=\prod_{v=0}^{n-1}\left(1-zq^{2v}\right) \tag{11.125}$$

The final answer is given by

$$\tilde{\mathscr{R}}^{j_1j_2}_{r_1-n,r_2+n;r_1,r_2}=\left[\left[\begin{matrix}r_1\\ n\end{matrix}\right]\right]_q q^{(-j_1+r_1-n)(-j_2+r_2)}\prod_{v=0}^{n-1}\left(1-q^{-4j_2}q^{2(r_2+v)}\right) \tag{11.126}$$

which, using (11.113), amounts to

$$\mathscr{R}^{j_1j_2}_{r_1-n,r_2+n;r_1,r_2}=\left[\left[\begin{matrix}r_1\\ n\end{matrix}\right]\right]_q q^{(j_1-r_1+n)(j_2-r_2)}\prod_{v=r_1-n}^{r_1-1}\left(1-q^{-4j_1+2v}\right) \tag{11.127}$$

or, equivalently,

$$\mathscr{R}^{j_1j_2}_{m_1+n,m_2-n;m_1m_2}=(1-q^{-2})^n q^{2(m_1+n)m_2}$$
$$\times\left[\left[\begin{matrix}j_1-m_1\\ n\end{matrix}\right]\right]_q\frac{[j_1+m_1+n]_q!}{[j_1+m_1]_q!} \tag{11.128}$$

11.5.2 Fusion rules and chiral vertex operators

The regular representations of the quantum group (with both highest and lowest weight states) are characterized by a vertex operator $V_j(z)$ with $2j\in\mathbb{N}$. The finite quantum group multiplet contains V_j and its screenings $\{e^j_n(z),\, 0\le n\le 2j\}$. We shall exploit the dual symmetry of the conformal weight (i.e. of the Casimir) under $j\leftrightarrow\tilde{\jmath}=-j-1$, which is just $\alpha\leftrightarrow 2\alpha_\circ-\alpha$ or, in a minimal (p,p') model,

$(m, m') \leftrightarrow (p-m, p'-m')$. This symmetry, accompanied by $m \leftrightarrow -m$, is also apparent in the expression for $Ee^j_n(z)$.

In order to compute dim $\{e^{\tilde{\jmath}}_n\}$, we just substitute $\tilde{\jmath}$ for j in (11.94) and find that $e^{\tilde{\jmath}}_n = (1 - q^{-4\tilde{\jmath}+2n-2}) \times \cdots$. Since $q^{2(k+2)} = 1$, dim $\{e^{\tilde{\jmath}}_n\} = k + 1 - 2j$ (see table 11.7).

For $j \geq (k+1)/2$, the quantum dual representations are not well defined, hence regular representations of $U_q(s\ell(2))$ are those with $0 \leq j \leq k/2$, with $q^{2(k+2)} = 1$. The contour representation of the quantum group when q is a root of unity shows us then that the quantum symmetry of the conformal field theory is not the full quantum group, but is, indeed, the quantum group modulo its center.

Having found the regular representations (irreps) of $U_q(s\ell(2))$, we must now check that the tensor product of any two irreps is a sum of irreps, i.e. that the irreps we have found close. Then the quantum group will be, effectively, semi-simple, and the fusion algebra will be consistent. We shall shortly define the tensor product of two irreps with the help of contour-like chiral vertex operators, but first let us state how we shall exploit this duality $j \leftrightarrow \tilde{\jmath}$ to obtain the correct truncated fusion algebra. The idea is to multiply naively

$$j_1 \times j_2 = \sum_{j \in A} j\,, \quad j_1 \times \tilde{\jmath}_2 = \sum_{j \in B} \tilde{\jmath}\,, \quad \tilde{\jmath}_1 \times j_2 = \sum_{j \in C} \tilde{\jmath} \tag{11.129}$$

and then ensure duality by requiring that all three products be consistent, i.e.

$$j_1 \otimes j_2 = \sum_{j \in A \cap B \cap C} j \tag{11.130}$$

Table 11.7 Dimensions of the quantum spaces $\{e^j_n\}$ and its dual $\{e^{\tilde{\jmath}}_n\}$ under reflection on the background charge. We impose that they both be strictly positive.

j	dim $\{e^j_n\}$	dim $\{e^{\tilde{\jmath}}_n\}$
0	1	$k+1$
$\frac{1}{2}$	2	k
1	3	$k-1$
$\vdots$	$\vdots$	$\vdots$
$\frac{k-2}{2}$	$k-1$	3
$\frac{k-1}{2}$	k	2
$\frac{k}{2}$	$k+1$	1

This expression means that the irreps j not in $A \cap B \cap C$ do not appear on the right-hand side. This discussion is similar to the one for the $c < 1$ theories in section 10.2.

To define the tensor product of two irreps j_1 and j_2, introduce a chiral vertex operator which acts on $\mathscr{F}_{j_1} \otimes \mathscr{F}_{j_2}$ to produce the irrep j out of j_1 and j_2 as follows. Given j_1 and j_2, screen $V_{j_1}(z_1)$ and $V_{j_2}(z_2)$ with r double-eight (Pochhammer) contours. The Pochhammer contour P is the double-eight contour familiar from the integral representation of hypergeometric functions, which surrounds each of $V_{j_1}(z_1)$ and $V_{j_2}(z_2)$ twice, once clockwise and once counter-clockwise (see figure 11.19). A branch cut joins z_1 and z_2, so we can relate the P screening of $j_1 \otimes j_2$ to a time ordered (say $|z_1| > |z_2|$) line integral from z_1 to z_2. Thus, the r Pochhammer contours can be deformed, as for (11.94), into r contours from z_1 to z_2:

$$\begin{aligned}\Phi^{r,0}_{j_1 j_2} &= \left[\prod_{i=1}^{r} \int_P dt_i S(t_i)\right] V_{j_1}(z_1) V_{j_2}(z_2) \\ &= \left(1 - q^{-4j_1}\right) \dots \left(1 - q^{-4j_1+2r-2}\right) \left(1 - q^{-4j_2}\right) \dots \left(1 - q^{-4j_2+2r-2}\right) \\ &\quad \times V_{j_1}(z_1) \left[\prod_{i=1}^{r} \int_{z_1}^{z_2} dt_i S(t_i)\right] V_{j_2}(z_2) \end{aligned} \tag{11.131}$$

We shall sometimes also use the notation $\Phi^{r,0}_{j_1 j_2} = \Phi^{j_3}_{j_1,j_2}(z_1, z_2)$ to make the dependence on the location of the vertices explicit.

It is easy to deform each of these straight contours from z_1 to z_2 into a pair of contours: one from z_1 to $i\infty$, the other from $i\infty$ to z_2 (we pick up q-phases from braiding in doing this). These straight contours from $i\infty$ to z_i can finally be recast as the defining analytic contours C, and the net result is a linear combination of $e^{j_1}_{n_1}(z_1) e^{j_2}_{n_2}(z_2)$.

This is not the whole story, however. We can also add n contours C around $\Phi^{r,0}_{j_1 j_2}$, i.e. act with F^n. The operator thus defined, $\Phi^{r,n}_{j_1 j_2}$, can be expanded similarly in terms of $e^{j_1}_{n_1} e^{j_2}_{n_2}$. Quite generally,

$$\Phi^{r,n}_{j_1 j_2} = K^{rn}_{j_1 j_2}(n_1, n_2) e^{j_1}_{n_1} e^{j_2}_{n_2} \tag{11.132}$$

We interpret the operator $\Phi^{r,n}_{j_1 j_2}$ as a chiral vertex operator for producing a state $|j, m\rangle$ at $i\infty$ out of $|j_1, m_1\rangle$ at z_1 and $|j_2, m_2\rangle$ at z_2 (recall $m = j - n$). What is j? Or, in other words, what selection rules can we derive from the simple contour manipulations leading to the expression (11.132)? To start with, the number of screening currents S did not change, hence $r + n = n_1 + n_2$. Secondly, since we

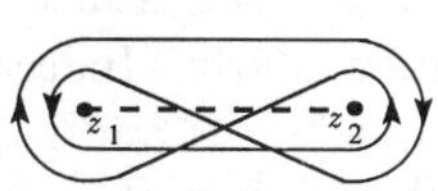

Fig. 11.19 The Pochhammer contour.

have already established $U_q(s\ell(2))$ as a good symmetry, the quantum isospin m must be conserved, hence

$$j - n = j_1 - n_1 + j_2 - n_2 \tag{11.133}$$

So the chiral vertex operator $\Phi^{r,n}_{j_1 j_2}$ creates the state

$$|j, m\rangle = |j_1 + j_2 - r, m_1 + m_2\rangle \tag{11.134}$$

Indeed, note that the state $\Phi^{r,n}_{j_1 j_2}(z_1, z_2)(V_{j_1} \otimes V_{j_2})$ vanishes unless

$$0 \leq r \leq \min(2j_1, 2j_2) - 1 \tag{11.135}$$

This requirement follows from the definition of $\Phi^{r,n}_{j_1 j_2}(z_1, z_2)$ and the fact that more than $2j_i$ contours annihilate the vertex V_{j_i}. If we compute now the scalar product

$$\left\langle V_{-1-j_3}(i\infty)\Phi^{r,n}_{j_1,j_2}(z_1, z_2)\left(V_{j_1} \otimes V_{j_2}\right)\right\rangle \tag{11.136}$$

it is easy to see, by deforming the contours as in figure 11.15, that this scalar product never vanishes provided $(k+2)$ is not rational, so the condition (11.135) is necessary and sufficient for $\Phi^{r,n}_{j_1 j_2}(z_1, z_2)(V_{j_1} \otimes V_{j_2})$ to be non-zero.

The charge conservation condition implies

$$2j_3 = 2j_1 + 2j_2 - 2r \tag{11.137}$$

which, together with (11.135), yields the "classical fusion rules"

$$|j_1 - j_2| \leq j_3 \leq j_1 + j_2 \tag{11.138}$$

with truncation from below but not from above. In the rational case ($k + 2 \in \mathbb{N}$), we obtain a vanishing scalar product (11.136) unless

$$0 \leq r \leq k - 2j_3 \tag{11.139}$$

We have supposed here that j_3 is an integrable representation of the $\widehat{SU(2)}_k$ theory. From (11.135), (11.137) and (11.139), we find the fusion rules of the rational conformal field theory:

$$|j_1 - j_2| \leq j_3 \leq \min(j_1 + j_2, k - j_1 - j_2) \tag{11.140}$$

which are also the decomposition rules of the restricted tensor product $(\mathscr{V}^{j_1} \otimes \mathscr{V}^{j_2})$.

Let us emphasize that the full selection rules are amazingly simple to derive. It is sufficient to consider the highest weight chiral vertex operators with $n = 0$. The only information we need is that the number r of double-eight contours is bounded, below because the Coulomb gas admits no "antiscreenings", and above from (11.131): $0 \leq r \leq \min(2j_1, 2j_2)$. Using these bounds in $j = j_1 + j_2 - r$, we find that $|j_1 - j_2| \leq j \leq j_1 + j_2$. The truncation below is customary, but duality also gives us the truncation above. Indeed, the chiral vertex operator $\Phi^{r,0}_{j_1 \tilde{j}_2}$ produces a state with $\tilde{j} = j_1 + \tilde{j}_2 - r$, and now the bounds on r from the tilded version of (11.131) read $0 \leq r \leq \min(2j_1, k - 2j_2)$, which translate into $j_2 - j_1 \leq j \leq \min(j_1 + j_2, k - j_1 - j_2)$. The third case is completely analogous, and the result from applying duality to the quantum chiral vertex operators is the correct fusion algebra for $\widehat{SU(2)}$ Kac–Moody at level k:

$$j_1 \otimes j_2 = \sum_{|j_1-j_2|}^{\min(j_1+j_2,k-j_1-j_2)} j \tag{11.141}$$

The actual computation of the quantum Clebsch–Gordan coefficients in some simple cases is left for the exercises.

11.5.3 On intertwiners: a clarification

Before ending this section, let us make an important general clarification on intertwiners with the help of the $\widehat{SU(2)}_k$ example at hand.

The Moore–Seiberg intertwiner $\Phi^{j_3}_{j_1 j_2}(z)$ or chiral vertex operator is a map

$$\Phi^{j_3}_{j_1 j_2}(z) : \mathscr{H}_{j_1} \otimes \mathscr{H}_{j_2} \to \mathscr{H}_{j_3} \tag{11.142}$$

among Verma modules $\mathscr{H}_{j_i}$ of the Kac–Moody algebra.

The Frenkel–Reshetikhin intertwiner is really an intertwiner of the Kac–Moody algebra :

$$\Phi^{j_1 j_2}_{j_3}(z) : \mathscr{H}_{j_3} \to V_{j_1}(z) \otimes \mathscr{H}_{j_2} \tag{11.143}$$

In general, an intertwiner of a Hopf algebra is a map

$$V_{\xi_3} \to V_{\xi_1} \otimes V_{\xi_2} \tag{11.144}$$

where the representation ξ_3 is contained in $\xi_1 \otimes \xi_2$.

Clearly, the Moore–Seiberg intertwiner is not an intertwiner of the Virasoro algebra. However, it can be interpreted as an intertwiner for the quantum group. Indeed, the equation (11.132) obtained in the contour picture allows us to view Moore and Seiberg's $\Phi^{j_3}_{j_1 j_2}$ as the projection of the highest weight vector of the quantum group module $V_q^{j_3}$ onto $V_q^{j_1} \otimes V_q^{j_2}$. In this notation, $V_q^{j_i}$ are the representation spaces of the quantum group, whose highest weight vectors are identified with the chiral algebra primary fields.

Intrinsic in the contour picture is the proposal to interpret the chiral vertex operators as intertwiners for the quantum group. They should not be confused with the intertwiners for the chiral algebra!

11.6 The quantum group of minimal models

We have seen how the quantum group of a (rational) conformal field theory emerges from its Coulomb gas version. Now, we only quote the results for the minimal models with central charge less than one; they follow directly from the Coulomb gas for minimal models discussed in chapter 10 and the techniques of this chapter.

Since minimal models have two screening operators, S_+ and S_-, their quantum group Q has six generators, $E_\pm$, $F_\pm$ and $K_\pm$, subject to the relations (11.70) with $\Omega_{\pm\pm} = 2\alpha_\pm^2$ and $\Omega_{-+} = \Omega_{+-} = -2$ [see (11.41)]. Each set of generators, $\{E_+, F_+, K_+\}$ and $\{E_-, F_-, K_-\}$, generates a Hopf subalgebra, respectively $U'_{q_+}(s\ell(2))$ and $U'_{q_-}(s\ell(2))$, with $q_\pm = \exp i\alpha_\pm^2$. It is easy to show that $U'_{q_+}(s\ell(2))$ and $U'_{q_-}(s\ell(2))$ commute since, for example,

$$\begin{aligned} K_+F_- &= \mathrm{e}^{-2\pi i\alpha_+ \oint \frac{\partial\phi}{4\pi}} \int_C S_-(t)dt \\ &= \mathrm{e}^{-2\pi i\alpha_+\alpha_-} \int_C S_-(t)dt\, \mathrm{e}^{-2\pi i\alpha_+ \oint \frac{\partial\phi}{4\pi}} = F_-K_+ \end{aligned} \tag{11.145}$$

Note that we have used the primed notation $U'_q(s\ell(2))$, introduced in section 2.5, to distinguish the quantum group whose Cartan subalgebra is generated by K from the quantum group $U_q(s\ell(2))$ whose Cartan part is generated by H, with $K = q^H$. Taking the "logarithm" of K_+, we see from (11.145) that it does not commute with F_-. This means that the quantum group Q of the Virasoro minimal models is not the tensor product $U'_{q_+}(s\ell(2)) \otimes U'_{q_-}(s\ell(2))$. The solution to this perplexing conclusion is that the Cartan generators of Q are related to H_+ and H_- ($[H_+, H_-] = 0$) in a slightly more elaborate fashion than we have naively guessed:

$$K_+ = q_+^{H_+}\,\mathrm{e}^{-i\pi H_-}\,, \qquad K_- = q_-^{H_-}\,\mathrm{e}^{-i\pi H_+} \tag{11.146}$$

This change of variables implies that the generators of $U_{q_+}(s\ell(2))$ and $U_{q_-}(s\ell(2))$ (i.e. $E_\pm$, $F_\pm$ and $H_\pm$) mix through the co-product, as in

$$\Delta_Q F_+ = F_+ \otimes \mathbf{1} + q_+^{-H_+}\,\mathrm{e}^{i\pi H_-} \otimes F_+ \tag{11.147}$$

The appearance of the factor $\mathrm{e}^{i\pi H_-}$ can be viewed as coming from a "Drinfeld twist" (section 6.11) of the original co-product of $U_{q_+}(s\ell(2)) \otimes U_{q_-}(s\ell(2))$, namely

$$\Delta_Q F_+ = T\,\Delta F_+\,T^{-1} = T\left(F_+ \otimes \mathbf{1} + q_+^{-H_+} \otimes F_+\right)T^{-1} \tag{11.148}$$

with

$$T = \exp\left[-i\frac{\pi}{2} H_- \otimes H_+\right] \tag{11.149}$$

In agreement with the above considerations, and recalling that the universal $\mathcal{R}$-matrix of $U_q = U_q(s\ell(2))$ belongs to $U_q \otimes U_q$ but not to $U'_q \otimes U'_q$ because of the factor $q^{H\otimes H}$, the $\mathcal{R}$-matrix of the quantum group Q of minimal models can be written as

$$\mathcal{R}_Q = \mathcal{R}_{+-}\mathcal{R}_{++}(q_+)\mathcal{R}_{--}(q_-)\mathcal{R}_{-+} \tag{11.150}$$

where $R_{\pm\pm}(q_\pm)$ is the usual $\mathcal{R}$-matrix for $U_{q_\pm}(s\ell(2))$, and

$$\mathcal{R}_{\pm\mp} = \exp\left[-i\frac{\pi}{2} H_\pm \otimes H_\mp\right] \tag{11.151}$$

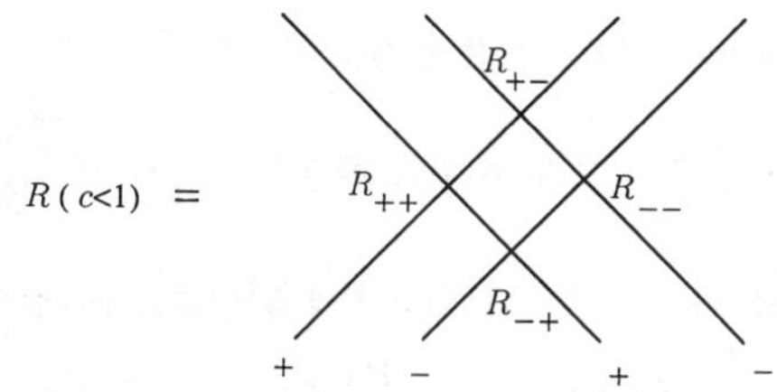

Fig. 11.20. The $\mathscr{R}$-matrix for the quantum group Q contains two dynamical factors and two twist contributions.

This $\mathscr{R}$-matrix is shown in figure 11.20. In conclusion, the quantum group of the minimal models of Belyavin, Polyakov and Zamolodchikov is $Q = U_{q_+}(s\ell(2))\widehat{\otimes}U_{q_-}(s\ell(2))$, where $\widehat{\otimes}$ denotes the twist by the element T.

Note that the pictorial representation in figure 11.20 assumes that we classify our states according to the untwisted quantum group. The $U_{q_+}(s\ell(2)) \otimes U_{q_-}(s\ell(2))$ labels for a primary field whose charge is $\alpha_{m,m'}$ are given by (j^+, j^-), where $m = 2j^+ + 1$ and $m' = 2j^- + 1$, so that $\alpha_{m,m'} = -j^+\alpha_+ - j^-\alpha_-$. For a minimal (p, p') model, the quantum deformation parameters $q_\pm$ are roots of unity, $q_+ = e^{i\pi p'/p}$, $q_- = e^{i\pi p/p'}$. With the help of standard quantum group techniques, it is easy to prove the truncation in the spectrum and fusion rules of rational conformal field theories.

The reader will find in the exercises the sketch of the construction of finite quantum groups associated with various conformal field theories whose free field realization is known. We end this chapter with the conjecture that the difficult problem of classifying all rational conformal field theories is equivalent to the classification of all finite-dimensional quasi-triangular Hopf algebras. More concretely, all the solutions of the polynomial equations which characterize the data of a rational conformal field theory should be given by the representation theory of finite quantum groups.

Exercises

Ex. 11.1 To prove Verlinde's theorem (9.100) or (11.15), consider the link of figure 11.21. Its braid group representation yields the following invariant:

$$W_{ijk} = \mathrm{tr}_{i\otimes j\otimes k}\left[\sigma\left(\mathscr{R}_{12}\right)\,\sigma\left(\mathscr{R}_{13}\right)\,\mathscr{R}_{13}\,\mathscr{R}_{12}\,q^{-H}\otimes q^{-H}\otimes q^{-H}\right] \tag{11.152}$$

The strategy is to compute W_{ijk} in two ways. First, use the quasi-triangularity of $\mathscr{R}$, namely

$$\begin{aligned}(\mathbf{1}\otimes\Delta)\,\mathscr{R} &= \mathscr{R}_{13}\,\mathscr{R}_{12}\\ (\mathbf{1}\otimes\Delta)\,\sigma(\mathscr{R}) &= \sigma(\mathscr{R})_{12}\,\sigma(\mathscr{R})_{13}\end{aligned} \tag{11.153}$$

to obtain

$$W_{ijk} = \mathrm{tr}_{i\otimes j\otimes k}\,(\mathbf{1}\otimes\Delta)\left(\sigma(\mathscr{R})\mathscr{R}\,q^{-H}\otimes q^{-H}\right) \tag{11.154}$$

Secondly, using (11.18) and (11.25), find that

$$W_{ijk} = \sum_m N_{jk}^{m} \mathrm{tr}_{i\otimes m}\sigma(\mathcal{R})\mathcal{R}\, q^{-H} \otimes q^{-H} = \sum_m N_{jk}^{m} \frac{S_{mi}}{S_{11}} \tag{11.155}$$

On the other hand, the trace (11.152) has the Markov property (D5.7), so that

$$\begin{aligned} W_{ijk} &= \frac{\mathrm{tr}_{i\otimes j}\left(\sigma(\mathcal{R})\mathcal{R}\, q^{-H} \otimes q^{-H}\right)\ \mathrm{tr}_{i\otimes k}\left(\sigma(\mathcal{R})\mathcal{R}\, q^{-H} \otimes q^{-H}\right)}{S_{1i}/S_{11}} \\ &= \frac{S_{ij}S_{jk}}{S_{11}S_{1i}} \end{aligned} \tag{11.156}$$

From this equation and (11.155), deduce

$$\frac{S_{ij}S_{jk}}{S_{1i}} = \sum_m N_{jk}^{m}\, S_{mi} \tag{11.157}$$

To complete the proof of Verlinde's theorem, you still need to use the fact that $S^2 = C$.

Ex. 11.2 Using equation (11.16), show that the modular group equation $(ST)^3 = C$ is equivalent to

$$\mathrm{e}^{2\pi i \Delta_j} S_1^j = \mathrm{e}^{i\pi c/4} \sum_k \mathrm{e}^{2\pi i \Delta_k} S_1^k S_k^j \tag{11.158}$$

Next, try to give a quantum group proof of this relation, relying on the completeness of the set of regular (standard, non-generic) irreducible representations of the quantum group at roots of unity.

Ex. 11.3 The contours in the screened vertex operators are a multi-contour version of the contour appearing in Hankel's formula of the Γ function:

$$\Gamma(z) = \frac{1}{-2i\sin(\pi z)} \int_C dt\, (-t)^{z-1}\, \mathrm{e}^{-t} = \int_0^{\infty} dt\, t^{z-1}\, \mathrm{e}^{-t} \tag{11.159}$$

where C surrounds the positive real axis and $z \notin \mathbb{Z}$. Comparing (11.159) with equation (11.32), we observe that the integral $\int_C dt\, (-t)^{z-1}\, \mathrm{e}^{-t}$ is the analog of

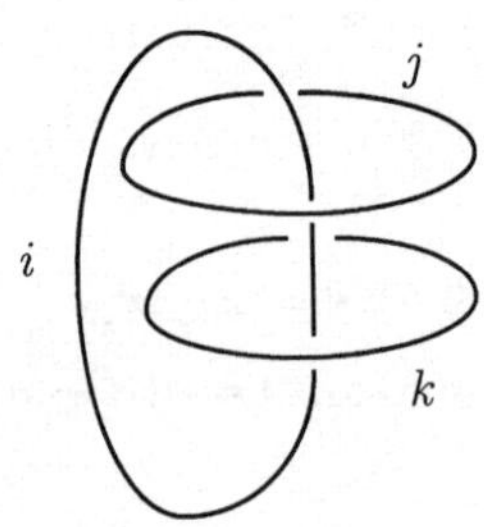

Fig. 11.21 Link used to prove Verlinde's theorem (11.15).

e_I^a while $\Gamma(z)$ is the analog of $\tilde{e}_I^a$. In strict analogy, the function $\sin \pi z$ in (11.159) reflects the existence of a cut. Find a contour representation for the function

$$B(x, y) = \frac{\Gamma(x)\Gamma(y)}{\Gamma(x + y)} \tag{11.160}$$

and rewrite the integrand as the operator product expansion of exponentials of a free field: this is the Veneziano amplitude for scattering of strings.

Ex. 11.4 In the derivation of the $\mathcal{R}$-matrix (11.128), the possible extra terms in $e_r^j(z)$ coming from integrals around the small circles near the point z were neglected. Include these terms to show that all the terms combine to yield the correct R-matrix for the braiding of screened vertex operators e_r^j.

Ex. 11.5 The Clebsch–Gordan coefficients for the regular representations of $U_q(s\ell(2))$ are obtained by inverting the (normalized) matrix K in (11.132). Let us sketch their computation in the simple case $\frac{1}{2} \otimes \frac{1}{2} = 0 \oplus 1$. In briefer notation,

$$\begin{aligned}
\Phi_0^0 &= V_1 \otimes V_2 = |+\rangle\,|+\rangle \\
\Phi_1^0 &= \int_P S\; V_1 \otimes V_2 = \left(1 - q^{-2}\right)\left(|+\rangle\,|-\rangle - q^{-1}\,|-\rangle\,|+\rangle\right) \\
\Phi_0^1 &= \int_C S\; V_1 \otimes V_2 = |-\rangle\,|+\rangle + q^{-1}\,|+\rangle\,|-\rangle \\
\Phi_0^2 &= \left(\int_C S\right)^2 V_1 \otimes V_2 = q^{-1}\,|-\rangle\,|-\rangle - q^{-1}q\,|-\rangle\,|-\rangle \\
\Phi_1^1 &= \int_C S \int_P S\; V_1 \otimes V_2 = \left(1 - q^{-2}\right)\left(|-\rangle\,|-\rangle - q^{-1}q\,|-\rangle\,|-\rangle\right) = 0
\end{aligned} \tag{11.161}$$

Clearly, $\Phi_0^3 = 0$ also. After normalizing the states on the right, we find

$$\begin{pmatrix} \Phi_0^0 = |11\rangle \\ \Phi_0^1 = |10\rangle \\ \Phi_1^0 = |00\rangle \\ \Phi_0^2 = |1-1\rangle \end{pmatrix} = \begin{pmatrix} 1 & 0 & 0 & 0 \\ 0 & \frac{q^{-1}}{\sqrt{1+q^{-2}}} & \frac{1}{\sqrt{1+q^{-2}}} & 0 \\ 0 & \frac{1}{\sqrt{1+q^{-2}}} & \frac{-q^{-1}}{\sqrt{1+q^{-2}}} & 0 \\ 0 & 0 & 0 & 1 \end{pmatrix} \begin{pmatrix} |+\rangle\,|+\rangle \\ |+\rangle\,|-\rangle \\ |-\rangle\,|+\rangle \\ |-\rangle\,|-\rangle \end{pmatrix} \tag{11.162}$$

- Invert this matrix and check that it yields the correct quantum Clebsch–Gordan coefficients.
- Compute the quantum Clebsch–Gordan coefficients for $\frac{1}{2} \otimes 1 = \frac{1}{2} \oplus \frac{3}{2}$ and check that they have the correct classical limit.

Ex. 11.6 Show that the chiral vertex operators for the chiral $\widehat{SU(2)}_k$ Wess–Zumino model, defined in terms of the contour of figure 11.19, can be expressed as

$$\Phi^{j_3}_{j_1,j_2}(z_1,z_2)\left(V_{j_1}\otimes V_{j_2}\right)=\prod_{x=0}^{r-1}\left(1-q^{-4j_1+2x}\right)\left(1-q^{-4j_2+2x}\right)$$
$$\times\int_{z_1}^{z_2}\cdots\int_{z_1}^{z_2}dt_1\cdots dt_r V_{j_1}(z_1)S(t_1)\cdots S(t_r)V_{j_2}(z_2) \tag{11.163}$$

Next, using the opening rule (R3) (figure 11.13), show that the quantum Clebsch–Gordan coefficients are given by

$$K^{j_1\ j_2\ 0}_{r_1\ r_2\ j_3}(q)=(-1)^{r_1}\left[\left[\begin{matrix} r\\ r_1\end{matrix}\right]\right]_q q^{r_1(r_1-1)}q^{-2j_1r_1}$$
$$\times\prod_{x=r_1}^{r-1}\left(1-q^{-4j_1+2x}\right)\prod_{y=r_2}^{r-1}\left(1-q^{-4j_2+2y}\right) \tag{11.164}$$

with the restriction $r=r_1+r_2$.

Ex. 11.7 Write out in full detail the relationships between the generators of the quantum group Q of minimal models, as well as their co-product, antipode and co-unit. Taking into account equation (11.150), write explicitly the $\mathcal{R}$-matrix for Q in terms of its generators.

Ex. 11.8 The conformal field theory realization of $U_q(\mathcal{G})$ is very similar to the case of $U_q(s\ell(2))$ discussed in detail in section 11.5, with an important difference: it is necessary to verify the Serre relations in the Chevalley basis. This requires a recurrence relation which can be sketched as follows. The light-house operators are given by the time ordered multiple integral

$$\underline{i_n\ldots i_1}=\mathrm{T}\left[\left(\prod_{i=1}^{n}\int_{i\infty}^{z}dt_i\right)S_{i_n}(t_n)\cdots S_{i_1}(t_1)V_{\vec{\lambda}}(z)\right] \tag{11.165}$$

where S_i are the screening operators of the free field representation of $\widehat{\mathcal{G}}_k$. In a somewhat abusive but transparent notation,

$$F_iV_{\vec{\lambda}}(z)=\left(1-q^{-2\lambda_i}\right)\underline{i} \tag{11.166}$$
$$F_iF_jV_{\vec{\lambda}}(z)=\left(1-q^{-2\lambda_j}\right)\left[\left(1-q^{-2\lambda_i+2\alpha_i\cdot\alpha_j}\right)\underline{ij}+q^{2\alpha_i\cdot\alpha_j}\left(1-q^{-2\lambda_i}\right)\underline{ji}\right]$$
$$=\left(1-q^{-2\lambda_i+2\alpha_i\cdot\alpha_j}\right)\underline{iF_jV_{\vec{\lambda}}}+q^{\alpha_i\cdot\alpha_j}\left(1-q^{-2\lambda_j}\right)\underline{jF_iV_{\vec{\lambda}}}$$

In general, the recurrence relation reads

$$F_{i_n}\cdots F_{i_1}V_{\vec{\lambda}}(z)=\sum_{\ell=1}^{n}q^{\sum_{m=\ell+1}^{n}\alpha_{i_\ell}\cdot\alpha_{i_m}}\left(1-q^{-2\lambda_{i_\ell}+2\sum_{p=1}^{\ell-1}\alpha_{i_\ell}\cdot\alpha_{i_p}}\right)$$
$$\times\underline{i_\ell F_{i_n}\cdots F_{i_{\ell+1}}F_{i_{\ell-1}}\cdots F_{i_1}V_{\vec{\lambda}}} \tag{11.167}$$

from which the Serre relations follow.

- In the contour picture, prove the following q-Serre relation of the quantum group $U_q(s\ell(3))$ hidden in the Wess–Zumino model $\widehat{SU(3)}_k$ ($q = \exp \pi i/(k+3)$):

$$F_1^2 F_2 - [|2|]_q F_1 F_2 F_1 + F_2 F_1^2 = 0 \tag{11.168}$$

- In the contour picture, prove the following q-Serre relation of the quantum group $U_q(G_2)$ hidden in the Wess–Zumino model $\widehat{G_2}_k$ ($q = \exp \pi i/(k+4)$):

$$F_1^4 F_2 - [|4|]_q F_1^3 F_2 F_1 + [|6|]_q F_1^2 F_2 F_1^2 + [|4|]_q F_1 F_2 F_1^3 - F_1 F_2^4 = 0 \tag{11.169}$$

Ex. 11.9 [R. Cuerno, *Phys. Lett.* **271B** (1991) 314.]

A straightforward generalization of the Virasoro algebra is provided by the so-called W-algebras. The free field representation of a Kac–Moody algebra $\hat{\mathscr{G}}_k$ (exercise 10.7) uses a set of free scalar fields $\vec{\phi}$ and another one of spin-1 β–γ systems. Consider the conformal field theory whose free field energy–momentum tensor consists of the scalar part only, $T(z) = -\partial\vec{\phi}(z) \cdot \partial\vec{\phi}(z) - (i/v)\vec{\rho} \cdot \partial^2\vec{\phi}(z)$, where $v^2 = k + h^*$ and $2\vec{\rho} = \sum_{\vec{\alpha} \in \Delta_+} \vec{\alpha}$. If the Kac–Moody algebra is $\widehat{SU(2)}_k$, this is just the free field representation of Virasoro. In the general case, primary fields are exponentials $V_{\vec{\lambda}}(z) = \mathrm{e}^{i\vec{\lambda}\cdot\vec{\phi}}$, and there are two screening operators $S_\pm(z) = V_{\vec{\alpha}_\pm}(z)$. Show that the quantum group Q hidden in the conformal field theory with chiral algebra $W(\mathscr{G})$ is $Q = U_{q_+}(\mathscr{G}) \hat{\otimes} U_{q_-}(\mathscr{G})$, where $q_\pm = \exp(i\pi\vec{\alpha}_\pm^2)$ and $\hat{\otimes}$ indicates the presence of a Drinfeld twist.

Ex. 11.10 [F. Jiménez, *Phys. Lett.* **252B** (1990) 577; **273B** (1991) 399.]

Show that the quantum group constructed from the free field realization of the $N = 1$ and $N = 2$ super-conformal field theories (see appendix G) is given by the twisted tensor products

$$U_{q_+}(osp(2,1)) \hat{\otimes} U_{q_-}(osp(2,1))\,, \qquad U_{q_+}(s\ell(2)) \hat{\otimes} U_{q_-}(osp(2,2))$$

respectively, where, for $N = 1$, the central charge is $c = \frac{3}{2}(1 - 16\alpha_0^2$ and $q_\pm = \exp i\pi\alpha_\pm^2$ ($\alpha_\pm = \alpha_0 \pm \sqrt{\alpha_0^2 + \frac{1}{2}}$), whereas, for $N = 2$, $c = 3(1 - 8\beta^2)$ and $q_+ = \exp 4\pi i\beta^2$, $q_- = \exp i\pi/(4\beta^2)$.

Ex. 11.11 Super-symmetry and the free fermion R-matrix. Find the trigonometric R-matrix for $U_q(\widehat{s\ell(2)})$ when $q^4 = 1$, and check that it is a special case of the six-vertex model, namely that it satisfies the free fermion condition

$$\left(R_{00}^{00}(u)\right)^2 + \left(R_{10}^{01}(u)\right)^2 = \left(R_{10}^{10}(u)\right)^2 \tag{11.170}$$

How are the free fermion condition and $N = 2$ super-symmetry related?

Ex. 11.12 In section 6.10, we introduced the affine quantum group $U_q(\widehat{\mathcal{G}})$ with the help of the Cartan matrix, as in (6.109). Obtain the following representation of $U_q(A_1^{(1)})$:

$$\begin{aligned}
&[T, Q_\pm] = \pm 2Q_\pm \\
&[T, \overline{Q}_\pm] = \pm 2\overline{Q}_\pm \\
&Q_\pm \overline{Q}_\pm - q^2 \overline{Q}_\pm Q_\pm = 0 \\
&Q_\pm \overline{Q}_\mp - q^{-2} \overline{Q}_\mp Q_\pm = \frac{1 - q^{\pm 2T}}{1 - q^2}
\end{aligned} \tag{11.171}$$

with the help of the isomorphism

$$\begin{aligned}
Q_+ &= E_1 q^{H_1/2} & Q_- &= E_0 q^{H_0/2} \\
\overline{Q}_+ &= F_0 q^{H_0/2} & \overline{Q}_- &= F_1 q^{H_1/2} \\
T &= H_1 & c &= H_0 + T
\end{aligned} \tag{11.172}$$

Recall that the central extension is $c = H_0 + H_1$, and consider the simplest case $c = 0$. Derive the Serre relations for the generators (11.171).

Ex. 11.13 Consider the sine-Gordon lagrangian of exercise 3.3. Define the following conserved currents:

$$\begin{aligned}
J_{\pm,z}(z, \overline{z}) &= \exp\left(\pm \frac{2i}{\beta} \phi\right) \\
J_{\pm,\overline{z}}(z, \overline{z}) &= \lambda \gamma \exp\left(\pm i \frac{2 - \beta^2}{\beta} \phi \mp i\beta \overline{\phi}\right) \\
\overline{J}_{\pm,z}(z, \overline{z}) &= \exp\left(\mp \frac{2i}{\beta} \phi\right) \\
\overline{J}_{\pm,\overline{z}}(z, \overline{z}) &= \lambda \gamma \exp\left(\mp i \frac{2 - \beta^2}{\beta} \phi \pm i\beta \overline{\phi}\right)
\end{aligned} \tag{11.173}$$

with $\gamma = \beta^2/(2 - \beta^2)$,

$$\begin{aligned}
\phi(x, t) &= \frac{1}{2}\left(\Phi(x, t) + \int_{-\infty}^{x} dy \partial_t \Phi(y, t)\right) \\
\overline{\phi}(x, t) &= \frac{1}{2}\left(\Phi(x, t) - \int_{-\infty}^{x} dy \partial_t \Phi(y, t)\right)
\end{aligned} \tag{11.174}$$

and Φ the sine-Gordon field.

Show that the associated charges $Q_\pm$, $\overline{Q}_\pm$ satisfy the relations

$$\begin{aligned}
Q_+ \overline{Q}_- - q^2 \overline{Q}_- Q_+ &= \frac{\lambda}{2\pi i} \gamma^2 \int dx \partial_x X^* \\
\overline{Q}_+ Q_- - q^{-2} Q_- \overline{Q}_+ &= -\frac{\lambda}{2\pi i} \gamma^2 \int dx \partial_x X
\end{aligned} \tag{11.175}$$

with $q = \exp(-2\pi i/\beta^2)$. Note that the classical limit is obtained when $\beta \to 2$.

Ex. 11.14 Consider the Knizhnik–Zamolodchikov equation (10.94) for a four-point correlation function ($n = 4$). Let $z_4 = \infty$. Following techniques similar to those in section 10.1 for minimal models, show that the four-point correlator can be written as

$$\Psi(z_1, z_2, z_3, z_4 = \infty) = (z_3 - z_1)^{(\Omega_{12}+\Omega_{13}+\Omega_{23})/(k+h^*)} \, G\left(\frac{z_2 - z_1}{z_3 - z_1}\right) \tag{11.176}$$

where the Ω_{ij} are given by (10.96) and $G(x)$ satisfies the differential equation

$$G'(x) = \frac{1}{k+h^*}\left(\frac{\Omega_{12}}{x} + \frac{\Omega_{23}}{x-1}\right) G(x) \tag{11.177}$$

It follows from this equation that we may choose two kinds of conformal blocks according to their asymptotic behavior near $x = 0$ or $x = 1$, which correspond to the associations (12)3 or 1(23):

$$G_{(12)3}(x) \underset{x\to 0}{\to} x^{\Omega_{12}/(k+h^*)}, \qquad G_{1(23)}(x) \underset{x\to 1}{\to} (1-x)^{\Omega_{23}/(k+h^*)} \tag{11.178}$$

Show that the fusion matrix ϕ defined by $G_{(12)3}(x) = G_{1(23)}(x)\phi$ satisfies all the axioms for quasi-triangular quasi-Hopf algebras. The R-matrix is just $R_{12} = \exp\left[i\pi\Omega_{12}/(k+h^*)\right] = q^{\Omega_{12}}$. Show that there exists a twist transformation (section 6.11) which takes this quasi-Hopf algebra to the Hopf algebra $U_q(\mathcal{G})$. Using the Coulomb gas formulation of Wess–Zumino models (section 10.4.2) and exercise 10.7, find the solution to the Knizhnik–Zamolodchikov equation for $\widehat{SU(2)}_k$ with four spin-$\frac{1}{2}$ irreps on the external legs. What is the contour representation of the conformal blocks found by Drinfeld? Using contour deformation techniques, compute the fusion matrix ϕ in the representation $V_{\frac{1}{2}} \otimes V_{\frac{1}{2}} \otimes V_{\frac{1}{2}}$.

Ex. 11.15 Consider a conformal field theory defined as the coset space $\widehat{G}/\widehat{H}$, where $\widehat{H} \subset \widehat{G}$ are Kac–Moody algebras. What can you say about the quantum group associated with this conformal field theory? In section 11.6, we found that the quantum group of the unitary minimal models is the twisted product of two $U_q(s\ell(2))$ algebras. Does this generalize?

Appendix G
Super-conformal field theories

In this appendix we will briefly describe some general aspects of super-conformal field theories. In general, super-symmetries are symmetries whose infinitesimal parameters are fermionic, and thus they mix bosonic and fermionic degrees of freedom. Spacetime super-symmetry is motivated as the square root of the Poincaré algebra, so that, in addition to the bosonic momentum operators P^μ, we have some fermionic generators (the super-symmetry charges) Q_α such that $\{Q_\alpha, Q_\beta\} = (\gamma_\mu)_{\alpha\beta}P^\mu$. Perhaps the most important motivation for the study of super-symmetric conformal field theories comes from string theory, about which we shall say nothing.

G11.1 Super-conformal transformations

Let $\widehat{\mathbb{C}}$ denote the super-complex plane with holomorphic co-ordinates $Z = (z, \theta)$ such that $\theta^2 = 0$. Since the bosonic co-ordinate z is accompanied by one fermionic co-ordinate θ, we are dealing with $N = 1$ super-symmetry. The standard derivations $\partial = \partial_z$ and $\overline{\partial} = \partial_{\overline{z}}$ on $\mathbb{C}$ are replaced by

$$D = \partial_\theta + \theta\partial_z \,, \qquad \overline{D} = \partial_{\overline{\theta}} + \overline{\theta}\partial_{\overline{z}} \tag{G11.1}$$

which can be thought of as a square root of the normal derivatives, since $D^2 = \partial$ and $\overline{D}^2 = \overline{\partial}$.

Super-holomorphic functions are characterized by the super-analyticity condition $\overline{D}F(Z) = 0$. Cauchy's residue theorem becomes

$$\frac{1}{2\pi i}\oint dZ_1 \frac{F(Z_1)\theta_{12}}{Z_{12}^{n+1}} = \frac{1}{n!}D^n F(Z) \tag{G11.2}$$

with

$$\theta_{12} = \theta_1 - \theta_2 \,, \qquad Z_{12} = z_1 - z_2 - \theta_1\theta_2 \,, \qquad dZ = dz d\theta \tag{G11.3}$$

Super-conformal transformations are generated by super-vector fields

$$v = \delta z + \theta\,\delta\theta \tag{G11.4}$$

The generator of super-conformal transformations is the super-energy–momentum tensor $\mathcal{T}(z)$. A super-conformal tensor of weights $(\Delta, \overline{\Delta})$ is characterized by the transformation law

$$\Phi(Z, \overline{Z}) \to (D\theta')^{\Delta}\left(\overline{D}\overline{\theta}'\right)^{\overline{\Delta}}\Phi(Z', \overline{Z'}) \tag{G11.5}$$

Infinitesimally, in terms of the vector field (G11.4), equation (G11.5) becomes

$$\delta\Phi(Z,\overline{Z}) = \left[v\cdot\partial + \frac{1}{2}(Dv)D + \Delta(\partial v)\right]\Phi(Z,\overline{Z}) \tag{G11.6}$$

Using (G11.6), we may associate to this transformation law the following operator product expansion:

$$\mathscr{T}(Z_1)\Phi(Z_2) = \Delta\frac{\theta_{12}}{Z_{12}^2}\Phi(Z_2) + \frac{1}{2}\frac{D_2\Phi(Z_2)}{Z_{12}} + \frac{\theta_{12}\partial_2\Phi(Z_2)}{Z_{12}} + \text{regular terms} \tag{G11.7}$$

The super-energy–momentum tensor transforms anomalously

$$\mathscr{T}(Z) = (D\theta')^3\mathscr{T}(Z') + \frac{\hat{c}}{4}S(Z,Z') \tag{G11.8}$$

due to the super-schwartzian derivative

$$S(Z,Z') = \frac{D^4\theta'}{D\theta'} - 2\frac{D^3\theta'}{D\theta'}\frac{D^2\theta'}{D\theta'} \tag{G11.9}$$

The operator product expansion version of (G11.8) is

$$\mathscr{T}(Z_1)\mathscr{T}(Z_2) = \frac{\hat{c}}{4}\frac{1}{Z_{12}^3} + \frac{3}{2}\frac{\theta_{12}}{Z_{12}^2}\mathscr{T}(Z_2) + \frac{1}{2}\frac{D_2\mathscr{T}(Z_2)}{Z_{12}} + \frac{\theta_{12}\partial_2\mathscr{T}(Z_2)}{Z_{12}} + \cdots \tag{G11.10}$$

This operator product expansion will be less compact if expressed in the components of $\mathscr{T}(z)$, but it will be more enlightening to do so. Since $\theta^2 = 0$, the Taylor expansion of the super-energy–momentum tensor

$$\mathscr{T}(z) = G(z) + \theta T(z) \tag{G11.11}$$

yields the familiar energy–momentum tensor $T(z)$ and the super-current $G(z)$. Equation (G11.10) encodes the following three operator product expansions:

$$\begin{aligned}
T(z)T(\zeta) &= \frac{c}{2}\frac{1}{(z-\zeta)^4} + \frac{2T(\zeta)}{(z-\zeta)^2} + \frac{\partial_\zeta T(\zeta)}{z-\zeta} + \cdots \\
T(z)G(\zeta) &= \frac{3}{2}\frac{G(\zeta)}{(z-\zeta)^2} + \frac{\partial_\zeta G(\zeta)}{z-\zeta} + \cdots \\
G(z)G(\zeta) &= \frac{2c}{3}\frac{1}{(z-\zeta)^3} + \frac{2T(\zeta)}{z-\zeta} + \cdots
\end{aligned} \tag{G11.12}$$

where the relationship between the super-symmetric central charge $\hat{c}$ and the usual bosonic central charge c is

$$c = \frac{3}{2}\hat{c} \tag{G11.13}$$

The identity **1**, the energy–momentum tensor $T(z)$ and the super-symmetry current $G(z)$ generate the $N = 1$ super-conformal chiral algebra. We may define their Laurent expansions:

$$\begin{aligned} T(z) &= \sum L_n z^{-n-2} \\ G(z) &= \sum G_n z^{-n-\frac{3}{2}} \end{aligned} \tag{G11.14}$$

In terms of the modes, (G11.12) becomes the super-Virasoro algebra:

$$\begin{aligned} [L_m, L_n] &= (m-n)L_{m+n} + \frac{c}{12}(m^3 - m)\delta_{m+n,0} \\ [L_m, G_n] &= \left(\frac{m}{2} - n\right) G_{m+n} \\ \{G_m, G_n\} &= 2L_{m+n} + \frac{c}{3}\left(m^2 - \frac{1}{4}\right)\delta_{m+n,0} \end{aligned} \tag{G11.15}$$

Now, the generator of translations on the complex plane is $L_{-1} = \partial_z$, so, in order to preserve super-symmetry, we must be able to set $m = n = -\frac{1}{2}$ in the last equation of (G11.15) to obtain

$$\left\{G_{-\frac{1}{2}}, G_{-\frac{1}{2}}\right\} = 2L_{-1} \tag{G11.16}$$

Thus, we conclude that, whereas the usual Virasoro modes L_n are integer-moded ($n \in \mathbb{Z}$), the super-current modes must be half-odd-moded ($n \in \mathbb{Z} + \frac{1}{2}$). This requirement allows us to have super-symmetry on the complex plane; it also makes $G(z)$ single valued. The algebra (G11.15) with half-odd modes for the super-current is known as the Neveu–Schwarz super-algebra.

A somewhat different super-algebra arises if we work on the cylinder instead of on the plane. Using the conformal map $z = e^w$ from the complex z plane to the w cylinder, we find that the generator of translations in the w co-ordinate is not L_{-1}, but rather L_0 or, taking into account the Casimir energy as we of course should, $L_0 - c/24$. On the cylinder, super-symmetry makes sense, and $G(w)$ is single-valued provided the third equation in (G11.15) becomes

$$\{G_0, G_0\} = 2\left(L_0 - \frac{c}{24}\right) \tag{G11.17}$$

By consistency, the super-current modes are now integer. The super-algebra (G11.15) with all indices integer is the first super-symmetric algebra ever written down, by Ramond.

G11.2 Representations

Let us consider separately the Ramond (R) and Neveu–Schwarz (NS) super-algebras. We only summarize the results for the unitary representations and leave their proofs as exercises. With the help of these representations, $N = 1$ super-conformal minimal models can be built.

A Neveu–Schwarz highest weight vector $|\phi\rangle$ is characterized by

$$\begin{aligned} L_n |\phi\rangle &= G_n |\phi\rangle = 0 \qquad n > 0 \\ L_0 |\phi\rangle &= \Delta |\phi\rangle \end{aligned} \tag{G11.18}$$

Unitary representations are those with central extension

$$c = \frac{3}{2}\left(1 - \frac{8}{m(m+2)}\right) \tag{G11.19}$$

and conformal weights

$$\Delta_{p,q} = \frac{[(m+2)p - mq]^2 - 4}{8m(m+2)} \qquad \begin{cases} 1 \leq p \leq m \\ 1 \leq q \leq m+2 \end{cases} \tag{G11.20}$$

The central extension (G11.19) can be reproduced with the Goddard–Kent–Olive construction (section 10.4.3) for the coset G/H, with

$$G = SU(2)_k \otimes SU(2)_2 \,, \qquad H = SU(2)_{k+2} \tag{G11.21}$$

provided we identify $m = k + 2$.

A Ramond highest weight vector is subject not only to (G11.18), but the value of G_0 must also be specified. Using (G11.17), we obtain

$$\langle \phi | \, G_0^2 \, | \phi \rangle = \Delta - \frac{c}{24} \tag{G11.22}$$

and therefore $\Delta \geq c/24$ for unitary representations. Whereas (G11.19) remains valid, the unitary weights are not given by (G11.20) but instead by

$$\Delta_{p,q} = \frac{[(m+2)p - mq]^2 - 4}{8m(m+2)} + \frac{1}{16} \qquad \begin{cases} 1 \leq p \leq m \\ 1 \leq q \leq m+2 \end{cases} \tag{G11.23}$$

G11.3 $N = 2$ super-conformal algebras

The $N = 2$ (global) super-symmetry algebra contains two super-charges $Q^{\pm}$ instead of one, in addition to the momentum P. The freedom of an internal relabeling among these two charges gives rise to a $U(1)$ abelian charge J_0. The defining relations between these generators are

$$\begin{aligned} &[J_0, Q^{\pm}] = \pm Q^{\pm} \,, \qquad \{Q^+, Q^-\} = 2P \\ &[P, Q^{\pm}] = [P, J_0] = (Q^+)^2 = (Q^-)^2 = 0 \end{aligned} \tag{G11.24}$$

The $N = 2$ super-conformal algebra is generated by the bosonic energy–momentum tensor $T(z)$, the two fermionic super-currents $G^{\pm}(z)$, and a bosonic $U(1)$ current $J(z)$. Using the Laurent expansions

$$\begin{aligned} T(z) &= \sum L_n z^{-n-2} \\ J(z) &= \sum J_n z^{-n-1} \\ G^{\pm}(z) &= \sum G_n^{\pm} z^{-n-\frac{3}{2}} \end{aligned} \tag{G11.25}$$

the $N = 2$ super-conformal algebra is

$$\begin{aligned}
[L_m, L_n] &= (m-n)L_{m+n} + \frac{c}{12}(m^3 - m)\delta_{m+n,0} \\
\left[L_m, G_n^{\pm}\right] &= \left(\frac{m}{2} - n\right) G_{m+n}^{\pm} \\
[L_m, J_n] &= -nJ_{m+n} \\
\left[J_m, G_n^{\pm}\right] &= \pm G_{m+n}^{\pm} \\
[J_m, J_n] &= \frac{c}{3}\, m\, \delta_{m+n,0} \\
\left\{G_m^-, G_n^+\right\} &= 2L_{m+n} - (m-n)J_{m+n} + \frac{c}{3}\left(m^2 - \frac{1}{4}\right)\delta_{m+n,0}
\end{aligned} \tag{G11.26}$$

As in the $N = 1$ case above, we may use the algebra (G11.26) to define $N = 2$ super-symmetry on the plane (NS) or on the cylinder (R). The super-symmetry charges $Q^{\pm}$ are accordingly

$$Q^{\pm} = \begin{cases} G^{\pm}_{-\frac{1}{2}} & \text{NS} \\ G_0^{\pm} & \text{R} \end{cases} \tag{G11.27}$$

A most relevant feature of the $N = 2$ super-conformal algebra, common to all $N = 2$ super-symmetric theories, is the invariance of the algebra under the spectral flow transformations

$$U_\alpha \quad : \quad L_n \to L_n + \alpha J_n + \frac{c}{6}\,\alpha^2 \delta_{n,0} \tag{G11.28a}$$

$$J_n \to J_n + \frac{c}{3}\,\alpha\,\delta_{n,0} \tag{G11.28b}$$

$$G_n^{\pm} \to G_{n\pm\alpha}^{\pm} \tag{G11.28c}$$

$$G_n^{-} \to G_{n-\alpha}^{-} \tag{G11.28d}$$

The importance of this spectral flow transformation is that it allows us to interpolate smoothly between the NS and R algebras. In string theory, the generators of spectral flow on the world-sheet become super-symmetry generators in spacetime.

G11.4 $N = 2$ irreps and the chiral ring

A highest weight vector $|\phi\rangle$ of the $N = 2$ NS super-conformal algebra is characterized by the conditions

$$\begin{aligned}
&L_n\,|\phi\rangle = J_n\,|\phi\rangle = 0 \qquad n > 0 \\
&G_n^{\pm}\,|\phi\rangle = 0 \qquad n > 0 \\
&L_0\,|\phi\rangle = \Delta\,|\phi\rangle \\
&J_0\,|\phi\rangle = q\,|\phi\rangle \\
&G^{\pm}_{-\frac{1}{2}}\,|\phi\rangle = \left|\psi^{\pm}\right\rangle
\end{aligned} \tag{G11.29}$$

The labels of the irreducible representation are the conformal weight Δ and the charge q. The last equation just defines, for convenience, the super-symmetric partners $\left|\psi^{\pm}\right\rangle$ of the highest weight state $|\phi\rangle$.

Unitary irreps are those with

$$\begin{aligned} c &= 3 - \frac{6}{k+2} && k \in \mathbb{N} \\ \Delta_{\ell,m} &= \frac{\ell(\ell+2) - m^2}{4(k+2)} && \ell = 0, 1, 2, \ldots, k \\ q_m &= \frac{m}{k+2} && m = -\ell, -\ell+2, \ldots, \ell-2, \ell \end{aligned} \tag{G11.30}$$

Although the central charge is the same as for $SU(2)_k$ Wess–Zumino models, the conformal weights are different.

The irreps (G11.30) can be associated, by standard procedure, with primary fields $\Phi_{\ell,m}$ satisfying

$$\begin{aligned} T(z)\Phi_{\ell,m}(\zeta) &= \frac{\Delta_{\ell,m}}{(z-\zeta)^2}\,\Phi_{\ell,m}(\zeta) + \frac{1}{z-\zeta}\,\partial\Phi_{\ell,m}(\zeta) \\ J(z)\Phi_{\ell,m}(\zeta) &= \frac{q_m}{z-\zeta}\,\Phi_{\ell,m}(\zeta) \\ G^{\pm}(z)\Phi_{\ell,m}(\zeta) &= \frac{1}{z-\zeta}\,\Psi^{\pm}_{\ell,m}(\zeta) \end{aligned} \tag{G11.31}$$

Some special primary fields are called chiral or antichiral; they are annihilated by one of the super-symmetry charges:

$$\begin{aligned} G^{+}_{-\frac{1}{2}}\Phi(z) &= 0 \qquad \text{chiral} \\ G^{-}_{-\frac{1}{2}}\Phi(z) &= 0 \qquad \text{antichiral} \end{aligned} \tag{G11.32}$$

Table G11.1 Unitary minimal models of the $N = 0$, $N = 1$ and $N = 2$ super-conformal algebras.
The level k is a positive integer ($k = 0$ is still allowed in all cases as the trivial theory), whereas N counts the number of super-symmetric charges in the theory. The third row refers to the highest weight field content, and the last row indicates the free fields needed for a Feigin–Fuks construction.

	$N = 0$	$N = 1$	$N = 2$
c_k	$1 - \frac{6}{(k+2)(k+3)}$	$\frac{3}{2}\left(1 - \frac{8}{(k+2)(k+4)}\right)$	$3\left(1 - \frac{2}{k+2}\right)$
$\Delta_{m,m'}$	$\frac{[(k+2)m-(k+1)m']^2-1}{4(k+2)(k+3)}$	$\frac{[(k+4)m-(k+2)m']^2-4}{8(k+2)(k+4)}$	$\frac{m(m+2)-m'^2}{4(k+2)}$ $\left(q=\frac{m'}{k+2}\right)$
m, m'	$1 \le m \le k+1$ $1 \le m' \le k+2$	$1 \le m \le k+2$ $1 \le m' \le k+4$	$0 \le m \le k$ $-m \le m' \le m$
Fields	one scalar field	one real super-field	one complex super-field

If $|\Phi\rangle$ is a chiral primary field, then

$$\langle\Phi| \left\{G^{+}_{-\frac{1}{2}}, G^{-}_{-\frac{1}{2}}\right\} |\Phi\rangle = 0 \tag{G11.33}$$

provided of course that $\left(G^{\pm}_{-\frac{1}{2}}\right)^{\dagger} = G^{\mp}_{\frac{1}{2}}$. Using now (G11.26) and (G11.29), we obtain

$$0 = \langle\Phi| \left\{G^{+}_{-\frac{1}{2}}, G^{-}_{-\frac{1}{2}}\right\} |\Phi\rangle = \langle\Phi|\Phi\rangle\,(2\Delta_{\Phi} - q_{\Phi}) \tag{G11.34}$$

Therefore, the weight and charge of chiral primary fields are related:

$$\Delta_{\Phi} = \frac{q_{\Phi}}{2} \tag{G11.35}$$

Using the operator product expansion

$$\Phi_1(z)\Phi_2(\zeta) = (z-\zeta)^{\Delta_3-\Delta_2-\Delta_1}\Phi_3(\zeta) + \cdots \tag{G11.36}$$

we may equip the set of chiral primary fields with the structure of a ring. Indeed, when $(z-\zeta) \to 0$, the product $\Phi_1(z)\Phi_2(\zeta)$ of two chiral primary fields yields a field Φ_3 whose weight is such that $\Delta_3 - \Delta_2 - \Delta_1 = 0$. Since the $U(1)$ charge is conserved, it follows that Φ_3 is both primary and chiral. We shall denote by $\mathscr{R}$ the ring of chiral primary fields of an $N = 2$ super-conformal field theory. The above two paragraphs apply equally well to antichiral fields with $2\Delta_{\Phi} = -q_{\Phi}$; we shall use $\overline{\mathscr{R}}$ to denote the ring of antichiral fields.

The chiral primary fields of the $N = 2$ minimal models (G11.30) are obtained by setting $\ell = m$, $\mathscr{R}_k = \{\Phi_{\ell,\ell} = \Phi_{\ell}, \ell = 0, 1, 2, \ldots, k\}$, and the multiplication in $\mathscr{R}_k$ is given by

$$\lim_{z\to\zeta} \Phi_{\ell_1}(z)\,\Phi_{\ell_2}(\zeta) = \Phi_{\ell_1+\ell_2}(\zeta) \tag{G11.37}$$

with the understanding that $\Phi_{\ell}(z) = 0$ if $\ell > k$. In this simple example, the generators of the ring $\mathscr{R}_k$ are just $\mathbf{1}$ and $x = \Phi_{1,1}$. The chiral ring $\mathscr{R}_k$ can be characterized as the quotient of the ring $P[x]$ of all polynomials in one variable by a k-dependent ideal, namely

$$\mathscr{R}_k = P[x]/\left\{\partial W^{(k)}(x)\right\} \tag{G11.38}$$

where $W^{(k)}(x) = x^{k+2}$ is the super-potential.

Exploiting the spectral flow transformation (G11.28), it is easy to see that chiral primary fields are associated with Ramond ground states, i.e. states for which

$$\Delta = \frac{c}{24} \quad \Longleftrightarrow \quad G_0\,|\Phi\rangle = 0 \tag{G11.39}$$

Alternatively, the chiral ring $\mathscr{R}$ can be viewed as a "co-homology" of Q^+. In fact, since by (G11.24) the super-symmetric charge Q^+ is nilpotent, the definition (G11.32) of chiral primary field yields immediately that $\mathscr{R} = \text{Ker}\ Q^+/\text{Im}\ Q^+$. Interpreting Q^+ as a Becchi–Rouet–Stora (BRS) operator (a field theoretic exterior differential) would indicate that some other theory exists whose spectrum is saturated by the chiral primary fields of an $N = 2$ super-conformal field theory. We turn our attention to such a theory next.

G11.5 $N = 2$ topological theories

Following Eguchi and Yang, we use the $U(1)$ current $J(z)$ to twist the energy–momentum tensor:

$$T^t(z) = T(z) - \frac{1}{2}\partial J(z) \tag{G11.40}$$

Taking modes, we find

$$L_0^t = L_0 - \frac{1}{2}J_0 \tag{G11.41}$$

and therefore $\Delta^t = \Delta - \frac{1}{2}q$ It follows that chiral primary fields have vanishing twisted conformal weight Δ^t.

On the other hand, the super-currents $G^{\pm}(z)$ have (twisted) conformal weights 1 and 2, whereby G^+ can be integrated to yield a charge, and G^- acts like a reparametrization ghost field. With the help of (G11.26), we may write the twisted energy–momentum tensor in terms of the super-currents as

$$T^t(z) = \oint d\zeta \; G^+(\zeta)\, G^-(z) \tag{G11.42}$$

In more compact notation, this reads as

$$T^t(z) = \{Q, G^-(z)\} \tag{G11.43}$$

with

$$Q = \oint G^+(\zeta)\, d\zeta \tag{G11.44}$$

Equation (G11.43), and the fact that Q is nilpotent (i.e. $Q^2 = 0$), constitute the soft underbelly of two-dimensional topological theories. Viewing Q as a BRS charge, then (G11.43) tells us that $T^t(z)$ is a "total BRS derivative". In general, Witten denoted by "topological" precisely those theories in which the metric is a gauge artifact.

We have already collected the ingredients we need to make the topological field theory associated with the $N = 2$ super-conformal field theory. The recipe reads as follows:

(i) elevate one of the super-symmetric charges, say Q^+, to the selective club of BRS charges;
(ii) identify the Hilbert space of the topological theory with chiral ring $\mathscr{R}$ at hand, i.e. with the BRS cohomology;
(iii) use as energy–momentum of the theory Eguchi and Yang's twisted energy–momentum (G11.40), which is a pure BRS operator due to (G11.43).

Only in the context of such topological theories do chiral rings become truly meaningful. In fact, from (G11.41), we know that the conformal weights of all chiral fields are zero and therefore that there is no co-ordinate dependence in the operator product expansions. The chiral ring is thus defined by a set of fields Φ_i with the multiplication rule given simply by

$$\Phi_i \Phi_j = c_{ij}^k \Phi_k \tag{G11.45}$$

For the topological theory arising from the minimal models of the previous section, if we write $\Phi_n = \Phi_{n,n}$, then we find $c_{nm}^{\ell} = \delta_{\ell,n+m}$

G11.6 Perturbed chiral ring

Let us sketch the perturbation of a topological field theory. A topological field theory associated with an $N = 2$ super-conformal field theory is defined by its chiral ring and the twisted energy–momentum tensor (G11.40). Relative to this energy–momentum tensor, chiral primary fields have conformal weight zero, and, therefore, all their correlation functions are independent of the co-ordinates:

$$\langle \Phi_{i_1}(z_1) \cdots \Phi_{i_n}(z_n) \rangle = F_{i_1 \cdots i_n} \neq F_{i_1 \cdots i_n}(z_1, \ldots, z_n) \tag{G11.46}$$

The $U(1)$ current $J(z)$ is anomalous with respect to the twisted energy–momentum:

$$\left[L_n^{(t)}, J_m \right] = -mJ_{m+n} - \frac{c}{6} n(n+1) \delta_{n+m,0} \tag{G11.47}$$

Hence, the $U(1)$ zero mode acts on the in- and out-vacua as follows:

$$J_0 \left|0\right\rangle = 0 \,, \quad \left\langle 0\right| J_0^{\dagger} = \left\langle 0\right| \frac{c}{3} \tag{G11.48}$$

Therefore, in order to have a non-vanishing correlation function (G11.46) on the sphere, we require the selection rule

$$\sum_{i=1}^{n} q_i = \frac{c}{3} \tag{G11.49}$$

This immediately yields the two-point function for chiral primary fields on the sphere. Quite generally, the two-point function $\langle \Phi_i \Phi_j \rangle \equiv \eta_{ij}$ can be interpreted as the metric on the chiral ring $\mathscr{R}$.

In the case of the $N = 2$ minimal models, the $U(1)$ charges of the chiral primary fields $\Phi_\ell = \Phi_{\ell,\ell}$ are

$$q_\ell = \frac{\ell}{k+2} \tag{G11.50}$$

and thus we obtain

$$\langle \Phi_i \Phi_j \rangle = \delta_{i+j,k} \tag{G11.51}$$

This matches the selection rule (G11.49) for the two-point functions, which reads

$$\frac{i+j}{k+2} = \frac{k}{k+2} \tag{G11.52}$$

Using now the structure constants (G11.45) of $\mathcal{R}$ and the metric η_{ij} in (G11.51), we may compute the three-point functions (on the sphere) for minimal models:

$$c_{ij\ell} = \langle \Phi_i \Phi_j \Phi_\ell \rangle = \sum_m c_{ij}^m \, \eta_{m\ell} = \delta_{i+j+\ell,k} \tag{G11.53}$$

Any other correlator can be obtained from the three-point function by factorization. For instance, the four-point function is

$$\langle \Phi_i \Phi_j \Phi_\ell \Phi_m \rangle = \sum_n c_{ij}^n \, c_{n\ell m} = \tag{G11.54}$$

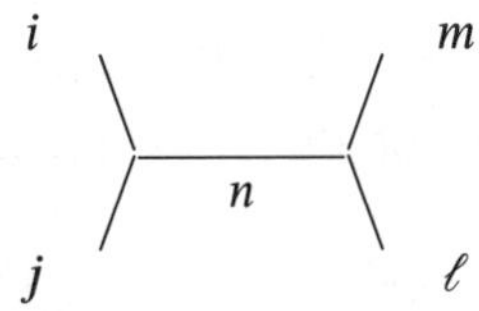

However, duality (which is a requirement for the consistency of factorizability) implies also

$$\sum_n c_{ij}^n \, c_{n\ell m} = \sum_n c_{j\ell}^n \, c_{nim} = \tag{G11.55}$$

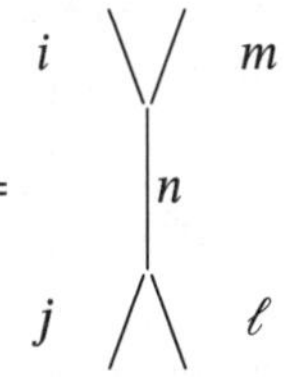

In this formalism, duality is guaranteed by the associativity of the chiral ring.

As shown in section G11.4, the chiral ring of minimal models can be represented by the factor of the ring of polynomials $P[x]$ in one variable x by the ideal generated by the derivative $W'(x) = x^{k+1}$ of the super-potential $W(x)$, after identifying x with Φ_1 and x^n with Φ_n. The metric and the structure constants can be represented by

$$\begin{aligned} \eta_{ij} &= \text{res}\left(\frac{\Phi_i \Phi_j}{W'}\right) = \frac{1}{2\pi i} \oint \frac{x^i x^j}{x^{k+1}} \\ c_{ij\ell} &= \text{res}\left(\frac{\Phi_i \Phi_j \Phi_\ell}{W'}\right) = \frac{1}{2\pi i} \oint \frac{x^i x^j x^\ell}{x^{k+1}} \end{aligned} \tag{G11.56}$$

After these preliminaries, we proceed now to perturb the chiral ring.

We define the perturbed correlators as

$$F_{i_1, i_2, \ldots, i_n}(t_0, \ldots, t_k) = \left\langle \Phi_{i_1} \Phi_{i_2} \cdots \Phi_{i_n} \, \exp \sum_{\ell=0}^{k} t_\ell \int \Phi_\ell^{(2)} \right\rangle \tag{G11.57}$$

where $\Phi_\ell^{(2)}$ is the $(1,1)$ field defined by

$$\Phi_\ell^{(2)} = G^-_{-\frac{1}{2}} \overline{G}^-_{-\frac{1}{2}} \Phi_\ell \tag{G11.58}$$

We think of $t = (t_0, t_1, \ldots, t_k)$ as $k+1$ coupling constants, measuring the strength of the perturbation. Our objective is to extend the factorization equations (G11.54) and (G11.55) to the perturbed correlators (G11.57). Note that the perturbation breaks super-conformal invariance, and thus mixes the left and right movers.

Let us consider the first order correction to the four-point function,

$$\left\langle \Phi_{i_1}\Phi_{i_2}\Phi_{i_3}\Phi_{i_4}\Phi_{\ell}^{(2)}\right\rangle = \sum_j c_{i_1i_2}^{j} c_{ji_3i_4,\ell} + \sum_j c_{i_1i_2,\ell}^{j} c_{ji_3i_4} \tag{G11.59}$$

with

$$c_{i_1i_2,\ell}^{j}\eta_{jm} = c_{i_1i_2m,\ell} = \left\langle \Phi_{i_1}\Phi_{i_2}\Phi_m \int \Phi_{\ell}^{(2)}\right\rangle \tag{G11.60}$$

Here, we have used the result that $\eta_{i_1i_2,\ell} = \left\langle \Phi_{i_1}\Phi_{i_2}\int \Phi_{\ell}^{(2)}\right\rangle$ is zero. This can be proved with the help of the Ward identities. Consistency of the factorization (G11.59) implies

$$\sum_j c_{i_1i_2}^{j} c_{ji_3i_4,\ell} + \sum_j c_{i_1i_2,\ell}^{j} c_{ji_3i_4} = \sum_j c_{i_2i_3}^{j} c_{ji_1i_4,\ell} + \sum_j c_{i_2i_3,\ell}^{j} c_{ji_1i_4} \tag{G11.61}$$

Comparing (G11.61) with (G11.55), we observe that the consistency conditions (G11.61) are satisfied if two requirements are met by the correlators

$$c_{i_1i_2i_3}(t) = \left\langle \Phi_{i_1}\Phi_{i_2}\Phi_{i_3} \exp\sum_{\ell=0}^{k} t_\ell \int \phi_{\ell}^{(2)}\right\rangle \tag{G11.62}$$

First, we must recover the super-conformal correlation functions when there is no perturbation, $c_{i_1i_2i_3,n} = \partial_n c_{i_1i_2i_3}(t)\Big|_{t=0}$. Secondly, $c_{i_1i_2i_3}(t)$ must satisfy (G11.55) for any value of t.

The perturbed chiral ring is obtained by taking as structure constants the $c_{ijm}(t)$. The relevance of this structure is that, at every order in perturbation theory, the perturbed correlation functions factorize.

Given a factorizable topological field theory, characterized by the structure constants $c_{ijm}(t)$, the associated Landau–Ginsburg super-potential $W(x,t)$ is obtained from the conditions

$$\begin{aligned} &\Phi_i(x,t) = -\partial_i W(x,t) \\ &\Phi_i(x,t)\Phi_j(x,t) = \sum_\ell c_{ij}^{\ell}(t)\Phi_\ell(x,t) \quad \text{mod } W'(x,t) \end{aligned} \tag{G11.63}$$

with $\partial_i = \partial/\partial t_i$ and $W'(x,t) = \partial W(x,t)/\partial x$.

G11.7 Landau–Ginsburg description

Let us come back to the representation (G11.38) of the chiral ring $\mathcal{R}$ as the quotient of the ring of polynomials by the relation $W'(x) = 0$. The function $W(x)$ is the Landau–Ginsburg super-potential associated to the corresponding $N = 2$ super-conformal field theory (recall exercise 8.2). Physically, this means that the infrared fixed points (under the renormalization group flow) of the $N = 2$ Landau–Ginsburg lagrangian with super-potential $W(x)$ are described, precisely, by the $N = 2$ super-conformal field theory at hand. This highly non-trivial statement is very difficult to prove, and in fact its domain of validity is unclear from a rigorous point of view; it works quite well in several interesting cases, however.

One of the main reasons that we consider the Landau–Ginsburg description of $N = 2$ theories is that, as already indicated in the previous section, it allows us to formulate a direct and simple understanding of the physical systems obtained by deforming an $N = 2$ super-conformal field theory: (super-)conformal invariance is broken as soon as we move away from the critical limit, but it may happen that integrability is preserved. We will restrict our attention to those deformations which preserve the $N = 2$ super-symmetry.

Formally, the deformed theory (recall exercise 8.8) is characterized by a lagrangian of the form

$$\mathscr{L} = \mathscr{L}_{N=2\,\mathrm{SCFT}} + \sum_{\ell=1}^{k} t_\ell \int \Phi_\ell^{(2)} d^2 z \tag{G11.64}$$

Consequently, the correlation functions of the deformed theory take the following form, with $A_i(z_i)$ any primary field of the (undeformed) $N = 2$ super-conformal field theory:

$$\left\langle \prod_i A_i(z_i) \exp \sum_\ell t_\ell \int \Phi_\ell^{(2)} d^2 z \right\rangle \tag{G11.65}$$

The Landau–Ginsburg description of the theory (G11.64) with correlators (G11.65) calls for a deformed super-potential, which depends explicitly on the deformation coupling constants t_ℓ. This deformed super-potential $W(x, t_\ell)$ can be derived in the context of the topological field theories sketched in the previous section. For some special deformations, $W(x, t_\ell)$ becomes a Morse function with non-degenerate critical points. Intuitively, the physics of such a system can be viewed in terms of solitons, i.e. the finite energy configurations which interpolate between boundary conditions (at spatial infinity on the right and on the left) fixed at the values of the various critical points. Let us concentrate on an example.

Imagine that we deform an $N = 2$ minimal model with central charge $c = 3k/(k+2)$ by the most relevant perturbation. The perturbed super-potential is

$$W(x,t) = \frac{x^{k+2}}{k+2} - tx \tag{G11.66}$$

The critical points $W' = 0$ are uniformly distributed around a circle in field space of radius proportional to the perturbation strength:

$$x_j = t^{1/(k+1)} \exp \frac{2\pi i j}{k+1} \qquad j = 1, \ldots, k+1 \tag{G11.67}$$

The solitons of this theory are the finite energy configurations of the field x such that $x(\pm\infty)$ take values in the set of critical points (G11.67). It is convenient to label the solitons according to the boundary conditions on the field at spatial infinity: we will speak of a $(j, j+r)$ soliton (with $r = 1, \ldots, k$ and $j = 1, \ldots, k+1$)

whenever $x(-\infty) = x_j$ and $x(-\infty) = x_{j+r}$. There are k different types of solitons, since any vacuum is essentially equivalent to any other vacuum and the only thing that matters is the distance between the vacua.

The general program of deforming a conformal field theory preserving integrability was initiated by Zamolodchikov in the 1980s. The goal is to reach (in a continuous fashion) integrable massive theories whose dynamics are completely characterized by a factorizable S-matrix. In the particularly interesting case of $N = 2$ super-conformal field theories, this general problem admits a very simple formulation: when does a Morse function (perturbed Landau–Ginsburg potential) describe an integrable deformation of an $N = 2$ super-conformal field theory? The obvious and painful approach is to construct explicitly in the deformed theory an infinity of conserved currents. We shall not even attempt to undertake such a tedious endeavor; rather, in the next section, we shall point out the interesting mathematical structure underlying some noteworthy Landau–Ginsburg super-potentials that is naturally tied to the integrability of the theory.

G11.8 Quantum groups and solitons

Let us consider a general perturbation of an $N = 2$ super-conformal field theory, described by a lagrangian of the form (G11.64) with $\Phi_\ell^{(2)}$ defined by (G11.58). The conservation laws $\bar{\partial} G^{\pm} = \partial \overline{G}^{\pm} = 0$ become now (see exercises 8.2 and 8.8)

$$\begin{aligned} \partial_{\bar{z}} G^{+} &= \sum t_\ell (1 - q_\ell)\, \partial_z \overline{G}^{-}_{-\frac{1}{2}} \Phi_\ell \\ \partial_{\bar{z}} G^{-} &= \sum t_\ell (1 - q_\ell)\, \partial_z \overline{G}^{+}_{-\frac{1}{2}} \Phi_\ell \\ \partial_z \overline{G}^{+} &= \sum t_\ell (1 - q_\ell)\, \partial_{\bar{z}} G^{-}_{-\frac{1}{2}} \Phi_\ell \\ \partial_z \overline{G}^{-} &= \sum t_\ell (1 - q_\ell)\, \partial_{\bar{z}} G^{+}_{-\frac{1}{2}} \Phi_\ell \end{aligned} \tag{G11.68}$$

where q_ℓ are the $U(1)$ charges of the chiral primary fields Φ_ℓ.

These new conservation laws extend the $N = 2$ super-symmetry algebra by topological central terms as found by Witten and Olive for $N = 1$ field theories.

We first write

$$\begin{aligned} Q_{\pm} &= \int G^{\pm}(z) dz - \sum t_\ell (1 - q_\ell)\, \overline{G}^{\mp}_{-\frac{1}{2}} \Phi_\ell(z, \bar{z}) d\bar{z} \\ \overline{Q}_{\pm} &= \int \overline{G}^{\pm}(z) dz - \sum t_\ell (1 - q_\ell)\, G^{\mp}_{-\frac{1}{2}} \Phi_\ell(z, \bar{z}) d\bar{z} \end{aligned} \tag{G11.69}$$

and use hermiticity to change notation to the more convenient

$$Q_+ = Q\,, \quad Q_- = Q^\dagger\,, \qquad \overline{Q}_+ = \overline{Q}\,, \quad \overline{Q}_- = \overline{Q}^\dagger \tag{G11.70}$$

It is a good exercise to check that the algebra satisfied by these charges (with of course their bosonic partners the hamiltonians $P = P^\dagger$, $\overline{P} = \overline{P}^\dagger$ and the fermion number $\mathscr{F} = \mathscr{F}^\dagger$) is then

$$\begin{array}{ll} Q^2 = \overline{Q}^2 = 0 & \left\{Q, \overline{Q}^\dagger\right\} = \{Q^\dagger, \overline{Q}\} = 0 \\ [\mathscr{F}, Q] = Q & \left[\mathscr{F}, \overline{Q}\right] = -\overline{Q} \\ \{Q, Q^\dagger\} = P & \left\{\overline{Q}, \overline{Q}^\dagger\right\} = \overline{P} \\ [Q, P] = \left[Q, \overline{P}\right] = & \left[\overline{Q}, P\right] = \left[\overline{Q}, \overline{P}\right] = 0 \\ \{Q, \overline{Q}\} = \Delta & [\mathscr{F}, P] = \left[\mathscr{F}, \overline{P}\right] = 0 \end{array} \tag{G11.71}$$

with $\overline{P} \neq P^\dagger$ and

$$\Delta = -2 \sum_{\ell} t_\ell \,(1 - q_\ell)\, [\Phi_\ell(+\infty) - \Phi_\ell(-\infty)] \tag{G11.72}$$

The right-hand side of (G11.72) can be written more compactly by using the functional $W(x,t)$ in (G11.63) associated to the perturbed chiral ring obtained from (G11.64). Imposing the boundary condition that the fields converge to a critical point of $W(x,t)$ at $\pm\infty$, we obtain

$$\{Q, \overline{Q}\} = 2\,(W_a - W_b) \tag{G11.73}$$

where W_a is the value of $W(x,t)$ at the critical point a (recall that a critical point of the super-potential is a value of x where $W' = 0$).

Similarly, the fermion number f_{ab} of a soliton interpolating between vacua a and b is determined by index theorems as

$$\exp\left(i\pi f_{ab}\right) = \text{phase of } \left[\frac{H(a)}{H(b)}\right] \tag{G11.74}$$

where $H(a) = \left(\det \partial_i \partial_j W\right)\Big|_a$ is the hessian of the super-potential at the critical point a.

From (G11.73), it follows that the modified super-symmetry charge obtained by including the perturbation (G11.64) becomes the topological number of a soliton configuration interpolating between two critical points of the super-potential $W(x,t)$. We shall encounter shortly some special solitons, called Bogomolny solitons, which are, in a sense, fundamental. They correspond to field configurations of finite energy, interpolating between different vacua or critical points, for which the value of the super-potential W evolves along a straight line in the W plane.

The centrally extended algebra (G11.71)–(G11.73) can be related easily to an old and trusted friend, namely the affine quantum group $U_q(\widehat{s\ell(2)})$ with $q^4 = 1$. The connection between the $N = 2$ super-algebra (G11.71) and $U_q(A_1^{(1)})$ for $q = e^{i\pi/2}$ is established by the map

$$\begin{array}{ll} Q = E_0 K_0^{-\frac{1}{2}} & Q^\dagger = K_1^{-\frac{1}{2}} E_1 \\ \overline{Q} = F_0 K_0^{-\frac{1}{2}} & \overline{Q}^\dagger = K_1^{-\frac{1}{2}} F_1 \end{array} \tag{G11.75}$$

so that the central extension terms are identified as

$$\Delta = \frac{1-K_0^{-2}}{2} \quad , \quad \overline{\Delta} = \frac{1-K_1^{-2}}{2} \tag{G11.76}$$

whereas the fermion number is essentially the (complex) spin:

$$(-1)^{\mathscr{F}} = K_1 = K_0^{-1} \tag{G11.77}$$

and the momenta are given by

$$\begin{aligned} P &= (E_0E_1 + E_1E_0)K_0^{-\frac{1}{2}}K_1^{-\frac{1}{2}} \\ \overline{P} &= (F_0F_1 + F_1F_0)K_0^{-\frac{1}{2}}K_1^{-\frac{1}{2}} \end{aligned} \tag{G11.78}$$

We have implicitly used nilpotent representations only, i.e. generic two-dimensional representations (chapter 7) with $E_i^2 = F_1^2 = 0$ $(i = 0, 1)$, to ensure that all four super-charges are nilpotent.

It is important to realize that this map from the $N = 2$ super-algebra to $U_q(\widehat{s\ell(2)})$ with $q^4 = 1$ is not an isomorphism. This can be seen easily by counting the independent labels of the irreducible representations, other than the momentum or rapidity. An irrep of the $N = 2$ super-algebra is characterized by three numbers: the fermion number f, the mass m and the central charge $\Delta = 2(W_a - W_b)$. On the other hand, two-dimensional nilpotent irreps of affine $SU(2)$ have only one label, λ. Thus, for a generic super-potential we should not expect the relationships between fermion number, mass and topological charge following from the identifications (G11.64)–(G11.66) to be valid. Nevertheless, for a very peculiar kind of super-potential, namely for A super-potentials perturbed by the most relevant operator, this is in fact the case.

Indeed, from the super-potential (G11.66), it is not hard to compute the fermion number of a $(j, j + r)$ soliton: it is given by $r/(k + 1)$. Similarly, the mass of a $(j, j + r)$ Bogomolny soliton is proportional to $\sin[\pi r/(k + 1)]$, which coincides, incidentally, with the spectrum of type A Toda theories. It is a simple but instructive exercise to verify that the Bogomolny $(j, j + r)$ solitons of such a perturbed type A Landau–Ginsburg $N = 2$ super-conformal field theory are in one to one correspondence with the nilpotent irreps of $U_{q=i}(\widehat{s\ell(2)})$ with label $\lambda = q^{2r}$. In particular, the fermion numbers, masses and topological extensions all match.

From this result, it is immediately obvious that the scattering matrix for solitons of the perturbed $N = 2$ super-conformal field theory is just the quantum group intertwiner of the nilpotent irreps, i.e. the trigonometric R-matrix of $U_{q=i}(\widehat{s\ell(2)})$.

For other, more general, super-potentials, the map from fundamental or Bogomolny solitons to irreps of $U_q(\widehat{s\ell(2)})$ with $q^4 = 1$ cannot be carried out, because their fermion number, mass and topological charge are not related in the simple way implied by the quantum group. For deformed super-potentials yielding integrable massive theories, some generalization of the affine quantum group should exist to support integrability. The reader is invited to elucidate further this point by investigating, for instance, the perturbed D type Landau–Ginsburg super-potentials.

Exercise

Ex. G11.1 Classification of the $N = 2$ Landau–Ginsburg potentials.
Given the chiral ring $\mathscr{R}$ with generators x_a of conformal weights $q_a/2$, consider quasi-homogeneous potentials $W(x_a)$ satisfying

$$W(\lambda^{q_a} x_a) = \lambda\, W(x_a) \tag{G11.79}$$

Prove that, in order to represent $\mathscr{R}$ as $P[x_a]/\{\partial W/\partial x_a\}$ with $P[x_a]$ the ring of polynomials in x_a, condition (G11.79) should be satisfied.

In his study of singularity theory, Arnold classified the quasi-homogeneous potentials as follows:

$$\begin{array}{ll} A_k & x^{k+1} \\ D_k & x^{k-1} + xy^2 \\ E_6 & x^3 + y^4 \\ E_7 & x^3 + xy^3 \\ E_8 & x^3 + y^5 \end{array} \tag{G11.80}$$

Find, for each of these "ADE" potentials, the associated chiral ring and conformal weights.

References

We have kept the bibliographic material below to a minimum. Its purpose is to direct the reader to some basic references (we have cited all the works used in the preparation of this book) and to provide some suggestions for further reading, chapter by chapter. Some of the exercises in the book also carry bibliographical notes on their specific subject. Current research articles on the topic of this book can be found in Paul Ginsparg's electronic database, hep-th@xxx.lanl.gov. The titles of all the references below have been translated into English, but we refer to the original published work. Most of the material in what were Soviet journals has subsequently appeared in English translations. We have made no attempt whatsoever to ensure historical completeness.

Chapter 1

• Integrability of (1+1)-dimensional relativistic particle systems

J.B. McGuire (1964). Study of exactly solvable N-body problems, *J. Math. Phys.* **5,** 622–36.

C.N. Yang (1967). Some exact results for the many-body problem in one dimension with repulsive delta-function interaction, *Phys. Rev. Lett.* **19,** 1312.

M. Karowski, H.J. Thun, T.T. Truong and P.H. Weisz (1977). On the uniqueness of purely elastic S-matrix in $(1 + 1)$ dimensions, *Phys. Lett.* **67B,** 321.

A.B. Zamolodchikov (1977). An exact two particle S-matrix for quantum solitons of the sine-Gordon model, *Pisma ZhETF* **25,** 499–502.

A.B. Zamolodchikov and Al.B. Zamolodchikov (1979). Factorized S-matrices in two dimensions as the exact solution of certain relativistic quantum field theory models, *Ann. Phys.* **120,** 253.

I.V. Cherednik (1980). On a method of constructing factorized S-matrices in elementary functions, *Teor. Mat. Fiz.* **43,** 117.

A.B. Zamolodchikov (1980). Factorized S-matrices in lattice statistical systems, *Sov. Sci. Rev.* **A2,** 1.

• Introduction to the Bethe ansatz

H.B. Thacker (1981). Exact integrability in quantum field theory and statistical systems, *Rev. Mod. Phys.* **53,** 253–85.

L.D. Faddeev (1982). Integrable models in $(1 + 1)$-dimensional quantum field theory, in *Proceedings, Les Houches XXXIX* (Elsevier, Amsterdam).

J.H. Lowenstein (1982). Introduction to Bethe ansatz approach in $(1 + 1)$-dimensional models, in *Proceedings, Les Houches XXXIX* (Elsevier, Amsterdam).

- **A first contact with the six-vertex model**

E.H. Lieb and F.Y. Wu (1972). Two-dimensional ferroelectric models, in *Phase Transitions and Critical Phenomena* **1**, eds. C. Domb and M.S. Green (Academic Press, New York).

R.J. Baxter (1982). *Exactly Solved Models in Statistical Mechanics* (Academic Press, London).

- **The form factor program**

F.A. Smirnov (1992). *Form Factors in Completely Integrable Models of Quantum Field Theory* (World Scientific, Singapore).

Chapter 2

- **Algebraic Bethe ansatz and its connection with quantum groups**

H. Bethe (1931). On the theory of metals: eigenvalues and eigenfunctions of linear atomic chains, *Z. Phys.* **71,** 205.

L.D. Faddeev, E.K. Sklyanin and L.A. Takhtajan (1979). Quantum inverse scattering transform method, *Teor. Mat. Fiz.* **40,** 194.

L.D. Faddeev and L.A. Takhtajan (1979). The quantum inverse problem method and the *XYZ* Heisenberg model, *Usp. Mat. Nauk* **34,** 13.

L.A. Takhtajan (1981). Quantum inverse scattering method and the algebraized matrix Bethe ansatz, *Zap. Nauchn. Sem. LOMI* **101,** 158.

L.D. Faddeev (1982). Integrable models in $(1 + 1)$-dimensional quantum field theory, in *Proceedings, Les Houches XXXIX* (Elsevier, Amsterdam).

L.A. Takhtajan (1989). Quantum groups and integrable models, *Adv. Studies Pure Math.* **13,** 19.

E.K. Sklyanin (1993). Quantum inverse scattering method: selected topics, in *Proceedings of the Fifth Nankai Workshop,* ed. Mo-Lin Ge (World Scientific, Singapore).

- **Higher-spin descendants in the spirit of this book**

P.P. Kulish, N.Yu. Reshetikhin and E.K. Sklyanin (1981). Yang–Baxter equation and representation theory I, *Lett. Math. Phys.* **5,** 393.

V.V. Bazhanov and Yu.G. Stroganov (1990). Chiral Potts model as a descendant of the six-vertex model, *J. Stat. Phys.* **59,** 799.

- **Introduction to the braid group**

J. Birman (1974). Braids, links and mapping class groups, *Ann. Math. Studies* **82,** 1.

Chapter 3

- **Applications of the Bethe ansatz**

C.N. Yang and C.P. Yang (1966). One-dimensional chain of anisotropic spin-spin interaction I, II and III, *Phys. Rev.* **150**, 321–7; 327–39; **151**, 258–64.

M. Takahashi and M. Suzuki (1972). One-dimensional anisotropic Heisenberg model at finite temperatures, *Prog. Theor. Phys.* **48,** 2187.

L.D. Faddeev and L.A. Takhtajan (1981). What is the spin of a spin wave?, *Phys. Lett.* **85A,** 375.

C. Destri and J.H. Lowenstein (1982). Analysis of the Bethe ansatz equations of the chiral Gross–Neveu model, *Nucl. Phys.* **B205,** 369.

J.H. Lowenstein (1982). Introduction to Bethe ansatz approach in $(1 + 1)$-dimensional models, in *Proceedings, Les Houches XXXIX* (Elsevier, Amsterdam).

L.A. Takhtajan (1982). The picture of low-lying excitations in the isotropic Heisenberg chain of arbitrary spins, *Phys. Lett.* **87A,** 479.

O. Babelon, H. de Vega and C.M. Viallet (1983). Analysis of the Bethe ansatz equations of the XXZ model, *Nucl. Phys.* **B220 [FS8],** 13.

H.M. Babujian (1983). Exact solutions of the isotropic Heisenberg chain with arbitrary spins: thermodynamics of the model, *Nucl. Phys.* **B215 [FS7],** 317.

H.M. Babujian and A.M Tsvelick (1986). Heisenberg magnet with an arbitrary spin and anisotropic chiral field, *Nucl. Phys.* **B265 [FS15],** 24.

N.M. Bogolyubov, A.G. Izergin and V.E. Korepin (1986). Critical exponents for integrable models, *Nucl. Phys.* **B275,** 687–705.

L.D. Faddeev and N.Yu. Reshetikhin (1986). Integrability of the principal chiral field model in (1+1) dimensions, *Ann. Phys.* **167,** 227.

N. Yu. Reshetikhin (1986). S-matrices in integrable models of isotropic magnetic chains, *J. Phys.* **A24,** 3299–310.

A.N. Kirillov and N.Yu. Reshetikhin (1987). Exact solution of the integrable XXZ Heisenberg chain model with arbitrary spin, *J. Phys.* **A20,** 1565–87.

V.V. Bazhanov and N.Yu. Reshetikhin (1989). Critical RSOS models and conformal field theory, *Int. J. Mod. Phys.* **A4,** 115–42.

- **Definition of K operators**

E.K. Sklyanin (1988). Boundary conditions for integrable quantum systems, *J. Phys.* **A21,** 2375.

P.P. Kulish and E.K. Sklyanin (1992). Algebraic structures related to the reflection equations, *J. Phys.* **A25,** 5963–76.

- **Spin chains with boundaries**

F.C. Alcaraz, M.N. Barber and M.T. Batchelor (1987). Conformal invariance and the spectrum of the XXZ chain, *Phys. Rev. Lett.* **58,** 771.

F.C. Alcaraz, M.N. Barber, M.T. Batchelor, R.J. Baxter and G.R.W Quispel (1987). Surface exponents of the quantum XXZ, Ashkin–Teller and Potts models, *J. Phys.* **A20,** 6397–409.

L. Mezincescu, R. Nepomechie and V. Rittenberg (1990). Bethe ansatz solution of the Fateev Zamolodchikov quantum spin chain with boundary terms, *Phys. Lett.* **147A,** 70–8.

- **Quantum group invariant chains**

V. Pasquier and H. Saleur (1990). Common structures between finite systems and conformal field theories through quantum groups, *Nucl. Phys.* **B330,** 523.

- **Valence bond models**

I. Affleck and F.D.M. Haldane (1987). Critical theory of quantum spin chains, *Phys. Rev.* **B36,** 5291.

I. Affleck, E.H. Lieb and H. Tasakishi (1988). Valence bound ground state in isotropic quantum antiferromagnets, *Commun. Math. Phys.* **115,** 477.

Chapter 4

- **Eight-vertex model**

J.D. Johnson, S. Krinsky and B.M. McCoy (1973). Vertical arrow correlation lengths in the eight-vertex model and the low-lying excitations of the XYZ hamiltonian, *Phys. Rev.* **A8,** 2526–47.

R.J. Baxter (1978). Solvable eight-vertex model on an arbitrary planar lattice, *Proc. Roy. Soc.* **289A,** 1359.

L.D. Faddeev and L.A. Takhtajan (1979). The quantum inverse problem method and the XYZ Heisenberg model, *Usp. Mat. Nauk* **34,** 13.

R.J. Baxter (1982). *Exactly Solved Models in Statistical Mechanics* (Academic Press, London).

- **Sklyanin algebra**

E.K. Sklyanin (1982). On certain algebraic structures connected with the Yang–Baxter equation, *Funk. Anal. Pril.* **16**, 27–34; 327–39; **17**, 34–48.

A.V. Odessky and B.L Feigin (1989). Sklyanin elliptic algebras, *Funk. Anal. Pril.* **23,** 45–54.

- **Elliptic functions**

E.T. Whittaker and G.N. Watson (1962). *Modern Analysis* (Cambridge University Press).

J.D. Fay (1975) *Theta Functions on Riemann Surfaces,* Lecture Notes in Mathematics **352** (Springer, Berlin).

Chapter 5

- **Face models**

A. Onsager (1944). Crystal statistics I: a two-dimensional model with an order-disorder transition, *Phys. Rev.* **65,** 117–49.

H.N.V. Temperly and E.H. Lieb (1971). Relations between the percolation and colouring problem and other graph theoretical problems associated with regular planar lattices: some exact results for the percolation problem, *Proc. Roy. Soc.* **A322,** 251.

G.E. Andrews, R.J. Baxter and P.J. Forrester (1984). Eight-vertex SOS model and generalized Rogers–Ramanujan-type identities, *J. Stat. Phys.* **35,** 193.

E. Date, M. Jimbo, A. Kuniba, T. Miwa and M. Okado (1987). Exactly solvable SOS models: local height probabilities and theta function identities, *Nucl. Phys.* **B290 [FS20],** 231–73.

E. Date, M. Jimbo, A. Kuniba, T. Miwa and M. Okado (1988). Exactly solvable SOS models II: proof of the star-triangle relation and combinatorial identities, *Adv. Studies Pure Math.* **16,** 17–122.

M. Jimbo, T. Miwa and M. Okado (1988). Local state probabilities of solvable lattice models: an $A^{(1)}_{n-1}$ family, *Nucl. Phys.* **B300 [FS22],** 74.

M. Jimbo, T. Miwa and M. Okado (1988). Solvable lattice models related to the vector representations of classical simple Lie algebras, *Commun. Math. Phys.* **116,** 507.

- **Approach in this book**

V. Pasquier (1987). two-dimensional critical systems labelled by Dynkin diagrams, *Nucl. Phys.* **B285 [FS19],** 162–74.

V. Pasquier (1988). Continuum limit of lattice models built on quantum groups, *Nucl. Phys.* **B295,** 491–510.

V. Pasquier (1988). Etiology of IRF models, *Commun. Math. Phys.* **118,** 355.

• **Towers of algebras**

F.M. Goodman, P.M. de la Harpe and V.F.R. Jones (1989). *Coxeter–Dynkin Diagrams and Towers of Algebras* (MSRI Publications/Springer Verlag, New York).

A. Ocneanu (1989). Quantized groups, string algebras and Galois theory for algebras, *London Math. Soc. Lecture Notes* **136,** 119.

• **Knots and statistical mechanics**

D. Rolfsen (1976). *Knots and Links* (Publish or Perish, Berkeley).

L. Kauffmann (1977). *On Knots* (Princeton University Press).

V.F.R. Jones (1985). A polynomial invariant for knots via von Neumann algebras, *Bull. Am. Math. Soc.* **12,** 103–12.

V.F.R. Jones (1987). Hecke algebra, representations of braid groups and link polynomials, *Ann. Math.* **126,** 335–88.

V.G. Turaev (1988). The Yang–Baxter equation and invariants of links, *Invent. Math.* **92,** 527.

Y. Akutsu, T. Deguchi and M. Wadati (1989). Exactly solvable models and knot theory, *Phys. Rep.* **180,** 247–332.

V.F.R. Jones (1989). On knot invariants related to some statistical mechanical models, *Pacific J. Math.* **137,** 311–34.

E. Witten (1989). Gauge theories and integrable lattice models, *Nucl. Phys.* **B322,** 629.

• **Face model fusion**

E. Date, M. Jimbo, T. Miwa and M. Okado (1986). Fusion of the eight-vertex SOS model, *Lett. Math. Phys.* **12,** 209.

M. Jimbo, T. Miwa and M. Okado (1988). Local state probabilities of solvable lattice models: an $A^{(1)}_{n-1}$ family, *Nucl. Phys.* **B300 [FS22],** 74.

M. Jimbo, T. Miwa and M. Okado (1988). Solvable lattice models related to the vector representations of classical simple Lie algebras, *Commun. Math. Phys.* **116,** 507.

• **Spin models and subfactors**

V.F.R. Jones (1983). Index for subfactors, *Invent. Math.* **72,** 1.

F.M. Goodman, P.M. de la Harpe and V.F.R. Jones (1989). *Coxeter–Dynkin Diagrams and Towers of Algebras* (MSRI Publications/Springer Verlag, New York).

Chapter 6

• **The basic reference remains**

V.G. Drinfeld (1987). Quantum groups, in *Proceedings of the 1986 International Congress of Mathematics,* ed. A.M. Gleason (American Mathematical Society, Berkeley).

• **Accessible introductions**

M. Jimbo (1989). Introduction to the Yang–Baxter equation, *Int. J. Mod. Phys.* **A4,** 3759–77; reprinted in *Braid Group, Knot Theory and Statistical Mechanics*, eds. C.N. Yang and M.L. Ge (World Scientific, Singapore).

L.A. Takhtajan (1990). Introduction to quantum groups and integrable massive models of quantum field theory, in *Nankai Lectures on Mathematical Physics,* eds. M.L. Ge and B.H. Zhao (World Scientific, Singapore).

M. Jimbo (1991). *Quantum Groups and the Yang–Baxter Equation* (Springer, Tokyo).

M. Jimbo (1992). Topics from representations of $U_q(\mathcal{G})$: an introductory guide to physicists, in *Proceedings, 1991 Nankai Workshop,* ed. M.L. Ge (World Scientific, Singapore).

- **Applications and representation theory**

M. Rosso (1988). Finite-dimensional representations of quantum analog of the enveloping algebra of a complex simple Lie algebra, *Commun. Math. Phys.* **117,** 581–93.

L.D. Faddeev, N.Yu. Reshetikhin and L.A. Takhtajan (1989). Quantum groups, in *Braid Group, Knot Theory and Statistical Mechanics,* eds. C.N. Yang and M.L. Ge (World Scientific, Singapore).

A.N. Kirillov and N. Yu. Reshetikhin (1989). Representations of the algebra $U_q(s\ell(2))$, q-orthogonal polynomials and invariants of links, in *Infinite-dimensional Lie Algebras and Groups,* ed. V.G. Kac (World Scientific, Singapore).

L.D. Faddeev, N.Yu. Reshetikhin and L.A. Takhtajan (1990). Quantum Lie groups and Lie algebras, *Leningrad Math. J.* **1,** 193–225.

- **Reprint volume**

M. Jimbo, ed. (1988). *Yang–Baxter Equation in Integrable Systems* (World Scientific, Singapore).

- **Affine quantum groups**

M. Jimbo (1986). Quantum R-matrix for the generalized Toda system, *Commun. Math. Phys.* **102,** 537.

V.G. Drinfeld (1988). A new realization of Yangians and quantized affine algebras, *Sov. Math. Doklady* **36,** 212–16.

I.B. Frenkel and N. Jing (1988). Vertex representation of quantum affine algebras, *Proc. Nat. Acad. Sci.* **85,** 9373.

V. Chari and A. Pressley (1991). Quantum affine algebras, *Commun. Math. Phys.* **142,** 261–83.

I. Frenkel and N.Yu. Reshetikhin (1992). Quantum affine algebras and holonomic q-difference equations, *Commun. Math. Phys.* **146,** 1–60.

- **Kac–Moody algebras**

V.G. Kac (1990) *Infinite Dimensional Lie Algebras,* 3rd edn (Cambridge University Press).

- **Particular aspects**

A.A. Belavin and D.G. Drinfeld (1983). Solutions of the classical Yang–Baxter equation for simple Lie algebras, *Funk. Anal. Pril.* **16,** 1.

H. Wenzl (1988). Hecke algebras of type A_n and subfactors, *Invent. Math.* **92,** 349–83.

V.G. Drinfeld (1990). Quasi-Hopf algebras and Knizhnik–Zamolodchikov equations, *Leningrad Math. J.* **1,** 1419–57.

- **A review on knots and quantum groups**

J. Birman (1983). New points of view in knot theory, *Bull. Am. Math. Soc.* **28,** 253–87.

Chapter 7

• Mathematical background

G. Lusztig (1989). Modular representations of quantum groups, *Contemp. Math.* **82,** 237–49.

C. de Concini and V.G. Kac (1990). Quantum group representations at q a root of 1, *Prog. Math.* **92,** 471.

C. de Concini, V.G. Kac and C. Procesi (1991). *The quantum co-adjoint action*, Pisa preprint.

• Some physical applications

D. Arnaudon and A. Chakrabarti (1991). Flat periodic representations of $U_q(\mathcal{G})$, *Commun. Math. Phys.* **139,** 605.

D. Arnaudon and A. Chakrabarti (1991). Periodic and partially periodic representations of $SU(N)_q$, *Commun. Math. Phys.* **139,** 461.

E. Date, M. Jimbo, K. Miki and T. Miwa (1991). Generalized chiral Potts model and minimal cyclic representations of $U_q(g\ell(n,\mathbb{C}))$, *Commun. Math. Phys.* **137,** 133–47.

E. Date, M. Jimbo, K. Miki and T. Miwa (1991). New R-matrices associated with cyclic representations of $U_q(A_2^{(2)})$, *Publ. Res. Inst. Math. Sci.* **27,** 639–55.

C. Gómez, M. Ruiz-Altaba and G. Sierra (1991). New R-matrices associated with finite dimensional representations of $U_q(s\ell(2))$ at roots of unity, *Phys. Lett.* **B265,** 95–8.

C. Gómez and G. Sierra (1992). A new solution to the star-triangle equation based on $U_q(s\ell(2))$ at roots of unity, *Nucl. Phys.* **B373,** 761.

M. Ruiz-Altaba (1992). New solutions to the Yang–Baxter equation from two-dimensional representations of $U_q(s\ell(2))$ at roots of unity, *Phys. Lett.* **B279,** 326–32.

• First $\mathbb{Z}_N$

models

D.A. Huse (1981). Simple three-state model with infinitely many phases, *Phys. Rev.* **B24,** 5180.

S. Ostlund (1981). Incommensurate and commensurate phases in asymmetric clock models, *Phys. Rev.* **B24,** 398.

V.A. Fateev and A.B. Zamolodchikov (1982). Self-dual solutions of the star-triangle relations on $\mathbb{Z}_L$ models, *Phys. Lett.* **92A,** 37.

• Super-integrable models

S. Howes, L.P. Kadanoff and M. den Nijs (1983). Quantum model for commensurate-incommensurate phase transitions, *Nucl. Phys.* **B215,** 169.

V. von Gehlen and V. Rittenberg (1985). $\mathbb{Z}_N$ symmetric quantum chains with an infinite set of conserved charges and $\mathbb{Z}_N$ zero modes, *Nucl. Phys.* **B257,** 351–70.

• Onsager algebra

B. Kaufmann and A. Onsager (1949). Crystal statistics III. Short range order in binary Ising lattice, *Phys. Rev.* **76,** 1244.

L. Dolan and M. Grady (1982). Conserved charges from self-duality, *Phys. Rev.* **D25,** 1587–604.

• **Chiral Potts model**

H. Au-Yang, B.M. McCoy, J.H.H. Perk, S. Tang and M.L. Yan (1987). Commuting transfer matrices in chiral Potts models: solutions of the star-triangle equations with genus > 1, *Phys. Lett.* **A123,** 219.

H. Au-Yang, R.J. Baxter and J.H.H. Perk (1988). New solutions of the star-triangle relations for the chiral Potts model, *Phys. Lett.* **A128,** 138.

V.V. Bazhanov and Yu.G. Stroganov (1990). Chiral Potts model as a descendant of the six-vertex model, *J. Stat. Phys.* **59,** 799.

V.V. Bazhanov, R.M. Kazaev, V.V. Mangazeev and Yu.G. Stroganov (1991). $(\mathbb{Z}_{N^\times})^{n+1}$ generalization of the chiral Potts model, *Commun. Math. Phys.* **138,** 393–408.

• **Spin chains based on non-standard irreps**

A. Berkovich, C. Gómez and G. Sierra (1993). Spin anisotropy, commensurate chains and $N = 2$ super-symmetry, *Nucl. Phys.* **B415,** 681–733.

Chapter 8

• **The basic references on conformal field theory are**

A.A. Belavin, A.M. Polyakov and A.B. Zamolodchikov (1984). Infinite conformal symmetry in two-dimensional quantum field theory, *Nucl. Phys.* **B241,** 333.

A.A. Belavin, A.M. Polyakov and A.B. Zamolodchikov (1984). Infinite conformal symmetry of critical fluctuations in two dimensions, *J. Stat. Phys.* **34,** 763.

• **A good collection of references is the reprint volume**

C. Itzykson, H. Saleur and J.B. Zuber, eds. (1988). *Conformal Invariance and Applications to Statistical Mechanics* (World Scientific, Singapore).

• **Good reviews**

J. Cardy (1987). Conformal invariance, in *Phase Transitions and Critical Phenomena* **11,** eds. C. Domb and J.L. Lebowitz (Academic Press, London).

C. Itzykson and J.M. Drouffe (1989). *Statistical Field Theory* (Inter Editions/Editions du CNRS, Paris).

P. Ginsparg (1990). Applied conformal field theory, in *Proceedings, Les Houches XLIX,* eds. E. Brézin and J. Zinn-Justin (North Holland, Amsterdam).

• **Scaling**

K.G. Wilson and J. Kogut (1974). The renormalization group and the ϵ expansion, *Phys. Rep.* **C12,** 75.

M.E. Fisher (1975). Critical phenomena, in *Proceedings, 51st Enrico Fermi Summer School at Varenna,* ed. M.S. Green (Academic Press, London).

S.K. Ma (1976). *Modern Theory of Critical Phenomena* (Benjamin, Reading).

• **Conformal hypothesis**

A.M. Polyakov (1970). Conformal symmetry of critical fluctuations, *Pisma ZhETF* **12,** 538.

• **Unitarity**

D. Friedan, Z. Qiu and S. Shenker (1984). Conformal invariance, unitarity and critical exponents in two dimensions, *Phys. Rev. Lett.* **52,** 1575.

D. Friedan, Z. Qiu and S. Shenker (1986). Details of the non-unitarity proof for highest weight representations of the Virasoro algebra, *Commun. Math. Phys.* **107,** 535.

• **Finite size effects**

H. de Vega and F. Woynarovich (1985). Method for calculating finite size corrections in Bethe ansatz systems: Heisenberg chain and six-vertex model, *Nucl. Phys.* **B251,** 439.

I. Affleck (1986). Universal term in the energy at the critical point and the conformal anomaly, *Phys. Rev. Lett.* **56,** 748.

H.W.J. Blöte, J.L. Cardy and M.P. Nightingale (1986). Conformal invariance, the central charge, and universal finite-size amplitudes at criticality, *Phys. Rev. Lett.* **56,** 742.

N.M. Bogolyubov, A.G. Izergin and N.Yu. Reshetikhin (1986). Finite-size effects and critical indices of 1D quantum models, *Pisma ZhETF* **44,** 405.

J. Cardy, ed. (1988). *Finite Size Scaling, Current Physics Sources and Comments* **2** (North Holland, Amsterdam).

Chapter 9

• **The basic works on duality in conformal field theory are**

R. Dijkgraaf and E. Verlinde (1988). Modular invariance and the fusion algebra, *Nucl. Phys. (Proc. Suppl.)* **5B,** 87–97.

G. Moore and N. Seiberg (1988). Polynomial equations for rational conformal field theories, *Phys. Lett.* **212B,** 451.

E. Verlinde (1988). Fusion rules and modular transformations in two-dimensional conformal field theory, *Nucl. Phys.* **B300 [FS22],** 360.

L. Alvarez-Gaumé, C. Gómez and G. Sierra (1989). Hidden quantum symmetries in rational conformal field theories, *Nucl. Phys.* **B319,** 155.

G. Moore and N. Seiberg (1989). Classical and quantum field theory, *Commun. Math. Phys.* **123,** 177–254.

• **More technical approach**

Y. Kanie and A. Tsuchiya (1987). Vertex operators in the conformal field theory on P^1 and monodromy representations of the braiding group, *Lett. Math. Phys.* **13,** 303.

J. Fröhlich (1988). Statistics of fields, the Yang–Baxter equation and the theory of knots and links, in *Non-perturbative Quantum Field Theory,* ed. 't Hooft (Plenum Press, New York).

Y. Kanie and A. Tsuchiya (1988). Vertex operators in the conformal field theory on P^1 and monodromy representations of the braiding group, *Adv. Studies Pure Math.* **16,** 297.

K.H. Rehren and B. Schroer (1989). Einstein causality and Artin braids, *Nucl. Phys.* **B312,** 715–50.

J. Fröhlich and H. Gabbiani (1990). Braid statistics in local quantum theory, *Rev. Math. Phys.* **2,** 251–353.

- **Virasoro characters**

A. Rocha-Caridi (1984). Vacuum vector representations of the Virasoro algebra, in *Vertex Operators in Mathematics and Physics,* eds. J. Lepowsky, S. Mandelstam and I.M. Singer (MSRI Publications/Springer, New York).

C. Gómez and G. Sierra (1991). Towers of algebras in rational conformal field theory, *Int. J. Mod. Phys.* **A6,** 2045.

- **Modular invariance**

J.L. Cardy (1986). Operator content of two-dimensional conformally invariant theories, *Nucl. Phys.* **B270 [FS16],** 186–204.

- **A review**

L. Alvarez-Gaumé, C. Gómez and G. Sierra (1990). Topics in conformal field theory, in *The Physics and Mathematics of Strings, Memorial Volume for Vadim Knizhnik,* eds. L. Brink, D. Friedan and A.M. Polyakov (World Scientific, Singapore).

Chapter 10

- **The Coulomb gas**

B.L. Feigin and D.B. Fuks (1982). Invariant skew-symmetric differential operators on the line and Verma modules over the Virasoro algebra, *Funk. Anal. Pril.* **16,** 47.

B.L. Feigin and D.B. Fuks (1983). Verma modules over the Virasoro algebra, *Funk. Anal. Pril.* **17,** 91.

V.S. Dotsenko and V.A. Fateev (1984). Conformal algebra and multi-point correlation functions in two-dimensional statistical models, *Nucl. Phys.* **B240 [FS12],** 312–48.

B.L. Feigin and D.B. Fuks (1984). Verma modules over the Virasoro algebra, *Lecture Notes in Mathematics* **1060,** 230.

J.L. Gervais and A. Neveu (1984). Novel triangular relation and absence of tachyons in Liouville string theory, *Nucl. Phys.* **B238,** 125–41.

V.S. Dotsenko and V.A. Fateev (1985). Four-point correlation functions and the operator algebra in 2d conformal invariant theories with central charge $c \leq 1$, *Nucl. Phys.* **B251 [FS13],** 691–734.

V.S. Dotsenko and V.A. Fateev (1985). Operator algebra of two-dimensional conformal theories with central charge $c \leq 1$, *Phys. Lett.* **154B,** 291–5.

B.L. Feigin and D.B. Fuks (1986). Representations of the Virasoro algebra, in *Representations of Infinite-Dimensional Lie Groups and Lie Algebras,* ed. V.G. Kac (Gordon and Breach, New York).

M. Wakimoto (1986). Fock representations of the affine Lie algebra $A_1^{(1)}$, *Commun. Math. Phys.* **104,** 604.

B. Nienhuis (1987). Coulomb gas formulation of two-dimensional phase transitions, in *Phase Transitions and Critical Phenomena* **11**, eds. C. Domb and J.L. Lebowitz (Academic Press, London).

- **Kac–Moody algebras**

P. Goddard and D. Olive (1985). Kac–Moody algebras, conformal symmetry and critical exponents, *Nucl. Phys.* **B257,** 226.

P. Goddard and D. Olive (1986). Kac–Moody and Virasoro algebras in relation to quantum physics, *Int. J. Mod. Phys.* **A1,** 303.
V.G. Kac (1990). *Infinite Dimensional Lie Algebras,* 3rd edn (Cambridge University Press).

• **Wess–Zumino models**

A.M. Polyakov and P.B. Wiegman (1983). Theory of nonabelian Goldstone bosons in two dimensions, *Phys. Lett.* **131B,** 121.
A.M. Polyakov and P.B. Wiegman (1983). Goldstone fields in two dimensions with multi-valued actions, *Phys. Lett.* **141B,** 223.
V.G. Knizhnik and A.B. Zamolodchikov (1984). Current algebra and Wess–Zumino model in two dimensions, *Nucl. Phys.* **B247,** 83.
E. Witten (1984). Nonabelian bosonization, *Commun. Math. Phys.* **92,** 455.

• **The Goddard–Kent–Olive construction**

P. Goddard, A. Kent and D. Olive (1985). Virasoro algebras and coset space models, *Phys. Lett.* **152B,** 88.
P. Goddard, A. Kent and D. Olive (1986). Unitary representations of the Virasoro and super-Virasoro algebras, *Commun. Math. Phys.* **103,** 105.

• **Connection with face models**

D.A. Huse (1984). Exact exponents for infinitely many new multi-critical points, *Phys. Rev.* **B30,** 3908.
M. Jimbo, T. Miwa and M. Okado (1988). Local state probabilities of solvable lattice models an $A_{n-1}^{(1)}$ family, *Nucl. Phys.* **B300 [FS22],** 74.

• ***q*-deformed Knizhnik–Zamolochikov equation**

I.B. Frenkel and N. Jing (1988). Vertex representations of quantum affine algebras, *Proc. Nat. Acad. Sci. US* **85,** 9373.
I. Frenkel and N.Yu. Reshetikhin (1992). Quantum affine algebras and holonomic q-difference equations, *Commun. Math. Phys.* **146,** 1–60.
B. Davies, O. Foda, M. Jimbo, T. Miwa and A. Nakayashiki (1993). Diagonalization of the XXZ hamiltonian by vertex operators, *Commun. Math. Phys.* **151,** 89–155.

• **Vertex operators**

M. Jimbo and T. Miwa (1983). Solitons and infinite-dimensional Lie algebras, *Publ. Res. Inst. Math. Sci.* **19,** 943–1001.
P. Goddard and D. Olive (1984). Algebras, lattices and strings, in *Vertex Operators in Mathematics and Physics,* eds. J. Lepowsky, S. Mandelstam and I.M. Singer (MSRI Publications/Springer, New York).

Chapter 11

• **Dominant ideology**

O. Babelon (1988). Extended conformal algebra and the Yang–Baxter equation, *Phys. Lett.* **B215,** 523–9.
L. Alvarez-Gaumé, C. Gómez and G. Sierra (1989). Hidden quantum symmetries in rational conformal field theories, *Nucl. Phys.* **B319,** 155.

L. Alvarez-Gaumé, C. Gómez and G. Sierra (1989). Quantum group interpretation of conformal field theory, *Phys. Lett.* **220B,** 142.
G. Moore and N.Yu. Reshetikhin (1989). A comment on quantum group symmetry in conformal field theory, *Nucl. Phys.* **B328,** 557.
L. Alvarez-Gaumé, C. Gómez and G. Sierra (1990). Duality and quantum groups, *Nucl. Phys.* **B330,** 347.
V. Pasquier and H. Saleur (1990). Common structures between finite systems and conformal field theories through quantum groups, *Nucl. Phys.* **B330,** 523.

• Contour picture of quantum groups

C. Gómez and G. Sierra (1990). Quantum group meaning of the Coulomb gas, *Phys. Lett.* **240B,** 149.
C. Ramírez, H. Ruegg and M. Ruiz-Altaba (1990). Explicit quantum symmetries of Wess–Zumino–Novikov–Witten theories, *Phys. Lett.* **B247,** 499–508.
C. Gómez and G. Sierra (1991). The quantum symmetry of rational conformal field theories, *Nucl. Phys.* **B352,** 791–828.
C. Ramírez, H. Ruegg and M. Ruiz-Altaba (1991). The contour picture of quantum groups in conformal field theories, *Nucl. Phys.* **B364,** 195–236.

• Other approaches and extensions to massive theories

A.C. Ganchev and V.B. Petkova (1989). $U_q(s\ell(2))$ invariant operators and minimal theory fusion rules, *Phys. Lett* **233B,** 374.
D. Bernard and A. LeClair (1990). Residual quantum symmetries of the restricted sine-Gordon theories, *Nucl. Phys.* **B340,** 721.
P. Bouwknegt, J. McCarthy and K. Pilch (1990). Quantum group structure in the Fock space resolutions of $\widehat{s\ell}(n)$ representations, *Commun. Math. Phys.* **131,** 125.
P. Furlan, A.G. Ganchev and V.B. Petkova (1990). Quantum groups and fusion rule multiplicities, *Nucl. Phys.* **B343,** 205.
J.-L. Gervais (1990). The quantum group structure of 2D gravity and minimal models, *Commun. Math. Phys.* **130,** 257.
N.Yu. Reshetikhin and F. Smirnov (1990). Hidden quantum group symmetry and integrable perturbations of conformal field theories, *Commun. Math. Phys.* **131,** 157.
E. Witten (1990). Gauge theories, vertex models and quantum groups, *Nucl. Phys.* **B330,** 285.
D. Bernard and A. LeClair (1991). Quantum group symmetries and non-local conserved currents in 2D QFT, *Commun. Math. Phys.* **142,** 99.
G. Felder and C. Wieczerkowski (1991). Topological representations of the quantum group $U_q(s\ell_2)$, *Commun. Math. Phys.* **138,** 583.
V.V. Schechtman and A.N. Varchenko (1991). Arrangement of hyperplanes and Lie algebra homology, *Invent. Math.* **106,** 139.
A.Yu. Alekseev, L.D. Fadeev, M. Semenov-Tian-Shansky (1992). Hidden quantum groups inside Kac-Moody algebra, *Commun. Math. Phys.* **149,** 335.
G. Felder and A. LeClair (1992). Restricted quantum affine symmetry of perturbed minimal conformal models, *Int. J. Mod. Phys.* **A7** **(***Suppl.* **1A),** 239.
G. Mack and V. Schomerus (1992). Quasi-Hopf quantum symmetry in quantum theory, *Nucl. Phys.* **B370,** 185.

S.D. Mathur (1992). Quantum Kac–Moody symmetry and quantum affine algebras, *Nucl. Phys.* **B369,** 433–60.

• **A review of $N = 2$ super-symmetry**

N. Warner (1993). $N = 2$ quantum field theories, in *Proceedings, Trieste Summer School 1992,* eds. R. Iengo, K. Narain, *et al.* (World Scientific, Singapore).

Index

$3j$ symbol, *see* Clebsch–Gordan coefficient
$6j$ symbol, 200–1, 319–23, 381

ADE partition functions, 338
affinization parameter, 56
Alexander–Conway polynomial, 167, 171, 176, 252, 271
anisotropy, 14, 21, 25, 45, 66, 74, 75, 77, 102, 110, 121, 129
anomalous dimension, 279
antiferromagnetic state, 72, 75, 78, 82
antipode, 185, 229, 396
auxiliary space, 34, 53, 62, 90

b–c system, 305
β–γ system, 305
background charge, 342
Baxter lattice, 15
baxterization, 269
Bethe ansatz
  algebraic, 41–7, 245
  coordinate, 12, 13, 20, 21, 107
  for descendants, 61–4
  for eight-vertex model, 111–20, 122
  for higher spin, 84, 85
  for nilpotent hamiltonians, 245–9
  for open chains, 92–6
  for XXX model, 80–3
  for XXZ chain, 74–8
Bethe equations, 44, 64, 96, 120
Bethe reference state, 10, 12, 20, 42, 63, 93, 118, 120
Bethe root, 75–83, 85, 104
Bethe string, 79–83, 103
bi-algebra, 40, 57, 185
bi-colored graph, 137, 160
Birman–Wenzl algebra, 173
bi-unitary matrix, 178
block spin renormalization, 275
Bogomolny soliton, 435

Boltzmann weight, 16, 24
  for boundaries, 88
  for descendants, 65
  for eight-vertex model, 110, 124, 130
  for face model, 134, 146
  for fused face model, 156
  for Ising model, 148
  and quasi-momenta, 43
  and $\mathcal{R}$-matrix, 35, 109
  for six-vertex model, 17, 23, 67
  as vector fields, 68
boost, 2, 160
bootstrap, 29–32
Borel subalgebra, 189, 394
boundary operator, 88, 92, 107
braid group, 48, 139, 165–96, 250, 323
braid limit, 49–53
braiding matrix, 203, 319–23, 349–53
Bratelli diagram, 137–40, 142, 144, 204, 322
Brauer representation, 161
BRS charge, 429
Burau representation, 171

c theorem, 121, 306
Cartan matrix, 207
Casimir, 194, 229
central charge, 285, 297, 300
central extension, 104, 106, 121, 122, 215
centralizer, 201, 379
charge conjugation, 31, 299, 321
checkerboard model, 264
Chevalley basis, 56, 207
chiral algebra, 289, 308
chiral Potts model, 252–68, 271
chiral primary field, 427
chiral ring, 429, 437
chiral vertex operator, 312, 336, 401, 418
Clebsch–Gordan coefficient, 197–200, 417
clock model, 252

co-adjoint action, 232
co-algebra, 185
co-associativity, 184, 219
co-cycle, 215
commuting square, 178, 181
co-multiplication, *see* co-product
conditional expectation, 142, 164, 168, 323
conformal block, 87, 197, 312
conformal invariance, 72, 272
conformal transformation, 68
conformal weight, 287, 297
contour deformation, 313, 393, 398–400
co-product, 40, 53, 55, 57, 97, 184, 191, 313, 392
corner transfer matrix, 158–60, 162–4, 226, 365, 371
correlation length, 272, 276
Coulomb gas, 340, 345, 379
co-unit, 185, 229, 396
coupling constant, 9, 102
Coxeter graph, 135, 143, 146–8, 339
Coxeter number, 147, 162, 381
CPT invariance, 7
critical exponent, 272, 276, 287
crossing unitarity, 91
crystal base, 227, 379
Curie point, 273
current conservation, 17

decoupling of null vectors, 291–9, 320, 332, 343, 358, 378
Dedekind's η function, 331
deformation parameter, 53
derivation, 215, 232
descendant field, 287
descendant model, 58
difference property, 3, 38, 39, 50, 76, 109, 258–60
dilogarithm, 301
Dirac sea, 76, 79, 83, 85, 98, 104
dispersion relation, 71, 72, 83, 300
Drinfeld's double, *see* quantum double
dual algebra, 190
dual Coxeter number, 356
dual lattice, 151
duality, 308, 319
Dynkin diagram, 135, 212–14, 339

eight-vertex model, 108–24, 136, 155
elementary excitations, 72
elliptic curve, 109, 110
elliptic exponential, 123, 124, 128
elliptic functions, 125–9
elliptic nome, 110, 150, 153, 328
energy–momentum tensor, 281, 288
entropy, 103
Euler's pentagonal identity, 337, 369
extended Yang–Baxter system, 173, 206

face model
  Boltzmann weights, 134
  chiral Potts, 255
  corner transfer matrix, 365
  and eight-vertex model, 115
  fusion, 155, 164
  as hidden symmetry in higher-spin chains, 87–8
  restricted, 145, 154
  Yang–Baxter equation, 136
factor, 177–82
factorization equations, 5, 8, 13
Fan–Wu condition, 122
Feigin–Fuks, *see* Coulomb gas
Fermat curve, 258
Fermi surface, 72
fermion number, 16, 435
ferromagnetic state, 12, 29, 72
Fibonacci number, 338
finite-size scaling, 286, 300–2
form factor, 28–33
four-Fermi interaction, 102
free energy
  in critical phenomena, 273
  finite size behavior, 300
  Kramers–Wannier duality, 153
  minimal model, 122
  one-dimensional Ising model, 277
  six-vertex model, 121
  spin-S chain, 103, 104
free fermion
  character, 337
  eight-vertex model, 123, 124
  Ising model, 31
  nilpotent irreps, 241
  sine-Gordon, 10
  supersymmetry, 418
  XX model, 36
free field, 340–1
  see also Coulomb gas
Freudenthal–de Vries strange formula, 368
Friedan–Shenker bundle, 401
fundamental construction, 141, 181, 327
fusion
  and quantum groups, 378–81
  and towers of algebras, 324–7
fusion algebra, 296–9
fusion matrix, 319–21, 352
fusion rules, 296, 338, 346

galilean group, 221

Gauss binomial formula, 237, 408
Gauss decomposition, 233, 359
gaussian model, 278, 302
Goddard–Kent–Olive construction, 364, 425

Haldane's conjecture, 98, 101
harmonic oscillator, 220, 250, 271
harmonic ratio, 68, 311
Hecke algebra, 69, 139, 170, 202, 378
hexagonal equation, 115, 321, 336
high temperature expansion, 153
higher-spin chain, 83, 104
higher-spin descendant, 155, 264
hole, 79, 81
HOMFLY polynomial, 167, 171
homotopy group, 165
Hopf algebra, 184, 220, 392
hopping parameter, 26
Hubbard model, 244, 270
hypergeometric function, 311, 370

incidence matrix, 134–7, 141, 146
inclusion matrix, *see* incidence matrix
inner derivation, 232
integrability, 1, 2, 14, 21, 23, 37, 47, 72, 91, 109, 160
integrable deformation, 434
integrable representation, 413
interaction round a face model, *see* face model
intertwiner, 58, 89, 234, 265, 313, 355, 413
IRF model, *see* face model
irreducible representations
  affine $A_1^{(1)}$, 56
  cyclic, 230, 265–7
  nilpotent, 231, 238
  $U_q(s\ell(2))$, 194
irrelevant operator, 270, 287
Ising model
  Boltzmann weights, 148
  conformal blocks, 348–50
  conformal weights, 299
  eight-vertex model, 121
  form factors, 31–2
  $N = 2$ chiral Potts, 252
  one-dimensional, 275
  Onsager algebra, 261
  partition function, 331, 337

Jacobi elliptic function, 78, 109, 127
Jacobi theta function, 126
Jacobi's triple product identity, 337, 369
Jones polynomial, 167, 170, 171, 206
Jordan–Wigner transformation, 26, 102, 123, 241

Kac determinant, 292
Kac formula, 292, 295, 346, 391
Kac–Moody algebra, 213, 216, 290, 363, 376
Kac–Moody character, 364
Killing vector, 38, 68, 258, 288
kink operator, 26
Knizhnik–Zamolodchikov equation, 148, 356, 367, 376, 382, 419
knot, 165
Kramers–Wannier duality, 31, 150–3, 260
Kronecker's theorem, 146

Landau–Ginsburg lagrangian, 304
Landau–Ginsburg super-potential, 432
lattice hamiltonian, 22, 23, 46, 65, 69, 84, 92
lattice momentum, 10, 22, 23, 46
Lee–Yang edge singularity, 298
Leibnitz's chain rule, 231
Leningrad formula, 86–7
level, 216, 291
Levi–Civita tensor, 106
Lie bracket, 215
link, 165
link invariant, 197
local state probabilities, 366
loop algebra, 212
Lorentz group, 221
Lorentz spin, 2, 287
low lying excitation, 71, 72, 78, 82, 98, 105
low temperature expansion, 151

magnetization, 273
Majorana fermion, 31
Majumdar–Ghosh hamiltonian, 106
Mandelstam variables, 318
marginal operator, 270, 287
Markov move, 168
Markov trace, 122, 139, 142, 144, 164, 168, 205, 327, 385
mass gap, 71, 72, 98, 102, 106, 261
maximal root, 223
minimal model
  central extension from finite size, 121
  fusion rules, 299
  Kac formula, 297
  and Landau–Ginsburg, 304
  $N = 0, 1, 2$ models, 428
  and quantum group, 413
  Rocha–Caridi formula, 330
  and XXZ chain, 106, 302

Möbius transformation, 285
modular invariance, 330, 339
modular matrix, 329, 338, 383
modular parameter, 328
modular transformation, 329
moduli space, 315
momentum operator, 21
monodromy invariance, 310
monodromy matrix, 38, 51, 61, 65, 89, 116, 245
monodromy transformation, 310
Morse function, 433
mutual locality index, 28

Néel state, 27, 102
Neveu–Schwarz super-algebra, 423
normal order, 188, 340
null descendant, 292
null vector, 291, 295, 343, 358, 387
number operator, 18

Ocneanu's string operator, 144
Onsager algebra, 261
operator formalism, 308
operator product expansion, 285, 293, 340, 350, 423
order and disorder operators, 32
order parameter, 273

parity symmetry, 53
partition function, 17, 151, 152, 159, 278
path representation, 225
Pauli exclusion principle, 11, 80, 270
pentagonal equation, 200
permutation map, 47, 49, 68, 195
Perron–Frobenius theorem, 73, 78, 145
perturbed chiral ring, 435
perturbed lagrangian, 307
phase shift, 12
phase space, 1
phase transition, first and second order, 273, 281
Pochhammer contour, 370, 410
Poincaré algebra, 224
Poisson bracket, 1, 132
polynomial equations, 321, 328
Potts model, 162, 253, 261, 298
primary field, 289, 291, 301, 341
projective representation, 285
propagator, 318, 328, 340
puncture, 314

q-intertwiner, 371
q-Knizhnik–Zamolodchikov equation, 371
q-number, 193

quantum determinant, 67, 123, 132
quantum dimension, 326, 337, 384
quantum double, 188, 221, 400
quantum group
  affine, 56, 90, 162, 214
  contour representation, 392–6
  Hopf algebra, 184–6
  knots, 171
  quasi-triangular Hopf algebra, 186
  and spin chain hamiltonian, 97, 107
  $U_q(s\ell(2))$, 52, 55, 189
  as Yang–Baxter algebra, 51
quantum space, 34, 53, 62, 90
quasi-Hopf algebra, 218, 227, 382, 420
quasi-momentum, 12

radial quantization, 283
Ramond ground state, 429
Ramond super-algebra, 425
rapidity, 2, 9, 27, 33, 45, 66, 75, 118, 255, 257
rational conformal field theory, 261, 297
reflection coefficient, 95
Reidemeister move, 166, 168, 198
relevant operator, 280, 287
renormalization group, 272, 275, 279
representation
  of braid group, 50, 138
  higher-spin, 84
  of Sklyanin algebra, 131
  of Temperley–Lieb–Jones algebra, 140, 145
  of $U_q(A_1^{(1)})$, 56
  of $U_q(s\ell(2))$, 55
  of Yang–Baxter algebra, 40, 41, 53, 58
Reshetikhin criterion, 70, 271
restricted face model, 120, 155
reversal symmetry, 17
ribbon Hopf algebra, 197, 251, 384
Riemann surface, 288, 314
Riemann–Hilbert problem, 320, 377
R-matrix
  affine, 217–18
  classical, 376
  in Coulomb gas, 396–400
  for descendants, 59
  eight-vertex, 112, 129
  an incomplete catalog, 208
  infinite-dimensional, 251
  as intertwiner, 54–5, 60, 234
  Jimbo's, 50
  nilpotent, 235–40
  permuted, 66
  six-vertex, 37, 45, 58, 91, 366
  symmetry properties, 90

R-matrix (*cont.*)
universal, 55, 66, 186, 193
in vertex models, 35
Rocha–Caridi formula, 330
root of unity
Bethe ansatz, 88, 107, 121
chiral Potts, 263–8
conformal field theory, 382
face models, 147, 154
quantum groups, 228–31
RSOS model, *see* face model
$\mathcal{R}TT$ equation, 38, 39, 52

scale invariance, 281
scale transformation, 272
scaling function, 50, 273
Schur's lemma, 230, 232
schwarzian derivative, 285, 304, 423
Schwinger term, 285
screened vertex operator, 386–92, 397
screening current, 345, 386, 402
self-dual chiral Potts model, 261
Serre relations, 207, 394, 418, 420
sewing construction, 314
sewing parameter, 315
shift operator, 10, 22, 69, 180
sine-Gordon, 9, 10, 27, 102, 366, 420
singularity theory, 437
six-vertex model, 15–17, 39, 45, 67, 74, 270
skein relations, 167, 210
Sklyanin algebra, 130, 132
Sklyanin equations, 91
$s\ell(2,\mathbb{C})$ vacuum, 285
$SL(2)$ invariance, 343
S-matrix, 4
analyticity, 29
crossing symmetry, 7, 8, 90
elliptic, 121
factorizability, 3, 4
real analyticity, 7
for solitons, 136
for spin wave, 25
unitarity, 7, 8, 69, 90
Smirnov's program, 28–33
solid on solid model, *see* face model
soliton, 9, 29, 137, 434
SOS model, *see* face model
Spec Z_ϵ, 233
specific heat, 301
spectral flow, 426
spectral parameter, 50, 112, 266
speed of sound, 104
spin-1 hamiltonian, 70, 98, 107, 242
spin chain, 10, 14
open, 92
spin model, 180, 252, 255
spin parity, 250
spin wave, 11, 12, 19, 72, 83, 93
star-triangle equation, 24, 183, 257
string hypothesis, 79, 86
string operator, 144
structure constant, 312, 430
structure function, 289
subfactors, 140
Sugawara construction, 355, 363
super-conformal algebra, 426
super-conformal field theory, 422
super-conformal transformation, 423
super-integrable chiral Potts model, 261
super-potential, 429, 432
super-symmetry, 420, 422, 425
super-Virasoro algebra, 423

T invariance, 90
Temperley–Lieb–Jones algebra
Brauer representation, 161
conformal field theory, 323–7, 378
definition, 139
face models, 145
fundamental construction, 140–4
Markov moves, 168
open XXZ chain, 98
relation to Hecke, 139
spin models, 181
thermal subalgebra, 297, 309, 331
thermodynamic limit, 17, 73, 75, 83, 103, 104, 153, 159
θ functions, 125
θ parameter, 119, 121, 122
time reversal, 31
topological field theory, 430, 432
torus, 328, 332
tower of algebras, 141, 323, 327, 337
transfer matrix
corner, *see* corner transfer matrix
descendant models, 61, 65
derivatives, 69
eigenvalues, 20, 44, 64
eight-vertex model, 117
face models, 135, 138
and monodromy matrix, 38
nilpotent models, 240
open chain, 90
and $\mathcal{R}$-matrix, 34, 35
six-vertex model, 18, 43
spin models, 180
twisted boundary conditions, 106
tricritical Ising model, 298
truncation
in Bethe ansatz, 87, 88

truncation (*cont.*)
in Coulomb gas, 413
in face models, 155
in minimal models, 297–9
in representation theory, 415
twist operator, 219, 414
twisted energy-momentum tensor, 429

uniformization
chiral Potts, 258
rapidity of solitons, 29
for six-vertex model, 23, 37, 45
for S-matrix, 9
universal enveloping algebra, 53, 189
universality class, 272, 287
unrestricted graph, 154
$U_q(A_1^{(1)})$, 56
$U_q(\mathcal{G})$, 207
$U_q(g\ell(1,1))$, 271
$U_q(s\ell(2))$, 52, 55, 57, 97, 107, 190, 196
$U_q'(s\ell(2))$, 57

vacancy, 81
valence bond solid, 99–101, 106
Verlinde's theorem, 339, 384, 415
Verma module, 287, 291, 295
vertex model, 15, 23
vertex operator, 341, 375
vertex–face map, 115, 132, 136, 147, 156
Virasoro algebra, 285, 286
Virasoro character, 328, 333, 337, 365, 370
Virasoro irreducible representation, 289, 291
von Neumann–Jones index, 177

Wakimoto construction, 362, 367
W-algebra, 419
Ward identities, 292, 294, 345
Wess–Zumino model, 88, 304, 336, 382, 401, 418
Wick theorem, 340
writhe number, 170

XX model, 25, 29, 241, 244
XXX model, 14, 25, 47, 70, 71, 86, 97, 111, 226, 242
XXZ model, 14, 22, 25, 27, 69, 73, 75, 93, 102, 110, 162, 163, 270, 406
XY model, 123, 124, 307
XYZ model, 14, 110

Yang–Baxter algebra, 39, 40, 51, 59, 116
Yang–Baxter equation
with affine parameter, 217
for Boltzmann weights, 24
classical, 210, 356, 367
from commuting squares, 179
for conformal blocks, 320
in the contour representation, 398
for descendants, 60
for eight-vertex model, 110, 115
for face models, 115, 135
for free fermions, 124
for fused face model, 156–8
for Ising model, 148
and knots, 166
for nilpotent irreps, 240
operator, 36, 37
for permutations, 69
permuted, 48, 49
S-matrix, 5
for spin models, 183
for spin waves, 13
universal, 187, 195
for vertex models, 24, 36
Yang–Baxter operator, 138, 140, 159, 323
yangian, 223
Young diagram, 170
Young tableau, 204

Z symmetry, 5, 91
Zamolodchikov algebra and operators, 6, 8, 28, 68
ζ function, 106